DATE DUE

MTP International Review of Science

Main Group Elements
Groups V and VI

MTP International Review of Science

Publisher's Note

The MTP International Review of Science is an important new venture in scientific publishing, which we present in association with MTP Medical and Technical Publishing Co. Ltd. and University Park Press, Baltimore. The basic concept of the Review is to provide regular authoritative reviews of entire disciplines. We are starting with chemistry because the problems of literature survey are probably more acute in this subject than in any other. As a matter of policy, the authorship of the MTP Review of Chemistry is international and distinguished; the subject coverage is extensive, systematic and critical; and most important of all, new issues of the Review will be published every two years.

In the MTP Review of Chemistry (Series One), Inorganic, Physical and Organic Chemistry are comprehensively reviewed in 33 text volumes and 3 index volumes, details of which are shown opposite. In general, the reviews cover the period 1967 to 1971. In 1974, it is planned to issue the MTP Review of Chemistry (Series Two), consisting of a similar set of volumes covering the period 1971 to 1973. Series Three is planned for 1976, and so on.

The MTP Review of Chemistry has been conceived within a carefully organised editorial framework. The over-all plan was drawn up, and the volume editors were appointed, by three consultant editors. In turn, each volume editor planned the coverage of his field and appointed authors to write on subjects which were within the area of their own research experience. No geographical restriction was imposed. Hence, the 300 or so contributions to the MTP Review of Chemistry come from many countries of the world and provide an authoritative account of progress in chemistry.

To facilitate rapid production, individual volumes do not have an index. Instead, each chapter has been prefaced with a detailed list of contents, and an index to the 10 volumes of the MTP Review of Inorganic Chemistry (Series One) will appear, as a separate volume, after publication of the final volume. Similar arrangements will apply to the MTP Review of Physical Chemistry (Series One) and to subsequent series.

Butterworth & Co. (Publishers) Ltd.

Inorganic Chemistry
Series One

Consultant Editor
H. J. Emeléus, F.R.S.
Department of Chemistry
University of Cambridge

Volume titles and Editors

1 MAIN GROUP ELEMENTS— HYDROGEN AND GROUPS I–IV
Professor M. F. Lappert, *University of Sussex*

2 MAIN GROUP ELEMENTS— GROUPS V AND VI
Professor C. C. Addison, F.R.S. and Dr. D. B. Sowerby, *University of Nottingham*

3 MAIN GROUP ELEMENTS— GROUP VII AND NOBLE GASES
Professor Viktor Gutmann, *Technical University of Vienna*

4 ORGANOMETALLIC DERIVATIVES OF THE MAIN GROUP ELEMENTS
Dr. B. J. Aylett, *Westfield College, University of London*

5 TRANSITION METALS—PART 1
Professor D. W. A. Sharp, *University of Glasgow*

6 TRANSITION METALS—PART 2
Dr. M. J. Mays, *University of Cambridge*

7 LANTHANIDES AND ACTINIDES
Professor K. W. Bagnall, *University of Manchester*

8 RADIOCHEMISTRY
Dr. A. G. Maddock, *University of Cambridge*

9 REACTION MECHANISMS IN INORGANIC CHEMISTRY
Professor M. L. Tobe, *University College, University of London*

10 SOLID STATE CHEMISTRY
Dr. L. E. J. Roberts, *Atomic Energy Research Establishment, Harwell*

INDEX VOLUME

Physical Chemistry
Series One
Consultant Editor
A. D. Buckingham
Department of Chemistry
University of Cambridge

Volume titles and Editors

1 THEORETICAL CHEMISTRY
 Professor W. Byers Brown, *University of Manchester*

2 MOLECULAR STRUCTURE AND PROPERTIES
 Professor G. Allen, *University of Manchester*

3 SPECTROSCOPY
 Dr. D. A. Ramsay, *National Research Council of Canada*

4 MAGNETIC RESONANCE
 Professor C. A. McDowell, *University of British Columbia*

5 MASS SPECTROMETRY
 Professor A. Maccoll, *University College, University of London*

6 ELECTROCHEMISTRY
 Professor J. O'M Bockris, *University of Pennsylvania,*

7 SURFACE CHEMISTRY AND COLLOIDS
 Professor M. Kerker, *Clarkson College of Technology, New York*

8 MACROMOLECULAR SCIENCE
 Professor C. E. H. Bawn, F.R.S., *University of Liverpool*

9 CHEMICAL KINETICS
 Professor J. C. Polanyi, F.R.S., *University of Toronto*

10 THERMOCHEMISTRY AND THERMODYNAMICS
 Dr. H. A. Skinner, *University of Manchester*

11 CHEMICAL CRYSTALLOGRAPHY
 Professor J. Monteath Robertson F.R.S., *University of Glasgow*

12 ANALYTICAL CHEMISTRY —PART 1
 Professor T. S. West, *Imperial College, University of London*

13 ANALYTICAL CHEMISTRY — PART 2
 Professor T. S. West, *Imperial College, University of London*

 INDEX VOLUME

Organic Chemistry
Series One
Consultant Editor
D. H. Hey, F.R.S.
Department of Chemistry
King's College, University of London

Volume titles and Editors

1 STRUCTURE DETERMINATION IN ORGANIC CHEMISTRY
 Professor W. D. Ollis, *University of Sheffield*

2 ALIPHATIC COMPOUNDS
 Professor N. B. Chapman, *Duke University, North Carolina*

3 AROMATIC COMPOUNDS
 Professor H. Zollinger, *Swiss Federal Institute of Technology*

4 HETEROCYCLIC COMPOUNDS
 Dr. K. Schofield, *University of Exeter*

5 ALICYCLIC COMPOUNDS
 Professor W. Parker, *University of Stirling*

6 AMINO ACIDS AND PEPTIDES
 Professor D. H. Hey, F.R.S. and Dr. D. I. Johns, *King's College, University of London*

7 CARBOHYDRATES
 Professor G. O. Aspinall, *University of Trent, Ontario*

8 STEROIDS
 Dr. W. D. Johns, *G. D. Searle & Co., Chicago*

9 ALKALOIDS
 Professor K. Wiesner, *University of New Brunswick*

10 FREE RADICAL REACTIONS
 Professor W. A. Waters, F.R.S., *University of Oxford*

 INDEX VOLUME

Inorganic Chemistry
Series One

Consultant Editor
H. J. Eméleus, F.R.S.

MTP International Review of Science

Volume 2

Main Group Elements
Groups V and VI

Edited by **C. C. Addison, F.R.S.,** and **D. B. Sowerby**
University of Nottingham

Butterworths · London
University Park Press · Baltimore

THE BUTTERWORTH GROUP

ENGLAND
Butterworth & Co (Publishers) Ltd
London: 88 Kingsway, WC2B 6AB

AUSTRALIA
Butterworth & Co (Australia) Ltd
Sydney: 586 Pacific Highway 2067
Melbourne: 343 Little Collins Street, 3000
Brisbane: 240 Queen Street, 4000

NEW ZEALAND
Butterworth & Co (New Zealand) Ltd
Wellington: 26–28 Waring Taylor Street, 1

SOUTH AFRICA
Butterworth & Co (South Africa) (Pty) Ltd
Durban: 152–154 Gale Street

ISBN 0 408 70248 6

UNIVERSITY PARK PRESS

U.S.A. and CANADA
University Park Press Inc
Chamber of Commerce Building
Baltimore, Maryland, 21202

Library of Congress Cataloging in Publication Data

Addison, Cyril Clifford, 1913–
  Main group elements: groups V and VI

  (Inorganic chemistry, series one, v. 2) (MTP
international review of science)
  Includes bibliographies
  1.   Chemistry, Inorganic. I. Sowerby, D. B.,
1934–   joint author. II. Title
QD151.2.15 vol. 2 546′.5 78–160324
ISBN 0–8391–1005–7

QD
151.2
I5
vol. 2

First Published 1972 and © 1972
MTP MEDICAL AND TECHNICAL PUBLISHING CO. LTD.
Seacourt Tower
West Way
Oxford, OX2 OJW
and
BUTTERWORTH & CO. (PUBLISHERS) LTD.

Filmset by Photoprint Plates Ltd., Rayleigh, Essex
Printed in England by Redwood Press Ltd., Trowbridge, Wilts
and bound by R. J. Acford Ltd., Chichester, Sussex

Consultant Editor's Note

The problem of keeping abreast of research literature on as broad a front as possible is one that confronts all chemists. In the past this difficulty has been met, in the main, by literature surveys and by several uncorrelated reviews of progress in certain subject areas. There are obvious inadequacies in this approach, which have become increasingly apparent in recent years. I was, therefore, grateful for the opportunity of helping to plan this new series, which has been designed to provide a comprehensive, critical survey of each of the main branches of chemistry.

This section of the MTP International Review of Science deals with progress in Inorganic Chemistry. The subject is developing at an astonishing rate and in many directions. Fortunately, however, it lends itself to a systematic treatment. Ten volumes have been prepared, three dealing with the main group elements and two with the general chemistry of the transition metals. Organometallic derivatives of the main group elements and lanthanides and actinides are covered separately, as is the subject of reaction mechanisms. The two remaining volumes on radiochemistry and solid state chemistry have been planned to avoid, as far as possible, overlap with those that have gone before.

It is a pleasure to thank the many experts who have collaborated as authors and volume editors in making this publication possible. While working to a pre-arranged over-all plan, they have been able to assess and interpret the literature in terms of their own experience in specialised fields. I believe that in this way they will not only provide a record of what has been done, but will stimulate further exploration in this fascinating branch of chemistry.

Cambridge

H. J. Emeléus

Preface

This volume, which reviews the chemistry of the elements in Groups V and VI of the Periodic Table, must necessarily be concerned with advances made in selected areas within these Groups over the last five years. In the space of one volume it would not be possible to represent faithfully five years' progress in the chemistry of all ten elements, especially in view of the importance and extensive chemistry associated with the lighter elements in each group. We have therefore concentrated on certain aspects of nitrogen, oxygen, phosphorus and sulphur chemistry.

For oxygen, we have concentrated on the formation of ternary oxides while for sulphur, systems involving the element and complexes formed from sulphur ligands are considered. The chemistry of phosphorus is represented by two aspects of its halogen chemistry. Two diverse aspects of nitrogen chemistry, i.e. reactions in molten nitrates and nitrites, and the complexes formed by certain nitrogen ligands, reflect the breadth of knowledge in this area. To present a more complete view, we will consider in later volumes other aspects of the chemistry of these four elements.

The remaining six elements have been studied less extensively in recent years, but to preserve a balance through the two groups, the chemical behaviour of arsenic, selenium and polonium is reviewed in this volume. Discussion of antimony, bismuth and tellurium chemistry has been deferred to a later volume.

C. C. Addison
D. B. Sowerby

Nottingham

Contents

1
Alkali Metal Ternary Oxides

M. G. BARKER
University of Nottingham

1.1   INTRODUCTION

The chemistry of ternary oxides has for many years mainly been concerned with the preparation, composition and thermal stability of solid phases. Very little attention has been given to the chemical reactions of the compound once it has been prepared, principally because many of these compounds are insoluble in aqueous media and in those cases where solution is possible no new species are formed. The nature of oxides in the molten state or in solution with molten salts is an area which has hardly been studied at all. Early investigations in the chemistry of ternary oxides concentrated on preparative aspects. Only in a few cases was structural work carried out. The more widespread use of powder x-ray crystallography enabled the structures of many compounds to be formulated but the obvious limitations of the technique left many problems unsolved, and prevented detailed descriptions of the nature of the bonding in these often complex structures.

The realisation that many ternary oxides had useful electrical properties initiated a new interest in these oxides and the preparation of new compounds in the search for commercially usable materials. The existence of such properties demonstrated the paucity of knowledge as to the detailed structures and bonding of ternary oxides. To the chemist, studies of physical properties of this type can be used to gain a real insight into the nature of the chemical bond in crystals in that they are the most sensitive method for the detection of small changes of composition and structure.

Improvements in the techniques for growing single crystals from high temperature melts has provided a means of preparation of large single crystals suitable both for physical property measurements and for structural studies. We are now in a period when the study of the structures of ternary oxides has become a major field of research and from this an understanding of the basic physical and chemical properties of ionic compounds is slowly emerging. This new insight into the basic metal–oxygen structural unit together with order–disorder phenomena, non-stoichiometry, and structural distortions will form a major section of this review. In parallel with our increased knowledge of existing compounds must be the search for new compounds. Once again the improvement in experimental techniques has led to new areas of chemistry being opened up. The preparation and characterisation of new ternary oxides must also form a sizeable portion of this review.

No single review can possibly cover all the developments that have taken place in the last 5 years in the chemistry of ternary oxides. Some form of selection must therefore be made and as a result many interesting areas left unmentioned.

The system of classification to be used in the review treats the material in terms of $A_2O–M_xO_y$ systems, where A is the alkali metal. For example, in the $A_2O–M_2O_5$ group will be found the compounds AMO_3, A_3MO_4, $A_4M_2O_7$ and bronzes of the type $A_xM_2O_5$. The principal advantage of this method is that compounds of the same formula type are grouped together irrespective of the metal involved and thus the isostructural aspects of such compounds may be brought out more clearly and concisely than if the elements M were treated as individual groups.

1.2 ALKALI METAL MONOXIDE—METAL(I) OXIDE SYSTEMS

Until 1968 the only known compounds in this system were the lithium and sodium oxoargentates(I). More recently however, several new compounds have been prepared and more significantly their structures determined. The compounds KAgO and CsAgO were prepared[1] from the corresponding oxides as single crystals; x-ray studies showed them to be isostructural with a tetragonal unit cell, $a = 9.91$, $c = 5.46$ Å (KAgO), and $a = 10.2$, $c = 6.17$ Å (CsAgO). The structure, which has the space groups $S_4^2 - I\bar{4}$, contains eight formula units in the single cell. The compounds RbAgO[2], NaCuO[3], RbCuO[3] and KCuO[4] were later found to adopt this same structure, and the first oxoaurate(I), CsAuO[5], was also an isotype of KAgO. The compound CsAuO was, however, thought to be orthorhombic and not isostructural with the previous compounds.

The major feature of this type of structure is the existence of nearly planar groups $[M_4O_4]$ with the oxygen atoms at the corners of a square. Down the

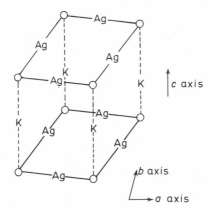

Figure 1.1 The packing of Ag_4O_4 groups in the compound KAgO

centre of the groups (along the c axis) are infinite channels with a diameter of 2.5 Å. The arrangement of the M_4O_4 groups showing the position of the channels is given in Figure 1.1.

The variation of the c-axis dimension with the ionic radius of the alkali metal ion is due to the arrangement of K—O—K or Cs—O—Cs bonds along the c axis.

The compounds KTlO and RbTlO[2] are not isostructural with KAgO having a monoclinic unit cell containing four and eight formula units respectively. These compounds are prepared by heating equimolar mixtures of Tl_2O and the alkali metal monoxide under argon at 500 °C for approximately 1 h. Both compounds are oxidised to the Tl^{3+} ternary oxide $ATlO_2$ when heated in oxygen. This type of oxidation reaction is characteristic of the AMO compounds.

CsAuO was prepared by the careful oxidation of the caesium gold com-

pound CsAu. Oxidation of the comparable NaTl and LiIn gave the metal
$3+$ compounds $NaTlO_2$ and $NaInO_2$.

The preparation of ternary oxides with the metal ions in the $1+$ state is a
relatively recent achievement. The production of single crystals has, however,
enabled the new structure type to be quickly established. It is still too early
to draw any conclusion as to the bonding of the metal $1+$ ion within the
planar groups M_4O_4 since few measurements of the physical properties of
these compounds have been performed. Since the alkali metals will form
aurides of the same types as CsAu the preparation of further oxoaurides(I)
should soon be carried out.

1.3   ALKALI METAL MONOXIDE-METAL(II) OXIDE SYSTEMS

Relatively few reactions have been studied between the alkali metal oxides
and metal(II) oxides. Isolated examples of $A_2M_2O_3$, A_2MO_2, A_4MO_3 and
A_6MO_4 type compounds have been reported for $M = Zn$ [6], Cd [7], Cu [8],
Ni [9], Pd [10] and Co [9], but the only complete series of compounds is the A_2HgO_2
series where A is Li, Na, K, Rb or Cs [10]. A recent study[11] of the A_2O-PbO
system, where $A = Li$ or Na, has established the existence of several new
compounds. PbO reacts with the alkali metal monoxides at temperatures
between 400 °C and 600 °C to form compounds which are extremely sensitive
towards moisture and aqueous or alcoholic solvents. The more reactive
monoxide is required since neither sodium hydroxide nor sodium carbonate
react with PbO between 650 °C and 800 °C. Mixtures of sodium oxide and
PbO were heated at temperatures between 450 °C and 600 °C in ratios from
$Na_2O:PbO = 0.2$ to $Na_2O:Pb = 3$. The compound $Na_2Pb_2O_3$ was prepared from a mixture of 1 mol of PbO with $\frac{1}{2}$ mol of Na_2O, Na_2PbO_2 from an
equimolar mixture and Na_4PbO_3 from a mixture of 1 mol PbO with 2.02 mol
Na_2O. No structural studies were carried out on the compounds but the
appearance of the x-ray powder patterns indicates a decreasing symmetry
with increasing alkali metal oxide content. The compounds were totally
hydrolysed by water, and treatment with absolute methanol yielded the
hydroxy methoxo compound. The Li_2O-PbO system yielded only two
new compounds namely $Li_2Pb_2O_3$ and Li_2PbO_2. Mixtures of lithium oxide
and PbO in ratios greater than two showed the presence of excess lithium
oxide with the compound Li_2PbO_2 as the product. One must conclude
therefore that no lithium analogue of Na_4PbO_3 exists under the conditions
used in this study. There seemed to be no obvious structural relationship
between the sodium and lithium compounds, the compound Li_2PbO_2
giving a much more complex x-ray pattern than Na_2PbO_2.

A study of the K_2O-PbO system[12] revealed the presence of four new compounds of Pb^{II}. The three alkali-rich oxides, K_4PbO_3, $K_6Pb_2O_5$ and K_2PbO_3
are white, extremely hygroscopic powders, being hydrolysed instantly by
atmospheric moisture to PbO and KOH. They have a rather low temperature
stability, decomposing under vacuum at 550 °C by the successive loss of
K_2O to $K_2Pb_2O_3$ which is the only compound stable at this temperature.
The x-ray powder patterns of the compounds revealed no structural analogues
with other compounds of the same general formula. One might have expected

some correspondence between K_2PbO_2 and K_2HgO_2 in view of the close similarity in ionic radii ($Pb^{2+} = 1.20$ Å $Hg^{2+} = 1.10$ Å) but none was found.

The compound $K_2Pb_2O_3$ was found to be the most stable of the potassium plumbites having a greater resistance to hydrolysis and a higher decomposition temperature of 600 °C.

$$K_2Pb_2O_3 \longrightarrow 2PbO + K_2O$$

$K_2Pb_2O_3$ has the caesium chloride structure ($a = 4.203$ Å) with ordering of the cations and oxygen vacancies.

The reactions of lithium oxide with tin monoxide at 400 °C did not prepare any compounds of tin(II) since disproportionation takes place with the formation of Li_2SnO_3, lithium oxide and elemental tin.

$$2Li_2O + 2Sn^{II}O \rightarrow Li_2Sn^{IV}O_3 + Li_2O + Sn^0$$

A similar reaction has been reported[13] for the reaction of FeO with lithium monoxide, namely

$$3FeO + Li_2O \rightarrow 2LiFeO_2 + Fe^0$$

being the disproportionation of 3 mol of Fe^{2+} to 2 mol of Fe^{3+} and 1 mol Fe^0.

The reaction of sodium oxide with Fe has been shown[14] to take place in two stages, with the formation of the compound Na_4FeO_3 as an intermediate. Mixtures of iron and sodium oxide in ratios greater than 1:2 were heated in vacuum at 600 °C. After 4 h a weight loss corresponding to the formation of Na_4FeO_3 was observed,

$$3Na_2O + Fe \rightarrow Na_4FeO_3 + 2Na$$

On further heating, the weight loss indicated that oxidation to the compound $NaFeO_2$ had taken place.

$$3Na_2O + Fe \rightarrow NaFeO_2 + Na_2O + Na$$

Prolonged heating of equimolar mixtures in vacuum gave weight losses corresponding to the formation of Na_2FeO_2 which suggests that the compound is stabilised by the presence of excess iron.

$$2Na_2O + 2Fe \rightarrow Na_2FeO_2 + Fe + Na$$

The compound Na_4FeO_3 has been prepared in a pure state by the reaction of Na_2O with iron in the presence of sodium vapour, or by the reduction of Fe_2O_3 with sodium vapour. Solution calorimetry gave a heat of formation $\Delta H_{298}^{\circ} = -25.07 \pm 0.55$ kcal mol^{-1} (-104.9 ± 2.3 kJ mol^{-1}). Using this data together with the ΔH for $NaFeO_2$ it could be shown that the reaction

$$3Na_2O + Fe \rightarrow Na_4FeO_3 + 2Na$$

has a $\Delta G^{\circ} = -2.2$ kcal mol^{-1} at 600 °C and so will proceed. The reaction

$$Na_4FeO_3 \rightarrow NaFeO_2 + Na_2O + Na$$

has $\Delta G = +19.8$ kcal mol^{-1} which gives a sodium activity for equilibrium to be about 10^{-5}; enough for a slow reaction in vacuum. A practical application of this work is the study of the corrosion of iron in liquid sodium containing dissolved oxygen since the compound Na_4FeO_3 was considered to be the likely corrosion product. The thermodynamic data indicate that for Na_4FeO_3 to be formed as a single phase, the sodium must contain at least 1000 p.p.m. oxygen. This oxygen level is far higher than occurs in practice so that from thermodynamic considerations the compound is unlikely to be as significant as was first thought.

1.4 ALKALI METAL MONOXIDE—METAL(III) OXIDE SYSTEMS

In the past 5 years very many compounds have been prepared with the general formula AMO_2. For this reason the treatment of the material in this section will differ from that used in the other systems. The compounds will be grouped in terms of their structural type, the principal structures adopted by AMO_2 compounds being
  (1) The α-$NaFeO_2$ hexagonal-rhombohedral structure
  (2) The α-$LiFeO_2$ tetragonal structure
  (3) The cubic NaCl type
  (4) The orthorhombic $KCuO_2$ type.

1.4.1 Compounds with the α-$NaFeO_2$ structure

Sodium ferrite has three crystal modifications; the α form is the stable form at room temperature and is prepared by heating sodium carbonate with γ-Fe_2O_3. Its structure is based on a variant of $CsCl_2I$. This structure represents a bridge between the CsCl and NaCl type of arrangements. The basic unit is that of a rhombohedron with one molecule in the unit cell but is often expressed on the basis of a hexagonal unit cell containing three molecules. α-$NaFeO_2$ has a rhombohedral angle α of $31°20'$ which places the structure as being based on the sodium chloride lattice rather than the caesium chloride structure. The sodium chloride lattice may be visualised as being distorted along the body diagonal to form the rhombohedral unit. The representative structure for this class of compounds with $\alpha \sim 30$ degrees is $NaHF_2$, but for oxides in general the α-$NaFeO_2$ structure is a better example. This may be visualised as having layers of oxygen atoms in a face-centred cubic arrangement with a sequence ABC ABC. Between the oxygen layers lie alternate layers of alkali metal ions and metal 3+ ions, each being in octahedral sites relative to the oxygen atom. The metal ions lie on the 111 plane of the basic NaCl lattice, alternate planes being occupied by the different ions. The structure is shown in Figure 1.2.

Compounds of the type AMO_2 having the α-$NaFeO_2$ structure are shown in Table 1.1.

Hagenmuller and co-workers[15, 23] have studied the K–Ln–O system where Ln is a lanthanide 3+ element. The compounds $KLnO_2$ were generally prepared by heating mixtures of Ln_2O_3 and potassium monoxide at $550°C$ in sealed capsules under vacuum. This method was not successful for the preparation of $KCeO_2$ or $KPrO_2$ [23] since the K_2O is in effect reduced to potassium metal by Ce_2O_3 according to the reaction:

$$Ce_2O_3 + K_2O \rightarrow 2CeO_2 + 2K$$

Some of the CeO_2 formed reacted with residual K_2O to form the metacerate.

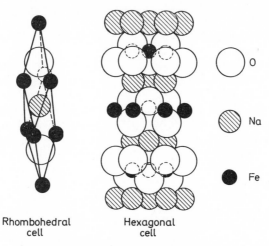

Rhombohedral   Hexagonal
cell   cell

Figure 1.2 The α-NaFeO$_2$ structure

Table 1.1 Crystallographic data for some compounds with the α-NaFeO$_2$ structure

Compound	a	c	Reference	Compound	a	c	Reference
KLaO$_2$	3.70	18.71	15	NaErO$_2$	3.377	16.573	16
KNdO$_2$	3.59	18.65	15	NaTmO$_2$	3.371	16.501	16
KSmO$_2$	3.55	18.55	15	NaYbO$_2$	3.350	16.530	16
KEuO$_2$	3.53	18.58	15	NaLmO$_2$	3.322	16.495	16
KGdO$_2$	3.51	18.62	15	NaCrO$_2$	2.968	15.94	17
KTbO$_2$	3.49	18.61	15	NaMoO$_2$	2.92	17.12	18
KDyO$_2$	3.47	18.57	15	NaAlO$_2$	2.868	15.88	19
KHoO$_2$	3.45	18.57	15	NaTlO$_2$	3.35	16.5	20
KErO$_2$	3.43	18.58	15	RbYO$_2$	3.48	19.5	21
KYbO$_2$	3.39	18.49	15	CsTlO$_2$	3.39	20.8	21
KYO$_2$	3.47	18.6	21	RbTlO$_2$	3.46	19.1	20
KTlO$_2$	3.43	18.3	20	RbErO$_2$	3.45	19.4	22
RbDyO$_2$	3.48	19.4	22	RbTmO$_2$	3.44	19.4	22
RbHoO$_2$	3.46	19.4	22	RbYbO$_2$	3.71	19.2	22
RbLuO$_2$	3.40	19.1	22	KCeO$_2$	3.66	18.66	23
KPrO$_2$							

K_2CeO_3. $KCeO_2$ and $KPrO_2$ were finally prepared by the action of potassium on the dioxides of Ce and Pr at 900 °C. The reaction mixture was quenched, and excess potassium removed by vacuum distillation at 320 °C.

The compounds $KLnO_2$ are all extremely hygroscopic and are rapidly oxidised in air. Apart from $KCeO_2$ which is red-brown, and $KPrO_2$ which is yellow-green, the compounds are white in colour.

Compounds of the $NaLnO_2$ series show[16] a much greater variation in structure type with only a few of the lanthanide elements forming compounds with the α-$NaFeO_2$ structure. The α-$NaFeO_2$ structure is only found for ratios of M^+ to M^{3+} in the range 1.02–1.5. At the lower limit of the ratio, the compounds show polymorphic behaviour with the formation of phases with a disordered structure of the NaCl β-$LiFeO_2$ type and compounds with the monoclinic $LiDyO_2$ type of structure.

The lanthanide compounds with Rb also show a varied structure type. Compounds $RbLnO_2$, where Ln is an element from lanthanum to gadolinium have the α-$LiFeO_2$ structure but for Ln = dysprosium to lutecium the α-$NaFeO_2$ structure predominates.

For the system $KLnO_2$, $NaLnO_2$ and $RbLnO_2$ a reasonably linear relationship exists for the variation of the cell parameter a with the ionic radius of the M^{3+} ion. This is not so surprising as it reflects the change in dimension within a layer and will be expected to change more rapidly than the dimension perpendicular to the layers in the unit cell.

$NaMoO_2$ has been prepared[18] for the first time by the reaction of sodium vapour with molybdenum dioxide at 400 °C; the compound is black and hygroscopic, decomposing at 550 °C in vacuum according to the equation:

$$3NaMoO_2 \rightarrow Na + Na_2Mo_3O_6$$

Magnetic measurements at temperatures between 295 K and 773 K showed feeble, temperature-independent, Pauli-type paramagnetism which could not be attributed to the presence of Mo^{3+} ions in the compound. A d^3 ion in an octahedral field should give a value for the magnetic moment of 3.87 BM. The compound has a metallic-type electrical conductivity which suggests the presence of delocalised electrons; the metal–metal distance of 2.92 Å is less than the Goodenough critical distance R_c (4d)[24]. The possibility of metal–metal bonding within the Mo_3O_{13} groups would explain the magnetic properties by analogy with the theory proposed for $LiVO_2$[24], but this seems less likely than the totally delocalised electron proposal.

A measurement of the heat of formation of $NaCrO_2$[17] has provided a value of $\Delta H_{298} = -208.9$ kcal mol^{-1}. Thermodynamic considerations of the reaction between sodium oxide and chromium metal to form $NaCrO_2$ show that this compound may be formed below 850 °C in liquid sodium containing less than 5 p.p.m. oxygen. This result represents a valuable contribution to the understanding of the behaviour of chromium steels in liquid sodium.

The compound $KPbO_2$ has been prepared[25] by an ingenious method. An equimolar mixture of PbO and PbO_2 is heated with potassium hydroxide at 400 °C or a mixture of PbO and the compound K_2PbO_3 at 600 °C.

$$2KOH + PbO + PbO_2 \rightarrow KPbO_2 + H_2O$$
$$K_2PbO_3 + PbO \rightarrow KPbO_2$$

$KPbO_2$ is a yellow, very hygroscopic, powder with a melting point of 650 °C. When heated in oxygen at 300 °C $KPbO_2$ oxidises, not to form $K_2Pb_2O_5$, but giving a mixture of K_2PbO_3 and $K_2Pb_3O_7$. The compound being diamagnetic and with insulating properties has lead in both the 2+ and 4+ oxidation states and may be formulated as $K_2[Pb^{II} \cdot Pb^{IV}]O_4$. Two structure types have been found for the compound. At high temperatures (above 640 °C) $KPbO_2$ has the α-$NaFeO_2$ structure with $a = 3.358$ Å and $c = 18.56$ Å while the low-temperature structure is not known. The high temperature phase is isostructural with $KTlO_2$ and mixed crystals of the type $K_2[Pb_x^{II}Pb_x^{IV}Tl_{2(1-x)}]O_4$ have been prepared with both the hexagonal and rhombohedral structures.

1.4.2 Compounds with the α-$LiFeO_2$ structure

When $LiFeO_2$ is annealed at 570 °C, a tetragonal structure is formed having an ordered distribution of metallic atoms. The structure may be related to the sodium chloride lattice having oxygen atoms in the chlorine positions and metal atoms in the sodium positions. The unit cell contains four molecules and the c-axis length is approximately twice the a-axis length, due to the

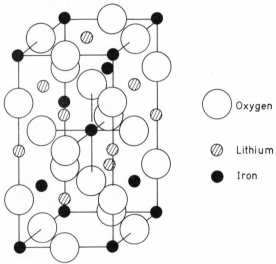

Oxygen

Lithium

Iron

Figure 1.3 The α-$LiFeO_2$ structure

Table 1.2

Compound	a	c	Reference	Compound	a	c	Reference
$LiFeO_2$	4.057	8.759	26,27	$NaSmO_2$	4.697	10.64	16
β-$LiTeO_2$	4.55	9.26	28	$NaEuO_2$	4.676	10.55	16
$LiInO_2$	4.30	9.32	29	$NaGdO_2$	4.660	10.52	16
$NaPrO_2$	4.763	10.966	16				
$NaNdO_2$	4.741	10.862	16				

need to accommodate ions of unequal size in the ordered form (Figure 1.3).

Compounds with the α-$LiFeO_2$ structure are listed in Table 1.2.

A linear relationship was observed[16] between the unit cell volume and atomic number for the $NaLnO_2$ compounds with the α-$LiFeO_2$ structure. The boundaries of the $LiFeO_2$ structure in these compounds lie between

0.74 and 1.06 for $M^+:M^{3+}$ ratios. The thermal and chemical stabilities of the sodium lanthanum oxide compounds with the $LiFeO_2$ structure are less than those with the α-$NaFeO_2$ structure.

1.4.3 ABO_2 compounds with cubic structures

Although the previous structures are both based on the cubic NaCl structure, very few ABO_2 compounds are found to adopt this basic structure. The compounds $NaHoO_2$ and $NaDyO_2$ [16] have a monoclinically distorted cubic lattice with little difference in the unit cell volume for both types. A second form of $NaHoO_2$ prepared by rapid heating of a mixture of $4NaNO_3$ and Ho_2O_3 at 1050 °C was cubic with $a = 4.767$ Å; conversion to the monoclinic form takes place below 950 °C. Sodium terbate $NaTbO_2$ was prepared[16] by heating a $NaNO_3$–Tb_2O_3 mixture in a current of hydrogen at 1000 °C for 1 h. This new compound, which is grey white and readily soluble in acids, is also cubic with $a = 4.819$ Å. The cubic structure of these compounds is isostructural with the cubic form of $LiFeO_2$.

Slow oxidation[30] of NaTl forms a new cubic form of $NaTlO_2$ with a cell dimension $a = 4.76$ Å. The structure is based on NaCl with $z = 2$ and with a random distribution of metal ions in the cation positions of the NaCl structure. At 500 °C it transforms to the known hexagonal form of $NaTlO_2$.

Caesium scandate also has the cubic structure, single crystal data[21] gives a unit cell of 17.42 Å containing 64 molecules.

1.4.4 ABO_2 compounds with orthorhombic structures

The well-established orthorhombic structure of β-$NaFeO_2$ is not found for many other AMO_2 compounds, only the sodium and lithium aluminates and gallates forming this structure. Recently[31], $KCuO_2$ has been found to have an orthorhombic structure with the cell constants $a = 4.37$ Å, $b = 11.75$ Å, $c = 5.42$ Å. Weissenberg and precession photographs gave the space group D_{2h}^{17}–$Cmcm$ and powder data showed that the compounds $RbCuO_2$ and $CsCuO_2$ were isotypic. The basic unit in this structure is the $[CuO_2]^-$ group which exists as one dimensional, infinite chains with the Cu^{3+} ion in a 4-coordinate planar configuration. The Cu—O and O—O bond lengths suggest that the Cu—O bond is very covalent in these compounds. The square-planar CuO_4 groups share edges and extend along the c-axis of the crystal structure.

$\longrightarrow c$ axis

The oxygen lattice in the structure extends along the a and c axes in the form of a series of trigonal pyramids with the K^+ ion at the centre of alternate prisms along the a-axis. The stacking sequence along the c-axis is ABAB.

The alkali aurates(III), $RbAuO_2$ and $CsAuO_2$, also have the $KCuO_2$ structure[32], but the potassium aurate, $KAuO_2$, has a structure which exhibits certain differences, namely the cell constant b, which is approximately half that found for $KCuO_2$. The unit cell of $KAuO_2$ contains a single molecule whilst that of $KCuO_2$ contains four molecules, but both contain the characteristic infinite one-dimensional chain, $^1_\infty[MO_{4/2}]^-$, along the c-axis. The atomic positions of the potassium ions ($\frac{1}{2}\frac{1}{2}0$) indicate that the alkali metal ion is placed at the body centre of a cubic array of oxygen ions rather than the triangular prisms as in $KLnO_2$. A plot of the b-axis dimension of the alkali metal aurates(III) against the ionic radius of the alkali metal proves to be linear, and the value of the b' axis of $KAuO_2$ lies on the same line if doubled.

The crystal structure of $NaCuO_2$ has been determined[33] to be triclinic ($a = 2.74$ Å, $b = 6.67$ Å, $c = 3.46$ Å, $\alpha = 76.2$ degrees, $\beta = 113.4$ degrees, $\gamma = 128.1$ degrees) with the space group C_i^1 $P\bar{1}$. The unit cell contains one formula unit and like $KCuO_2$ a chain arrangement $[CuO_2]^-$ has been found. The c-axis dimension is extremely close to that of $KCuO_2$ and $RbCuO_2$, indicating the alignment of the infinite chain along the c-axis. The sodium ion is placed at the body centre of the unit cell and is octahedrally coordinated with oxygen ions.

1.4.5 Alkali-rich compounds in the A_2O-M_2O_3 system

Relatively few compounds are known in the alkali-rich ($A_2O:M_2O_3 > 1:1$) region of this system, particularly where M is a transition element. In the past 3 years, however, Hoppe and his co-workers have established the existence of three compounds in this category. Na_3CuO_3 may be prepared[33] as bright black, diamagnetic, single crystals by the reaction of sodium peroxide with CuO at 600 °C under oxygen. According to single-crystal work, the compound has triclinic symmetry with the cell constants $a = 18.25$ Å, $b = 16.78$ Å, $c = 11.12$ Å, $\alpha = 136.2$ degrees, $\beta = 106.9$ degrees, $\gamma = 90.6$ degrees; the unit cell contains 36 formula units. The compound most certainly does not contain a peroxy grouping as it may also be prepared by the thermal decomposition of $NaCuO_2$:

$$3NaCuO_2 \rightarrow Na_3CuO_3 + 2CuO + \tfrac{1}{2}O_2$$

The compounds Li_3AuO_3, Li_5AuO_4 and Na_3AuO_3 have also been prepared by Hoppe et al.[32]. Li_3AuO_3 has a tetragonal structure ($a = 9.111$ Å, $c = 3.576$ Å) with four molecules in the unit cell, Li_5AuO_4 is orthorhombic ($a = 3.673$ Å, $b = 9.505$ Å, $c = 2.940$ Å) with one molecule in the unit cell, and Na_3AuO_4 is also orthorhombic ($a = 8.56$ Å, $b = 10.5$ Å, $c = 9.00$ Å) but with eight molecules in the unit cell.

The compound Li_5AuO_4 may be regarded as an isotype of Li_2PdO_2 and Li_2CuO_2 having the same space group D_{2h}^{25}-$Immm$. The compound may be written as $Li_4(Li_{0.5}Au_{0.5})_2O_4$ since the atomic position 0,0,0 is occupied by both Li^+ and Au^{3+} ions. The structure also contains the one-dimensional infinite chains, $[AuO_{4/2}]^-$, running along the c-axis, which were found in the compounds $KAuO_2$, $RbAuO_2$ and $CsAuO_2$. The b-axis dimension also lies on the straight-line plot for the *b*-axis dimension of these AMO_2 compounds. Li_2AuO_3 differs from the other alkali metal aurates(III) in having $[Au_2O_6]^{6-}$ groups instead of the now familiar $[AuO_{4/2}]$ chains.

1.4.6   General comments on compounds of the type ABO_2

The efforts of various workers to correlate the structure type of the AMO_2 compounds with their chemical composition have not produced a totally viable theory. A linear relationship of the lattice parameter *a* with the ionic radius of the metal 3+ ion has been shown to hold for the $K-LnO_2$ series of compounds but has not been rigorously applied to other systems. Figure 1.4

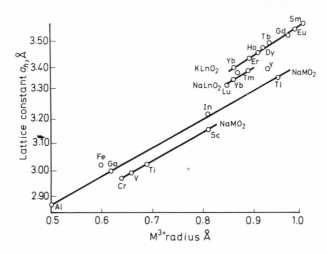

Figure 1.4   The variation of the lattice constant a_h for compounds with the α-NaFeO$_2$ structure with the ionic radius of the M^{3+} ion

shows such variation of the lattice parameter *a* with the M^{3+} radius for compounds with the αNaFeO$_2$ type of structure. Several points of interest arise from the diagram.

(1) A linear relationship is found for the compounds $NaMO_2$, where M = Al, Ga, In and Tl.

(2) In the sodium-lanthanide-oxygen compounds only a few lanthanide elements form the α-NaFeO$_2$ structure and a stability range was postulated with the ionic radius of the M^{3+} ion as the limiting factor.

(3) In the sodium–transition element–oxygen series of compounds, a linear relationship is found for the elements Cr, Ti and Sc, but the values for $NaVO_2$ and $NaFeO_2$ are anomalous. Whilst it may seem reasonable for the Fe^{3+} ion to show anomalous behaviour to the other M^{3+} ions there is no obvious reason for the V^{3+} ion to be different from either Cr^{3+} or Ti^{3+}. A re-calculation of the lattice parameters of $NaVO_2$ show that the data of Rudorff and Becker are probably incorrect and that the compound $NaVO_2$ has the approximate parameters $a = 3.01$ Å, $c = 16.07$ Å. These values show a much better correlation with the values obtained for $NaCrO_2$ and $NaTiO_2$.

Whilst α-$NaFeO_2$ is usually the representative compound in this series, it ought to be borne in mind that the chemical behaviour of this compound is not as simple as might first be thought. α Sodium ferrite is prepared by the reaction of stabilised γ-Fe_2O_3 with either sodium carbonate or sodium hydroxide; the use of α-Fe_2O_3 in the reaction leads to the formation of β-$NaFeO_2$. The $\alpha \rightarrow \beta$-phase transition temperature of 760 °C is less than the temperature used in the preparation, so that the reason for the formation of the two forms must lie elsewhere.

1.5 ALKALI METAL MONOXIDE–METAL(IV) OXIDE SYSTEMS

1.5.1 Alkali metal titanates

A study[34] of the reaction of sodium hydroxide with titanium dioxide by systematic x-ray powder analysis has been used to identify the titanates formed and to gain some insight into the process of their formation. The use of sodium hydroxide in place of the more usual sodium carbonate gave more vigorous reactions at lower temperatures, an important factor when studying the mechanisms of formation of compounds of limited stability. Titanates with the empirical formula, $Na_2O \cdot nTiO_2$, were prepared, where $n = 1$, 5/4, 2, 3 and 6, and a titanate of indefinite composition was found with n lying between 1 and 5/4 and having the interplanar spacings of Na_2TiO_3. An attempt to prepare the compound with $n = 2$, Na_4TiO_4, was not successful, as reaction was not complete even in a fused mass at 900 °C. All the products contained excess sodium hydroxide and the x-ray pattern obtained for the titanate was not deemed reliable.

The dititanate $Na_2O \cdot 2TiO_2$, the existence of which has been questioned by several previous workers, may be prepared by heating mixtures of sodium hydroxide and titanium dioxide (Na_2O:TiO_2 = 1:2) at 665–670 °C. At temperatures above 670 °C the compound decomposes with the formation of $4Na_2O \cdot 5TiO_2$ at 800 °C and $Na_2O \cdot 3TiO_2$ at higher temperatures. A great number of lines were observed in the pattern at 800 °C which could not be characterised but which disappeared completely on raising the temperature to 900 °C. Heating the same mixture (Na_2O:TiO_2 = 1:2) at lower temperatures showed that at 250 °C the metatitanate is the primary product together with the anatase form of titanium dioxide. On raising the temperature to 650 °C, the formation of the metatitanate increases whilst lines due to anatase decrease. At 665–670 °C the dititanate becomes the main phase. The small temperature stability of the dititanate is the reason why many investigators have not been able to obtain the phase by the reaction of sodium carbonate and titanium dioxide, since the reaction only proceeds sufficiently fast at 800 °C at which temperature the dititanate is unstable.

The sequence of reaction for the system Na_2O–TiO_2 was investigated for all the compounds formed. In every case the metatitanate is the first product, the formation of other titanates being determined by both the molar ratio of the reactants and the temperature involved.

A second study[35] of the Na_2O–TiO_2 system has shown that in the reaction of sodium carbonate with titanium dioxide with molar ratios greater than

1:3, an intermediate titanate is formed. The compound was not isolated in a pure state, but chemical analyses and quantitative x-ray powder diffraction methods on mixtures containing this phase showed it to be the compound tetrasodium trititanate ($2Na_2O \cdot 3TiO_2$). The compound decomposes with the formation of $Na_2O \cdot 3TiO_2$ and $4Na_2O \cdot 5TiO_2$ so that it becomes extremely difficult to obtain in the pure state.

Two potassium-rich titanates have been prepared[36] by the stoichiometric reaction of potassium monoxide and anatase at 400 °C. The orthotitanate, K_4TiO_4, is a white, extremely hygroscopic, powder and has triclinic symmetry. Although isostructural with the orthostannate, K_4SnO_4, the orthotitanate reacts differently with aqueous media, being completely hydrolysed with the formation of the gel $TiO_2 \cdot xH_2O$; K_4SnO_4 reacts with water to form the hexahydroxy stannate, $K_2Sn(OH)_6$.

The metatitanate, K_2TiO_3, may also be prepared by the thermal degradation of K_4TiO_4 which has limited stability, decomposing at 650 °C according to the equation:

$$K_4TiO_4 \rightarrow K_2TiO_3 + K_2O$$

The compound is hygroscopic, and on exposure to the atmosphere a hydrated form having the x-ray pattern previously reported for the anhydrous compound is produced. Complete hydrolysis yields the gel $TiO_2 \cdot xH_2O$.

1.5.2   The alkali-rich zirconates and hafnates

Crystal structure determinations[37] have shown both Li_2ZrO_3 and Li_2HfO_3 to have a monoclinic unit cell containing four formula units. The use of MAPLE (Madelung Part of the Lattice Energy) calculations enabled the positions of the lithium atoms to be confirmed. The use of such calculations as a means of structure determination will be dealt with later in the review. The alkali rich compounds Li_4MO_4 and Li_8MO_6 may be prepared[38] by the reaction of lithium oxide with zirconia and hafnia, but no corresponding

Table 1.3

Oxy compound	Reaction temperature °C	Most alkali rich compound
K_2CO_3	950	$K_2Zr_3O_7$
KOH	600	α-$K_2Zr_2O_5$
KNO_3	800	K_2ZrO_3
K_2O	450	K_4ZrO_4
K_2CO_3, KOH and KNO_3	No reaction with ThO_2	—
K_2O	550	K_2ThO_2

thorates may be prepared with alkali contents greater than Li_2ThO_3.

The action of some oxycompounds of potassium on zirconia and hafnia has revealed[39] some interesting trends in the relative reactivity of the oxy compounds (Table 1.3).

In these reactions, the less acid the gas formed, the richer in potassium monoxide is the most basic compound obtained. The reaction temperature is also an important parameter since many of the potassium zirconates possess limited temperature stability. In the case of potassium carbonate and potassium nitrate, the compounds start to decompose by loss of potassium monoxide at the temperature at which they are formed. Only by careful choice of starting composition (by making allowance for losses due to volatilisation), temperature, and heating time may single phases be formed. The thermal decomposition of the potassium zirconates is made more complex by the temperature-dependence of the decomposition products of K_2ZrO_3. Between 570 °C and 850 °C the metazirconate forms α-$K_2Zr_2O_5$, but from 720 °C this compound starts to dissociate with the formation of monoclinic zirconia. Above 850 °C the final decomposition product is always monoclinic zirconia, but two intermediate phases occur, the dizirconate, β-$K_2Zr_2O_5$, and the trizirconate, $K_2Zr_3O_7$.

1.5.3 Structural aspects of compounds in the K_2O–MO_2 system

The large number of compounds in the potassium oxide–metal dioxide system enables a comparison of their various crystallographic characteristics to be made. Few comparisons may be made for compounds of the 1:3 (K_2O:MO_2) type since the governing factors in determining the structure type are the size and configuration of the M^{4+} ion. Thus $K_2Zr_3O_7$ and $K_2Hf_3O_7$ are isotypes being tetragonal, $K_2Pb_3O_7$ and $K_2Sn_3O_7$ are orthorhombic and again are isotypic. Of the 1:2 compounds known, only β-$K_2Zr_2O_5$ and $K_2Hf_2O_5$ are isotypes with the orthorhombic structure. The 1:1 compounds are, however, much more regular in their structure types; only the metagermanate, K_2GeO_3, and the metathorate, K_2ThO_3 are exceptions. The compounds of the type K_2MO_3, where M = Ti, Zr, Hf, Sn and Pb, all crystallise with orthorhombic symmetry in the space group D_{2h}^{16}, and the ionic radius of the M^{4+} ion in these compounds, which changes from 0.68 Å to 0.84 Å, has little influence. The structure of K_2ThO_3 is monoclinic, and is isomorphous with Na_2ZrO_3, while K_2GeO_3 has an x-ray pattern completely different from either of the two others. All phases that correspond to a K_2O/MO_2 ratio of 2 are isomorphous (K_4GeO_4, K_4TiO_4, K_4SnO_4, K_4ZrO_4 and K_4PbO_4) and crystallise in the triclinic system.

1.5.4 The lithium monoxide–molybdenum dioxide system

A recent study[40] of the lithium oxide–molybdenum dioxide system revealed the existence of the compound $Li_4Mo_5O_{12}$ and a solid solution with the composition range Li_2MoO_3 to $Li_6Mo_2O_7$. The three definite compounds were all indexed in the monoclinic (C_{2h}^6) symmetry and were isotypic with the compound Li_2SnO_3. Since the compounds Na_2SnO_3 and Na_2PbO_3, which are isostructural with Li_2SnO_3, also possess a second cubic form it was thought that the molybdenum compounds might also possess this structural modification. The use of lithium hydroxide instead of lithium oxide enabled the cubic α-forms of Li_2MoO_3 and $Li_6Mo_2O_7$ to be prepared at temperatures between 400 °C and 700 °C. Both compounds were found to be isostructural with the cubic form of Li_2TiO_3 $(O_h^7 \ F_{d3m})$. The unit cell contains the formula unit $Li_{64/3} \ Mo_{32/3} \ O_{32}$ with both Li atoms and molybdenum atoms occupying the d positions in the ratio 1Li:2Mo in Li_2MoO_3, and 1Li:1Mo in $Li_6Mo_2O_7$. The structurē is built up of planes of oxygen atoms with planes of metal atoms between the oxygen layers. Alternating between planes containing lithium alone are planes with lithium and molybdenum atoms in random configuration. In the solid-solution region proceeding from Li_3MoO_3 to $Li_6Mo_2O_7$, a progressive enrichment of lithium takes place in the planes containing both metal ions with the formation of vacancies in the oxygen layers. The transformation, cubic to monoclinic, takes place at 700 °C and is irreversible.

1.6 ALKALI METAL MONOXIDE—METAL(V) OXIDE SYSTEMS

1.6.1 Phase diagrams

This system is probably the most intensively investigated system in the whole area of ternary oxides in that it contains the compounds of the type ABO_3 which are probably the best characterised compounds in inorganic chemistry.

Phase diagrams for $Li_2O \cdot V_2O_5$ [41], $Li_2O \cdot Nb_2O_5$ [42] and $Li_2O \cdot Ta_2O_5$ [43] are well established as are the phase diagrams for $Na_2O \cdot Nb_2O_5$ [44, 46], $Na_2O \cdot Ta_2O_5$ [43], $K_2O \cdot V_2O_5$ [47], $K_2O \cdot Nb_2O_5$ [46], $K_2O \cdot Ta_2O_5$ [48], $Rb_2O \cdot Nb_2O_5$ [49] and $Cs_2O \cdot Nb_2O_5$ [50]. Recently, two further systems have been investigated namely the $Na_2O \cdot V_2O_5$ [51] and $Cs_2O \cdot V_2O_5$ [52] systems. Five compounds were found in the $Na_2O \cdot V_2O_5$ system, i.e. NaV_6O_{15}, $Na_8V_{24}O_{63}$, $NaVO_3$, $Na_4V_2O_7$ and Na_3VO_4, and interplanar spacings were quoted for all five compounds. Phase changes were observed for $NaVO_3$ (540 °C) and $Na_4V_2O_7$ (426 °C), but x-ray data were quoted for one form in each case. The x-ray data and phase transition temperatures for $NaVO_3$ are not in good agreement with the earlier data by Lukacs and Strusievici[53] who studied the phase change of $NaVO_3$ by differential thermal analysis, infrared spectroscopy and x-ray powder diffraction. The transition temperature reported by these workers is between 403 °C and 405 °C; the irreversibility of the transition is, however, not in doubt.

The phase diagram for the system $Na_2O–NbO_2$ [54] has shown the existence of two sodium niobates(IV), namely Na_2NbO_3 and Na_4NbO_4. These new compounds were prepared by reaction of a mixture of sodium carbonate and a mixed oxide of the transition metal in the 5+ oxidation state in N_2 followed by reduction with hydrogen at 1000 °C. This method was used since direct reaction of sodium carbonate with the dioxide gave oxidised products, presumably due to the carbon dioxide liberated in the reaction.

The phase diagram of binary systems from alkali metal metavanadates[55] gave the melting points of KVO_3 520 °C, $RbVO_3$ 560 °C and $CsVO_3$ 640 °C with phase transition temperatures for $RbVO_3$ at 518 °C and $CsVO_3$ at 402 °C. The same authors[52] reported the melting point of $Cs_4V_2O_7$ at 892 °C with a phase transition at 342 °C and the Cs_3VO_4 melting point 1200 °C and a phase transition at 460 °C.

1.6.2 A_3MO_4 compounds

Le Flem[56] has shown that the compound Na_3VO_4 has two forms depending on the thermal treatment of the sample. The low-temperature form is difficult to prepare in a crystalline state, requiring lengthy heat treatment over a period of 6 days from 500 °C to 600 °C. The high-temperature form is obtained by heating at 1150 °C in oxygen for 3 days. The x-ray data quoted by Fotiev and Slobodin[51] have been shown by Le Flem to be due to a mixture of the high and low temperature forms of Na_3VO_4. The phase change does not appear to be reversible and the compounds themselves are highly hygroscopic.

Bouillaud[57] has shown that the previous x-ray data[44-46] for the compound Na_3NbO_4 may be explained on the basis of two forms of this compound, namely a cubic form and a monoclinic form. Chemical analysis gives the same formulation for each form yet the phase diagrams give no indication of a phase change up to the melting point. The interrelationship of the two forms is not fully understood, but the monoclinic form may be converted to the cubic form[58] by prolonged heating above 800 °C. Heating at 1200 °C causes the loss of sodium oxide with the formation of the niobate $NaNbO_3$. The densities of the cubic and monoclinic forms were found to be 3.83 and 3.84 g cm^{-3} respectively, which would explain why separation of the two phases by floatation is not possible. The closeness of the density values would also suggest that the structural relationship between the two compounds is very close, and that the thermal change at the transition temperature would be low, even if the transition were an abrupt one. The loss of sodium oxide from these compounds at high temperatures poses certain problems in the stoichiometry of the alkali-rich niobates and, in view of the apparent slowness of the phase transition, it might be that the two forms are not in fact isomorphs, but are the result of small deviations from the ideal stoichiometry.

A third niobate reported by Bouillaud could not be obtained in a pure state and only tentative chemical analyses were obtained. The conclusions drawn were that the compound was the hitherto unknown $Na_4Nb_2O_7$ and single-crystal crystallography gave a triclinic unit cell.

Several studies of alkali metal niobates and tantalates have been reported as a result of the interest in liquid metal corrosion studies. The compounds K_3NbO_4 [59] and K_3TaO_4 [60] are formed as corrosion products of Nb and Ta in liquid potassium, and are also formed in the reaction of the pentoxides[61] with liquid potassium. The compound Na_3UO_4 has also been prepared in the reaction of uranium oxides with liquid sodium[62], and by the reaction of uranium dioxide with liquid sodium saturated with oxygen[63]. The idea of using the alkali metals in the liquid state as a preparative medium is an interesting one and may result in the preparation of several new ternary oxides. The existence of different phase modifications of the potassium orthoniobates and tantalates has been confirmed by x-ray diffraction[59-61] and by observation of the behaviour on cooling of single crystals drawn from the melt[64]. Single crystals of K_3TaO_4 and K_3NbO_4 cracked on cooling at 900 °C and 500 °C, and it seems unlikely that these compounds may be prepared as single crystals in atmospheric conditions.

Isostructural aspects of A_3MO_4 compounds have been studied by Tarte[65] and Blasse[66]. Li_3PO_4, Li_3AsO_4 and Li_3VO_4 all form isostructural phases at low temperatures, having an orthorhombic unit cell with the space group $Pmn2_1 - C_{2v}^7$. Li_3PO_4 and Li_3AsO_4 have isomorphic high-temperature forms, closely related to the low temperature modifications (b dimension doubled in high-temperature forms). The close similarity in structure as derived from the x-ray powder patterns, i.r. spectra, and the reversible nature of the transition, suggest that both structures are derived from non-deformed MO_4 tetrahedra and fourfold coordination of lithium. No phase change was reported for Li_3VO_4 by quenching, but a high-temperature x-ray study[27] has shown the identical behaviour of this compound also.

The preparation of alkali chromate(V) compounds has been the subject of several[68, 69] publications by Scholder, and all the alkali metal chromates(V) are now characterised by chemical analysis. Detailed x-ray structural studies have yet to be carried out on the heavy alkali metal chromates(V) but the lithium compound, Li_3CrO_4, appears to be isostructural with the corresponding vanadate and manganate.

1.6.3 AMO_3 compounds—physical data

A recent book[70] is to be recommended as giving an up-to-date presentation of the preparation and general chemistry of compounds with the perovskite structures.

The thermodynamics of the AMO_3 compounds has received little attention considering the wide application of their physical properties. An approximate calculation[71] of the heats of formation of the metaniobates, $LiNbO_3$, $NaNbO_3$ and $KNbO_3$, gave values of $\Delta H°$ $LiNbO_3$ = -315 kcal mol^{-1}, $\Delta H°$ $NaNbO_3$ = -306 kcal mol^{-1} and $\Delta H°$ $KNbO_3$ = -306 kcal mol^{-1}. The similarity in the heats of formation of sodium and potassium niobates is due to the large heat of reaction of K_2O with Nb_2O_5, which compensates for the fact that the heat of formation of K_2O is less than that of Na_2O. The heat of formation of $NaVO_3$ has been re-measured[72] using an improved value for V_2O_5, and gives $\Delta H°$ = -274.2 kcal mol^{-1}. Using the known entropy value[73], the free energy of formation for $NaVO_3$ is ΔG = -254.4 kcal mol^{-1}. The relative thermal stability of the alkali metatantalates was derived[74] from their reduction with hydrogen at 1200 °C. The metatantalates were much more stable than metaniobates, sodium and potassium metatantalates being more stable than the lithium, rubidium and caesium metatantalates.

1.6.4   Lithium metaniobate and tantalate

The compounds $LiNbO_3$ and $LiTaO_3$ have recently received a great deal of attention since single crystals of these compounds are of interest for a wide variety of applications, e.g. electro-optic modulators[75, 76], parametric oscillators[77], and harmonic generators[78]. These compounds, which are ferroelectrics, may readily be grown as large single crystals[79, 80] but certain variations in physical properties have been observed with changes in the method of preparation. In particular, the variation in the Curie temperature of lithium metatantalate was found to be dependent upon both temperature[81] and melt stoichiometry[82]. Changes were even observed[81] along the length of a single crystal. Although work on the measurement of the physical properties and applications has proceeded at a rapid rate, much has yet to be done on the understanding of the mechanisms responsible for these properties. The techniques of absorption spectroscopy[83], Raman spectroscopy[84], nuclear magnetic resonance[85, 86] and x-ray diffraction have been used to gain some information on the chemical bonding, so as to provide a starting point for theories of the various physical properties.

The spontaneous polarisation of $LiTaO_3$, previously thought to increase with temperature up to 450 °C, is now considered to decrease with temperature[87]. The Curie temperatures for $LiNbO_3$ and $LiTaO_3$ are now considered to be at 1140 °C and 665 °C respectively. The results of high-temperature x-ray studies[88, 89] on these compounds have been used to explain the observed changes in physical properties. The temperature dependence of the cell constants a_h and c_h of $LiTaO_3$ [88] at 31 temperatures between 20 °C and 750 °C showed that a_h increased smoothly with temperature whilst c_h, which coincides with the direction of spontaneous polarisation, had a more complicated behaviour. Some five phase transitions were detected from the variation in c_h, but the unit cell maintained its hexagonal symmetry throughout the temperature range. Several points emerge from this work.

(1) The ferroelectric phase transformation at 665 °C takes place without a change in unit cell symmetry. This result contradicts the mechanism predicted by Megaw[90] according to which the unit cell should change to perovskite above the transition temperature.

(2) There is a regular decrease in spontaneous polarisation with increasing temperature above 110 °C, but at 510 °C and 615 °C discontinuous changes take place. This contradicts the earlier observation[91] of an increase in spontaneous polarisation with temperature.

(3) The mechanism of the ferroelectric transformation in $LiTaO_3$ is similar to that proposed for $LiNbO_3$; namely a displacement of the Li^+ ions towards the nearest oxygen layer accompanied by a displacement of the Ta^{5+} ions, such that they adopt a symmetric position. This causes a removal of the dipole moment of the unit cell and the formation of the para-electric phase.

(4) The gradual decrease of the spontaneous deformation of the unit cell leads to the belief that $LiTaO_3$ is in fact a hard ferroelectric rather than a frozen-in[90] ferroelectric.

A high-temperature x-ray diffraction study[89] of the isostructural niobate $LiNbO_3$ showed eleven phase transformations between 20 °C and 1170 °C. The phase transition temperature explained the change in dielectric constant and relative expansion anomalies observed by Smolenski et al.[92], and the conclusions drawn as to the fall in spontaneous polarisation with increasing temperature were confirmed by the results of Savage[93].

The behaviour of $LiTaO_3$ and $LiNbO_3$ are therefore closely related and the mechanisms put forward to explain the changes in physical properties are analogous. A more complete structural study of $LiNbO_3$ [94] has shown that it is indeed possible for the Li^+ ion to slip through the triangular face of its octahedron into the adjacent octahedral site, with a movement of the Nb^{5+} ion within its own octahedron parallel to the triad axis. The result is a change in the size of the niobium octahedra, and interchange in shape between the original Li octahedron and the adjacent empty octahedron. A complete structural study at the higher temperatures is now required in order to fully understand the numerous changes in the cell dimension c_h and to explain the observed changes in physical properties.

1.6.5   Sodium metaniobate

The heats of transition for the complex series of phase changes in $NaNbO_3$ have been re-measured[95]. No heat changes were observed at 420 °C or 478 °C but the transition enthalpy and entropy were measured for the changes at 372°, 527°, 576° and 640 °C. Below 372 °C the compound is antiferro-electric, while above 372 °C its electrical state is still in dispute. Substitution of some of the sodium atoms by lithium atoms[96] promotes instability of the antiparallel dipole in $NaNbO_3$, and makes easier the conversion from the antiferroelectric phase to the ferroelectric phase by application of an electric field. The system $(Na_{1-x}Li_x)NbO_3$ exhibits single-phase solid solution for values of x up to 0.14, with a multiple perovskite structure based on a mono-clinic sub-cell at low temperature. A pseudo cubic high-temperature phase was observed, the transition temperature decreasing with increasing lithium content. For the composition with $x = 0.04$, the transition temperature was at a minimum at 350 °C.

Sodium metaniobate can be deficient in either oxygen or sodium oxide content[97]. The reduced phase, prepared by prolonged heating at 1200 °C in a reducing atmosphere, is black and paramagnetic, and has a lower resistivity than the stoichiometric phase; no change in the x-ray powder pattern is however observed.

Re-oxidation of the reduced phase is accompanied by a weight gain ($\sim 2\%$) and establishes the composition of the reduced phase as $NaNbO_{2.98}$.

The loss of sodium oxide from $NaNbO_3$ gives a single phase region of composition $Na_{1-x}NbO_{3-\frac{x}{2}}$ with a limit at the composition $Na_{0.90}NbO_{2.95}$; further loss of sodium oxide is accompanied by the formation of $NaNb_3O_8$. The loss of sodium oxide, which has the effect of diminishing the monoclinic deformation of the pseudo cubic cell, is achieved by thermal treatment under vacuum.

1.6.6   Alkali-rich vanadates

The system KVO_3–V_2O_5 has in the past been the subject of much contro-versy. In an excellent series of papers, Pouchard[98–100] has re-investigated this area and removed many of the points of confusion. These were due, in many cases, to the use of high temperatures leading to oxygen loss and the for-mation of the compound, $K_{2-x}V_{3+2x}O_{8+2x}$. This has an extremely large, two-dimensional, composition range and exhibits all the aspects of a bronze-type compound with a monoclinic structure derived from KV_5O_{13}. The complex phase diagram for the K_2O–V_2O_5–V_2O_4 system has now been thoroughly investigated and a large number of new phases characterised.

The compound KV_6O_{15}, prepared[101, 102] by the reaction of 6 mol V_2O_5 with 1 mol of K_2CO_3, is stable to oxygen and air up to 580 °C. Under slow cooling, however, the phase decomposes to V_2O_5 and $KV_4O_{10.5-x}$ below 528 °C and may only be obtained by quenching. The compound $KV_4O_{10.5-x}$ is stable at low temperatures only, decomposing on melting at 528 °C. Similar behaviour is observed[99, 101] for KV_3O_8 which melts and decomposes at 486 °C, a factor which undoubtedly led to previous investigators not observing its formation.

A study[103] of the sodium vanadium bronzes $Na_xV_2O_5$, where x is less than 2, has revealed the existence of four new compounds. The compounds $Na_xV_2O_5$ are prepared by heating a mixture of V_2O_5 with a mixture of VO_2 or V_2O_3 and the meta- or pyro-vanadate of sodium in a sealed tube at 600 °C in argon. The general equations are:

$$(1-x)V_2O_5 + xVO_2 + xNaVO_3 \rightarrow Na_xV_2O_5 \qquad x \leqslant 1$$
$$(1-\tfrac{3x}{4})V_2O_5 + \tfrac{x}{4}V_2O_3 + xNaVO_3 \rightarrow Na_xV_2O_5 \qquad x \leqslant 3$$
$$(1-\tfrac{x}{2})V_2O_5 + \tfrac{x}{4}V_2O_3 + \tfrac{x}{4}Na_4V_2O_7 \rightarrow Na_xV_2O_5 \qquad x \leqslant 2$$

The compound $Na_2V_2O_5$ ($x = 2$) could not be prepared by the above method. The products of reaction were always $Na_4V_2O_7$ and V_2O_3, and for values of x between 1.82 and 2.00 these compounds co-existed with the bronze $Na_{1.8}V_2O_5$. The compounds which exist in the $Na_xV_2O_5$ system are: a phase α with values of x from 0 to 0.02, isotypic with the corresponding α-$Li_xV_2O_5$ compound; β-$Na_xV_2O_5$ where x has the range 0.22–0.40, which has a monoclinic cell; α'-$Na_xV_2O_5$ with x from 0.70 to 1.00 and having orthorhombic symmetry; η-$Na_xV_2O_5$ with x from 1.28 to 1.45; and finally $Na_xV_2O_5$ with x from 1.68 to 1.82, having rhombohedral symmetry.

The α-phase and α'-phase are closely related to the oxide V_2O_5 in their structures and may be regarded as insertion compounds. The α-phase represents the insertion of only a very small concentration of sodium into the V_2O_5 lattice with correspondingly little change in the size of the unit cell. The environment of the vanadium is unchanged and all the vanadium atoms are in equivalent bipyramids. The phase α' represents the insertion of a much larger concentration of sodium; the unit cell volume increases from 179 Å^3 for V_2O_5, to 196 Å^3 in the α'-phase, and the c axis increases from 4.369 Å to 4.791 Å. The structure also differs in the environment of the vanadium atoms in that, in the α'-phase, there exist two different types of vanadium atoms resulting in a change of space group from D_{2h}^{13} to C_{2v}^7. This is largely due to the presence of V^{4+} ions as well as V^{5+} ions in the bronze. The V^{4+} ions occupy larger sites than do the smaller V^{5+} ions and the vanadium–oxygen distances are 1.88 Å and 1.82 Å.

The electrical conductivity, Hall voltage, Seebeck effect and e.s.r. properties of the β-phase $Na_{1.33}V_2O_5$ have been extensively studied in order to elucidate the nature of the electron transport mechanism in the compound. Numerous theories have been proposed, but no single theory was able to fit all the experimental data. Recent work[104] has shown that the Hall effect data give the number of electron carriers in good agreement with the sodium concentration, and the mobility being low, points to a hopping mechanism. Magnetic susceptibility measurements indicate that the activated conduction process does not involve an oxygen intermediary, but that the electron jumps from one vanadium site to another. The metal-metal separation in $Na_{0.33}V_2O_5$ is 3.07 Å, which is larger than the critical value calculated for overlap of the 3d orbitals of vanadium. Thus the Goodenough theory also predicts a hopping mechanism for this compound.

1.6.7 Hypovanadates

A study[105] of the non-stoichiometric hypovanadates (Na_2O-VO_2 system) showed three phases to be present in the composition range between vanadium dioxide and the hypothetical compound $Na_2V_2O_5$. The first phase, having a composition range between $Na_2V_{12}O_{25}$ and $Na_2V_8O_{17}$, was found to have an x-ray powder pattern corresponding to the α' phase NaV_2O_5 in the $Na_xV_2O_5$ system. The cell constants of the orthorhombic unit cell remained practically constant between the whole of the range from the α'-NaV_2O_5 compound to $Na_2V_8O_{17}$ or $Na_2V_{12}O_{25}$. However, the change in density with V/Na ratio enabled two mechanisms to be postulated depending on the value of the ratio V:Na. The change in composition from NaV_2O_5 to the hypovanadates dictates a change in oxidation state of the vanadium from 4.5 to 4.0 and a change in the ratio V:Na from 2 to either 4 or 6. The postulated mechanism is concerned with the filling of the vacancies caused by the loss of sodium with either vanadium or oxygen atoms. For low values of the ratio V/Na (between 2 and 2.5) it seems that a small fraction only of the vacancies resulting from the loss of sodium atoms are occupied by oxygen, but that almost all the octahedral sites formed are occupied by vanadium atoms. For higher values of the ratio V:Na, the available vacancies are progressively occupied by oxygen atoms and the rate of insertion of vanadium atoms decreases. The observed densities may then be explained by the substitution, atom by atom, of sodium by oxygen for the ratio V:Na approximately 4:5. The progressive change of phases from the bronze type to the hypovanadates might also be interpreted as the partial, or total, occupation of the sites made available by the loss of sodium by chains of composition $(V_mO)_n$ with m varying between 4 and 1.2.

The remaining two compounds in the Na_2O-VO_2 system possess much smaller ranges of composition, but also corresponded crystallographically to the vanadium bronze type of compound rather than a ternary oxide of vanadium 4+, e.g. the compound $Na_2V_3O_7$ was not isostructural with the corresponding compound of titanium, $Na_2Ti_3O_7$, but was isostructural with the bronze $Na_xV_2O_5$ where $x = 1.68$ to 1.82.

Two intermediate compounds have been established in the $V_2O_5-LiVO_3$ system by x-ray diffraction and optical phase analysis[106], namely LiV_6O_{15} and $LiV_{2.5}O_{6.75-x}$. The defect nature of the oxidation state in the compound LiV_6O_{15} is directly related to the quantity of oxygen liberated in the crystallisation of the compound.

$$MV_6O_{16}(liq) \rightleftharpoons MV_6O_{15}(s) + \tfrac{1}{2}O_2(g)$$

Whereas the sodium vanadium bronze, which crystallises from the melt in an atmosphere of air, is stable in the range of low temperatures, the lithium and potassium vanadium bronzes of the same composition, and under the sáme conditions, are metastable compounds. Slow cooling of a melt (close to equilibrium) with molar ratio $Li_2O:V_2O_5 = 1:6$ is accompanied by the formation of two phases below 585 °C.

$$LiV_6O_{16}(liq) \rightarrow V_2O_5(s) + LiV_{2.5}O_{6.75-x}(s) + xO_2(g)$$

This process is made up of a series of intermediate steps, including the oxygen loss reaction, to form LiV_6O_{15} and the subsequent interaction of this compound with oxygen to form V_2O_5 and $LiV_{2.5}O_{6.75-x}$.

No evidence was found for the existence of $Li_{0.04}V_2O_5$ or $Li_{0.21}V_2O_5$, the only phases found in this region of the system being V_2O_5, LiV_6O_{15} and $LiV_{2.5}O_{6.75-x}$.

1.7 ALKALI METAL MONOXIDE–METAL(VI) OXIDE SYSTEMS

1.7.1 A_2O–MO_3 system

A re-examination[107] of the A_2MO_4–MO_3 system, where A is Na or K and M molybdenum or tungsten, has shown that the trimolybdate of sodium does not exist, nor the tritungstate. Solid solution was found to take place between the tetramolybdates and MoO_3, and the tetratungstates and WO_3, from the composition $A_2M_4O_{13}$ to $A_2M_4O_{13}\cdot2MO_3$ in the case of A being sodium, and $A_2M_4O_{13}\cdot4MO_3$ for A being potassium. A study of the variations in density and in the intensity of the x-ray lines, led to the belief that the solid solutions are doubly deficient in both alkali metal oxide and transition metal oxide.

The alkali rich portion of the A_2O–MO_3 system has not been studied until very recently[66]. A comprehensive examination of the system[108] showed the following phases to be formed in the $Li_2O\cdot MoO_3$ and $Li_2O\cdot WO_3$ systems. For ratios $Li_2O:MO_3$ between 1 and 2, a mixture of Li_2MO_4 and Li_4MO_5 compounds were obtained. The compounds Li_4MoO_5 and Li_4WO_5 were prepared in a pure state by the reaction of a 2:1 molar mixture of Li_2O and MoO_3 at 350 °C; and a 2:1 mixture of Li_2O and WO_3 at 300 °C, in a sealed tube. An excess of Li_2O had no reaction with Li_4MoO_5 up to 800 °C–the maximum temperature studied, but lithium oxide reacted with Li_4WO_5 at 500 °C to form the new compound Li_6WO_6. These compounds are white, or slightly yellow, hygroscopic powders and are soluble in water. The compound Li_4WO_5 has two modifications with an irreversible transformation at 690 °C. On the evidence of lines attributed to a superstructure, the unit cell dimension of the low-temperature phase of Li_4WO_5 was doubled to 8.29 Å. The high temperature phase closely resembles that of Li_2TiO_3, the dimensions b and c being identical. The dimension a is, however, almost five times larger, suggesting an ordering of the lithium and tungsten atoms, with three atoms of lithium and two atoms of tungsten in the unit cell. The compound Li_4MoO_5 has four allotropic modifications, with transformations α to β at 450 °C, $\beta \rightarrow \gamma$ at 680 °C, and $\gamma \rightarrow \delta$ at 840 °C; all the transformations are irreversible. The α-form of Li_4MoO_5 has a cubic cell $a = 4.14$ Å and the δ-form is isostructural with the high-temperature form of Li_4WO_5. The β- and γ-forms are both of orthorhombic symmetry with 32 molecules of

Li_4MoO_5 in the unit cell. The tungstate Li_6WO_6, which indexes in the orthorhombic system, decomposes on heating to 1000 °C in oxygen to Li_2O and Li_2WO_4.

The reaction of sodium oxide with the trioxides at 450 °C gave three new phases, Na_4MoO_5, Na_4WO_5 and Na_6WO_6. Na_4WO_5 was isostructural with β-Li_4WO_5 but Na_4MoO_5 and Na_6WO_6 had no structural analogues with the corresponding lithium compounds.

1.7.2 Molybdenum and tungsten bronzes

The tungsten bronzes were first prepared in 1824 but the molybdenum bronzes proved to be much more difficult to characterise. The unusual electrical properties of A_xMO_3 compounds continue to create interest in the preparative[109, 110, 111] and structural aspects of these compounds. Wold et al. first reported the remarkable difference in electrical conductivity between the red and blue potassium molybdenum bronzes. The red bronze had a behaviour typical of a semiconductor being almost insulating at room temperature, whilst the blue bronze had much lower resistivity with an apparent change from semiconductor to metallic behaviour above −100 °C. The chemical formulae of the red and blue bronzes have been the subject of some discussion; the original formulation was undoubtedly in error and present data[112] give the red bronze as $K_{0.33}MoO_3$ and the blue bronze as $K_{0.30}MoO_3$. The basic unit of the red bronze is a cluster of six distorted MoO_6 octahedra connected by edge-sharing, giving a stoichiometry Mo_6O_{18}. Two K^+ ions are associated with each cluster and two electrons are therefore incorporated in the cluster. The basic unit of the blue bronze is a cluster, $Mo_{10}O_{30}$, made up of ten edge-sharing MoO_6 octahedra with three electrons in the cluster and three K^+ ions. In both bronzes the clusters are interconnected by corner sharing of octahedra.

The existence of isolated cluster orbitals is used to explain the observed low, temperature-independent magnetic susceptibility of the red bronze. The two electrons are probably spin-paired because of the occupancy of the same molecular orbital with accessible excited states in which the spins might be uncoupled. Such an arrangement of isolated cluster orbitals would also explain the observed low conductivity corresponding to the activated hopping of electrons from cluster to cluster. The lower magnetic susceptibility and good conductivity at room temperature of the blue bronze suggests a metallic model, in which the electrons are delocalised and mostly paired as a Pauli paramagnetic, degenerate, electron gas. However, both the Seebeck effect and Hall effect indicate p-type carriers at room temperature, and n-type carriers as the temperature is lowered. A satisfactory model for the blue bronze is therefore a low-temperature conductivity dominated by electron hopping, or excitation from donor levels into a narrow, low mobility, conduction band, and a high-temperature conductivity dominated by hole conduction in a high-mobility valence band.

The electronic energy level diagram for the molybdenum bronzes has been derived[113] from the u.v. and visible spectra of the blue and red bronzes. An e.s.r. signal was obtained from the blue bronze, being derived from the

Mo^{5+} or M^+/Mo^{5+} pairs providing donor levels approximately less than 1 eV below the conduction band.

The tungsten bronzes have, since their discovery, been the subject of many investigations. In some of the early studies an anomaly in the conductivity of the bronze Na_xWO_3 was found at a value of $x = 0.75$. This was later thought to be the result of inhomogeneity in the crystal used. Recent measurements[114] of the electrical resistivities, Hall coefficients, and Seebeck coefficients of metallic Na_xWO_3 as a function of the sodium concentration and of temperature have shown a definite anomaly at $x = 0.75$. At $x = 0.75$ the sodium atoms in the bronze are ordered but the small magnitude of the anomalies suggests that only partial ordering takes place. The Hall coefficients and Seebeck effect[115] measurements also indicate that there is not a 1:1 correspondence between the number of conduction electrons and sodium atoms. Whilst there is no doubt that the sodium atoms are all completely ionised, and that the number of free electrons is equal to the number of sodium atoms, the free-electron model for Na_xWO_3 is perhaps an over simplification.

1.8 CONCLUDING REMARKS

In the past few years a change of emphasis has become apparent in the study of the chemistry of ternary oxides. Very few investigations are concerned only with the preparation of new compounds, most papers now contain details of the structural aspects of the compounds. In many cases, however, details of the structure adopted by a compound comprises only a small part in our understanding of the whole chemistry of a system. The work of Hoppe and his co-workers attempts to extend the structural information available by considering why any particular structure is favoured when several possibilities exist. The MAPLE calculation has been used, with considerable success, to determine the most probable structure in a series of structures which could not be resolved by x-ray crystallographic methods. The theories of bonding in oxide compounds have not advanced at the same rate as in other areas, e.g. coordination compounds. The long-range order in ionic compounds poses greater problems in calculation than in the isolated covalent molecule and our attempts to understand such effects are still in their infancy. The ideas used to explain the magnetic and electrical properties of binary oxides are now being increasingly applied to ternary oxides, the major disadvantage being in many cases the lack of complete structural information which is essential before exact calculations of the bonding scheme may be performed.

The increasing interest in ternary oxides for electronic devices will, however, provide the necessary driving force for basic studies into the cause of the magnetic and electrical properties of oxides.

Parallel with an interest in the compounds themselves is an increase in the number of studies into the mechanism of reaction. Frequently these studies are associated with the reactivity of the alkali metal oxo compound used as a starting material. The use of alkali metal carbonates as sources of alkali metal oxides was almost universal until advances in practical methods enabled the much more reactive alkali metal oxides to be used themselves. It became apparent that not only were reaction temperatures lowered but that the choice of oxo compound could influence the phase formed in the reaction. An excellent example has already been quoted, namely the reactions of zirconia with the oxo compounds of potassium, where changing the alkali metal compounds caused a variation in the type of product formed.

References

1. Sabrowsky, H. and Hoppe, R. (1968). *Z. Anorg. Allg. Chem.*, **358**, 241
2. Sabrowsky, H. (1969). *Z. Anorg. Chem.*, **365**, 146
3. Hoppe, R., Hestermann, K. and Schenk, F. (1969). *Z. Anorg. Allg. Chem.*, **367**, 275
4. Hestermann, K. and Hoppe, R. (1968). *Z. Anorg. Allg. Chem.*, **360**, 113
5. Wasel-Nielen, H. D. and Hoppe, R. (1968). *Z. Anorg. Allg. Chem.*, **359**, 36
6. Velhaber, E. and Hoppe, R. (1965). *Z. Anorg. Allg. Chem.*, **338**, 209
7. Hoppe, R. and Velhaber, E. (1964). *Naturwissenschaften*, **51**, 103
8. Klemm, W., Wehrmeyer, G. and Bade, H. (1959). *Z. Electrochem.*, **63**, 56
9. Woltersdorf, G. (1943). *Z. Anorg. Allg. Chem.*, **252**, 126
10. Sabrowsky, H. and Hoppe, R. (1966). *Naturwissenschaften*, **53**, 501
11. Hoppe, R. and Rohrborm, H. J. (1961). *Z. Anorg. Allg. Chem.*, **312**, 87
12. Fouassier, C. (1968). *Bull. Soc. Chim. Fr.*, 1338
13. Bauer, H. (1956). *Dissertation*, Karlsruhe.
14. Gross, P. and Wilson, G. L. (1970). *J. Chem. Soc. A*, 1913
15. Clos, R., Devalette, M., Hagenmuller, P., Hoppe, R. and Paletta, E. (1967). *Compt. Rend.*, **265**, 801
16. Spitsyn, V. I., Muraveva, I. A. Kovba, L. M. and Korchak, I. I. (1969). *Russ. J. Inorg. Chem.*, **14**, 759
17. Gross, P., Wilson, G. L. and Gutteridge, W. A. (1970). *J. Chem. Soc. A*, 1908
18. Ringenbach, C., Kessler, H. and Hatterer, A. (1969). *Compt. Rend.*, **269**, 1394
19. Reid, A. F. and Ringwood, A. E. (1968). *Inorg. Chem.*, **7**, 443
20. Hoppe, R. and Werding, G. (1961). *Z. Anorg. Allg. Chem.*, **307**, 174
21. Hoppe, R. and Sabrowsky, H. (1968). *Z. Anorg. Allg. Chem.*, **387**, 202
22. Seeger, K. and Hoppe, R. (1969). *Z. Anorg. Allg. Chem.*, **365**, 22
23. Devalette, M., Fouassier, C. and Hagenmuller, P. (1970). *Mater. Res. Bull.*, **5**, 179
24. Goodenough, J. B. (1967). *J. Phys. B.*, **17**, 304
25. Fouassier, C. and Hagenmuller, P. (1968). *Bull. Soc. Chim. Fr.*, 1340
26. Hoppe, R. (1965). *Bull Soc. Chim. Fr.*, 1115
27. Anderson, J. C. and Schieber, M. (1964). *J. Phys. Chem. Solids*, **25**, 961
28. Hoppe, R. and Werding, G. (1961). *Z. Anorg. Allg. Chem.*, **307**, 174
29. Hoppe, R. and Schepers, B. (1958). *Z. Anorg. Allg. Chem.*, **295**, 233
30. Hoppe, R. and Rohrborn, H. J. (1964). *Z. Anorg. Allg. Chem.*, **327**, 199
31. Hoppe, R. and Hestermann, K. (1969). *Z. Anorg. Allg. Chem.*, **367**, 249
32. Wasel-Nielen, H. D. and Hoppe, R. (1970). *Z. Anorg. Allg. Chem.*, **375**, 43
33. Hestermann, K. and Hoppe, R. (1969). *Z. Anorg. Allg. Chem.*, **367**, 270
34. Batygin, V. G. (1967). *Russ. J. Inorg. Chem.*, **12**, 762
35. Belyaev, E. K., Panasenko, N. M. and Linnik, E. V. (1970). *Russ. J. Inorg. Chem.*, **15**, 336
36. Devalette, M. and Tournoux, M. (1965). *Bull. Soc. Chim. Fr.*, 2337
37. Dittrich, G. and Hoppe, R. (1969). *Z. Anorg. Chem.*, **371**, 306
38. Scholder, R., Rade, D. and Schwarz, H. (1968). *Z. Anorg. Allg. Chem.*, **362**, 149
39. Devalette, M., Fouassier, C., Le Flem, G., Tournoux, M. and Hagenmuller, P. (1968). *Science of Ceramics*, **4**, 381
40. Reau, J. M., Fouassier, C. and Gleitzer, C. (1967). *Bull. Soc. Chim. Fr.*, 4294
41. Kohlmuller, R. (1961). *Bull. Soc. Chim. Fr.*, 748
42. Reisman, A. and Holtzberg, F. (1958). *J. Amer. Chem. Soc.*, **80**, 6503
43. Reisman, A. (1962). *J. Phys. Chem.*, **66**, 15
44. Reisman, A., Holtzberg, F. and Banks, E. (1958). *J. Amer. Chem. Soc.*, **80**, 37

45. Shafer, M. W. and Roy, R. (1959), *J. Amer. Ceram. Soc.*, **42**, 482
46. Whiston, C. D. and Smith, A. J. (1965). *Acta Crystallogr.*, **19**, 169
47. Holtzberg, F., Reisman, A., Berry, M. and Berkenblit, M. (1956). *J. Amer. Chem. Soc.*, **78**, 1536
48. Reisman, A., Holtzberg, F., Berkenblit, M. and Berry, M. (1956). *J. Amer. Chem. Soc.*, **78**, 4514
49. Reisman, A. and Holtzberg, F. (1960). *J. Phys. Chem.*, **64**, 748
50. Reisman, A. and Mineo, S. (1961). *J. Phys. Chem.*, **65**, 996
51. Slobodin, B. V. and Fotiev, A. A. (1965). *Zh. Prikl. Khim.*, **38**, 801–6
52. Belyaev, I. N. and Golovanova, T. G. (1964). *Russ. J. Inorg. Chem.*, **9**, 125
53. Lukacs, I. and Strusievici, C. (1962). *Z. Anorg. Allg. Chem.*, **315**, 323
54. Ball, M. C. (1970). *Trans. Brit. Ceram. Soc.*, **69**, 157
55. Belyaev, I. N. and Golovanova, T. G. (1970). *Zh. Prikl. Khim.*, **43**, 892
56. Le Flem, G., Olazcuaga, R. and Hagenmuller, P. (1968). *Bull. Soc. Chim. Fr.*, 2769
57. Bouillaud, Y. (1967). *Bull. Soc. Chim. Fr.*, 3879
58. Bouillaud, Y. (1965). *Bull. Soc. Chim. Fr.*, 519
59. Stecura, S. (1971). *J. Less-Common Metals*, in press
60. Hickam, C. W. (1968). *J. Less-Common Metals*, **14**, 315
61. Addison, C. C., Barker, M. G. and Lintonbon, R. M. (1970). *J. Chem. Soc. A*, 1465
62. Addison, C. C., Barker, M. G., Lintonbon, R. M. and Pulham, R. J. (1969). *J. Chem. Soc. A*, 2457
63. Pepper, R. T., Stubbles, J. R. and Tottle, C. R. (1964). *Appl. Mater. Res.*, 203
64. Loiacono, G. M. and Michelman, B. (1968). *J. Amer. Ceram. Soc.*, **51**, 542
65. Tarte, P. (1967). *J. Inorg. Nucl. Chem.*, **29**, 915
66. Blasse, G. (1964). *Z. Anorg. Allg. Chem.*, **331**, 44
67. Hooper, A. J. (1971). Ph. D. Thesis University of Nottingham
68. Scholder, R., Schwochow, F. and Schqarz, H. (1968). *Z. Anorg. Allg. Chem.*, **363**, 10
69. Scholder, R. and Schwartz, H. (1963). *Z. Anorg. Chem.*, **326**, 1
70. Galasso, F. S. (1969). 'Structure, properties and preparation of perovskite-like compounds.' *International series of monographs in solid state physics* 5, (Oxford: Pergamon Press)
71. Voskrensenskaya, N. K. and Budova, G. P. (1969). *Russ. J. Inorg. Chem.*, **14**, 1566
72. Bertrand, G. L. and Hepler, L. G. (1967). *J. Chem. Eng. Data*, **12**, 412
73. Kelly, K. K. and King, E. G. (1961). *U.S. Bur. Mines Rep.*, 592
74. Kapitskii, A. V. and Artamonova, E. P. (1962). *Russ. J. Inorg. Chem.*, **7**, 986
75. Lenzo, P. V., Spencer, E. G. and Nassau, K. (1966). *J. Opt. Soc. Am.*, **56**, 633
76. Denton, R. T., Chen, F. S. and Ballman, A. A. (1967). *J. Appl. Phys.*, **38**, 1611
77. Giordmaine, J. A. and Miller, R. C. (1965). *Phys. Rev. Lett.*, **14**, 973
78. Miller, R. C., Boyd, G. D. and Savage, A. (1965). *Appl. Phys. Lett.*, **6**, 77
79. Nassau, K., Levirstein, H. J. and Loiacone, G. M. (1966). *J. Phys. Chem. Solids*, **27**, 983
80. Hill, V. G. and Zimmerman, K. G. (1968). *J. Electrochem. Soc.*, **115**, 978
81. Glass, A. M. (1968). *Phys. Rev.*, **172**, 564
82. Ballman, A. A., Levinstein, H. J., Capio, C. D. and Brown, H. (1967). *J. Amer. Ceram. Soc.*, **50**, 657
83. Valyashko, E. G., Rashkovich, L. N. and Timoshenkov, V. A. (1968). *Optika Spektroskopiya*, **24**, 637
84. Schantele, R. F. (1966). *Phys. Rev.*, **152**, 705
85. Golenischev-Kut, V. A. and Kopvillem, U. K. (1968). *Fiv. Tverd. Tela.*, **10**, 759
86. Peterson, G. E., Bridenbaugh, P. M. and Green, P. (1967). *J. Chem. Phys.*, **46**, 4009
87. Shapiro, Z. I., Tedalov, S. A., Veneskv, N. and Rigerman, L. G. (1966). *Kristallografiya*, **10**, 725
88. Ismailzade, I. G., Nesterenko, V. I. and Bairamov, A. A. (1968). *Kristallografiya*, **13**, 188
89. Ismailzade, I. G., Nesterenko, V. I. and Mirishli, F. A. (1968). *Kristallografiya*, **13**, 25
90. Megaw, H. D. (1957). *Ferroelectricity in crystals*, 103, (London: Methuen)
91. Matthias, B. T. and Remerka, J. P. (1949). *Phys. Rev.*, **76**, 1886
92. Smolenskii, G. A., Kraimik, N. N., Khuchua, N. P., Zhdahova, V. V. and Mylinikova, I. E. (1966). *Phys. Status. Solidi*, **13**, 309
93. Savage, A. (1966). *J. Appl. Phys.*, **37**, 3071
94. Megaw, H. D. (1968). *Acta Crystallogr.*, **24**, 583.
95. Tennery, V. J. and Hang, K. W. (1968). *J. Amer. Ceram. Soc.*, **51**, 469

96. Nitta, T. (1968). *J. Amer. Ceram. Soc.,* **51,** 626
97. Bouillaud, Y. (1969). *Bull. Soc. Fr. Mineral. Crist.,* **92,** 347
98. Pouchard, M., Galy, J., Rabordel, L. and Hagenmuller, P. (1968). *Compt. Rend.,* **264,** 1943
99. Pouchard, M. (1967). *Bull. Soc. Chim. Fr.,* 4271
100. Pouchard, M., Casalot, A. and Villeneuve, G. (1968). *Bull Soc. Chim. Fr.,* 81
101. Glazyrin, M. P. and Totiev, A. A. (1968), *Russ. J. Phys. Chem.,* **42,** 1288
102. Fotiev, A. A., Glazyrin, M. P. and Alyamovskii, S. I. (1967). *Russ. J. Inorg. Chem.,* **14,** 1325
103. Pouchard, M., Casalot, A., Galy, J. and Hagenmuller, P. (1967). *Bull. Soc. Chim. Fr.,* 4343
104. Perblein, J. H. and Sienko, M. J. (1968). *J. Chem. Phys.,* **48,** 174
105. Pouchard, M., Casalot, A., Rabardel, L. and Hagenmuller, P. (1968). *Bull. Soc. Chim. Fr.,* 2742
106. Fotiev, A. A., Glazyrin, M. P. and Bausova, N. V. (1968). *Russ. J. Inorg. Chem.,* **13,** 1007
107. Caillet, P. (1967). *Bull. Soc. Chim. Fr.,* 4750
108. Reau, J. M., Fouassier, C. and Hagenmuller, P. (1967). *Bull. Soc. Chim. Fr.,* 3873
109. Chamberland, B. L. (1969). *Inorg. Chem.,* **8,** 1183
110. Conroy, L. E. and Podolsky, G. (1968). *Inorg. Chem.,* **7,** 614
111. Gier, T. E., Pease, D. C., Sleight, A. W. and Bither, T. A. (1968). *Inorg. Chem.,* **7,** 1646
112. Bouchard, G. H., Perlstein, J. and Sienko, M. J. (1967). *Inorg. Chem.,* **6,** 1652
113. Dickens, P. G. and Nield, N. J. (1968). *Trans. Faraday Soc.,* **64,** 13
114. Muhlestein, L. D. and Danielson, G. C. (1967). *Phys. Rev.,* **158,** 825
115. Muhlestein, L. D. and Danielson, G. C. (1967). *Phys. Rev.,* **160,** 562

2
Chemistry of Molten Nitrates and Nitrites

D. H. KERRIDGE
University of Southampton

2.1 INTRODUCTION

Molten nitrates display one of the most interesting and varied chemistries so far known for any class of fused salt solvents. However, this variety is still being made apparent and is by no means fully understood as yet, for the majority of the known chemistry has been reported over the last 10 years. The explanation of this situation is that when the revival of interest in molten salts began in about 1950, attention was concentrated on the physical properties of fused salts with the hope that a theory of the interactions between ions in a purely ionic medium would enable progress to be made on the difficult problem of understanding concentrated aqueous solutions.

With this aim the initial choice of salts of simple spherical ions with single, or at worst identical, charges on cation and anion, and with small covalent interactions, was perfectly natural. However, this has resulted in concentrating much attention on the alkali metal halide melts which necessarily display a less varied chemistry. Moreover, this class of melts has the practical disadvantage of comparatively high melting points (e.g. LiCl 613 °C, KCl 776 °C, LiF 870 °C, KF 880 °C), and even the use of eutectic mixtures (e.g. LiCl/KCl (59/41 mol%) m.p. 352 °C and LiF/NaF/KF (46.5/11.5/42 mol%) m.p. 654 °C) creates the necessity of developing special apparatus and techniques to overcome the attendant handling and corrosion problems.

There are, therefore, very real advantages in using salts with polyatomic anions, which have very much lower melting points, for an investigation into the chemistry of molten salts. Of the possible salts, the alkali metal nitrates have some of the lowest melting points (e.g. $LiNO_3$ 255 °C; KNO_3 334.5 °C; $LiNO_3/KNO_3$ (43/57 mol%) m.p. 132 °C), are stable over a reasonable temperature range and have desirable physical properties, i.e. they are colourless, have low volatility, a viscosity and surface tension similar to aqueous solutions etc. For these reasons a number of investigators have been drawn to these melts and have contributed to our understanding of the range of chemistry (acid–base and oxidation–reduction reactions, coordination equilibria etc.) which they can display.

In this connection molten nitrites show promise of even more varieties of behaviour, since possibilities exist of coordination through either the nitrogen or the oxygen of the nitrite ion. In addition, it is possible for these melts to show a more strongly reducing character as well as, in other circumstances, to provide an oxidising environment. Despite these possibilities the range of chemistry is at present only sparsely covered. However, some useful and interesting comparisons with the presently better-known fused nitrate chemistry will emerge from the review of recent work with which this chapter concludes.

2.2   CHEMISTRY OF MOLTEN NITRATES

2.2.1   Nature of the acidic species in nitrate melts

One of the most vexing problems in the interpretation of reactions in molten nitrates is the nature of the acidic and basic species involved. At an early stage Duke studied the acid–base equilibrium in pure sodium–potassium nitrate eutectic and in solutions containing added Lux–Flood acids and bases (these acids and bases are defined as compounds capable of removing, or donating, oxide ions from, or to, the solution). Duke then postulated the equilibrium

$$NO_3^- \xrightleftharpoons{K} NO_2^+ + O^{2-} \tag{2.1}$$

which was followed by the reaction

$$NO_2^+ + NO_3^- \xrightarrow{K} N_2O_5 \longrightarrow 2NO_2 + \tfrac{1}{2}O_2 \tag{2.2}$$

The extent of the ionisation, equation (2.1), was very small, $K_1 = 2.7 \pm 0.3 \times 10^{-26}$ at 250 °C and $5.7 \pm 0.1 \times 10^{-24}$ at 300 °C in equimolar sodium–potassium nitrate[1], but the concentration of the acidic species, the nitryl ion, could be considerably increased by the addition of Lux–Flood acids. A compound which functioned as a strong acid was potassium pyrosulphate

$$S_2O_7^{2-} + NO_3^- \rightleftharpoons NO_2^+ + 2SO_4^{2-} \tag{2.3}$$

the value of the equilibrium constant (K_3) being 7.2×10^{-3} at 250 °C and 50.8×10^{-3} at 300 °C[1]. In view of later results it must be mentioned that when required the basic species (oxide ion) was generated by electrolytic reduction of the nitrate in the melt, rather than added directly because of the difficulty of obtaining and handling pure sodium monoxide.

On varying the sodium:potassium ion ratio in the melt the value of K_3 was found to increase with the potassium concentration and the rate of reaction $2(k_2)$ to decrease[2]. This was rationalised in terms of the more strongly polarising sodium ion weakening the nitryl cation–nitrate anion interaction, and thus incidentally assuming the first stage of reaction (2.2) to be an equilibrium. Unfortunately neither dinitrogen pentoxide, nor any other nitryl compound, can be added directly to the melt to test this hypothesis, since thermal decomposition is rapid at melt temperatures (e.g. nitryl perchlorate is appreciably decomposed within minutes at 127 °C[3]).

These postulated ionisations and reactions have been severely criticised by Topol et al.[4], who claimed to have found no evidence for the nitryl ion in sodium–potassium nitrate at 280–350 °C by chronopotentiometry, though necessarily they were unable to establish the sensitivity of this method. They also quoted a private communication to the effect that an infrared spectrum had shown no signs of nitryl ion. Although no details of this investigation were given, a 2 M solution of potassium pyrosulphate at 300 °C would, on Duke's value of the equilibrium constant, only be expected to produce a maximum of 0.4 mol% of nitryl ion in the solution, and since the diagnostic absorption frequency (2360 cm^{-1}) is not strong it is not unexpected that a negative result was obtained.

Topol *et al.* suggested, as an alternative to Duke's hypothesis, that the acidic species was nitrogen dioxide itself. The electrochemical results, showing nitrogen dioxide to be present in all acidic solutions, were buttressed with brief reports of the stoichiometries of a number of reactions. Many of these are not easy to understand and in some cases are in conflict with later evidence (see dichromate–iodide reaction p. 42). The authors unfortunately made no suggestions as to the pathway whereby pyrosulphate could react to produce the ultimate decomposition products, nitrogen dioxide and oxygen. However, it may be noted that in every case acid–base reaction pathways can be more easily explained postulating a nitryl ion than using nitrogen dioxide alone. Possibly as a result the Topol hypothesis, which was invoked by a number of investigators in 1968–1969, has now become less popular. As an instance, Bartholomew and Donigan[5] showed Duke's bromide oxidation kinetics to be compatible with nitrogen dioxide as the acidic species, and explicitly supported the Topol hypothesis in 1968, while in 1970 Kozlowski, Bartholomew and Garfinkel[6] proposed a mechanism for another reaction involving the nitryl ion. It must also be noted that the distinction made by Topol *et al.* is not necessarily clear-cut, since nitrogen dioxide can dimerise and the dimers have been shown to ionise in two ways, unsymmetrically

$$2NO_2 \rightleftharpoons N_2O_4 \rightleftharpoons NO^+ + NO_3^- \qquad (2.4)$$

and less readily, symmetrically

$$2NO_2 \rightleftharpoons N_2O_4 \rightleftharpoons NO_2^+ + NO_2^- \qquad (2.5)$$

Though many reactions involving these ionisations in low-temperature solvents have been suggested, the validity of their extension to nitrate melts is much less certain. On the one hand, the highly ionic medium might well facilitate the ionisation, as do solvents of high dielectric constant, but on the other hand the concentration of dimers at melt temperatures would be very low indeed. However, in this connection it is customarily found that reactions involving nitrogen dioxide as such, are quite slow, though much of this slowness can be attributed to the difficulty of transport across the gas–melt interface.

2.2.2 Nature of the basic species in nitrate melts

Despite the disagreement on the nature of the acidic species, both Duke and Topol agreed that oxide ions were the basic species in nitrate melts. Furthermore, Topol was able to use both alkali metal carbonates and hydroxides as a source of oxide ions, i.e. the equilibria

$$CO_3^{2-} \rightleftharpoons O^{2-} + CO_2 \qquad (2.6)$$
$$2OH^- \rightleftharpoons O^{2-} + H_2O \qquad (2.7)$$

lie well to the right, (cf. K_{18} and discussion on p. 37).

Against this, Francini and Martini[7], from polarographic measurements on the sodium–potassium melt at 230–280 °C, found that sodium monoxide and sodium hydroxide were only equivalent when water was present in the melt, i.e. that equilibrium (2.7) lies to the left.

Equilibrium (2.7) is also considered to lie far to the left by Zambonin[8]. Further, on the basis of a series of electrochemical investigations, Jordan and Zambonin[9–12] have postulated three equilibria in basic nitrate melts,

$$NO_3^- + O^{2-} \rightleftharpoons NO_2^- + O_2^{2-} \qquad (2.8)$$

for which the equilibrium constant $K_8 \sim 3$ at 229 °C, in an equimolar sodium–potassium melt[10]

$$2NO_3^- + O_2^{2-} \rightleftharpoons 2NO_2^- + 2O_2^- \qquad (2.9)$$

for which the equilibrium constant was 6.7×10^{-11}, and in an atmosphere containing oxygen

$$O_2^{2-} + O_2 \rightleftharpoons 2O_2^- \qquad (2.10)$$

for which the equilibrium constant is 3.5×10^5. They further asserted that the failure of previous workers to find that 'oxide ions are not capable of existence at appreciable concentration levels' was due to the presence of silica and/or water in their melts. To avoid these contaminants, besides using melts which were 'painstakingly dried', Zambonin and Jordan used a platinum-lined vessel and platinum 'tube protectors' on the glass tubing of their rotating disc and reference electrodes, though it is not clear to what extent the melt may have penetrated between the platinum and the glass tubes.

The hypothesis that peroxide is the stable oxygen-containing basic species in nitrate melts is supported by the one-electron-transfer (RT/F) Nernst slope found for the oxygen electrode by Zambonin and Jordan. Previous authors[1, 13–15] had all found two-electron-transfer $(RT/2F)$ slopes corresponding to the equilibrium

$$O^{2-} \rightleftharpoons \tfrac{1}{2}O_2 + 2e \qquad (2.11)$$

Shams El Din[14, 15] did, however, report that unbuffered melts (i.e. not containing a Lux–Flood acid) with added sodium peroxide did give one-electron-transfer slopes, though whether this was due to dissolved peroxide ions or to hydroxide ions (assuming both the presence of water and equilibrium (2.7) to lie to the left) was not established.

A recent finding not supporting the work of Zambonin and Jordan is that of Kust[16], who reported that permanganate titration of an aqueous solution of a sodium–potassium melt containing carbonate, which had been heated under vacuum for 48 h, (i.e. which would have been expected to have consisted of a basic solution, initially of oxide, but, at equilibrium, of peroxide, and even superoxide if allowed access to oxygen) in fact gave no indication of either peroxide or superoxide.

As one final comparison between these authors, it may be noted that while Zambonin and Jordan found oxygen could be reduced in the nitrate melt, and are supported in this by Duke and co-workers, both Topol et al. and Swofford and McCormick[17] found that oxygen could not be reduced.

The true nature of the basic species discussed above is therefore at present unresolved, but the situation is further complicated by other claims that the basic species are in fact oxide ions combined with one or two nitrate ions. Kohlmuller[18] claimed that orthonitrate ions (NO_4^{3-}) were formed on addition of sodium monoxide to sodium nitrate, on the basis of some rather unconvincing evidence for differences in thermal decomposition, solubilities in liquid ammonia and reactions with carbon dioxide and oxygen. Moreover, his cryoscopic measurements could not distinguish between this compound and a mixture. A similar compound was found with potassium but no evidence was obtained for the lithium analogue. Fresh attention has been focused on this work since the formation of orthonitrate has been invoked by both Duke[1] and Kust[13] to explain the deviation of equilibrium constants from calculated values.

A further compound containing oxide ion, pyronitrate ($N_2O_7^{4-}$), has been claimed by Shams El Din and El Hosary[19, 20] on the basis of a number of potentiometric titrations. Some support for the new compound was claimed from x-ray powder photography of chilled melts of potassium hydroxide and potassium nitrate heated together for 6 h at 350 °C, which thus implies that equilibrium (2.7) lies to the right.

It will now be sadly evident that both with the acidic species and the basic species it is difficult to draw firm conclusions as to the actual ions existing in nitrate melts. It is to be hoped that this complex situation may be speedily resolved by the acquisition of additional data, preferably by other than purely electrochemical methods, for example by the study of the kinetics and products of reaction of the various postulated basic or acidic species. Some attempts[21-23] have already been made in this direction and will be reviewed later under the heading of the individual reactant element (manganese, copper, mercury), but it has not, so far, proved possible to categorically exclude any hypothetical species.

For simplicity, in the major part of the review concerned with the discussion of the chemistry of particular solutes we shall refer to nitryl and oxide ions as being present in acidic and basic solutions respectively. It will, however, be apparent to the reader that in most cases it is possible to substitute the alternative species discussed above.

2.2.3 Thermal decomposition of nitrate melts

It has been known for many years that alkali metal nitrates lose oxygen and form nitrites at high temperatures

$$NO_3^- \rightleftharpoons NO_2^- + \tfrac{1}{2}O_2 \tag{2.12}$$

While the precise value of the equilibrium constant (K_{12}) is still the subject of discussion[24-26], it has recently become apparent that significant concentrations of nitrite are present at the temperatures customarily used for the study of nitrate melts[26, 27]. Thus, this nitrite concentration and the presence of oxygen are two further factors which have to be taken into account, along with the possible acidic and basic species, when endeavouring to explain the course of reactions in nitrate melts.

2.3 REACTIONS IN MOLTEN NITRATES

The chemistry of a considerable number of compounds has now had some study in molten nitrate solutions, and it is proposed to review the work of the last 5 years by arranging the solutes in the order in which the metal, or central non-metal atom, appears in the Periodic Table. The review commences with the Main Group elements, proceeds to the transition metals and ends with the lanthanide and actinide elements.

2.3.1 Main Group elements

2.3.1.1 Alkali metals

These cations were not found to be separated by chromatography on glass fibre paper impregnated with zirconium phosphate, or on glass fibre alone, in lithium–potassium nitrate at 160 °C [28]. The cations were separable, however, on zeolites when in sodium nitrate at 330 °C [29]. Chabazite was found to be the best exchanger, the selectivity being in the same order as in aqueous solution, i.e., $Na^+ < Rb^+ < Cs^+ < Ba^{2+}$, which implies complex or ion-pair formation as the counterpart of the hydration spheres assumed in order to explain the aqueous results.

Sodium metal was found to react violently with lithium–potassium nitrate eutectic at its melting point [30]. The amalgam reacted less readily, and at 250 °C produced nitrous oxide and nitrogen (in the proportions 3:1) together with a precipitate of sodium monoxide, and nitrite in solution. In contradiction to this result, Arvía et al. claim that lithium [31], sodium [32] and potassium metal [33] exist in equilibrium with their nitrates in the cathode compartments of reversible electrochemical cells at 370, 350 and 380 °C respectively. However, many other authors have reported that sodium oxide is formed at the cathode.

2.3.1.2 Alkaline earth metals

Beryllium(II) compounds have been found to be moderately strong acids in equimolar sodium–potassium nitrate at 240 °C, reacting with the liberation of nitrogen dioxide [34]. In contrast, compounds containing magnesium(II), calcium(II), strontium(II) and barium(II) ions have been found to be stable in nitrate melts. The latter three ions have not been found to be separated chromatographically on glass fibre [35]. In an infrared study the strength of the bond between the metal and oxygen of nitrate was correlated with the extent of splitting of the $v_3(E')$ vibrational mode of the free nitrate ion, and were found to be in the order

$$Mg^{II} > Cd^{II} > Ca^{II} > Pb^{II} > Sn^{II} > Ba^{II}$$

when dissolved in sodium and potassium nitrate melts, and which, except for tin, is the order of their ionic potentials. In support of the correlation, the addition of potassium chloride was found to restore the potassium nitrate spectrum in the case of the calcium and cadmium solutes, i.e. to displace coordinated nitrate by chloride, but not to do so with magnesium[36].

Calcium nitrate has been found to be quite soluble (~ 7.5 M) in equimolar sodium–potassium nitrate at 240 °C [37].

Magnesium metal reacted at 420 °C in the pure lithium–potassium eutectic giving nitrite and a precipitate of magnesium oxide, together with a gaseous mixture of nitrous oxide, nitrogen dioxide and nitrogen (in the proportions 67:26:7). Acidic melts reacted at lower temperatures. Thus 0.8 M solutions of potassium dichromate (a weak acid in fused nitrates) reacted at 310 °C and a 0.1 M solution of potassium pyrosulphate (a strong acid) reacted at 160 °C, evolving nitrogen dioxide and producing a solution of magnesium(II) ions[30]. Calcium metal reacted similarly in the pure melt at 230 °C, though the same gaseous products were now found to be in the ratio 30:<1:70, and a trace of hyponitrite was found in the melt. Again the acidic melts reacted at lower temperatures, 220 °C and 160 °C respectively[30].

It is suggested that these products show evidence of electron transfer to nitryl ion particularly in the acidic melts, i.e.

$$NO_2^+ + M \rightarrow M^+ + NO_2 \qquad (2.13)$$

and, in addition, there is reduction of nitrate by direct reaction of oxygen with the metal. The products would then be successively,

$$NO_3^- + M \rightarrow MO + NO_2^- \qquad (2.14)$$

$$NO_2^- + M \rightarrow MO + [NO^-]$$
$$\qquad\qquad\qquad \downarrow$$
$$\qquad\qquad \tfrac{1}{2}N_2O_2^{2-} \qquad (2.15)$$

$$N_2O_2^{2-} + M \rightarrow MO + [N_2O^{2-}]$$
$$\qquad\qquad\qquad\qquad \downarrow$$
$$\qquad\qquad\qquad N_2 + O^{2-} \qquad (2.16)$$

The nitrous oxide, and perhaps also some nitrogen, could be produced by thermal decomposition of hyponitrite, though it was found that sodium hyponitrite solutions in this melt gave nitrous oxide as the only product.

2.3.1.3   Group III elements

The only report of reaction of a boron compound is that of sodium or potassium tetrafluoroborate in sodium or potassium nitrate, on soda-lime glass at 350–400 °C. The reaction corresponded to the stoichiometry

$$4SiO_2 + 6BF_4^- + 2NO_3^- \rightarrow 4SiF_6^{2-} + 2NO_2 + \tfrac{1}{2}O_2 + 3B_2O_3 \qquad (2.17)$$

and was faster in the sodium melt and in 1% sodium carbonate solutions, but showed a large decrease in rate in pyrosulphate solutions[38]. These rate-variations correspond with the concentration of oxide ions and suggest it must be involved in the rate-determining step.

Aluminium nitrate is reported to behave as a Lux–Flood acid[4, 34]. The metal reacts in pure melts to give passivating films[39, 40], and as usual at lower temperatures in melts containing acids, giving a solution of aluminium(III) ions with 0.1 M pyrosulphate[30]. In melts with oxide present in excess, aluminates are formed, which are stated to be insoluble in sodium–potassium nitrate at 300 °C [39, 40], but soluble to the extent of 10^{-3} M in the same melt at 240 °C [34].

No information is available on thallium(III), but thallium(I) compounds are stable and have been studied by phase diagram methods[41-43]. The metal is reported to be surface-oxidised in lithium–potassium eutectic at 143 °C [44].

2.3.1.4 Group IV elements

Carbon, in the form of granulated charcoal, reacts at 250 °C in lithium–potassium nitrate to form carbon dioxide, nitric oxide and nitrogen dioxide[30], and as a graphite anode in sodium–potassium nitrate at 220–300 °C reacts stoichiometrically to form carbon dioxide, nitrogen dioxide and two electrons[45]. Carbon dioxide has a small solubility in sodium nitrate at 300–400 °C, which decreases with temperature, thus indicating some interaction with the ions of the melt[46]. This solubility cannot be attributed to the formation of carbonate with an oxide impurity since the experimental method measured the gas evolved on 'chilling' the melt, and sodium carbonate although undergoing the dissociation

$$CO_3^{2-} \rightleftharpoons CO_2 + O^{2-} \tag{2.18}$$

has a very small equilibrium constant at these temperatures. Kust has found this constant (K_{18}) to be 5×10^{-5} at 300 °C in sodium–potassium nitrate[13] and the rate of reaction has been found to be low below 1000 °C [16, 46]. The solubility of sodium carbonate in the lithium–potassium eutectic at 160 °C was itself found to be low (1.1×10^{-2} M) though potassium cyanide was very much more soluble (2.7 M) [47].

Organic compounds appear to have a limited stability in nitrate melts, particularly at higher temperatures. However, phenolphthalein has been shown to act as an acid–base indicator in lithium–potassium nitrate at 210 °C, and to form coloured species very similar to those well known in the aqueous system[48]. A number of phase diagram studies of urea[49] and the sodium salts of carboxylic acids[50] with alkali metal nitrates have appeared, the latter with no reports of any decomposition reaction. Two other workers have reported reactions at slightly higher temperatures, but disagree on the products and the stoichiometry.

Shams El Din, by reacting the compounds in potassium nitrate at 350 °C, deduced the following reactions on the basis of electrochemical titration of the products[51]

$$2HCOONa + KNO_3 \rightarrow Na_2CO_3 + KNO_2 + CO_2 + H_2 \tag{2.19}$$
$$4CH_3COONa + 5KNO_3 \rightarrow 4(Na,K)_2CO_3 + KNO_2 + CH_3NO_2 \tag{2.20}$$
$$(NaOOC)_2 + KNO_3 \rightarrow Na_2CO_3 + KNO_2 + CO_2 \tag{2.21}$$

Kozlowski and Bartholomew, on the other hand, as a result of cryoscopic, voltametric and chronopotentiometric measurements on the reaction products in sodium nitrate at 320–350 °C, suggested the stoichiometries[52]

$$2HCOONa + 2NaNO_3 \rightarrow Na_2CO_3 + 2NaNO_2 + H_2O + CO_2 \quad (2.22)$$
$$2CH_3COONa + 8NaNO_3 \rightarrow Na_2CO_3 + 8NaNO_2 + 3H_2O \quad (2.23)$$
$$4C_2H_5COONa + 14NaNO_3 \rightarrow 7Na_2CO_3 + 4NaNO_2 + 10H_2O$$
$$+ 5CO_2 + 4N_2 \quad (2.24)$$

and reported that explosions could occur with high concentrations of the organic salt (30 wt %) at 350 °C.

Some interesting nitration reactions have been carried out in a lithium–sodium–potassium nitrate melt containing pyrosulphate at 250–300 °C producing the mononitro-derivatives of a variety of organic compounds of aromatic type[53].

The only recent report on tin indicates that the metal is relatively inert to a 0.1 M pyrosulphate solution in the lithium–potassium eutectic, with no surface oxidation occurring until 250 °C [30]. In contrast, lead compounds have continued to be the subject of numerous physical and electrochemical investigations. These have recently included the stability constants of chloro complexes[54] and of bromo[55] and sulphato[56] complexes. Lead carbonate has been shown to decompose to 'white lead' (2PbCO₃·PbO) in potassium nitrate at 350–415 °C in a zero-order reaction with an activation energy of 38 kcal mol^{-1} [57]. Lead(II) ions have been titrated with electrolytically-generated oxide ions[35] to give the insoluble oxide, while the metal and the amalgam were both surface-oxidised in lithium–potassium nitrate at 143 °C [44], and cathodically to form lead(II) oxide at 250 °C in sodium–potassium nitrate[58].

2.3.1.5   Group V elements

Nitrogen gas has been shown to have a low solubility (10^{-6} mol (mol solvent)$^{-1}$ barr^{-1}) in lithium and sodium nitrates[59], though ammonia has a much larger solubility (10^{-2} mol (mol solvent)$^{-1}$ barr^{-1}) in lithium–sodium–potassium and lithium–potassium nitrates which decreased with temperature and increased markedly with lithium ion concentration, suggesting an interaction with the solvent[60].

Potassium azide has been shown to be soluble in lithium–potassium and sodium–potassium eutectic, and to precipitate the heavy metal azides from solutions of the metals[61]. These azides have been shown to be rather more stable than those precipitated from aqueous solution, and in the case of lead azide to have a decomposition temperature some 30 °C higher than the usual product.

An interesting series of physical measurements, including proton magnetic resonance, viscosity, density and electrical conductivity, have been made on nitric acid solutions in the lithium–sodium–potassium eutectic at 140–220 °C. These have been interpreted to indicate that in dilute solutions each nitric acid molecule is associated with a nitrate ion (O_2N—O—H—O—NO_2^- ?), while in solutions with a one-third mole fraction of nitric acid the hydrogen appears to have more than two neighbouring oxygens. As the mole fraction reaches one half, the properties of the solutions approach those of concentrated nitric acid. In all cases only a single p.m.r. line was found, indicating that exchange between forms was rapid[62].

Ammonium nitrate has been found to decompose thermally in nitrate melt solutions at rates very similar to those found for the pure salt. The effect of 15 different solutes on the rate of decomposition in the sodium–potassium eutectic has been measured. Some correlation was found with the polarising

Table 2.1 Titration of bases with sodium metaphosphate in potassium nitrate at 350°C

Base	Forward titration		Backward titration	
	acid:base ratio	product	acid:base ratio	product
NaOH Na$_2$O$_2$ Electrolytically- produced O^{2-}	2:1 ↓ 1:1	P$_2$O$_7^{4-}$ PO$_4^{3-}$	1:2 ↓ 2:1 ↓ 3:1	(PO$_4^{3-}$·O^{2-})* P$_2$O$_7^{4-}$ P$_3$O$_{10}^{5-}$
HCOONa (COONa)$_2$ Na$_2$CO$_3$ K$_2$CO$_3$ NaHCO$_3$	3:1 ↓ 2:1 ↓ 1:1	P$_3$O$_{10}^{5-}$ P$_2$O$_7^{4-}$ PO$_4^{3-}$	1:1 ↓ 2:1 ↓ 3:1	PO$_4^{3-}$ P$_2$O$_7^{4-}$ P$_3$O$_{10}^{5-}$
CH$_3$COONa			2:1 ↓ 3:1	P$_2$O$_7^{4-}$ P$_3$O$_{10}^{5-}$
Li$_2$CO$_3$	1:1	PO$_4^{3-}$	1:1 ↓ 3:2	PO$_4^{3-}$ (PO$_4^{3-}$·P$_2$O$_7^{4-}$)*
CaCO$_3$	2:3 ↓ 4:9	(2PO$_4^{3-}$·O^{2-})* [4(PO$_4^{3-}$·O^{2-})·CO$_3^{2-}$]*	1:2 ↓ 3:4	(PO$_4^{3-}$·O^{2-})* (3PO$_4^{3-}$·O^{2-})*
SrCO$_3$	4:3 ↓ 2:3	(2PO$_4^{3-}$·P$_2$O$_7^{4-}$)* (2PO$_4^{3-}$·O^{2-})*	2:3 ↓ 3:4	(2PO$_4^{3-}$·O^2)* (2PO$_4^{3-}$·P$_2$O$_7^{4-}$)*
BaCO$_3$	2:1 ↓ 1:1	P$_2$O$_7^{4-}$ PO$_4^{3-}$	1:1 ↓ 5:1	PO$_4^{3-}$ (2P$_2$O$_7^{4-}$·PO$_3^-$)*

*Indicates a compound previously unreported.

power of the solute cation, but the solute anions produced considerable variation in their effects on the rates of reaction[37].

Recent progress in the chemistry of phosphorus has added very considerable detail to the earlier observation that phosphorus pentoxide in potassium nitrate at 350 °C acted as a Lux–Flood acid, forming metaphosphate ions. These ions, produced in this way or added as such, could then be titrated with additional oxide ions (added in the form of sodium peroxide) firstly to form

pyrophosphate and secondly orthophosphate[63]. Coumert *et al.* have now
reported that sodium trimetaphosphate, tripolyphosphate and pyrophosphate (when dissolved in potassium nitrate at 350 °C) all react with the correct
ratios of sodium peroxide to form orthophosphate. The trimetaphosphate
and tripolyphosphate could also be titrated with orthophosphate in the
correct ratio to form pyrophosphate, the products being confirmed by polarography or chromatography[64]. These titrations could also be performed in
sodium nitrate solvent at 350 °C, but in these cases the acid:oxide ratios were
less exact. A smaller quantity of oxide was in fact required and it was suggested that this could possibly indicate that sodium nitrate itself displayed
some basicity. (Usually however, the acidity of alkali metal cations is considered to increase with decreasing size.)

Since then Shams El Din *et al.*[51, 65] have reported a very much more numerous series of products, some of them novel compounds, as apparently
being formed when sodium metaphosphate is titrated against various bases
in potassium nitrate at 350 °C. Confusingly, different products were sometimes found as a result of the 'forward' (i.e. base added to a solution of acid)
and 'backward' (i.e. acid added to a solution of base) titrations. The results
are summarised in Table 2.1.

It was suggested that the seven new compounds could be 'phosphate
glasses' analogous to the silicates, with bridging and non-bridging oxide ions.
It may be noted that phosphorus bonded to five oxygens is well known in
organic compounds, though not so far reported in an inorganic system.

From the results of Table 2.1 Shams El Din[65] arranged the cations of the
bases in a series

$$K^I, Na^I > Ba^{II} > Sr^{II} > Li^I > Ca^{II}$$

of decreasing phosphorus:oxygen, or phosphorus:charge, ratios (the ratios
being calculated on the initial product of the forward titrations), and suggested that these ratios were related to the decreasing basicity of the cations.
The series is in fact a plausible order for increasing acidity, though it is
extremely difficult to obtain any other experimental evidence which relates
the alkali metal and alkaline earth cations. A disadvantage of this theory is
that it implies no exchange of cations between base and melt, and thus a
very much more covalent view of the nature of the solute bases than has
been held hitherto.

Of the remaining members of the Group V elements there are only two
recent reports. The electrochemical titration of meta-arsenate with sodium
hydroxide, or sodium peroxide, indicated that pyroarsenate (acid:base
ratio 2:1) was formed initially, followed by the basic ion $AsO_4^{3-} \cdot O^{2-}$ (acid:base
ratio 1:2) which is again a new species[66]. There is also a brief report that bis-
muth(III) ions acted as a Lux–Flood acid at 150–200 °C in lithium–potassium
nitrate containing dissolved iodide, and liberated iodine and nitrogen dioxide
with the formation of bismuthyl iodide[67].

2.3.1.6   Group VI elements

The extent to which water is soluble in molten nitrates continues to receive
attention. The recent determination by Zambonin *et al.*[68] gives a value of the

same order of magnitude as those previously obtained for molten nitrate systems[69-71], and is similar to the value recently obtained for the potassium nitrate–sodium nitrite melt[72]. It is generally agreed that simple solubility occurs and that no dissociation is involved, though there is evidence of strong cation–water bonding[69, 72].

Sulphur dioxide, potassium sulphite and potassium sulphide all reduce sodium or potassium nitrate at 310–350 °C, the stoichiometries being determined as[6]

$$SO_2 + 2NaNO_3 \rightarrow Na_2SO_4 + 2NO_2 \qquad (2.25)$$

$$K_2SO_3 + KNO_3 \rightarrow K_2SO_4 + KNO_2 \qquad (2.26)$$

$$K_2S + 4KNO_3 \rightarrow K_2SO_4 + 4KNO_2 \qquad (2.27)$$

Elemental sulphur also reduces molten lithium–potassium nitrate at 250 °C forming sulphate, nitrous oxide and a trace of nitrogen dioxide[30].

Potassium thiocyanate is stable and very soluble (88.9 mol %) at 160 °C in the lithium–potassium eutectic[47].

2.3.1.7 Group VII elements

Although chlorine is reported to oxidise nitrate melts slowly with the formation of nitrogen dioxide[73], the solubility of chlorine, together with those of bromine and iodine, has been measured by Delimarskii and Shilina[74], who found there was a small increase in solubility with increasing atomic weight. The solubility of the three potassium halides is rather larger and decreases sharply with increasing atomic weight[47]. All three halide ions have been titrated against silver nitrate potentiometrically[75].

Varying views have again been expressed regarding the stability of the halide ions in nitrate melts. Swofford and Propp[75] maintain that all three ions are stable to 250 °C in sodium–potassium nitrate, whereas all three are said to react at 150 °C in the lithium–sodium–potassium melt[74]. In lithium–potassium nitrate iodide is said to commence reaction at 147 °C though it is only appreciably oxidised at temperatures of 217 °C and above[76]. The stoichiometry of oxidation has been established at 220 °C as[77]

$$2I^- + NO_3^- \rightarrow NO_2^- + O^{2-} + I_2 \qquad (2.28)$$

The equilibrium constants for the formation of the tri-iodide ion have been measured[76] as $K_{29} = 88 \pm 8$ mol kg^{-1}.

$$I_{2(soln)} + I^- \rightleftharpoons I_3^- \qquad (2.29)$$

The spectrum of the tri-iodide solution has been shown to be similar to that in aqueous solution[77]. Bromide solutions in lithium–potassium nitrate did not show any reaction until 300 °C and chloride solutions until 470 °C, in both cases with the formation of nitrite and oxygen[77].

The halide ions reacted at lower temperatures in acidic solutions. For example all three ions were oxidised by nitrogen dioxide at 160 °C in lithium–potassium nitrate[77], though in the sodium–potassium melt bromide and iodide reacted[4] but chloride and fluoride did not[5]. The kinetics of the nitrogen dioxide–iodide reaction in the latter solvent showed that the reaction rate was controlled by transport across the gas–melt interface[5].

In dichromate solutions, iodide was oxidised to iodine in potassium nitrate at 350 °C but was stated to produce no nitrogen dioxide or nitric oxide[5], though in sodium–potassium nitrate iodine and nitric oxide were reported as the products[4]. The dichromate:iodide ratio required for complete reaction was stated[4] to be 4:1 which differs both from the 1:3 ratio reported earlier[78], and from the 4:3 ratio which was established in the lithium–potassium melt on the basis of a quantitative estimation of the products, the stoichiometry being[77]

$$6I^- + 8Cr_2O_7^{2-} \rightarrow 3I_2 + Cr_2(CrO_4)_3 + 11CrO_4^{2-} \tag{2.30}$$

Bromide and chloride also react with dichromate solutions at lower temperatures (300 °C and 380 °C) than in the pure lithium–potassium melts, and all three halides reacted at an even lower temperature (160 °C) in pyrosulphate solutions evolving free halogen[77].

Potassium bromate, chlorate and iodate reacted in the pure lithium–potassium melt, at 230, 390 and 500 °C respectively, with the production of oxygen, while the bromate and iodate produced the free halogen at 400 and 600 °C. In the case of iodate this reaction was attributed to nitrite produced by thermal decomposition, and iodate was found to react in nitrite solutions at 320 °C, and above, with the stoichiometry[77]

$$^-IO_3^- + 3NO_2^- \rightarrow I^- + 3NO_3^- \tag{2.31}$$

In dichromate solutions of lithium–potassium nitrate, reaction occurred at progressively lower temperatures (210, 250 and 200 °C) with bromate, chlorate and iodate with liberation of free halogen[77], while in sodium–potassium nitrate solutions the halates again liberated the halogen. The reaction of chlorate has been found to be catalysed by the chloride ion[79, 80], while with bromate the stoichiometry has recently been found to depend on the dichromate: chromate ratio. At low dichromate concentrations the reaction was

$$BrO_3^- + Br^- + Cr_2O_7^{2-} \rightarrow 2CrO_4^{2-} + Br_2 + O_2 \tag{2.32}$$

which was suggested to proceed via the following stages

$$BrO_3^- + Cr_2O_7^{2-} \rightarrow 2CrO_4^{2-} + BrO_2^+ \tag{2.33}$$
$$BrO_2^+ + 2Br^- \rightarrow BrO_2^- + Br_2 \tag{2.34}$$
$$BrO_2^- \rightarrow Br^- + O_2 \tag{2.35}$$

while at low chromate concentrations the stoichiometry approached

$$BrO_3^- + 5Br^- + 3Cr_2O_7^{2-} \rightarrow 6CrO_4^{2-} + 3Br_2 \tag{2.36}$$

which is analogous to that found in acidic aqueous solution[81].

2.3.1.8   Rare gas elements

The solubility of helium and argon has been measured[59] and is of the same order of magnitude ($10^{-4} - 10^{-5}$ mol (mol solvent)$^{-1}$ barr^{-1}) as that previously found for xenon[82].

2.3.2 Transition metals

2.3.2.1 Titanium group

The only additions to the meagre number of observations so far reported for these metals is that titanium is surface-oxidised at 250 °C by 0.1 M pyrosulphate solution in lithium–potassium nitrate[30], and that zirconium hydrogen phosphate $(Zr(HPO_4)_2 \cdot H_2O)$, used as an ion-exchange medium, was stable up to 350 °C in sodium–potassium nitrate when alkali metal ions are exchanged for hydrogen. Breakdown and release of phosphate did not commence until 500 °C [83].

2.3.2.2 Vanadium group

A report has appreared of a phase-diagram study of vanadium pentoxide with sodium nitrate and with potassium nitrate[84], which categorically states that no gases were evolved at temperatures up to 700 °C. The only reaction observed was the formation of 3:2 compounds.

This report runs completely counter to the observation by Shams El Din et al.[85] that vanadium pentoxide reacts rapidly with potassium nitrate at 350 °C, forming metavanadate which behaved as a strong Lux–Flood acid.

Vanadium metal has been found to react with a 0.1 M pyrosulphate solution in lithium–potassium nitrate at 160 °C forming vanadium pentoxide[30].

2.3.2.3 Chromium group

Potassium chromate, dichromate and trichromate solutions in lithium–potassium nitrate at 160 °C have been examined by spectroscopy[86]. The absorptions listed in Table 2.2 are quite similar to the charge-transfer bands observed in aqueous solution.

Table 2.2

Solute	Maximum absorption	Absorptivity
K_2CrO_4	27 100 cm^{-1}	3620 l mol^{-1} cm^{-1}
$K_2Cr_2O_7$	26 900	2350
$K_2Cr_3O_{10}$	26 700	5550

Reaction of trichromate and dichromate with the nitrate melt was observed at 200 and 400 °C respectively, the compounds functioning as Lux–Flood acids with the ultimate formation of chromate[86]

$$2Cr_3O_{10}^{3-} + 2NO_3^- \rightarrow 3Cr_2O_7^{2-} + 2NO_2 + \tfrac{1}{2}O_2 \qquad (2.37)$$

$$Cr_2O_7^{2-} + 2NO_3^- \rightarrow 2CrO_4^{2-} + 2NO_2 + \tfrac{1}{2}O_2 \qquad (2.38)$$

The dichromate reaction (2.38) could be reversed by the passage of nitrogen dioxide through the melt[4] or by adding potassium pyrosulphate[86]. However, the former reactant required some 2000 times the stoichiometric quantity required for the latter reactant, thus again indicating the extreme slowness of reactions involving nitrogen dioxide. This may eventually be significant in the controversy over the nature of the acidic species in nitrate melts, but may also be attributed to the difficulty of transport across the gas–melt interface.

Like metaphosphate, potassium dichromate has been titrated electro-chemically with a great variety of bases, and also revealed a complicated behaviour with differing forward and backward titration products[51, 57]. The bases used, acid:base ratios and the stoichiometries of the products are listed in Table 2.3.

Table 2.3

Base	Forward titration		Backward titration	
	Acid:base ratio	Product	Acid:base ratio	Product
NaOH Na$_2$O$_2$ Electrolytically-produced O^{2-}	1:1	CrO$_4^{2-}$	1:2 ↓ 1:1	(CrO$_4^{2-}$·O^{2-})* CrO$_4^{2-}$
Na$_2$CO$_3$ NaHCO$_3$ (NaOOC)$_2$ NaOOCH Li$_2$CO$_3$ K$_2$CO$_3$ CaCO$_3$ SrCO$_3$			1:1	CrO$_4^{2-}$
BaCO$_3$			2:3	(2BaCrO$_4$·BaCO$_3$)*
PbCO$_3$	1:1 ↓ 1:2	PbCrO$_4$+K$_2$CrO$_4$ (PbCrO$_4$·PbCO$_3$)*	2:1 ↓ 3:2	K$_2$CrO$_4$+(PbCrO$_4$·PbCO$_3$)* +PbO K$_2$CrO$_4$+2PbCrO$_4$+PbO

*Indicates new compound.

The relative acidity of the base cations was stated to be

$$Pb^{II} > Ca^{II} > Sr^{II} > Ba^{II}$$

and $$Li^I, Na^I > K^I$$

Chromium(VI) oxide was found to be an even stronger Lux–Flood acid than the above oxyanions, and reacted at the melting point of the lithium–potassium eutectic to form trichromate and nitrogen dioxide[86]. In potassium nitrate at 350 °C it was found to form dichromate, nitrogen dioxide and oxygen[87].

Dichromate solutions in lithium–potassium nitrate could be reduced to insoluble chromium(III) oxide by sodium amalgam[30], but this oxide was re-oxidised by the melt at 250 °C [86], the stoichiometry being

$$Cr_2O_3 + 5NO_3^- \rightarrow 2CrO_4^{2-} + NO_2^- + 4NO_2 \tag{2.39}$$

Chromium(III) nitrate was found to be a strong Lux–Flood acid reacting in sodium–potassium nitrate at 275 °C, giving off nitrogen dioxide[4]. Chromium(III) chloride was said to be oxidised to dichromate in potassium nitrate at 350 °C, and to evolve nitryl chloride and nitrosyl chloride if additional chloride ions were present in the melt. No evidence was given for the presence of these gases which were previously supposed to decompose at considerably lower temperatures[88].

Chromium metal reacted in pure lithium–potassium nitrate at 300 °C and in a 0.1 M pyrosulphate solution at 250 °C [30]. The former reaction has been shown to have the stoichiometry[86]

$$2Cr + 7NO_3^- \rightarrow 2CrO_4^{2-} + 3NO_2^- + 3NO_2 + NO \qquad (2.40)$$

The phase diagrams of molybdates[89, 90] and tungstates[90, 91] with alkali metal nitrates have been determined, with no reports of any reactions. Kust has, however, found a slow dimerisation equilibrium in sodium–potassium nitrate at 320–370 °C [92],

$$MoO_4^{2-} \rightleftharpoons Mo_2O_7^{2-} + O^{2-} \qquad (2.41)$$

and has found that the basicity decreases in the order[93]

$$WO_4^{2-} > MoO_4^{2-} > CrO_4^{2-}$$

Molybdenum(VI) oxide has been found to display Lux–Flood acidity in potassium nitrate at 350 °C, reacting to form the trimolybdate. This product and tungsten(VI) oxide, which was found to be stable in the pure melt, could both be titrated against sodium peroxide with the formation of molybdate and tungstate ions[87].

2.3.2.4 Manganese group

Potassium permanganate dissolves in lithium–potassium nitrate at 160 °C to form a purple solution (absorption maximum 18 800 cm^{-1}) which decomposes slowly over 36 h to a black precipitate in accordance with the equation[94]

$$2KMnO_4 + 2Li^+ \rightarrow (Li_{0.92}/K_{0.08})_2Mn_2O_5 + \tfrac{3}{2}O_2 + 2(Li_{0.08}/K_{0.92})^+ \qquad (2.42)$$

Potassium manganate(VI) disproportionates in the same solvent to form a purple solution, a black precipitate of another manganate(IV) and a mole of oxide,

$$2K_2MnO_4 \rightarrow 2KMnO_4 + K_2MnO_3 + K_2O \qquad (2.43)$$

before reacting completely to form an insoluble manganate(IV) again containing a high proportion of lithium. The overall equation in this case is[94]

$$K_2MnO_4 + 2Li^+ \rightarrow (Li_{0.96}/K_{0.04})_2MnO_3 + \tfrac{1}{2}O_2 + 2(Li_{0.04}/K_{0.96})^+ \qquad (2.44)$$

These reactions are not identical with those reported earlier in sodium–potassium nitrate[95], where permanganate first decomposed to the green manganate(VI) before decomposing further to manganese dioxide and oxygen.

Manganese dioxide in the lithium–potassium eutectic has been found to be in equilibrium with manganese(II) ions at low temperatures (e.g. at 160 °C with 1.6 mol % Mn^{2+}) [21]. Addition of acid produces a larger concentration of manganese(II). Once again pyrosulphate in solution is found to dissolve considerably more manganese dioxide than gaseous nitrogen dioxide.

Manganese(II) compounds react with the pure lithium–potassium melt to form manganese dioxide, the reaction being complete at 350 °C and above [94].

$$Mn^{2+} + 2NO_3^- \rightarrow MnO_2 + 2NO_2 \qquad (2.45)$$

Manganese metal reacts in the pure melt at 300 °C to give the dioxide, nitrous oxide and nitrogen, but with acidic melts to give manganese(II) ions [30].

The only recent report on rhenium compounds involved the solvent extraction of perrhenate from lithium–potassium nitrate eutectic at 150 °C into a polyphenyl phase containing tetraheptyl ammonium nitrate [96].

2.3.2.5 Iron group

Little chemistry has been reported in the last 5 years for this group of metals. Iron(III) is stated to be immobile on various chromatographic supports [28]. The metal has been found to be generally inert, not forming an oxide film at 250 °C in sodium–potassium nitrate [58], and not until 500 °C in the lithium–potassium melt [30]. A less pure iron did however react in the latter melt at 400 °C, the gases evolved consisting of carbon dioxide (82 %), nitrogen (10 %), nitrous oxide (8 %) and nitrogen dioxide (<1 %). In acidic melts reaction occurred at lower temperatures (0.1 M $K_2S_2O_7$ 180 °C, 0.8 M $K_2Cr_2O_7$ 320 °C) with the formation of iron(III) [30].

Ruthenium and osmium metals are reported to be oxidised to the (VII) and (VIII) states respectively, but no further details have so far been published [97].

2.3.2.6 Cobalt group

Cobalt(II) compounds are stable at 160 °C in lithium–potassium nitrate, and some solubility values have been reported [47]. The cation has been separated chromatographically, being eluted with an ammonium nitrate solution from glass fibre paper impregnated with zirconium phosphate [28]. The electronic spectrum has been re-examined and is now attributed to dodecahedral coordination [98] rather than to octahedral. Evidence of complex formation with halide ions has again been found, in this case in lithium–sodium–potassium nitrate at 145 °C by polarography [55]. The metal has been shown by potential measurements to form an oxide film in lithium–potassium nitrate at 143 °C [44], though this was not observed visually until 500 °C [30].

Rhodium(III) chloride dissolved slowly in the sodium–potassium eutectic at 300 °C, giving a yellow solution with a spectrum similar to that of a hexanitrorhodate(III) complex which decomposed over the course of two days giving a black precipitate of rhodium(III) oxide. The yellow solution became pink on addition of chloride ions, with a spectrum attributed to the hexa-chlororhodate(III) anion[99]. A rhodium cathode on electrolysis in potassium nitrate at 350 °C gave first a yellow film and then a green solution which it was suggested also contained rhodium(III) ions[100].

2.3.2.7 Nickel group

The nickel(II) oxidation state is stable in nitrate melts at reasonably low temperatures. Recently solubility[47], chromatographic[28] and polarographic studies[55] have been carried out, the latter producing stability constants for the chloro and bromo complexes. A spectroscopic study in lithium–sodium–potassium nitrate eutectic[101] has confirmed the earlier octahedral assignment and, because the spectrum was found to be identical with the 1:3 nickel(II) nitrate complex formed in dimethylsulphone, it was suggested that the octahedral environment was formed by three bidentate nitrate groups. As a result of a potentiometric investigation, nickel metal was reported to form a film of nickel(II) oxide at 250 °C in sodium–potassium nitrate[58], though a visible film was not found in lithium–potassium nitrate until 500 °C [30].

Platinum, when acting as a cathode in sodium nitrate at 340 °C, was said to give a red or orange melt and then to form a green platinum(II) nitrosyl compound. A similar green compound was formed in potassium nitrate at 350 °C [100].

2.3.2.8 Copper group

The stability of copper(II) compound in lithium–potassium nitrate is indicated by the determination of the solubilities of the chloride and nitrate at 160 °C [47]. However, other workers using the same melt reported the precipitation of the black copper(II) oxide at 143 °C [44]. The ion was also stated to oxidise iodide ions with the precipitation of brown copper(I) iodide, and silver metal with the formation of silver(I) ions and black copper(II) oxide, both reactions taking place at 150 °C [87]. The latter reaction is difficult to rationalise in view of the general stability of silver towards acidic melts at low temperatures. A kinetic study of the reaction of copper(II) ions in sodium–potassium nitrate at 310–350 °C has recently been reported. The reaction

$$Cu^{2+} + 2NO_3^- \rightarrow CuO + 2NO_2 + \tfrac{1}{2}O_2 \qquad (2.46)$$

was found to be autocatalysed by the copper(II) oxide surface. Ferric oxide was also found to act as a catalyst for this reaction[102].

Copper metal, like cobalt was found by potential measurements to be surface-oxidised at 143 °C in lithium–potassium nitrate[44], though again like cobalt and nickel, no visible film was observed below 500 °C [30]. In sodium–potassium nitrate copper(I) oxide may be formed at 250 °C [58].

Silver(I) ions have been found to be stable in nitrate melt solutions and have been used extensively for various electrochemical measurements[103]. The ion has recently been shown to have a higher mobility than the potassium ion over the whole concentration range at 300 °C [104]. The solubility products of the sparingly soluble silver(I) chromate[105] and oxide[106] have recently been determined in sodium–potassium and potassium nitrates, as have those of the halides and cyanide[107] in lithium–potassium nitrate at 150 °C. Measurements of the stability constants of the various halide complexes continue to be reported[96, 108, 109].

Silver metal was slowly corroded at 325 °C in sodium–potassium nitrate with the production of silver(I) ions, reaction being faster if the more acidic lithium ions were present, but being slower in the presence of nitrite or oxalate. When anodised in lithium–sodium–potassium nitrate at 150 °C, the metal formed a black film of silver(I) oxide which was quickly removed by the addition of nitrite ions[110]. With a silver cathode in potassium nitrate at 380 °C a brown melt was produced which precipitated silver(I) oxide on solution in water[100].

2.3.2.9 Zinc group

Little new work has been carried out on zinc compounds in recent years, though evidence continues to accumulate that zinc(II) is both soluble in nitrate melts and can, under suitable conditions, react as a Lux–Flood acid with precipitation of oxide[111]. The metal has been shown to oxidise in sodium–potassium nitrate at 250 °C [58]. In lithium–potassium nitrate potentiometric measurements showed a film was formed at 143 °C [44], although a visible layer was only observed at 450 °C [30]. Visible reaction occurred at lower temperatures in 0.8 M dichromate (350 °C) with the formation of insoluble zinc chromate, and in 0.1 M pyrosulphate (160 °C) with the formation of a solution of zinc(II) ions[30].

Cadmium(II) complexes have again been studied extensively, using a variety of experimental methods, solvent melts and temperatures, as well as ligand species. Among the complexes examined recently are those with fluoride[55, 56], chloride[55, 112–115], bromide[55, 115–118], iodide[55, 115, 118], cyanide[55] and sulphate[56]. Cadmium metal has been reported to be oxidised in lithium–potassium nitrate at 143 °C, and in sodium–potassium nitrate at 250 °C.

Mercury(II) compounds have attracted somewhat less attention but have been the subject of phase diagram[119] and biamperimetric titration studies[67]. In the latter work the ion was found to form stable complexes with iodide. At higher temperatures (250 °C) the chloride and bromide were found to volatilise quickly from sodium–potassium nitrate[120].

The equilibrium

$$Hg + Hg^{2+} \rightleftharpoons Hg_2^{2+} \tag{2.47}$$

was found to have a rather larger equilibrium constant in lithium–potassium nitrate ($K_{47} = 224$ at 177 °C) than in aqueous solution, and this was attributed to the greater ionic strength of the melt solution[121]. The solubility of the mercury(I) halides in this melt was found to be low[122, 123], but both coulometric[122] and potentiometric titration[123] showed more soluble anionic complexes to be formed with excess halide. Similar behaviour was found with the sparingly soluble mercury(I) chromate which also formed soluble complexes[123]. Titration of the basic species produced by coulometric reduction of the melt with mercury(I) ions was observed to produce a black precipitate of mercury metal and mercury(II) oxide which sometimes became red on standing[123]. It was suggested that either Zambonin's postulated equilibria (equations (2.8), (2.9) and (2.10) above) were displaced in the presence of mercury(II) ions, or that possibly water was present and hydroxide was formed (i.e. equilibrium (2.7) above moved to the left) and that these ions reacted in accordance with the equation

$$Hg^{2+} + 2OH^- \rightarrow Hg + HgO + H_2O \qquad (2.48)$$

In earlier work[44] it had been found that certain solutions did give small quantities of a yellow precipitate, and this was attributed to the water present forming basic mercury(I) nitrate.

Mercury metal was found to react in pure lithium–potassium nitrate melts at 400 °C forming mercury(II) oxide, and at successively lower temperatures in 0.8 M dichromate (300 °C) and in 0.1 M pyrosulphate (150 °C) to give soluble mercury(II) ions and insoluble mercury(II) sulphate respectively[30]. An Ellingham diagram, obtained by calculation from the best available and estimated data, indicated that mercury(II) oxide would not be expected to be formed by oxidation of mercury by sodium or potassium nitrate. However, the oxygen produced by thermal decomposition of the nitrates might very well react when it was present at appreciable partial pressure[124].

2.3.2.10 Lanthanide elements

Several methods of separating lanthanide ions have been studied using nitrate melts as solvents. Solutions in lithium–potassium nitrate have been solvent-extracted with a polyphenyl solution containing tri-n-octyl phosphine oxide[125], the results showing that the metal concentration in the organic phase increased in the order

$$Nd^{III} < Gd^{III} < Er^{III}$$

Neodymium has been extracted from the sodium–potassium melt by absorption on a zeolite[126], and europium in lithium–potassium nitrate solution has been separated on glass fibre paper[28]. Several metal ions [Nd^{III}, Pr^{III}, Sm^{III}, Er^{III}] have also been studied polarographically in potassium nitrate at 360 °C [127].

These reports, which may all be taken as an indication of the general stability of the ions in the various melts, only serve to emphasise the extent of our ignorance of the detailed chemistry of the lanthanide metals. The same may be said of the intriguing report that several measurements of physical properties (spectra, density and e.m.f.) all suggest that a change in the coordination of neodymium(III) ions occurs at around 1 M concentration[128].

2.3.2.11 Actinide elements

The only recent work to be reported on this series of elements is by Carnall et al.[129], who showed uranyl(VI) nitrate to be considerably more thermally stable in lithium–sodium nitrate eutectic solution, where it commenced to decompose above 270 °C to form uranate anions, than as the pure solid which decomposed at 170 °C to form uranium(VI) oxide. Diuranates with a high proportion (96%) of sodium cations were formed when bromate was added to the solution between 210 and 270 °C.

Neptunyl(VI) nitrate dissolved in this melt giving a solution with a spectrum very similar to that of the aqueous solution, and was reduced over the course of 10 min to the neptunyl(V) ion. Addition of bromate to either solution produced a dineptunate precipitate with a similar proportion of sodium ions. This method of preparing diactinates at relatively low temperatures is potentially most useful since the customary method involves heating the highest oxides (Np_3O_8 and UO_3) to 700 or 800 °C with sodium carbonate.

2.4 CHEMISTRY OF MOLTEN NITRITES

As with the molten nitrates, many investigations have made use of the lower working temperatures made available by the eutectic mixtures, though some have preferred to use a pure alkali metal nitrite. The melting points of the pure alkali metal nitrites have been shown to increase markedly with atomic weight ($LiNO_2$ m.p. 222 °C, $NaNO_2$ 284 °C, KNO_2 438 °C) and to a lesser extent in the binary eutectics ($LiNO_2/KNO_2$ (45:55 mol%) m.p. 98 °C [130]; $LiNO_2/NaNO_2$ (62:38 mol%) m.p. 151 °C [131]; $NaNO_2/KNO_2$ (63:35 mol%) m.p. 220 °C [132]).

Very much less investigation has taken place so far with these melts than with the corresponding nitrates, so that comparisons of the effects of varying the melt cations are in most cases not at present possible. Similarly, despite the more extensive chemistry potentially possible in molten nitrites than in molten nitrates (see introduction p. 30), in most cases these potentialities have barely begun to be explored, and this review is limited in the case of many groups of elements to a random scattering of facts often found by chance during the course of other investigations.

2.4.1 Nature of the acidic and basic species in nitrite melts

The species characteristic of, and actually present in, nitrite melts have been the subject of very little curiosity and of no controversy comparable to that recorded earlier for nitrate melts.

It has been found that compounds which functioned as Lux–Flood acids (i.e. metal ions and condensed anions) in molten nitrates behaved in a very similar fashion in molten nitrites, and it had therefore been assumed by several authors that an ionisation similar to that of equation (2.1) above did exist in molten nitrites, and that possibly the concentrations of the acidic and basic species were rather larger than in molten nitrates. However, the first quantitative work on the equilibrium

$$NO_2^- \rightleftharpoons NO^+ + O^{2-} \tag{2.49}$$

has only been quoted very recently[133], the value of the equilibrium constant (K_{49}) being given as 1.3×10^{-9} at 300 °C, though so far the experimental details of this investigation have not been published. It has also been assumed that the nitrosyl ion has a similarly short-lived existence (again probably as an ion-pair with a melt anion) as has the nitryl ion in nitrate melts, and would therefore decompose to form nitrogen dioxide and nitric oxide,

$$NO^+ + NO_2^- \rightleftharpoons [N_2O_3] \rightarrow NO_2 + NO \qquad (2.50)$$

The latter product has been frequently identified from melts held under vacuum or inert gas, though of course for melts in contact with oxygen the former product is the only one that can be isolated.

Although the authors quoted above, and many others, have made the simple assumption that oxide ions were the basic species in nitrite melts, Kohlmuller[18] has claimed the existence of the compounds Na_3NO_3 and K_3NO_3 (i.e. an oxide ion in combination with nitrite) though the evidence was only stronger than in the case of the corresponding nitrate compounds because the cryoscopic values (four ions per molecule in lithium nitrate solvent) did seem to rule out the possibility that the oxide ions were merely solvated.

2.4.2 Thermal reactions of nitrite melts

When allowed access to oxygen, nitrite melts should of course be converted to nitrates in accordance with the equilibrium depicted above in equation (2.12).

However, the evidence indicates that this equilibrium was only slowly established at temperatures below 600 °C, and that the rate-determining step again occurred at the gas–melt interface. Freeman[24] has also shown that the decomposition of sodium nitrite commenced at 650–700 °C with the production of sodium monoxide, some sodium nitrate and nitrogen. The formation of sodium monoxide precipitates has also been observed by Kozlowski and Bartholomew[134] but at the considerably lower temperature of 325 °C. They also implied support for the observations of Oza[135] that the gaseous products were nitrogen dioxide, nitric oxide and nitrogen. Lithium nitrite has also been found to display a number of decomposition equilibria at 250–350 °C, and the products are claimed to include nitrate, oxide, nitrogen dioxide, dinitrogen trioxide, nitric oxide, nitrous oxide, and nitrogen[136].

Batches of sodium–potassium nitrite eutectic have been analysed for oxide and nitrate, and have been shown to have concentrations which varied from $2.1–3.6 \times 10^{-2}$ M and $1.8–3.5 \times 10^{-2}$ M respectively[137]. The amounts in a particular batch were found to depend on the previous history of purification and on the nature of the thermal treatment it had undergone. The influence of these impurities, and of possible acidic and basic species, has to be borne in mind when considering the products and possible mechanisms of reactions in molten nitrite, though in some cases the reaction products can be attributed directly to the nitrite ions, no doubt partly because they are present in the melts in very much larger concentrations (13 M in the sodium–potassium eutectic).

2.5 REACTIONS IN MOLTEN NITRITES

As with the nitrate melts, recent publications will be reviewed arranging the solutes in the order in which the metal, or central non-metal, appears in the

Periodic Table, beginning with the Main Group elements (s and p blocks) and proceeding to the transition metals. Since the amount of material is so very much less it is not longer necessary to break up the review with sub-headings for the various groups of elements. In the case of nitrite melts no studies have so far been made with lanthanide or actinide compounds.

2.5.1 Main Group elements

Sodium metal, in cathode compartments, has been claimed to be unreactive towards sodium nitrite at 280–330°C by Calandra and Arvía[138]. However, Lundén[27] points out that he, and all other previous workers, had found that sodium monoxide was formed at the cathode. In the course of cryoscopic measurements Kozlowski and Bartholomew[134] deduced that sodium monoxide was soluble to less than 10^{-3} M in sodium nitrite at 320 °C.

Calcium(II) and strontium(II) ions appear to be stable in nitrite melts, since a number of phase diagrams of these nitrites with alkali nitrites have been published[139, 140]. Calcium metal has been reported to react in lithium–potassium nitrite at 230–240 °C with the formation of a precipitate of calcium oxide, a trace of dissolved hyponitrite and the evolution of nitrogen and nitrous oxide in the proportions 70:30 [30], possibly via reactions (2.15) and and (2.16) above. Barium sulphate has been shown to have a very low solubility in sodium nitrite at the melting point, but barium perchlorate was more reactive since it evolved gas if present at concentrations higher than 2×10^{-2} M [134].

Thallium(I) ions are stable in all the alkali metal nitrites and phase diagrams have been published[141].

The reactions of several sodium salts of organic acids have been investigated and their stoichiometries established. On the basis of cryoscopic, infrared, voltametric and chronopotentiometric measurements, the oxidation of sodium formate and acetate in sodium nitrite at 320–350 °C was found to be in accordance with the equations[142]

$$2HCOONa + 2NaNO_2 \rightarrow 2Na_2CO_3 + H_2O + N_2O \qquad (2.51)$$
$$2CH_3COONa + 6NaNO_2 \rightarrow 4Na_2CO_3 + 3H_2O + N_2O + 2N_2 \quad (2.52)$$

A kinetic study of the oxidation of sodium oxalate in this same melt has enabled the following stoichiometry to be deduced[143]

$$aC_2O_4^{2-} + (a+\alpha)NO_2^- = [(a-\alpha)/2]CO_2 + [(3a+\alpha)/2]CO_3^{2-}$$
$$+ 2\alpha NO + [(a-\alpha)/2]N_2O \qquad (2.53)$$

It was also suggested that the sequence of reactions involved hyponitrite-type intermediates. In contrast to these reported oxidations, the phase diagrams of sodium formate, acetate, proprionate and isobutyrate with sodium nitrite have recently been reported with no mention of any reaction having been observed[50]. Whilst in contradiction to the apparent stability of sodium carbonate as indicated by the above products, appreciable concentrations ($>10^{-1}$ M) in sodium nitrite melts have been said to evolve a gas and to precipitate sodium monoxide[134].

The only information on other elements of the carbon group is a brief statement that lead salts were found to be unstable in molten sodium nitrite[134].

In the nitrogen group, perhaps the most important reaction so far noted is that of nitrogen dioxide which has been found to oxidise nitrite to nitrate

$$NO_2 + NO_2^- \rightarrow NO_3^- + NO \qquad (2.54)$$

It is not clear, however, as to whether this reaction involves the direct oxidation of nitrite ions or whether the unsymmetrical ionisation of the dinitrogen tetroxide dimers (equation (2.4) above) is involved. This appears to be a possibly fruitful area for some tracer experiments, which might provide evidence as to whether the product nitrate originated from the nitrite or from the nitrogen dioxide. The above reaction (equation (2.54)) is significant both because it increases the nitrate concentration in the melt solutions, and also because in thermogravimetric analysis the weight loss of the gaseous reaction products is considerably reduced[137]. All the evidence indicates that this reaction is quite rapid. The fact that weight losses frequently correspond with the loss of nitric oxide rather than nitrogen dioxide is evidence that the solubility of oxygen and/or its rate of solution across the gas–melt interface is low.

Recently, it has been claimed that nitric oxide also reacts with sodium nitrite melts at about 330 °C, with the production of some nitrate and nitrogen dioxide[6], possibly via the reaction

$$2NO \rightarrow NO_2 + \tfrac{1}{2}N_2 \qquad (2.55)$$

though it was not clear why the nitrogen dioxide did not then react further, and why nitrogen was not the ultimate product of all these reactions. In this somewhat confusing situation it must of course be remembered that the pure melt has been claimed to decompose yielding nitrogen dioxide, nitric oxide and nitrogen[134, 135], so that some, or all, of the products discussed above may arise solely from thermal decomposition reactions.

The solubility of water has been determined in sodium nitrite, and the value of the Henry's law constant shown to be similar to the values found for nitrate melts[69-71], and more recently for the potassium nitrate–sodium nitrite melt[72]. The only remaining significant reports of Main Group element chemistry are those of some sulphur compounds. Sodium sulphate has been shown to be both soluble and stable in sodium nitrite, giving a cryoscopic depression equivalent to only one ion. Sodium thiocyanate was considerably less stable and evolved gas when the concentration exceeded 6×10^{-2} M [134]. Sulphur dioxide was quite reactive in the same melt and formed sulphate ions together with nitrogen dioxide, nitric oxide and nitrous oxide, according to the stoichiometry[6]

$$2SO_2 + 4NaNO_2 \rightarrow 2Na_2SO_4 + NO_2 + NO + N_2O \qquad (2.56)$$

though again the first two gaseous reaction products might have been presumed to have reacted further, at least to some extent.

No reports of halogen chemistry have been published, except the brief observations that sodium chloride formed solid solutions with sodium nitrite and hence was presumably stable at 285 °C, and that barium per-chlorate reacted at appreciable concentrations[134].

2.5.2 Transition metals

On progressing through the transition metals in the usual order, the first element for which some chemistry in molten nitrites has been reported is chromium.

Potassium chromate has been found to be stable in sodium–potassium nitrite eutectic, giving a yellow solution with an absorption curve (as far as this can be obtained, i.e. beyond the region of solvent absorption) which was very like those of chromate in other solvents[137]. Potassium dichromate was found to commence reaction below the melting point of the eutectic, and the stoichiometry was determined as

$$Cr_2O_7^{2-} + 2NO_2^- \rightarrow 2CrO_4^{2-} + NO_2 + NO \tag{2.57}$$

The ready reaction of this Lux–Flood acid may be compared with the rather weak acidic character displayed in molten nitrates (e.g. reacting at 400 °C in the lithium–potassium eutectic[86]). A fast reaction of dichromate in sodium nitrite has also been observed by Bartholomew[5]. The nitrogen dioxide produced in these reactions was able to react further to form nitrate and nitric oxide (equation (2.54)). The ratios of the two gaseous products of the dichromate–sodium nitrite reaction at 300 °C have been the subject of a separate study[144]. The ratio of nitric oxide to nitrogen dioxide was found to vary with dichromate concentration etc., but to be always greater than unity, and to be lowest under reduced pressure, where presumably the least opportunity for reaction (2.54) was provided. The amount of nitrate produced in the melt and the weight of gaseous products have been shown to be satis-factorily correlated[137].

Potassium trichromate and chromium(VI) oxide have also been found to commence reacting below the melting point of the solvent eutectic[137]. Again, the ultimate product of the reaction with nitrite was a yellow solution of chromate, together with related quantities of nitric oxide, nitrogen dioxide and nitrate. The overall stoichiometries were found to be

$$Cr_3O_{10}^{2-} + 4NO_2^- \rightarrow 3CrO_4^{2-} + 2NO_2 + 2NO \tag{2.58}$$
and
$$CrO_3 + 2NO_2^- \rightarrow CrO_4^{2-} + NO_2 + NO \tag{2.59}$$

It was reported that at an intermediate stage of these reactions yellow-green or brown-black solutions were obtained, the former colour being produced if the chromium concentration was low (<0.1 M), when the solution had an absorption at 16 700 cm^{-1}, characteristic of chromium(III), as well as the chromate absorption band. On solution in water these melts gave yellow solutions of chromate and green precipitates of chromium(III) hydroxide. The ratio of chromium(VI) to chromium(III) was always found to be less than three. On the other hand, the brown-black solutions were only obtained if the chromium concentration was higher (>0.1 M). On solution in water these melts produced a brown-black precipitate of chromium(III) chromate ($Cr_2(CrO_4)_3$) and a yellow chromate solution. In addition, in these melts the $CR^{VI}:Cr^{III}$ ratio was always much larger (>9). It seems clear that with these reactants nitrite melts are displaying reducing as well as Lux–Flood base behaviour, though the amount of reduction, and hence the nature of the intermediate products, was very sensitive to concentration. What was not clear was the nature of the oxidation product of the nitrite formed when the chromium(VI) was reduced. This might have been expected to be some type of nitrogen oxyanion, but was unlikely to have been nitrate as this would have been expected to remain inert during the subsequent oxidation of chromium(III) and such an excess of nitrate was not found.

Chromium(III) chloride was also found to react at a low temperature in the sodium–potassium eutectic to give intermediate products which were yellow-green solutions containing chromium(III) if dilute (<0.1 M), and brown-black melts if more concentrated. The eventual product was again chromate, and the stoichiometry determined was[137]

$$CrCl_3 + 5NO_2^- \rightarrow CrO_4^{2-} + 3Cl^- + NO_2 + 4NO \qquad (2.60)$$

though some nitrate was also formed in accordance with equation (2.54). Chromium(III) oxide has been reported to react in sodium nitrite at a lower temperature than in sodium nitrate[145]. Thermogravimetric analysis indicated that at high concentrations of chromium the reaction was close to

$$Cr_2O_3 + 4NaNO_2 \rightarrow 2Na_2CrO_4 + 3NO + \tfrac{1}{2}N_2 \qquad (2.61)$$

Molybdate and tungstate ions are presumed to be stable in molten nitrites since phase diagrams have been published[89, 90].

Iron(III) and iron(II) ions behaved as strong Lux–Flood acids in the sodium–potassium eutectic, the chlorides reacting at the melting point and 280 °C respectively according to the equations[146]

$$FeCl_3 + 6NO_2^- \rightarrow Fe_2O_3 + 6Cl^- + 3NO_2 + 3NO \qquad (2.62)$$

and

$$FeCl_2 + 4NO_2^- \rightarrow Fe_2O_3 + 4Cl^- + NO_2 + 3NO \qquad (2.63)$$

The rather high reaction temperature for the lower chloride was attributed to its very low solubility. Once more, as with reactions in molten nitrates, it proved easier to write reaction pathways postulating nitrosyl ions as the acidic species, rather than invoking the nitrogen oxide analogous to nitrogen dioxide in nitrates (in nitrites this is dinitrogen trioxide equivalent to nitrogen dioxide plus nitric oxide).

Cobalt(II) chloride reacted similarly in the sodium–potassium eutectic at the melting point[146]

$$3CoCl_3 + 6NO_2^- \rightarrow Co_3O_4 + 6Cl^- + 2NO_2 + 4NO \qquad (2.64)$$

but cobalt metal, acting as a cathode in sodium nitrite at 302 °C, was reported to form a dark-red melt which gave cobalt(III) ions on solution in water[100]. It would thus appear that in this melt cobalt metal ions do not have powerful acidic properties.

Nickel(II) chloride was found to behave as a strong Lux–Flood acid, reacting at 220 °C in sodium–potassium nitrite according to the equation[146]

$$NiCl_2 + 2NO_2^- \rightarrow NiO + 2Cl^- + NO_2 + NO \qquad (2.65)$$

with some nitrate formed via reaction (2.54).

No reports are available for the reaction of nickel metal, but platinum, when acting as a cathode at high current density in sodium nitrite at 302 °C, was found to form a red melt which turned green due to a platinum nitrosyl compound, with evolution of a gas which was not identified[100]. No platinum compounds have been reported to form in sodium or sodium–potassium nitrite at 240–340 °C when platinum was used as an anode[147].

The reactions of several silver compounds have been reported[148]. Silver(I) oxide was found to be insoluble in sodium–potassium nitrite at 230 °C (cf. low solubility of silver(I) oxide in sodium–potassium nitrate at 250 °C [106]) and was slowly reduced to metallic silver with formation of nitrate.

$$Ag_2O + NO_2^- \rightarrow 2Ag + NO_3^- \qquad (2.66)$$

This appeared to be a surface reaction, rather than a solution mechanism, involving the formation and thermal decomposition of silver nitrite. Silver nitrite itself was found to be very soluble in this melt and to decompose (with maximum rate at 300 °C) forming metallic silver and nitrogen dioxide; with some of the latter reacting further to form nitrate (equation (2.54)). These silver nitrite solutions exhibited greater thermal stability than the pure silver nitrite solid which commenced decomposing well below 200 °C. This rather greater stability in solution was considered to be due to some oxygen bonding to silver in the melt, in contrast to the solid in which x-ray diffraction has shown nitrogen to be nearest the silver[149].

Silver nitrate exhibited comparatively higher thermal stability, decomposing with a maximum rate at 320 °C according to the equation

$$AgNO_3 + NO_2^- \rightarrow Ag + NO_3^- + NO_2 \qquad (2.67)$$

It was suggested that the additional stability might indicate either that more nitrite was now coordinated via oxygen or that some nitrate ions were still bonded to silver. These hypotheses were not disproved by the observation that infrared examination of the partially reacted melt showed an absorption at $852 \, cm^{-1}$, indicative of silver nitrite, as well as the $808 \, cm^{-1}$ band of silver nitrate.

Silver chloride was found to be only slightly soluble in this melt ($\sim 7 \times 10^{-3}$ M) which corresponded closely with the value obtained in sodium nitrate at 300 °C (5.8×10^{-3} M) [150]. It reacted slowly in accordance with the equation

$$AgCl + NO_2^- \rightarrow Ag + Cl^- + NO_2 \qquad (2.68)$$

with a maximum reaction rate at 420 °C, this high temperature being attributed to the low solubility. The reaction was presumed to occur via the interaction of silver cations with nitrite anions, followed by their mutual reduction and oxidation, though in this case infrared examination of the quenched melt showed no indication of silver nitrite.

Silver metal, when acting as a cathode in sodium nitrite at 302 °C, has been reported to form an insoluble yellow compound, and the melt on solution in water gave silver ions and a yellow precipitate of silver hyponitrite[100]. This appears to be a reduction analogous to reaction (2.15) in nitrate melts, and might well be supposed to occur preferentially in the presence of silver because of the insolubility of the silver salt but for the fact that silver ions are apparently unstable in nitrite melts (as discussed above[148]). The true explanation of these observations is therefore in doubt at the present time.

Cadmium salts have been reported to be unstable in sodium nitrite at 280 °C, though no details of the products formed have been given[134].

2.6 CONCLUSION

It will have been noted that in a number of cases rather different reactions or products have been found when the proportions of alkali metal cations are varied in a melt. In several cases these differences are considerably greater than would have been expected from the variation of ion size alone. This, together with the companion observation that the reactions of supposedly ionic solute anions can vary with the nature of alkali metal cation with which they are combined, suggests that a note of caution is necessary in the customary interpretation of nitrate and nitrite melts as simple assemblages of ions, and in the implication that reactions in different melts should necessarily be strictly comparable.

These complications will, however, add to the rich variety and considerable interest of the chemistry of these melts, and suggests that the discoveries of future years may be even more exciting, as well as more extensive, than those of the past.

Note added in proof

The e.m.f. of the $Ag^{(I)}$/Ag electrode in sodium nitrite at 309 °C has again been reported, by Boxall and Johnson[151] (cf. ref. 100) with the consequent implication of the stability of the $Ag^{(I)}$ ion in molten nitrite. This conflicts with the findings of the extensive investigation of the decomposition of silver(I) compounds in sodium–potassium nitrite[148] and the observation by Calandra and Arvía[147] that decomposition occurred when sodium nitrite came into contact with a solution of $Ag^{(I)}$ in molten nitrate, and suggests that this discrepancy merits further investigation.

References

 1. Kust, R. N. and Duke, F. R. (1963). *J. Amer. Chem. Soc.,* **85,** 3338
 2. Luthy, J. A. and Duke, F. R. (1963). *U.S.A.E.C. Report* 1S-742
 3. Glasner, A., Pelly, I. and Steinberg, M. (1969). *J. Inorg. Nucl. Chem.,* **31,** 3395
 4. Topol, L. E., Osteryoung, R. A. and Christie, J. H. (1966). *J. Phys. Chem.,* **70,** 2857
 5. Bartholomew, R. F. and Donigan, D. W. (1968). *J. Phys. Chem.,* **72,** 3545
 6. Kozlowski, T. R., Bartholomew, R. F. and Garfinkel, H. M. (1970). *J. Inorg. Nucl. Chem.,* **32,** 401
 7. Francini, M. and Martini, S. (1968). *Electrochim. Acta,* **13,** 851
 8. Zambonin, P. G. (1970). *J. Electroanal. Chem.,* **24,** 365
 9. Zambonin, P. G. and Jordan, J. (1967). *J. Amer. Chem. Soc.,* **89,** 6365
10. Zambonin, P. G. and Jordan, J. (1969). ibid., **91,** 2225
11. Zambonin, P. G. (1969). *Anal. Chem.,* **41,** 868
12. Zambonin, P. G. (1970). *J. Electroanal. Chem.,* **24,** 25A
13. Kust, R. N. (1964). *Inorg. Chem.,* **3,** 1035

14. Shams El Din, A. M. (1962). *Electrochim. Acta*, **7**, 285
15. Shams El Din, A. M. and Gerges, A. A. A. (1963). *Proc. First Aust. Conf. Electrochem.*, p. 462. (Oxford: Pergamon)
16. Kust, R. N. (1969). *J. Electrochem. Soc.*, **116**, 1137
17. Swofford, H. S. and McCormick, P. G. (1965). *Anal. Chem.*, **37**, 970
18. Kohlmuller, R. (1959). *Ann. Chim.*, **4**, 1183
19. Shams El Din, A. M. and El Hosary, A. A. (1966). *J. Inorg. Nucl. Chem.*, **28**, 3043
20. Shams El Din, A. M. and El Hosary, A. A. (1967). *Electrochim. Acta*, **12**, 1665
21. Kerridge, D. H. and Habboush, D. A. (1968). *J. Inorg. Nucl. Chem.* **30**, 2870
22. Kust, R. N. and Burke, J. D. (1969). *Inorg. Chem.*, **8**, 1748
23. Mazzochin, G-A., Bombi, G. G. and Sacchetto, G. A. (1969). *J. Electroanal. Chem.*, **21**, 345
24. Freeman, E. S. (1954). *J. Phys. Chem.*, **60**, 1487; Freeman, E. S. (1957). *J. Amer. Chem. Soc.*, **79**, 838
25. Bartholomew, R. W. (1966). *J. Phys. Chem.*, **20**, 3442
26. Kust, R. N. and Burke, J. D. (1970). *Inorg. Nucl. Chem. Lett.*, **6**, 333
27. Lundén, A. (1969). *Electrochim. Acta*, **14**, 1068
28. Alberti, G., Conte, A. and Allulli, S. (1965). *J. Chromatog.*, **8**, 564
29. Callahan, C. M. and Kay, M. A. (1966). *J. Inorg. Nucl. Chem.*, **28**, 233
30. Brough, B. J. and Kerridge, D. H. (1965). *Inorg. Chem.*, **4**, 1353
31. Arvía, A. J., Calandra, A. J. and Martins, M. E. (1966). *Electrochim. Acta*, **11**, 963
32. Arvía, A. J., Calandra, A. J. and Triaca, W. E. (1964). *Electrochim. Acta*, **9**, 1417
33. Arvía, A. J., Calandra, A. J. and Martins, W. E. (1967). *Electrochim. Acta*, **12**, 347
34. Bombi, G. G. and Fioroni, M. (1965). *Talanta*, **12**, 1053
35. Alberti, G. and Grassini, G. (1960). *J. Chromatog.*, **4**, 425
36. Hester, R. E. and Krishnan, K. (1967). *J. Chem. Phys.*, **46**, 3405; **47**, 1747
37. Barclay, K. S. and Crewe, J. M. (1967). *J. Appl. Chem.*, **17**, 21
38. Ray, N. H., Stacey, M. H., Thomas, R. E. and Webster, S. J. (1967). *Glass Technol.*, **8**, 78
39. Conte, A. and Campanella, L. (1968). *Electrochim. Mettalorum.*, **3**, 183
40. Campanella, L. and Conte, A. (1969). *J. Electrochem. Soc.*, **114**, 144
41. Sinistri, C., Franzosini, P., Timidei, A. and Rolla, M. (1965). *Z. Naturforsch.*, **20a**, 561
42. Franzosini, P., Sinistri, C., Rolla, M. and Timidei, A. (1966). ibid., **21a**, 595
43. Sinistri, C., Franzosini, P. and Flor, G. (1967). *Gazz. Chim. Ital.*, **97**, 275
44. Mazzochin, G-A., Bombi, G. G. and Fiorani, M. (1966). *Ric. Sci.*, **36**, 338
45. Arvía, A. J. and Triaca, W. E. (1965). *Electrochim. Acta*, **10**, 1188
46. Bratland, D. and Krohn, C. (1969). *Acta Chem. Scand.*, **23**, 1839
47. Brough, B. J. and Kerridge, D. H. (1966). *J. Chem. Eng. Data*, **11**, 260
48. Brough, B. J., Kerridge, D. H. and Mosley, M. (1966). *J. Chem. Soc. A*, 1556
49. Shul'ga, N. A. and Bergman, A. G. (1969). *Russ. J. Inorg. Chem.*, **14**, 1034
50. Sokolov, N. M., Tsindrik, N. M. and Khritin, M. Y. (1970). ibid., **15**, 433
51. Shams El Din, A. M. and El Hosary, A. A. (1968). *Electrochim. Acta*, **13**, 135
52. Kozlowski, T. R. and Bartholomew, R. F. (1968). *Inorg. Chem.*, **7**, 2247
53. Temple, R. B., Fay, C. and Williamson, J. (1967). *Chem. Commun.*, 966
54. Grigorenko, F. F. and Suprunenko, A. A. (1968). *Visn. Kiiv. Univ. Ser. Khim.*, **9**, 67
55. Inmann, D., Lovering, D. G. and Narayan, R. (1968). *Trans. Faraday. Soc.*, **64**, 2476
56. Isbell, R. E., Wilson, E. W. and Smith, D. F. (1966). *J. Phys. Chem.*, **70**, 2493
57. Shams El Din, A. M. and El Hosary, A. A. (1968). *J. Electroanal. Chem.*, **16**, 551
58. Conte, A. and Casadio, S. (1966). *Ric. Sci.*, **36**, 433
59. Cleaver, B. and Mather, D. E. (1970). *Trans. Faraday Soc.*, **66**, 2469
60. Allulli, S. (1969). *J. Phys. Chem.*, **73**, 1084
61. Egghart, H. C. (1967). *Inorg. Chem.*, **6**, 2121
62. Siew, L. C. and Sundheim, B. R. (1969). *J. Phys. Chem.*, **73**, 4135
63. Shams El Din, A. M., El Hosary, A. A. and Gerges, A. A. A. (1964). *J. Electroanal. Chem.*, **8**, 312
64. Coumert, N., Porthault, M. and Merlin, J-C. (1965). *Bull. Soc. Chim. France*, 910
65. Shams El Din, A. M., Taki El Din, H. D. and El Hosary, A. A. (1968). *Electrochim. Acta*, **13**, 407
66. Shams El Din, A. M. and El Hosary, A. A. (1968). *J. Electroanal. Chem.*, **17**, 238
67. Mazzochin, G-A., Bombi, G. G. and Sacchetto, G. A. (1970). ibid., **24**, 31

68. Zambonin, P. G., Candetta, V. C. and Signorile, G. (1970). *J. Electroanal. Chem.,* **28,** 237
69. Duke, F. R. and Doan, A. S. (1958). *Iowa State College J. of Science,* **32,** 451
70. Frame, J. P., Rhodes, E. and Ubbelodhe, A. R. (1961). *Trans. Faraday Soc.,* **57,** 1075
71. Peleg, M. (1967). *J. Phys. Chem.,* **71,** 4553
72. Hull, H. S. and Turnbull, A. G. (1970). *J. Phys. Chem.,* **74,** 1783
73. Bombi, G. C., Sacchetto, G. A. and Mazzochin, G-A. (1970). *J. Electrochem.,* **24,** 23
74. Delimarskii, Iu. K. and Shilina, G. V. (1965). *Electrochim. Acta,* **10,** 973
75. Swofford, H. S. and Propp, J. H. (1965). *Anal. Chem.,* **37,** 974
76. Sacchetto, G. A., Bombi, G. G. and Fiorani, M. (1969). *J. Electroanal. Chem.,* **20,** 89
77. Habboush, D. A. and Kerridge, D. H. (1970). *Inorg. Chim. Acta,* **4,** 81
78. Novik, R. M. and Lyalikov Iu. S. (1958). *J. Anal. Chem. USSR,* **13,** 783
79. Schlegel, J. M. (1965). *J. Phys. Chem.,* **69,** 3638
80. Schlegel, J. M. (1967). ibid., **71,** 1520
81. Schlegel, J. M. (1969). ibid., **73,** 4152
82. Protsenko, P. I. and Bergman, A. G. (1950). *Zh. Obshch. Khim.,* **20,** 1365
83. Alberti, G., Allulli, S. and Cardini, G. (1969). *J. Chromatog.,* **45,** 298
84. Mokhosoev, M. V., Aleikina, S. M. and Fedorov, P. I. (1966). *Russ. J. Inorg. Chem.,* **11,** 644
85. Shams El Din, A. M. and El Hosary, A. A. (1964). *J. Electroanal. Chem.,* **7,** 464
86. Brough, B. J., Kerridge, D. H. and Tariq, S. A. (1967). *Inorg. Chim. Acta,* **1,** 267
87. Shams El Din, A. M. and El Hosary, A. A. (1965). *J. Electroanal. Chem.,* **9,** 349
88. Spink, M. W. Y. (1966). *Diss. Abs.,* **26,** 4274
89. Shurdumov, G. K. and Khokonova, T. N. (1969). *Zh. Neorg. Khim.,* **14,** 1706
90. Shurdumov, G. K. and Khokonova, T. N. (1970). ibid., **15,** 843
91. Korobka, E. I., Kishova, A. I. and Bergman, A. G. (1967). *Zh. Neorg. Khim.,* **12,** 3207
92. Kust, R. N. (1967). *Inorg. Chem.,* **6,** 157
93. Kust, R. N. (1967). ibid., **6,** 2239
94. Kerridge, D. H. and Tariq, S. A. (1968). *Inorg. Chim. Acta,* **2,** 371
95. Bennett, R. M. and Holmes, O. G. (1963). *Canad. J. Chem.,* **41,** 108
96. Gal, I. J., Méndez, J. and Irvine, J. W. (1968). *Inorg. Chem.,* **7,** 985
97. Walden, J. A. (1970). *Diss. Abs.,* **30B,** 3540
98. Fung, K. W. and Johnson, K. E. (1970). *Canad. J. Chem.,* **47,** 4699
99. Ogilvie, F. B. and Holmes, O. G. (1966). *Canad. J. Chem.,* **44,** 447
100. Bartlett, H. E. and Johnson, K. E. (1967). *J. Electrochem. Soc.,* **114,** 64
101. Liu, C. H., Hasson, J. and Smith, G. P. (1968). *Inorg. Chem.,* **7,** 2244
102. Kust, R. N. and Burke, J. D. (1969). *Inorg. Chem.,* **8,** 1748
103. Bowcott, J. E. L. and Plunkett, B. A. (1969). *Electrochim. Acta,* **14,** 363, 883
104. Okada, M. and Kawamora, K. (1970). ibid., **15,** 1
105. Brookes, H. C. and Flengos, S. N. (1970). *Canad. J. Chem.,* **48,** 55
106. Shams El Din, A. M., Gouda, T. and El Hosary, A. A. (1968). *J. Electroanal. Chem.,* **17,** 137
107. Bombi, G. G., Fiorani, M. and Mazzochin, G-A. (1965). *J. Electroanal. Chem.,* **9,** 457
108. Lindgren, R. M. (1968). *Diss. Abs.,* **29B,** 569
109. Elding, I. and Leden, I. (1969). *Acta Chem. Scand.,* **23,** 2430
110. Conte, A. and Ingram, M. D. (1968). *Electrochim. Acta,* **13,** 1551
111. Hills, G. J. and Power, P. D. (1969). *Trans. Faraday Soc.,* **65,** 3042
112. Bombi, G. G., Mazzochin, G-A. and Fiorani, M. (1966). *Ric. Sci.,* **36,** 573
113. Flaherty, T. P. and Braunstein, J. (1967). *Inorg. Chim. Acta,* **1,** 335
114. Liquornik, M. and Irvine, J. W. (1970). *Inorg. Chem.,* **9,** 1330
115. Inman, D. (1965). *Electrochim. Acta,* **10,** 11
116. Braunstein, J. and Minano, A. S. (1966). *Inorg. Chem.,* **5,** 942
117. Inman, D. and Braunstein, J. (1966). *J. Phys. Chem.,* **70,** 2726
118. Inmann, D. and Bockris, J. O. M. (1961). *Trans. Faraday Soc.,* **57,** 2308
119. Sinistri, C., Flor, G., Franzosini, P. and Rolla, M. (1967). *Z. Naturforsch,* **22,** 53
120. Swofford, H. S. and Holifield, C. L. (1965). *Anal. Chem.,* **37,** 1513
121. Mazzochin, G-A., Bombi, G. G. and Fiorani, M. (1968). *J. Electroanal. Chem.,* **17,** 95
122. Fiorani, M., Bombi, G. G. and Mazzochin, G-A. (1966). *Ric. Sci.,* **36,** 580
123. Mazzochin, G-A., Bombi, G. G. and Sacchetto, G. A. (1969). *J. Electroanal. Chem.,* **21,** 345
124. Bartlett, H. E. and Johnson, K. E. (1961). *Canad. J. Chem.,* **44,** 2119

125. Vittori, O. and Porthault, M. (1969). *Compt. Rend. Acad. Sci. Paris Ser. C.*, **269**, 967
126. Liquornik, M. (1971). *J. Inorg. Nucl. Chem.*, **33**, 283
127. Val'tsev, V. K. and Didora, N. F. (1968). *Izv. Sibirsh. Otdel Akad. Nauk. SSSR Ser. Khim. Nauk*, **4**, 26
128. Padova, J., Seleg, M. and Soriano, J. (1967). *J. Inorg. Nucl. Chem.*, **29**, 1895
129. Carnall, W. T., Neufeldt, S. J. and Walker, A. (1965). *Inorg. Chem.*, **4**, 1808
130. Protshenko, P. I. and Shisholina, R. P. (1963). *Russ. J. Inorg. Chem.*, **8**, 1438 (2244)
131. Shisholina, R. P. and Protshenko, P. I. (1963). ibid., **8**, 1436 (2741)
132. Polyakov, V. D. and Berul, S. I. (1955). *Izv. Sektora Fiz. Khim. Anal. Inst. Obschei; Neorg Khim Akad Nauk SSSR*, **26**, 164
133. Kust, R. N. and Fletcher, R. W. to be published, quoted in Reference 26
134. Kozlowski, T. R. and Bartholomew, R. W. (1967). *J. Electrochem. Soc.*, **114**, 937
135. Oza, T. M. (1945). *J. Indian Chem. Soc.*, **22**, 173; Oza, T. M. and Walarwalkar, B. R. (1945). ibid., **22**, 243
136. Lee, A. K. K. (1969). *Diss. Abs.*, **29B**, 4578
137. Kerridge, D. H. and Tariq, S. A. (1969). *Inorg. Chim. Acta*, **3**, 667
138. Calandra, A. J. and Arvia, A. J. (1967). *Electrochim. Acta*, **12**, 95
139. Protshenko, P. I. and Medvedev, B. S. (1963). *Russ. J. Inorg. Chem.*, **8**, 1434 (2733)
140. Protshenko, P. I. and Shurdumov, G. K. (1964). ibid., **9**, 916 (1692)
141. Protshenko, P. I. and Brykova, N. A. (1965). ibid., **10**, 659 (1220)
142. Kozlowski, T. R. and Bartholomew, R. W. (1968). *Inorg. Chem.*, **7**, 2247
143. Burke, J. D. and Kust, R. N. (1970). *Inorg. Chim. Acta*, **4**, 309
144. Kust, R. N. and Fletcher, R. W. (1969). *Inorg. Chem.*, **8**, 687
145. Sugiyama, K. and Takahashi, T. (1967). *J. Chem. Soc. Japan (Ind. Chem. Sect.)*, **70**, 830
146. Kerridge, D. H. and Tariq, S. A. (1970). *Inorg. Chim. Acta*, **4**, 499
147. Calandra, A. J. and Arvía, A. J. (1966). *Electrochim. Acta*, **11**, 1173
148. Kerridge, D. H. and Tariq, S. A., *Inorg. Chim. Acta*, **4**, 452
149. Ketelaar, J. A. A. (1936). *Z. Krist.*, **95A**, 383
150. Seward, R. P. and Field, P. E. (1964). *J. Phys. Chem.*, **68**, 210
151. Boxall, C. G. and Johnson, K. E. (1971). *J. Electroanal. Chem.*, **30**, 25

3
Halides of Phosphorus

R. SCHMUTZLER and O. STELZER
Technische Universität, Braunschweig

3.1   INTRODUCTION

A halide of phosphorus, within the scope of this review, is a compound of
phosphorus containing at least one phosphorus–halogen bond. In view of
the enormous number of reports pertinent to such compounds that have
been published within the short span of the 2 years covered, 1969–1970, we
have narrowed the scope of this article by putting emphasis on completely
or largely inorganic compounds containing a phosphorus–halogen bond.
As an ordering principle, we have arranged the various types of compound
according to the coordination number around phosphorus, i.e. 2 to 6.

In spite of the large number of publications it cannot be said that strikingly
novel developments have taken place in the area, and the period was one

of consolidation rather than of great innovation. Particular effort has been spent on the increased application of physical methods to the investigation of phosphorus halides, vibrational and nuclear magnetic resonance spectroscopy being particularly worth mentioning. Nuclear quadrupole resonance spectroscopy and photo-electron spectroscopy have also made their entry into phosphorus–halogen chemistry.

Among the classes of compound studied in most detail, the coordination compounds of trivalent phosphorus halides, nitrogen–phosphorus–halogen compounds, and 5-coordinate phosphorus halides may be singled out. As to preparative methods in the area of phosphorus halide chemistry, the cleavage reactions of element–trimethylsilyl compounds (element = N, P, O, S, etc.) with phosphorus halides have found further widespread application, providing ready access to the derivative chemistry of phosphorus halides.

A literature search for the 2 year period has been made, using *Chemical Abstracts*. Inevitably, complete coverage of all developments in the area has proved impossible, as it would far exceed the limits of this chapter. We had to be selective and, perhaps, subjective, in order to keep the size of the review within bounds. We feel, however, that we have covered what is essential in phosphorus halide chemistry in 1969–1970. As will be seen, we have in some instances referred to work prior to 1969–1970 if we felt this was required for the full description of a particular topic.

3.2 PHOSPHORUS HALIDES CONTAINING BI-COORDINATE PHOSPHORUS

Only a few organic compounds of bi-coordinate phosphorus are known[1-3] while stable phosphorus halides with bi-coordinate phosphorus are not known to exist. According to spectroscopic investigations, ionic, fluorine-bridged structures of type $[PF_2]^+[MF_6]^-$ have been postulated to be present in the systems PF_3/MF_5 (M = As, Sb)[4]. At high temperatures and low pressures tetrafluorodiphosphine, P_2F_4, is decomposed with formation of PF_2 radicals[5, 20]. These radicals are also formed upon electron impact on P_2F_4 [20], or on γ-irradiation of $[ND_4][PF_6]$ [6], or by β-bombardment of phosphorus trifluoride in a matrix of sulphur hexafluoride[7].

The e.s.r. spectrum of PF_2 (generated from P_2F_4 or PF_2H by pyrolysis above 200 °C and trapped at 20 °K) shows 12 lines[8, 9]. The e.s.r. data suggest that the radical electron is more strongly located on the central atom than in the case of the NF_2 radical.

The radicals PCl_2 and PBr_2 have been identified by i.r. spectroscopy in an argon matrix; they were generated in the reaction of phosphorus trichloride or tribromide with lithium atoms. The bond angle, X—P—X, is probably 120 degrees[10].

In the photolysis of PCl_3 the two radicals, PCl_4 and PCl_2, are produced. Formation of the former may be suppressed if xenon is employed as a diluent[11],

$$PCl_3 \xrightarrow{h\nu} \cdot PCl_2 + Cl\cdot \tag{3.1}$$

$$Cl\cdot + PCl_3 \rightarrow \cdot PCl_4 \tag{3.2}$$

Photolysis of methyldichlorophosphine furnishes the radical, CH_3—P—Cl [11] which was identified by its e.s.r. signals. PCl_2 radicals formed according to (3.1) may react with cycloalkanes. Thus, cyclohexyldichlorophosphine is formed upon irradiation of phosphorus trichloride and cyclohexane with ^{60}Co γ-rays[12]. The yields in these reactions are determined by the dose of γ-irradiation employed[13].

In the mass spectrum of phosphorus trichloride the positive ion, PCl_2^+, is most abundant, while in the mass spectrum of the negative ions the Cl^- ion is predominant[14, 15]. Formation of PCl_2^+ takes place within an ion-pair, in accord with equation (3.3).

$$PCl_3 + e \xrightarrow{\;12 \pm 0.5\,eV\;} Cl^- + PCl_2^+ + e \qquad (3.3)$$

3.3   HALIDES OF TRI-COORDINATE PHOSPHORUS CONTAINING PHOSPHORUS–PHOSPHORUS BONDS

In contrast to halogen compounds of types PX_3 and $R_{3-n}PX_n$, halides of tricoordinate phosphorus involving a P—P bond are not very well known. The iodide, P_2I_4, however, is an exception, in that it has been known since the nineteenth century[16, 17]. The compounds P_2F_4 [18], F_2PPH_2 [19] and $P(PF_2)_3$ [20] have been reported only recently. Diphosphorus tetrachloride, P_2Cl_4, was first described as early as 1910 [21], but no useful preparation was known until 1963 [22].

We propose to cover this type of halide of phosphorus in some detail, as there has been much recent development. Inevitably, we will have to include references to work prior to the period 1969–1970, but we feel that such a treatment is justified for this area.

3.3.1   Preparation

3.3.1.1   *Compounds of type* P_2X_4

Diphosphorus tetrafluoride, P_2F_4, was first obtained in 1966 by the reaction of PF_2I with mercury[18].

$$2PF_2I + 2Hg \rightarrow P_2F_4 + Hg_2I_2 \qquad (3.4)$$

In the reaction of phosphine with difluoroiodophosphine small amounts of phosphinodifluorophosphines, $P_2H_2F_2$ [19], and PF_2H, are formed.

$$PH_3 + PF_2I \rightarrow F_2PPH_2 + HI \qquad (3.5)$$

The methods for the preparation of P_2Cl_4 reported in the literature almost invariably involve electrical discharges in a PCl_3/H_2 mixture in the gas phase, they usually differ only with regard to the kind of electrode material employed and in the partial pressures of both hydrogen and phosphorus trichloride[21, 23, 24]. The ratio of the two partial pressures seems to be a critical factor in determining the yields of P_2Cl_4 [22]. A further method of preparation of P_2Cl_4 consists of passing an electric discharge through the gas phase above a solution of white phosphorus in phosphorus trichloride[22].

The formation of diphosphorus tetrabromide, P_2Br_4, in the neutron irradiation of phosphorus tribromide has been reported[25]. Its formation was proved chemically, by an investigation of its hydrolysis products. An organic derivative of P_2Br_4, 1,2-dibromo-1,2-diphenyl-diphosphine, was obtained by a number of routes, such as the reaction of the cyclopoly-phosphine, $(PC_6H_5)_5$ with $C_6H_5PBr_2$ or with a solution of bromine in carbon disulphide, also through halogen exchange in 1,2-di-iodo-1,2-diphenyldiphosphine and silver bromide[26].

No preparative method for P_2I_4 has been described recently. Reference is made to the review article by Payne[27] which describes the older preparations. P_4I_2 is claimed to be formed in the chain reaction between white phosphorus and iodine[28] which gave P_2I_4.

3.3.1.2 Compounds of type $P(PX_2)_3$

The unusual compound tris(difluorophosphino)phosphine[20], $P(PF_2)_3$, has been obtained in the pyrolysis of P_2F_4,

$$P_2F_4 \xrightarrow[3\,\text{mm}]{900\,°C} P(PF_2)_3 \qquad (3.6)$$

Identification was from the mass spectrum, a parent peak at $m/e = 238$ being observed, and by ^{19}F and ^{31}P n.m.r. spectroscopy.

$P(PF_2)_3$ is the only well-established compound of the type $P(PX_2)_3$. It seems possible that a closer investigation of the decomposition of P_2X_4 (X = Cl, I) may provide evidence for similar chlorides or iodides, respectively.

3.3.1.3 Compounds of type $(PX)_n$

Compounds of this type exist either in the form of short-lived, reactive species or as polymeric products. Thus, in the thermal decomposition of the adduct $P_2I_4 \cdot AlI_3$, besides $AlI_3 \cdot PI_3$, a polymeric compound of composition $(PI)_n$ [29] is formed. The radical PCl was detected in the isothermal flash photolysis of phosphorus trichloride vapour[30].

3.3.2 Physical and chemical properties

Physical properties of the compounds mentioned are listed in Table 3.1.

The chemistry of these compounds has been studied but little, so far. P_2F_4, upon hydrolysis, gives F_2POPF_2 and PF_2H [18]. It has been concluded from e.s.r. studies that P_2F_4 undergoes dissociation into PF_2 radicals[18, 20],

$$P_2F_4 \rightleftharpoons 2 \cdot PF_2 \qquad (3.7)$$

Only a few reactions of P_2Cl_4 have been investigated. Thus, at $-23\,°C$

no complex formation with boron trihalides takes place while decomposition of P_2Cl_4 commences at 0 °C. The decomposition products have not been completely identified but have been reported to include phosphorus trichloride, elemental phosphorus[36] and a sub-halide, $(PCl)_n$ [37]. Of all com-

Table 3.1   Physical properties of halides of tri-coordinate phosphorus containing phosphorus–phosphorus bonds

Compound	B.P.(°C)	M.P.(°C)	d_4^{20}	Reference
P_2F_4	−6.2			18
P_2Cl_4		−28	1.701	21, 22, 31, 24
		−29 to −28		
		−34		
P_2I_4		126–7		32, 33
		122		34
		124.5		17
		125.5	4.178	35
$P_2H_2F_2$			±0.002	19
$P(PF_2)_3$		−68		20

pounds of type P_2X_4 the longest known, P_2I_4. has been studied most thoroughly. Its known reactions are summarised in Figure 3.1.

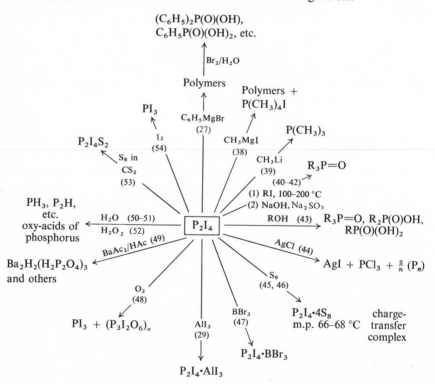

Figure 3.1 (References in parenthases)

It should be noted that some of the reactions listed in Figure 3.1 were reported a long time ago, and re-investigation using modern physical methods seems to be necessary.

While the ligand properties of diphosphines, R_2PPR_2 (R = alkyl, per-fluoroalkyl, or aryl) have been investigated in detail [55-57], similar studies, involving P_2X_4 have hardly been undertaken, although it would be expected that they might form complexes of similar stability as phosphorus trihalides.

Diphosphorus tetrachloride, P_2Cl_4, in large excess, reacts with $Ni(CO)_4$ as a monofunctional base and, depending on the stoichiometric ratio of the reactants, the complexes $Ni(CO)_2(P_2Cl_4)_2$, $Ni(CO)(P_2Cl_4)_3$, and $Ni(P_2Cl_4)_4$ are obtained. If a large excess of $Ni(CO)_4$ is employed, P_2Cl_4 behaves as a bifunctional base and forms $(CO)_3NiP_2Cl_4Ni(CO)_3$ [37].

Mention should be made, finally, of the reaction of 1,2-diaryldi-iododi-phosphines with alkyl halides in which 1,2-diaryl-tetra-alkyl-diphosphonium-iodopentaiodides are formed[58].

3.3.3 Spectroscopic investigations

N.M.R. investigations (^{19}F, ^{31}P) of P_2F_4 have been conducted, and the molecule has been treated as an XX'AA'X''X''' spin system (X = F, A = P) [59]. The observation of two different vicinal F—F coupling constants suggests that one rotamer predominates (see Figure 3.2).

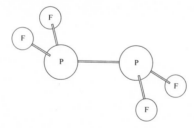

Figure 3.2   Molecular configuration of P_2F_4 as derived from n.m.r. spectra

I.R. and Raman spectroscopic investigations have also shown that the *trans*-rotamer shown in Figure 3.2 is most likely while a different order of stability of the rotamers of P_2F_4 was concluded from HMO-LCAO cal-culations[60],

$$gauche > cis > trans$$

Nearly first-order ^{19}F and ^{31}P n.m.r. spectra were observed for $P(PF_2)_3$ [20].

From the mass spectroscopic investigation of P_2Cl_4 a value for the enthalpy of formation of -106 kcal mol^{-1} and of 58 kcal mol^{-1} for the P—P bond energy have been derived[14, 61]. I.R. and Raman spectra of P_2Cl_4 have been studied in detail and potential fields and force constants were deter-mined[62]. It is suggested that the molecule has the *trans* structure[62, 63].

The u.v. spectrum of P_2I_4 shows a band at 228 nm (extinction not speci-fied)[64]. Two i.r. absorptions in P_2I_4 at 301 and 329 cm^{-1} were assigned to P—I stretching vibrations[53]. The *gauche* structure originally postulated for

P_2I_4 on the basis of dipole moment[35] and i.r. measurements[65] could not be confirmed in more recent work[62, 65a, 66]. The *trans* structure was established for P_2I_4 in solution (carbon disulphide, benzene, dichloromethane, carbon tetrachloride, methyl iodide) as well as for the solid[66].

The dissociation energy for the P—P bond in P_2I_4, D (P—P), has been obtained from electron impact studies as 64.8 kcal mol^{-1} [61]. The enthalpy of formation, $\Delta H_f^0 = -113.1 \pm 5.0$ kJ mol^{-1}, has been determined from the thermochemical investigation of the equilibrium,

$$P_2I_4 + I_2 \rightleftharpoons 2 PI_3 \qquad (3.8)$$

By the same method a mean P—P bond energy of 232 kJ mol^{-1} has been obtained[54].

3.3.4 Structural investigations

Accurate structural data are available only for P_2I_4. The crystals are triclinic with the space group $P\bar{1}$. The unit cell is occupied by one molecule of P_2I_4. The molecule belongs to the point group $C_{2h} - 2/m$ [34]. Figure 3.3 shows the spatial arrangement of the PI_2 groups, relative to each other; the corres-

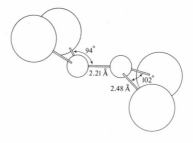

Figure 3.3   X-ray structure of P_2I_4

ponding structural parameters are also listed. The P—P length of 2.21 Å, the bond angles, and the results of the mass spectroscopic investigation also suggest a simple σ-bond between the phosphorus atoms. The P—I bond lengths differ only slightly from that in PI_3 [34].

3.4 HALIDES OF TRI-COORDINATE PHOSPHORUS OF TYPE PXYZ

3.4.1 Preparation

3.4.1.1 Halides of type PX_3

No fundamentally new methods of preparation for phosphorus trihalides, PX_3 (X = F, Cl, Br, I) were developed during the last few years. Preparative methods known up to about 1964 have been reviewed in detail by Kolditz[67] and Payne[68]. Attention is also drawn to the review articles by Schmutzler[69],

Drozd[70] and Nixon[71] in which the chemistry of phosphorus fluorides, including PF_3, is covered.

3.4.1.2 Preparation of mixed phosphorus halides, $PX_{3-n}Y_n$ and PXYZ

Syntheses of PF_2Cl, $PFCl_2$, PF_2Br, $PFBr_2$ and PF_2I, as well as of PFClBr, are described in the above-mentioned review articles[69-71], and will not be discussed here. Mention should be made of three methods of preparation for these compounds which are based on the cleavage of the P—N bond in aminofluorophosphines by means of hydrogen halides[72-74].

$$R_2N—PF_2 + 2HX \rightarrow (R_2NH_2)X + X—PF_2 \qquad (3.9)$$

on the equilibrium of phosphorus trihalides or partial flourination[75-78],

$$PX_3 + PY_3 \rightleftharpoons PX_2Y + PY_2X \qquad (3.10)$$
$$PZ_2Y + PX_2Y \rightleftharpoons 2PXYZ \qquad (3.11)$$

No new syntheses of the remaining seven mixed halides of phosphorus, PCl_2Br, $PClBr_2$, PCl_2I, $PClI_2$, PBr_2I, $PBrI_2$ and PClBrI, have been reported. The formation of the compounds PCl_2Br, $PBrCl_2$, PBr_2I and $PBrI_2$ could be established by spectroscopy in the systems $PCl_3–PBr_3$ and $PBr_3–PI_3$ (cf. Ref. 68 and literature quoted in that reference).

3.4.1.3 Amino- and alkoxy-halophosphines, $(R_2N)_n PX_{3-n}$ and $(RO)_n PX_{3-n}$

Both classes of compound are very important with regard to their synthetic potential because amino and alkoxy groups are good 'leaving groups' in reactions with compounds of type HX^{72-74}. In addition to the well established syntheses of these compounds, such as reaction of phosphorus trihalides with primary or secondary amines[79] or alcohols[80] and equilibration of mixtures of $PX_3—P(NR_2)_3$ [81, 82] or $PX_3—P(OR)_3$ [83, 84], the cleavage of Si—N and Si—O bonds in silylamines[85] and silyl ethers[86] has been described as a preparative method, e.g.

$$R_2NSiMe_3 + PCl_3 \rightarrow R_2NPCl_2 + Me_3SiCl \qquad (3.12)$$

Alkylamino-bis-dichlorophosphines, $RN(PCl_2)_2$, have been obtained in the reaction of hydrochlorides of primary amines with phosphorus trichloride[72]:

$$(RNH_3)Cl + 2PCl_3 \rightarrow RN(PCl_2)_2 + 3HCl \qquad (3.13)$$

The corresponding fluorophosphines were obtained by fluorination of the chlorophosphines with antimony trifluoride[72]. Aminodifluorophosphines are formed, also, in the unusual reaction of trichloromethyldifluorophosphine with secondary amines[87].

$$CCl_3PF_2 + R_2NH \rightarrow CCl_3H + R_2NPF_2 \qquad (3.14)$$

Hydroxylamino- and hydrazino-chlorophosphines were obtained in the reaction of various substituted hydroxylamines or hydrazines with phosphorus trichloride; the corresponding fluorides were prepared by chlorine–fluorine exchange with antimony trifluoride or sodium fluoride[88, 89]. Partial chlorine–fluorine exchange in R_2NPCl_2 with formation of R_2NPClF is also possible[90].

3.4.1.4 Organohalophosphines, R_nPX_{3-n}, and hydridohalophosphines, PH_nX_{3-n}

Russian workers have reported a number of new syntheses of organohalophosphines. Thus, reaction of p-dichlorobenzene with white phosphorus and phosphorus trichloride at 340 °C in the presence of an iodine catalyst was found to give p-bis-(dichlorophosphino)benzene[91]:

$$(3.15)$$

If o-xylylene dichloride is employed in this reaction, a cyclic phosphine is obtained[92]:

$$(3.16)$$

A synthesis of organodichlorophosphines from phosphorus pentachloride and aliphatic and aromatic hydrocarbons under the catalytic influence of iodine has been described[93]. If terminal olefins are employed, vinyldichlorophosphines are obtained upon reduction of the intermediate tetrachlorophosphorane complexes with methyldichlorophosphite[94].

$$R^1R^2CH{=}CH_2 + 2PCl_5 \rightarrow [R^1R^2C{=}CHPCl_3]^+[PCl_6]^- + HCl$$
$$\downarrow CH_3OPCl_2$$

$$R^1R^2C{=}CHPCl_2 \qquad (3.17)$$

A very interesting method of synthesis of organochlorophosphines is based on the cleavage of the aromatic carbon–phosphorus bond by means of phosphorus trichloride, a reaction which in principle has been known for a long time[95, 96]. Thus, the very interesting cyclic chlorophosphine, C_4H_8PCl, and a series of its derivatives, is now readily accessible, as follows,

$$(3.18)$$

The reaction is conducted at 280 °C under autogenous pressure.

The preparation of the t-butylfluorophosphines, Bu^tPF_2 and Bu^t_2PF, and of the pentafluorophenylfluorophosphines, $C_6F_5PF_2$ and $(C_6F_5)_2PF$, has been accomplished by chlorine–fluorine exchange on the respective chlorophosphines with sodium fluoride in an acetonitrile medium[97, 98]. As in the case of perfluoroalkylfluorophosphines[99], these compounds do not spontaneously disproportionate, as has been observed for alkylfluorophosphines[100]:

$$2RPF_2 \nrightarrow \tfrac{1}{n}(RP)_n + RPF_4 \qquad (3.19)$$

$$R_2PF \nrightarrow R_2PPR_2 + R_2PF_3 \qquad (3.20)$$

Hydrogen-containing halophosphines of type PH_nX_{3-n} have been described, so far, for the most and the least electronegative halogen, X = F and I.

PF_2H is formed, in moderate yield, upon reaction of equimolar amounts of hydrogen iodide and difluoroiodophosphine in the presence of mercury[101]:

$$PF_2I + HI + 2Hg \rightarrow PF_2H + Hg_2I_2 \qquad (3.21)$$

If, in this reaction, hydrogen iodide is replaced by phosphine, PH_3, the yield is increased substantially[19].

The transient formation of the iodophosphines, PH_2I and PHI_2, has been detected in the reaction of white phosphorus with hydrogen iodide[102], or of phosphine with phosphorus tri-iodide or iodine. Also, formation of PH_nI_{3-n} has been observed upon reaction of phosphonium iodide with diphosphorus tetraiodide. All attempts to actually isolate the compounds PHI_2 and PH_2I in a pure state have failed, but their existence could be established by i.r. and n.m.r. spectroscopy[103].

A convenient synthesis of iodophosphines, $ArPI_2$, by halogen exchange in the corresponding aromatic dichlorophosphines with lithium iodide has been reported[104, 105].

3.4.2 Physical and chemical properties

It is impossible, in view of the very large number of publications, to give complete coverage of all the reactions of halides of tri-coordinate phosphorus. In the following, therefore, some reactions are described which are believed to be fairly typical. Furthermore, in view of the recently published extensive review article by Nixon[71], the chemistry of fluorophosphines will not be covered in great detail.

A study has been made of the reaction between phosphorus trifluoride and trimethyl phosphite[106]. Simple ligand exchange and Arbuzov rearrangements, giving rise to the formation of phosphoryl compounds, have been observed[106]. Orthoformate esters were found to react vigorously with phosphorus trihalides, such as PCl_3 and PBr_3 [107], successive substitution and Arbuzov reaction of the alkoxyphosphorus halides by alkyl halide formed in the reaction taking place.

Phosphorus trifluoride is a very weak Lewis acid and forms an adduct with trimethylamine only at low temperatures[108]. In accord with this weak

Lewis acid character, formation of adducts between PF_3 and amino-fluorophosphines, such as

$$\begin{array}{c}>\!\ddot{P}-N\!< \end{array} \longrightarrow PF_3$$

or

$$\begin{array}{c}>\!\ddot{N}-P\!< \end{array} \longrightarrow PF_3$$

is not observed[109].

Exothermic reactions have been observed between tri-n-alkylphosphines and phosphorus trichloride, phenyldichlorophosphine, and diphenylchloro-phosphine[110]. The trialkylphosphines are converted to the corresponding chlorophosphoranes, and polymeric phosphorus or, perhaps, sub-halides of type $(PX)_n$ are formed. The cyclopolyphosphine, $(C_6H_5P)_5$, and tetra-phenyldiphosphine result from the reactions of the phenylchlorophosphines. Formation of a stable adduct has been noted in the reaction of trialkylphos-phines with dimethylchlorophosphine,

$$R_3P + Me_2PCl \rightarrow [R_3\overset{(+)}{P} - PMe_2]^+Cl^- \tag{3.22}$$

The above-mentioned reactions may proceed via a similar intermediate which is, however, higher in energy.

If, instead of trialkylphosphines, tertiary phosphines containing phenyl groups are allowed to react with phosphorus trichloride, exchange of chlorine for phenyl groups takes place[95, 96]. In the reaction of phosphorus tri-iodide with triphenylphosphine, diphosphorus tetraiodide and iodotri-phenylphosphonium iodide or tri-iodide are formed[111].

Ammonolysis of phenyldichlorophosphine in liquid ammonia in the presence of potassium amide[112] furnishes the dipotassium salt of 1,2-diamido-1,2-diphenyl-diphosphazane, KNH—P(Ph)—NH—P(Ph)—NHK. Phosphorus trifluoride reacts with primary amines with formation of alkylaminodifluorophosphines and bis-(alkylamino)fluorophosphines; form-ation of bis-(alkylamino)difluorophosphoranes in this reaction has also been noted[113].

Phosphorus trichloride has been found to undergo 1,4-addition to α, β-unsaturated ketones[114]. Thus, from 4-cholesten-3-one (1) in the presence of benzoic acid, 3-oxo-5α-cholestan-5-yl-phosphonic acid (3) is formed under the action of alcoholic potassium hydroxide (via the phostonyl chloride (2)). A mechanism has been proposed which involves electrophilic attack of phosphorus in PCl_3 on the carbonyl oxygen atom.

A number of recent reports are concerned with reactions of phosphorus trihalides of the Friedel–Crafts type. Thus, in the reaction of aliphatic-aromatic silanes of type $R_{3-n}X_nSiCH_2C_6H_5$ with phosphorus trichloride/ aluminium chloride, followed by treatment with phosphorus oxychloride, aryldichlorophosphines, $R_{3-n}X_nSiCH_2C_6H_4PCl_2$ (X = halogen, are formed[115]. The aluminium chloride-catalysed reactions of ferrocene with phosphorus trihalides have been studied, and the following order of re-activity has been observed:

$$PCl_3 \ll R_2NPCl_2 > (R_2N)_2PCl > (R_2N)_3P$$

C_8H_{17}

$$(3.23)$$

It is suggested that the reactive intermediate is formed by the coordination of aluminium chloride to nitrogen and not to phosphorus[116].

Phosphorus trichloride will react with t-butylbenzene when aluminium chloride is employed as a catalyst. Upon hydrolysis of the phosphinic chloride p-t-butylphenyl-t-butylphosphinic acid is formed[117]. In contrast to this observation, migration of a t-butyl group from carbon to phosphorus takes place during the $AlCl_3$-catalysed reaction of 1,3,5-tri-t-butyl-benzene. Hydrolysis of the reaction mixture furnishes t-butyl-3,5-di-t-butylphenylphosphinic chloride[118],

$$(3.24)$$

Nucleophilic displacement of bromine in bromodifluorophosphine by the anion of the 2-halo(or pseudohalo)-hexafluoropropoxide anion has given rise to a series of μ-oxo-difluorophosphino compounds[119], e.g.

$$(3.25)$$

Other μ-oxodifluorophosphine derivatives were formed upon a reaction of the mixed halofluorophosphines, PF_2X, with the silver salts of carboxylic acids[120]:

$$RCOOAg + XPF_2 \rightarrow AgX + RCOOPF_2 \qquad (3.26)$$
$$(X = Cl, Br, I; R = Me, CF_3, C_2F_5, C_3F_7)$$

The reducing properties of phosphorus trifluoride towards a number of oxides and sulphides have been investigated in a number of instances, e.g. with P_2S_5, P_2O_5 [121], As_2O_3, Sb_2O_3, and Bi_2O_3 [122], MoO_3 and WO_3 [123], and oxides of sulphur, selenium and tellurium [124]. Phosphorus trichloride also exhibits reducing properties towards oxides, and its reaction with SO_2 with formation of $POCl_3$ and SO has been reported[125].

The addition of cyanogen bromide to dimethylaminodifluorophosphine has been reported to give the interesting 5-coordinate phosphorus compound, Me_2NPF_3CN, and the bromophosphine, Me_2NPFBr [126].

3.4.3 Spectroscopic investigations

Spectroscopic investigations on fluorides of tri-coordinate phosphorus have been covered in detail in Nixon's review[71]. We will, accordingly, put emphasis here on compounds containing the heavier halogens, but will make occasional reference to P—F compounds for comparative purposes.

3.4.3.1 I.R. and Raman spectroscopy

A number of reports are concerned with the analysis and determination of molecular force fields of phosphorus–halogen compounds[76, 127–129]. Potential energy and mean amplitudes of vibration have been calculated for PF_3, POF_3, PSF_3, PCl_3, $POCl_3$ and $PSCl_3$, and PBr_3, $POBr_3$ and $PSBr_3$ [130, 131]. It has been concluded that the behaviour of the free electron pair in compounds PX_3 corresponds to that of oxygen or sulphur in $P(Y)X_3$ (Y = O, S). The importance of the free electron pair on phosphorus in the trihalides, PX_3, in determining the intramolecular potential functions has been stressed by Kim and Overend[132].

Based on depolarisation measurements on the Raman lines of the PCl_2 group in phenyldichlorophosphine, C_s symmetry has been derived for the molecular skeleton, although the phenyl vibrations correspond to C_{2v} symmetry[133].

A detailed analysis of the i.r. and Raman spectra of the series of compounds, $C_6H_5PX_2$(X = H, F, Me, Br) and $(C_6H_5)_2PX$ (X = Me, CD_3, Cl) has been carried out; particular attention has been given to the vibrations of the phenyl ring[134]. Vibrational spectra of phosphines of type R–PX_2 have also been studied[135].

3.4.3.2 Nuclear magnetic resonance

The largest number of publications dealing with n.m.r. investigations of tri-coordinate phosphorus halides are concerned with compounds containing a P—F bond.

An extensive compilation of ^{31}P n.m.r. data has been given by Latscha[136]. Lucken and Williams [137] have investigated the anisotropy of ^{31}P chemical shifts. If in phenylphosphines of type $PhPX_2$ (X = Cl) hydrogen atoms of the phenyl group are replaced by fluorine, a shift of δ_P to high field is noted. The effect of m-substituents is more pronounced than that of substituents in the p-position. This suggests that, to a first approximation, fluorine exerts an inductive substituent effect[138].

Fast chlorine exchange, inversion at phosphorus at high temperatures, slow rotation around the P—N bond, and slow inversion at low temperature cause a marked temperature dependence of the n.m.r. spectra of aminochlorophosphines[139, 140]. Restriction of the free rotation around the P—N bond is ascribed to $d\pi$–$p\pi$ bonding[141].

Ionov and Ionova[142], using a semi-empirical MO/LCAO method, have calculated effective nuclear charges on phosphorus and have given an estimate of the mean excitation energies in compounds of types R_nPX_{3-n} (R = Me, Et; X = F, Cl) and PX_3 (X = F, Cl, Br, I). The relative changes of δ_P within the series, R_nPX_{3-n}, and the trend for the series PX_3 (X = F, Cl, Br, I) are represented qualitatively by the Karplus–Das–Pople approximation[142].

^{35}Cl nuclear quadrupole resonance data have been recorded for a series of chlorophosphines, and have been compared to those of the corresponding phosphoryl and thiophosphoryl compounds[143]. The resonance frequencies are given by the following series,

$$\nu_{P=S} > \nu_{P=O} \gg \nu_{P(III)}$$

3.4.4 Structural investigations

The F—P—F bond angle in PF_3 has been re-determined by both microwave spectroscopy and electron diffraction[144, 145], giving values of 96.3 and 97.8 degrees. These values are rational within the series of bond angles in the series of molecules PX_3 (X = F, Cl, Br, I)[145] and do not require the assumption of double bond character for the P—F bond[146, 147]. The role of d orbitals in the bonding of PF_3 is still uncertain [148–150]. According to an *ab initio* SCF—MO calculation for PH_3, PF_3, and PMe_3[149, 150], a considerable π component, due mainly to 3d–2p and, to a lesser extent, 3p–2p interactions, in addition to the σ-component of the P—F bond in PF_3 has been noted. The agreement between the calculations and experimental results was satisfactory. The effect of d orbital participation was found to have only a minor effect upon the correlation of orbital energies, and of ionisation potentials. The participation of d orbitals may be related to the 'π donor capacity' of groups bonded to phosphorus.

Electron diffraction studies on the aminohalophosphines, R_2NPX_2 (R = Me; X = Cl, F), have provided evidence for a planar arrangement of ligands around the nitrogen atom in (4)–(6)[151, 152],

(4)

(5)

(6)

3.5 HALIDES OF TETRA-COORDINATE PHOSPHORUS

3.5.1 Halides of tetra-coordinate phosphorus containing a bond

The chemistry of nitrogen–phosphorus–halogen compounds involving a $P{=}N$ double bond has been the subject of intensive study during the past few years; in particular in connection with the investigation of the ammonolysis of phosphorus pentachloride[153, 154] and of chlorophosphoranes[155]. While a great variety of compounds belonging to this class has been made, considerable confusion exists with regard to a consistent nomenclature for such systems, thus making it difficult to compare series of related compounds. In the absence of generally accepted rules for the nomenclature of these compounds we have decided to use the nomenclature employed by the authors of the original papers.

3.5.1.1 Preparative methods

Derivatives of trifluorophosphazene, $F_3P{=}NH$, have been obtained in the reaction of trifluorodichlorophosphorane with primary acid amides[156], e.g.

$$F_3PCl_2 + FSO_2NH_2 \rightarrow F_3P{=}N{-}SO_2F + 2HCl \qquad (3.27)$$
$$(7)$$
$$F_3PCl_2 + F_2P(Y)NH_2 \rightarrow F_3P{=}N{-}P(Y)F_2 + 2HCl \qquad (3.28)$$
$$(Y = O, S) \qquad\quad (8)$$

The products thus obtained are liquids which, with the exception of $F_3P{=}N{-}P(\!:\!S)F_2$, gradually decompose at room temperature. If a compound such as $F_2P(\!:\!S){-}N{=}PF_3$ is treated with excess PF_3Cl_2, a PF_3 group may be converted to a PF_2Cl group[157], i.e.

$$F_2P(\!:\!S){-}N{=}PF_3 \xrightarrow{PF_3Cl_2} F_2P(\!:\!S){-}N{=}PF_2Cl \qquad (3.29)$$
$$(9)$$

Partial ammonolysis of $Cl_2P(\!:\!S){-}N{=}PF_3$ and $FClP(\!:\!S){-}N{=}PF_3$ yields the amides (10) and (11)[158]:

$$Cl_2P(\!:\!S){-}N{=}PF_3 + 2NH_3 \rightarrow Cl_2P(\!:\!S){-}N{=}PF_2NH_2 + NH_4F \qquad (3.30)$$
$$(10)$$
$$ClFP(\!:\!S){-}N{=}PF_3 + 2NH_3 \rightarrow ClFP(\!:\!S){-}N{=}PF_2NH_2 + NH_4F \qquad (3.31)$$
$$(11)$$

The related compound $F_2P(\!:\!S){-}N{=}PF_2NH_2$ was obtained upon ligand exchange on the thiophosphoryl group of (11).

The cleavage of the Si–N bond in aminosilanes with phosphorus halides provides a facile method of attaching amino groups on to phosphorus, and has been widely employed in the study of the derivative chemistry of phosphorus halides. A case in point is the facile cleavage of the Si–N bond in dimethylaminotrimethylsilane or in hexamethyldisilazane through N-tri-halophosphorylphosphoranylidene compounds[159, 160]:

$$Me_3SiNMe_2 + F_2P(:S)—N{=}PF_3 \rightarrow Me_3SiF + F_2P(:S)—N{=}PF_2NMe_2$$
$$(12) \qquad\qquad\qquad (3.32)$$

$$(Me_2Si)_2NH + F_2P(:S)—N{=}PF_3 \rightarrow F_2P(:S)—N{=}PF_2NHSiMe_3 + Me_3SiF$$
$$(13) \qquad\qquad\qquad (3.33)$$

In all these reactions substitution always takes place in the $N{=}PF_3$ but not in the thiophosphoryl-dihalide group. For compounds of type $R—N{=}PX_3$ ($R = FSO_2$, $ClSO_2$, $P_3N_3F_5$, and $X = F$, Cl) the following order of decreasing reactivity of specific groupings has been found[160]:

$$—N{=}PCl_3 > —N{=}PCl_2X > —N{=}PF_2Cl > —N{=}PF_3 \gg —ClSO_2 >$$
$$—FSO_2 \ (X = F, NMe_2)$$

With regard to the synthesis and properties of a great variety of further nitrogen–phosphorus–halogen compounds with $—N{=}PX_3$ units, reference is made to the review articles by Bermann[161] and Becke–Goehring[154]. We will mention only a few interesting members of this class of compound here.

Thus, trifluoromethylsulphonamide undergoes a Kirsanov reaction with PCl_5 [162],

$$CF_3SO_2NH_2 + PCl_5 \rightarrow CF_3SO_2N{=}PCl_3 + 2HCl \qquad (3.34)$$

In the reaction of N-fluorosulphonylsulphoximide with phosphorus pentabromide, tribromophosphazosulphonyl fluoride could be obtained as a colourless solid of m.p. 57–58 °C [163]:

$$FSO_2—N{=}S{=}O + PBr_5 \rightarrow FSO_2N{=}PBr_3 + SOBr_2 \qquad (3.35)$$

The synthesis of long-chain (oligomeric or polymeric) $—P{=}N$ systems has been described by several workers[164, 165]. The properties of polymeric phosphazenes have been reviewed in detail by Kireev et al.[166] who have proposed a new scheme of classification,

(1) Homopolyphosphazenes, containing the recurring unit $RR'P{=}N—$.

(2) Heteropolyphosphazenes, $—N{=}\overset{\mid}{\underset{\mid}{P}}—E$.

(3) Polymers containing a phosphazene grouping in the side chain.

E stands for an atom or group of atoms other than a phosphazo group, e.g. $—SiR_2—$, $—C_6H_4—$.

Numerous reports are concerned with the synthesis of cyclic nitrogen–phosphorus–halogen compounds in which the $P{=}N$ grouping forms part of a five-, six-, or eight-membered ring. Thus, from dicyanodiamide and phosphorus pentachloride (in a 1:2 molar ratio) 1,3,5-trichloro-1-tri-chlorophosphazo-1-phospha-2,4,6-triazine (14) is formed[167]. Similarly, from sodium dicyanimide and phosphorus pentachloride, the 1,1,3,5-tetra-chloro-1-phospha-2,4,6-triazine ring system (15) is obtained[167].

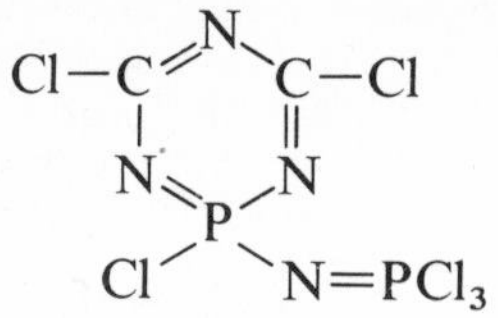

(14) 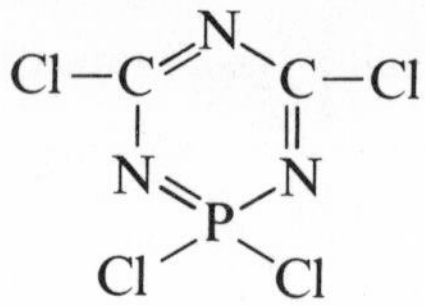   (15) 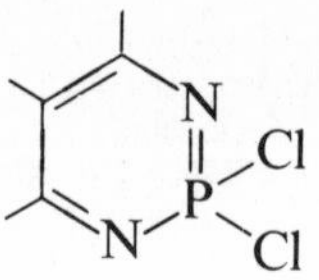   (16)

$$(CN)_2C{=}CR{-}N{=}PPhCl_2$$
(17)

The phosphorylation of enamines[168] has been found to give rise to compounds of type (16) and (17).

The heterocyclic ring system (18) has been synthesised in the pyrolysis of the reaction product obtained from $[Cl_3P{-}N{=}PCl_3][PCl_6]$ with amidosulphonic acid, $HOSO_2NH_2$ [169],

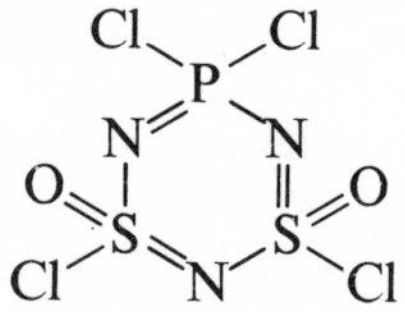

(18)

The ammonolysis of dialkylaminochlorophosphoranes[155] has been found to consist of a complex sequence of steps, involving condensation, substitution, formation of ionic products, etc. Dialkylamino-substituted cyclotriphosphazenes of type (19) are obtained.

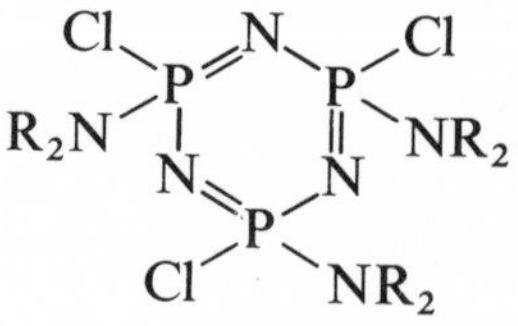

(19) 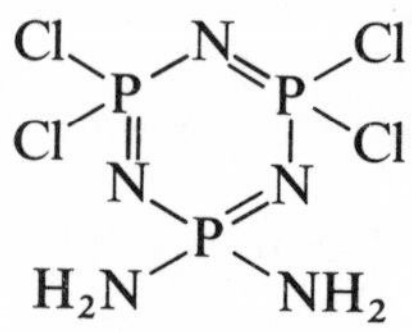   (20)

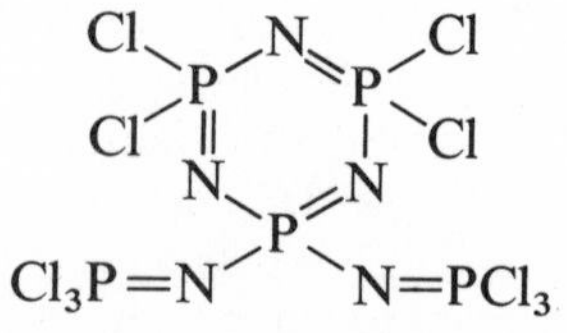

(21)

A Kirsanov reaction on the geminal diamide (20)[170, 171] yields the bis-phosphazo compound (21) which, on reaction with heptamethyldisilazane undergoes ring closure with formation of the spiro compound (22)[172]:

$$
\begin{array}{c}
\text{Cl} \quad \text{Cl} \qquad \text{Cl} \quad \text{Cl} \\
\diagdown \diagup \qquad\qquad \diagdown \diagup \\
\text{P}\!-\!\text{N} \quad\ \text{N}\!=\!\text{P} \\
\diagup\!\!\diagup \qquad\diagdown \diagup \qquad \diagdown \\
\text{N} \qquad\quad \text{P} \qquad\quad \text{N}\!-\!\text{Me} \\
\diagdown \qquad\quad \diagup \qquad \diagup \\
\text{P}\!=\!\text{N} \quad\ \text{N}\!=\!\text{P} \\
\diagup \diagdown \qquad\quad \diagup \diagdown \\
\text{Cl} \quad \text{Cl} \qquad \text{Cl} \quad \text{Cl}
\end{array}
$$

(22)

3.5.1.2 Reactions, chemical and physical properties of halides of tetra-coordinated phosphorus containing a P—N bond

A number of reports have been concerned with the study of substitution reactions on phosphonitrilic halides. In trimeric phosphonitrilic fluoride, controlled ammonolysis gives rise to substitution of one fluorine atom[173]. Similarly, reaction of $(PNF_2)_3$ with primary or secondary amines[174, 175], or with dimethylaminotrimethylsilane[176] yields the corresponding alkylamino-pentafluorophosphazenes,

$$(PNF_2)_3 \xrightarrow[\text{or } Me_2NSiMe_3 \ (-Me_3SiF)]{HNR_2} \quad F\text{-ring}\text{-}NMe_2 \qquad (3.36)$$

(R = H or hydrocarbon group)

Both methods may also be employed in the synthesis of derivatives of $(PNF_2)_3$ multiply-substituted with amino groups.

Amino-substituted phosphonitrilic fluoride derivatives are of considerable preparative interest, as their reaction with HCl or HBr provides a means of selectively replacing fluorine for chlorine or bromine on the ring system[174, 176].

$$P_3N_3F_5NR_2 + 2HX \rightarrow P_3N_3F_5X + [NR_2H_2]X \qquad (3.37)$$

Such compounds were hitherto accessible only via the uncontrolled fluorination of $(PNX_2)_3$ (X = Cl, Br) [177, 178].

The whole series of possible alkyl-mercaptofluorotriphosphazenes has been prepared by the reaction of sodium ethyl mercaptide with $(PNF_2)_3$ in the appropriate molar ratio[179]:

$$(PNF_2)_3 + nNaSEt \rightarrow N_3P_3(SEt)_nF_{6-n} + nNaF \qquad (3.38)$$
$(n = 1\text{--}5)$

Substitution of two chlorine atoms in cis- (23a) and trans- (23b) -1,3-bisdimethylaminotetrachlorophosphonitrile with KSO_2F gives rise to the corresponding cis- (24a) and trans- (24b) geminal 5,5-difluorides. By contrast, if SbF_3 is employed as a fluorinating agent, the 1,3-fluorinated products (25a) and (25b) are obtained. Further fluorination with KSO_2F gives the completely fluorinated compounds (26a) and (26b)[180, 181].

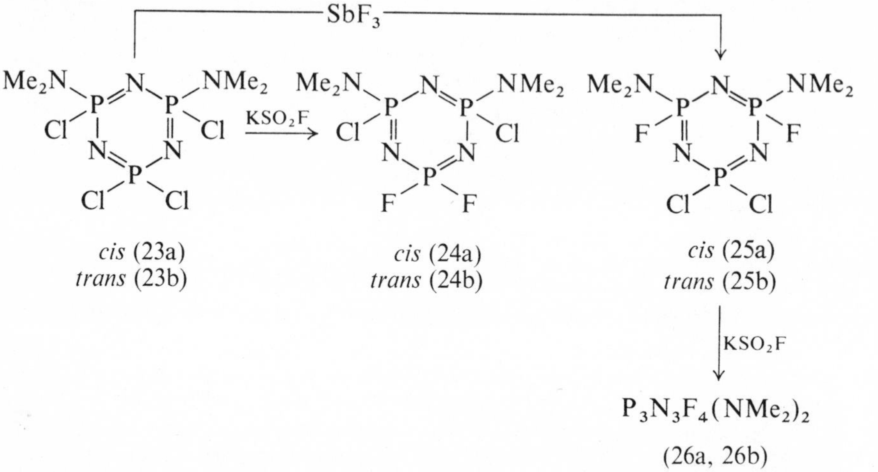

cis (23a)   *cis* (24a)   *cis* (25a)
trans (23b)   *trans* (24b)   *trans* (25b)

$$P_3N_3F_4(NMe_2)_2$$

(26a, 26b)

(3.39)

It is evident from this work that the electron distribution and, as a consequence thereof, the reactivity of the chlorine atoms, are strongly dependent on the substituent position. The same type of reaction may be extended to the tetrameric series. Thus, from dimethylamine and $(PNCl_2)_4$ in ether at $-18\,^{\circ}C$, $P_4N_4Cl_5(NMe_2)_3$ and $P_4N_4Cl_4(NMe_2)_4$ are obtained. The corresponding fluorides, $P_4N_4F_5(NMe_2)_3$ and $P_4N_4F_4(NMe_2)_4$, were formed upon fluorination with SbF_3 [182]. It is somewhat surprising that fluorination of PCl_2 groups by SbF_3 takes place readily here while in the case of $P_3N_3Cl_4(NMe_2)_2$ only the two chlorine atoms of the $PCl(NMe_2)$ groups are fluorinated under similar conditions[180, 181]. The difference is probably caused by the known greater general reactivity and basicity of the tetrameric compound[182]. The formation of addition compounds between various chloro-isopropyl-aminocyclotriphosphazatrienes has been reported[183].

Intermediate products of the ammonolysis of chlorophosphazenes, which may be carried through to complete replacement of all chlorine atoms, could be isolated[184]. Thus, a number of products of the partial ammonolysis of $(PNCl_2)_4$ could be synthesised and identified[185, 186], e.g.

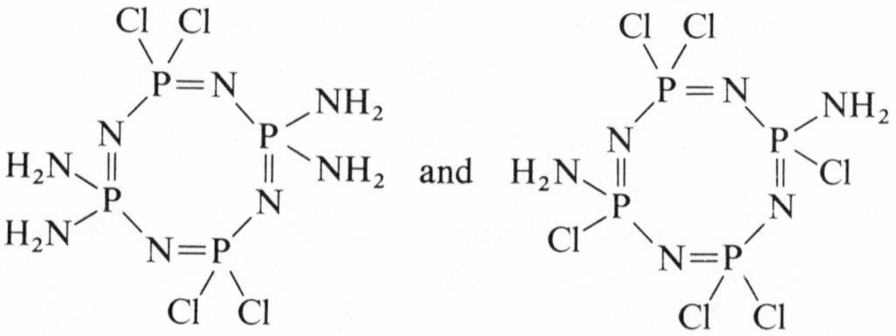

A novel anionic derivative of $(PNF_2)_3$ was obtained in the reaction of $(PNF_2)_3$ with caesium fluoride[187]:

$$CsF + (PNF_2)_3 \rightarrow Cs^+[(PNF_2)_3F]^- \qquad (3.40)$$

The structure of the anion is still uncertain, both an open-chain and a cyclic structure have been discussed.

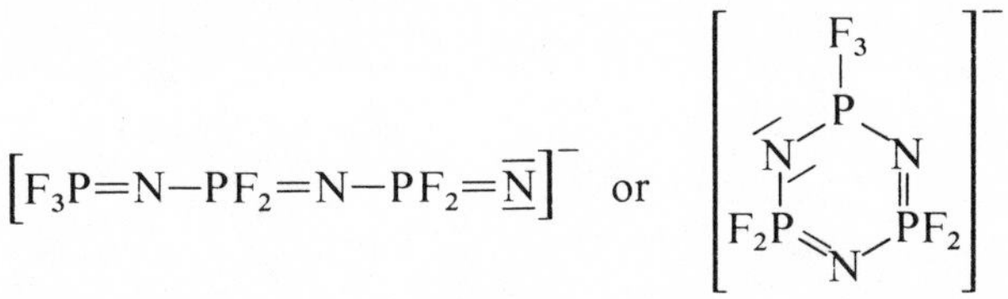

$$\left[F_3P{=}N{-}PF_2{=}N{-}PF_2{=}\overline{N}\right]^- \quad \text{or} \quad \left[\begin{array}{c} F_3 \\ | \\ P \\ N \quad\quad N \\ | \quad\quad || \\ F_2P \quad PF_2 \\ N \end{array}\right]^-$$

Biddlestone and Shaw[188-190] investigated the reaction between $(PNCl_2)_3$ and Grignard reagents. In the course of this reaction the ring is cleaved, arylated or alkylated, and then closed.

The lattice energies of $N_6P_6Cl_{12}$ and $N_4P_4Cl_8$ have been determined $(25.3 \pm 0.42$ and 29.4 ± 0.47 kcal mol^{-1} respectively) by measuring the temperature-dependence of their vapour pressure[191].

The phosphonitrilic halides, $(PNCl_2)_3$ and $(PNBr_2)_3$ form addition compounds, $(PNCl_2)_3 \cdot AlBr_3$, $(PNBr_2)_3 \cdot AlBr_3$ and $(PNBr_2)_3 \cdot 2AlBr_3$ when allowed to react with aluminium bromide in carbon disulphide solution. According to i.r. evidence $AlBr_3$ is coordinated to nitrogen[192], while in complexes of the type $(NPF_2)_n \cdot 2SbF_5$ $(n = 3-6)$ coordination is through the fluorine atoms of the ring[193].

3.5.1.3 Spectroscopic investigations

The i.r. spectra of N-carbacylphosphine imide derivatives, $R{-}N{=}PCl_{3-n}R'_n$ (R = CF_3CO, $PhCO$, CCl_3CO, p-ClC_6H_4CO, p-$NO_2C_6H_4CO$; CF_3CCl_2, CCl_3CCl_2; $R'{=}$), containing ^{14}N or ^{15}N have been investigated[194]. Assignments for $v_{P=N}$, v_{C-N}, and v_{P-Cl} have been made. N-acyl-P,P,P-trichloro- and -P,P-dichloro-P-methylphosphine imides with ^{14}N show strong absorption in the region 1290–1396 cm^{-1} (P$=$N). Substitution with ^{15}N causes a shift of these bands by 10–18 cm^{-1}. I.R. spectra of a related series of compounds of the type $R_{Hal}N{=}PCl_3$ (R_{Hal} = CCl_3, CCl_3CCl_2, $(CCl_3)_2$ CCl, CF_3CCl_2, $(CF_3)_2CCl$, and $(CF_3)_3C$) have also been studied[195]. An absorption in the region 1400–1500 cm^{-1} has been assigned to v^o_{PNC}, the out-of-phase vibration of the PNC-group. For the trichloromethyl derivatives the difference between the out-of-phase and in-phase PNC vibrations, i.e. $v^o_{PNC}-v^e_{PNC}$, increases in the order shown above as a result of the steric hindrance which leads to an increase in the PNC angle[196]. The increase in bond angle is accompanied by a high-field shift of δ_P, and by a decrease of the dipole moment.

The results of these i.r. studies have been discussed in terms of a $p\pi d\pi$ contribution in the P$=$N bond. An increase in the PNC bond angle should cause a higher multiplicity of the P—N bond, increased participation of the lone electron pair on nitrogen and an s–p re-hybridisation. The strengthening of the P—N bond, and the increase in d electron density on phosphorus are borne out in the i.r.- and n.m.r.-spectroscopic findings[197].

A similar trend has been noted in the series of the cyclic phosphonitrilic halides, $(PNCl_2)_n$ ($n = 3$, 4, or 5)[198, 199].

Table 3.2

Parameter	$(PNCl_2)_3$	$(PNCl_2)_4$	$(PNCl_2)_5$
PNP angle	119.0°	131.3°	148.6°
d_{P-N} (Å)	1.595	1.570	1.521
δ_P (p.p.m.) (85% H_3PO_4 ref.)	−20	+7	+17
$\nu_{P=N}$(cm^{-1})	1218	1310	1355

A correlation has been established between the partial charges, Δ_P, on phosphorus in $(PNX_2)_3$ (as calculated from the electro negatives of X) and the P—N stretching frequency, ν_{P-N}[200],

$$\nu_{P-N} = 756\Delta_P + 1017$$

Excellent agreement was found between calculated and experimental values for several of the $(PNX_2)_3$ compounds.

^{31}P n.m.r. spectra of a series of substituted derivatives of $(PNCl_2)_3$ have been investigated[201]. Spectra of higher order which had to be evaluated by analysis were encountered in some cases, e.g. type AB_2 for gem-$P_3N_3Cl_4$ $(NH_2)_2$, or A_2BX for gem-$H_2N(N=PPh_3)P_3N_3Cl_4$ [201].

Nuclear quadrupole resonance measurements (^{35}Cl) have been conducted on some chlorides of phosphorus[202]. Thus, for the compound cyclo-tri-μ-nitrido-dichlorophosphoro-bis(oxychloro)sulphur, $Cl_2PN(NSOCl)_2$, the molecular symmetry C_s could be established for the solid state[169].

^{35}Cl n.q.r. frequencies of $Cl_3P=N—P(:O)Cl_2$, $Cl_3P=N—SO_2Cl$, and $S_2PN_3O_2Cl_4$ at 77 K were determined and were employed to assess the nature of bonding in the P—Cl bond[314]. $Cl_3P=N—P(:O)Cl_2$, for example, showed five signals at 26.82 MHz and ten at 30.06 MHz. No ^{35}Cl n.q.r. signals were seen in the case of $[Cl_3PNPCl_3][PCl_6]$, $[Cl_3PNPCl_2PCl_3]$ $[PCl_6]$, and $Cl_3P=N—P(:S) Cl_2$.

3.5.1.4 Structural investigations

The crystal and molecular structure of fluorodiphenyl-N-methylphosphine imide, $Ph_2FP(NMe)$, has been determined[203]. The compound is monomeric, the observed P—N bond length indicates considerable multiple bond character and a resonance hybrid may be employed to describe this,

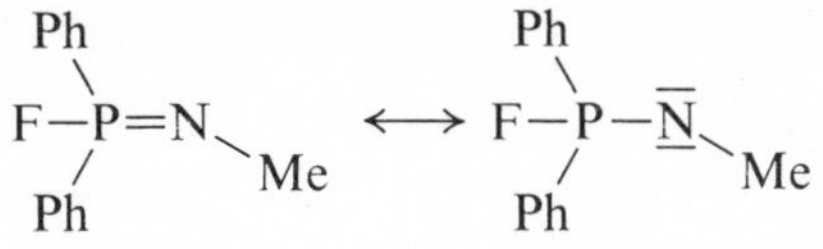

$$
\begin{array}{ccc}
\text{Ph} & & \text{Ph} \\
\backslash & & \backslash \\
\text{F—P=N} & \longleftrightarrow & \text{F—P—}\overline{\text{N}} \\
\diagup \qquad \diagdown \text{Me} & & \diagup \qquad \diagdown \text{Me} \\
\text{Ph} & & \text{Ph} \\
(27) & & (28)
\end{array}
$$

By contrast, the related difluoride, $F_2PhPNMe$, is clearly dimeric[204].

In the x-ray crystal structure determination of $N_4P_4Cl_4Ph_4$ (29)[205] the following mean bond distances and angles were observed, d(P—N) 1.57 Å, d(P—Cl) 2.03 Å, d(P—Ph) 1.79 Å, $\angle$ NPN 120 degrees, $\angle$ XPX 104 degrees.

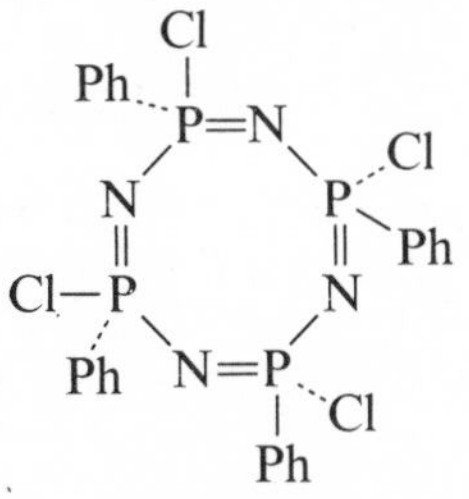

$$
\begin{array}{c}
\qquad\qquad \text{Cl} \\
\text{Ph}\underset{\cdot\cdot}{}\;\;| \\
\qquad\text{P=N} \\
\diagup\qquad\qquad\diagdown\text{Cl} \\
\text{N}\qquad\qquad\quad\text{P}^{\diagdown} \\
\|\qquad\qquad\qquad\|\;\text{Ph} \\
\text{Cl—P}\qquad\qquad\text{N} \\
\diagdown\qquad\diagup \\
\text{Ph}\quad\text{N=P}_{\cdot\cdot} \\
\qquad\qquad|\;\;\text{Cl} \\
\qquad\quad\text{Ph}
\end{array}
$$

$$(29)$$

The ring has chair configuration of approximate C_{2h} symmetry[205].

A re-determination of the bond parameters of trimeric phosphonitrilic chloride, $(PNCl_2)_3$, by electron diffraction has yielded the following values[206]: d(P—N) 1.585 ± 0.010 Å, d(P—Cl) 2.006 ± 0.007 Å, $\angle$ NPN 119.7 ± 0.3 degrees, $\angle$ ClPCl 101.8 ± 1.2 degrees. The ring shows slight deviation from planarity, the symmetry is C_{3v}.

3.5.2 Halides of tetra-coordinate phosphorus containing P=O or P=S bonds

In view of the tremendous number of publications concerned with preparation, properties, and spectroscopic investigations of the title compounds we have selected a limited amount of information which we believe is sufficiently illustrative of the developments.

3.5.2.1 Preparation

Several halophosphoric acids and their thio analogues have been synthesised for the first time. The synthesis of monofluorophosphorus acid has been realised by careful hydrolysis of PCl_3 with 40% hydrofluoric acid in an ether medium[207].

$$PCl_3 + 2H_2O + HF \rightarrow HP(:O)(F)(OH) + 3HCl \qquad (3.41)$$

Substitution of the NH_2 group in diamidothiophosphate by a fluoride ion gives rise to salts of the oxo-thio-fluoroamidophosphoric acid[208].

$$NH_4POS(NH_2)_2 + KF \rightarrow 2NH_3 + [FP(:O)(S)(NH_2)]^- K^+ \quad (3.42)$$

Difluorodithiophosphoric acid has been synthesised by the reaction of iodothiophosphoryl difluoride with hydrogen sulphide, or through the action of sulphur on difluoromonothiophosphoric acid[209]:

$$F_2P(:S)I + H_2S \rightarrow F_2P(:S)SH + HI \quad (3.43)$$

$$F_2P(:O)SH \xrightarrow{S_8} F_2P(:S)SH, P(:S)F_3, \text{ etc.} \quad (3.44)$$

Dipotassium difluorodiphosphate, amongst other products, is formed in the reaction of potassium hexafluorophosphate, potassium metaphosphate or P_4O_{10} with potassium fluoride[210-212]:

$$P_4O_{10} + 4KF \rightarrow 2K_2[P_2O_5F_2] \quad (3.45)$$

Highly volatile complexes are formed by the difluorodithiophosphate ion, $[PS_2F_2]^-$, with certain transition metals such as Co, Ni, Pd and Pt. The derivative of Co^{2+} exhibits a vapour pressure of 20 mm at 95 °C. The complexes from Ag^+ and Cu^+ have a polymeric structure and their volatility is small[213].

A series of organometallic derivatives of $[PS_2F_2]^-$ were prepared in the reaction of difluorodithiophosphoric acid or of its caesium salt with organometallic hydrides or halides. The mode of coordination of $[PS_2F_2]^-$ depends upon the reaction conditions, and upon the relative lability of the atoms or groups bonded to the metal[214], e.g.

$$HPS_2F_2 + \pi\text{-}C_5H_5M(CO)_xX \rightarrow \pi\text{-}C_5H_5M(CO)_{x-1}PS_2F_2 + HX + CO \quad (3.46)$$

$$HPS_2F_2 + M'(CO)_yX \rightarrow M'(CO)_{y-1}PS_2F_2 + HX + CO \quad (3.47)$$

$$HPS_2F_2 + \pi\text{-}C_5H_5M(CO)_xH \rightarrow \pi\text{-}C_5H_5M(CO)_xSP(S)F_2 + H_2 \quad (3.48)$$

M = Fe (only Fe in (3.46)), Mo, W; M' = Mn; X = Cl, Br; x = 2, 3; y = 5.

Dithiodiphosphoryltetrafluoride has been obtained by a number of routes[215]:

$$Me_2NP(:O)F_2 + 2HPS_2F_2 \rightarrow F_2P(:S)OP(:S)F_2 + [Me_2NH_2]^+[PS_2F_2]^- \quad (3.49)$$

$$2Me_2NP(:S)F_2 + 4HPO_2F_2 \rightarrow F_2P(:S)OP(:S)F_2 + 2[Me_2NH_2]^+[PO_2F_2]^- + F_2P(:O)OP(:O)F_2 \quad (3.50)$$

$$2NaPOSF_2 + H_2SO_4 \rightarrow F_2P(:S)OP(:S)F_2 + Na_2SO_4 + H_2O \quad (3.51)$$

The best yields of relatively pure product are obtained via method (3.49).

A series of dithio-diphosphoryl fluorides bridged via oxygen, sulphur, or nitrogen were prepared in accord with[216]:

$$2IP(:S)F_2 + \tfrac{1}{2}O_2 \rightarrow F_2P(:S)\text{---}O\text{---}P(:S)F_2 + I_2 \quad (3.52)$$

$$2HPS_2F_2 + Me_2NP(:E)F_2 \rightarrow F_2P(:S)\text{---}E\text{---}P(:S)F_2 + [Me_2NH_2][PS_2F_2] \quad (3.53)$$

$$MeNHP(:E)F_2 + ClP(:E)F_2 + NMe_3 \rightarrow$$
$$F_2P(:E)\!-\!NMe\!-\!P(:E)F_2 + [Me_3NH]Cl \quad (3.54)$$
$$H_2NP(:E)F_2 + ClP(:E)F_2 + 2NMe_3 \rightarrow$$
$$[F_2P(:E)\!-\!NH\!-\!P(:E)F_2]\cdot NMe_3 + HCl\cdot NMe_3 \quad (3.55)$$
$$(E = O, S)$$

The product formed in the last reaction is obtained in the form of an oil as a 1:1 adduct with trimethylamine. Related P—Cl compounds have been obtained in the reaction of $POCl_3$ with $MeNHP(:O)Cl_2$ and with hepta-methyldisilazane[217]:

$$MeNHP(:O)Cl_2 + P(:O)Cl_3 + Et_3N \rightarrow$$
$$Cl_2P(:O)\!-\!NMe\!-\!P(:O)Cl_2 + [Et_3NH]Cl \quad (3.56)$$
$$(Me_3Si)_2NMe + P(:O)Cl_3 \rightarrow$$
$$Cl_2P(:O)\!-\!NMe\!-\!SiMe_3 + Me_3SiCl \quad (3.57)$$
$$Cl_2P(:O)\!-\!NMe\!-\!SiMe_3 + P(:O)Cl_3 \rightarrow$$
$$Cl_2P(:O)\!-\!NMe\!-\!P(:O)Cl_2 + Me_3SiCl \quad (3.58)$$

The new phosphorus-hydrido derivatives, $HP(:X)F_2$ (X = S, Se), have been prepared by the reaction of PF_2I with the hydrides H_2X in the presence of mercury[218]:

$$PF_2I + H_2S \xrightarrow[\text{Hg}]{\text{Hg}} HP(:S)F_2 + HI \quad (3.59)$$
$$PF_2I + H_2Se \xrightarrow{\text{Hg}} HP(:Se)F_2 + HI \quad (3.60)$$

The syntheses of a series of fluorophosphoryl compounds, based on the fluorochlorides, $P(:O)FCl_2$ and $P(:O)ClF_2$, have been reported[219]. The mixed halides are accessible via the reaction of the appropriate fluorophosphoric acid with phosphorus pentachloride[220]:

$$(HO)_2P(:O)F + 2PCl_5 \rightarrow Cl_2P(:O)F + 2POCl_3 + 2HCl \quad (3.61)$$
$$HOP:O)F_2 + PCl_5 \rightarrow ClP(:O)F_2 + POCl_3 + HCl \quad (3.62)$$

3.5.2.2 *Chemical properties and reactions of phosphoryl and thiophosphoryl halides*

The behaviour of phosphoryl and thiophosphoryl halides towards poly-dentate phosphines has been studied[221]. From ethylene-1,2-bisdiphenyl-phosphine and $POCl_3$ a salt-like intermediate is obtained which, in the presence of moisture, loses PCl_3 giving ethylene-1,2-bisdiphenylphosphine oxide,

$$Ph_2PC_2H_4PPh_2 + 2POCl_3 \xrightarrow{POCl_3} \left[\begin{array}{c} Ph \quad\;\; Ph \\ | \quad\quad | \\ Cl_2P(:O)\!-\!PC_2H_4P\!-\!P(:O)Cl_2 \\ | \quad\quad | \\ Ph \quad\;\; Ph \end{array} \right]^{2+} 2Cl^-$$

$$(3.63)$$

$$\downarrow {}_{-PCl_3}\Big|{}^{H_2O}$$

$$Ph_2P(:O)C_2H_4P(:O)Ph_2$$

The corresponding phosphine sulphide is formed directly if $PSCl_3$ instead of $POCl_3$ is employed in the same reaction[221],

$$Ph_2PC_2H_4PPh_2 + 2PSCl_3 \xrightarrow{25\,°C}$$
$$Ph_2P(:S)C_2H_4P(:S)Ph_2 + 2PCl_3 \quad (3.64)$$

In the reaction of phosphoryl halides with triphenyl-phosphine or -arsine exchange of one halide ion for phosphine or arsine invariably takes place, and salt-like compounds of the type $[Ph_3MP(:O)X_2]X$ (M = P, As; X = Cl, Br) are formed[222, 223].

Adducts between $POCl_3$ or $POBr_3$ and tertiary amines of 1:3 stoichiometry ($POX_3:R_3N$) have been reported[224], R_3N may be pyridine, β- and γ-picoline, quinoline, isoquinoline, or piperidine. The complexes of $POBr_3$ with pyridine and piperidine have a 1:4 stoichiometry[224].

The cleavage of silicon–nitrogen or silicon–oxygen bonds in silyl amines[225, 226] or in silyl ethers[227] with phosphoryl or thiophosphoryl halides has furnished a series of interesting element-phosphorus compounds which may serve as useful precursors for further synthesis, e.g.

$$(Me_3Si)_2NH + P(:E)F_3 \rightarrow Me_3Si-NH-P(:E)F_2 + Me_3SiF \quad (3.65)$$
$$(Me_3Si)_2NH + P(:S)FCl_2 \rightarrow Me_3Si-NH-P(:S)FCl + Me_3SiCl \quad (3.66)$$
$$(Me_3Si)_2O + P(:O)F_2X \rightarrow Me_3Si-O-P(:O)F_2 + Me_3SiX \quad (3.67)$$
$$(E = O, S; X = Cl, Br)$$

The electrical conductivity of solutions of $SbCl_5$ or BCl_3 in nitrobenzene is increased upon addition of phosphoryl halides, and the effect was found to increase in the following order of increasing donor strength[228]:

$$POCl_3 < PhP(:O)F_2 < PhP(:O)FCl <$$
$$PhP(:O)Cl_2 < Ph_2P(:O)F < Ph_2P(:O)Cl$$

The ionisation is thought to be a consequence of the autoionisation of the adducts in accord with [228]:

$$2D \cdot ACl_n \rightleftharpoons [D_2ACl_{n-1}]^+[ACl_{n+1}]^- \quad (3.68)$$
$$(D = P{=}O \text{ compound}; A = Sb, B; n = 3, 5)$$

The enthalpy of formation, ΔH^0, and the enthalpy of dissolution in water of the adduct $Cl_3Al \cdot OPCl_3$ have been determined to be -18.6 ± 0.9 and -146.4 ± 0.3 kcal mol^{-1}, respectively[229]. The halogen exchange between aluminium halides and $POCl_3$ has been studied using ^{27}Al n.m.r. spectroscopy[230]. The solutions of Al_2X_6 (X = Cl, Br, I) in $POCl_3$, were found to contain, besides $[AlCl_4]^-$ ions, the hexa-coordinate cations, $[Al(OPCl_3)_6]^{3+}$. Since, in the case of the solvents of Al_2Br_6 or Al_2I_6, the solvent is the only source of chloride ions, halogen exchange between $POCl_3$ and Al_2X_6 must take place,

$$Al_2X_6 + 6POCl_3 \rightleftharpoons 2AlCl_3 + 6POCl_2X \quad (3.69)$$
$$4AlCl_3 + 6POCl_3 \rightleftharpoons 3[AlCl_4]^- + [Al(OPCl_3)_6]^{3+} \quad (3.70)$$

From a study of the phase diagram, formation of a 1:1 complex between gallium tribromide and $POCl_3$ has been established which, in the molten state, conducts electricity. This finding may be rationalised in terms of either of the following equilibria:

$$2POCl_3 \cdot GaBr_3 \; \rightleftharpoons \; [GaBr_4]^- \; + \; \begin{bmatrix} Cl_3PO & & Br \\ & Ga & \\ Cl_3PO & & Br \end{bmatrix}^+ \qquad (3.71)$$

or

$$POCl_3 \cdot GaBr_3 \; \rightleftharpoons \; [POCl_3 \cdot GaBr_2]^+ \, Br^- \qquad (3.72)$$

The available experimental data do not, as yet, allow a distinction between these alternatives[231].

The thermal decomposition of the addition compounds, $Br_3P(:S) \cdot AlBr_3$ and $Br_3P(:S) \cdot SO_3$ has been investigated[232]. The adduct, $Br_3P(:S) \cdot AlCl_3$ is unstable at room temperature. In a series of further metal halide adducts of $POBr_3$ and $PSBr_3$ evidence for the coordination of the phosphoryl or thiophosphoryl halide via oxygen or sulphur has been found[232].

While a large number of adducts of $POCl_3$ [233] and $POBr_3$ [234] are known, similar adducts of POF_3 have been studied but little. A recent report is concerned with adducts of POF_3 with BF_3, AsF_5, and SbF_5. In the case of $F_3PO \cdot AsF_5$ coordination via oxygen has been established[235].

The reactions between thiophosphoryl halides and Grignard reagents, especially of t-butyl derivatives, have been investigated, and have furnished interesting examples for n.m.r. spectroscopic studies, using the INDOR technique[236, 237].

Thiophosphoryl compounds involving both the P=S and the P=N bond were obtained in the reactions of amides of dihalothiophosphoric acids, $H_2NP(:S)X_2$ (X = F, Cl), with PF_3Cl_2, Ph_3PCl_2, Ph_2PCl_3, and $PhPCl_4$. Some of these compounds contain the —N=PF_3 grouping, and thus represent the first monomeric members of this class of compound[157]. The reaction of decaborane with $POCl_3$ has been studied. No substitution product was formed from $B_{10}H_{14}$ and $POCl_3$, while a yellow compound, $B_{10}H_{13}P(:O)Cl_2$ was obtained in the reaction of the sodium derivative, $NaB_{10}H_{13}$, with $POCl_3$ [238].

Thiophosphoryldifluoride isocyanate, the isomer of phosphoryl difluoride isothiocyanate, represents yet another case where a Si—N cleavage reaction was usefully employed[239]:

$$4PSF_3 + Si(NCO)_4 \xrightarrow[\substack{\text{autogenous} \\ \text{pressure}}]{200\,°C} 4P(:S)F_2NCO + SiF_4 \qquad (3.73)$$

Tertiary alkyl- and aryl-phosphine oxides have been found to react with phosphoryl and thiophosphoryl halides with formation of stable compounds containing P—O—P bridges[240]:

$$R_3P(:O) + X_3P(:E) \rightarrow [R_3\overset{(+)}{P}\!\!-\!\!O\!\!-\!\!P(:E)X_2]X^- \qquad (3.74)$$
(E = O, S; X = Cl, Br)

3.5.2.3 Spectroscopic investigations

The bond order of the P=O bond in POX$_3$ (X = F, Cl, Br) decreases in the order F > Cl > Br. The P—X bond order for X = F, Cl, Br is 1.35, 0.91, and 0.85, respectively, as has been concluded from i.r. and Raman studies on POX$_3$ and ^{18}POX$_3$ [241]. Force constants and thermodynamic functions of the P=S bond in PSX$_3$ (X = F, Cl, Br) have been determined[242].

The i.r. spectra of acylphosphonamidic dichlorides, RCONHP(:O)Cl$_2$, and of N-acyl-P-methyl-phosphonamidic chlorides, RCONHP(:O)MeCl, have been studied, using ^{14}N and ^{15}N-containing molecules. P—N, C—N, C—O, and N—H stretching frequencies have been assigned[243].

The splitting of the P—O or P=S bands in the i.r. spectra of the compounds (MeO)$_2$P(:S)Cl and HC≡C—CH$_2$—O—P (:S)Cl$_2$ has been rationalised in terms of restricted rotation around the P—O bond[244]. The low temperature Raman spectra of MeOP(:S)F$_3$ have shown the presence of at least two configurations. Since the band intensities do not vary the isomers must have nearly the same energies[245].

The ^{19}F and ^{31}P n.m.r. spectra of diphosphoryl tetrafluoride, F$_2$P(:O)OP (:O)F$_2$, and of F$_2$P(:S)OP(:S)F$_2$ have been analysed in terms of an AA'XX' A''A''' system[246]. Fast rotation around the P—O bond has been established[246]. Based on ^{19}F n.m.r. studies on compounds of type X$_4$P$_2$S$_2$, the structure X$_2$P(:S)—SPX$_2$ (X = F) with a P—S—P bond was suggested[247]. ^{19}F and ^{31}P n.m.r. spectra for a variety of P—F compounds with P=O and P=S bonds have been reported[248]. It proved impossible to derive simple, general substituent rules for either chemical shifts or coupling constants in compounds of type F$_n$P(:E)R$_{3-n}$ (E = O, S; R = OR', alkyl, halogen, NR'$_2$, where R' = alkyl or aryl)[248].

Quantum mechanical calculations concerned with the participation of 3d π orbitals in the bonding in POX$_3$ were carried out, and the exponents for 3s and 3p in an electrostatic field were calculated for X = F, Cl, C, H [249]. The environment of phosphorus in POF$_3$ gives rise to contraction of the d orbitals on phosphorus having π symmetry which are directed to oxygen, and thus permits their interaction with the 2p π orbitals of oxygen[249].

The photoelectron spectra of POF$_3$ (and of PF$_3$) have been discussed, based on ab initio SCF/MO calculations. The highest occupied orbital of POF$_3$ (7e), the acceptor orbital of PF$_3$, has 13% phosphorus 3d character, in accord with an increase of 3d population upon coordination[150].

3.5.2.4 Structural investigations

The crystal structure of POBr$_3$ has been investigated[250]. The crystals are orthorhombic, space group Pna2$_1$. The molecules form infinite chains via intermolecular Br—O bonds (d$_{Br-O}$ 3.08 Å); mean P—Br distance 2.14 Å; P=O distance 1.44 Å.

The structures of POCl$_3$ and PSCl$_3$ have been investigated by electron diffraction by the sector microphotometric method and a comparison of bond lengths was made. In the series PSCl$_3$, POCl$_3$, Me$_2$NPCl$_2$, Me$_2$NP

(:O)Cl$_2$, and PCl$_3$, d(P—Cl) was found to increase in the order,

$$POCl_3 < PSCl_3 < Me_2NP(:O)Cl_2 < PCl_3 < Me_2NPCl_2$$

while the Cl—P—Cl bond angles decrease in the same direction[251].

3.5.3 Transition metal complexes of phosphorus halides and halophosphines

The research activity in this area has been considerable during the past few years. Reference is made to the reviews by Nixon[71], Drozd[70] and Schmutzler[69] which are concerned with the coordination chemistry of fluorophosphine. While Nixon's article covers the literature up to part of 1970, Kruck[252] has reviewed the coordination chemistry of phosphorus trifluoride with transition metals up to 1967.

3.5.3.1 Preparation

The standard methods of synthesis, such as displacement of CO in metal carbonyls[253, 255, 256], or of olefinic ligands[254, 257, 258] by the fluorophosphines have been widely employed. In addition, some new methods have become known.

A series of PF$_3$–transition metal complexes has been obtained by co-condensation of the respective metal vapour and phosphorus trifluoride, or chlorodi fluorophosphine, and of phosphorus trifluoride and phosphine at $-196\,°C$[259, 260]. The following examples illustrate the scope of the synthesis, Cr(PF$_3$)$_6$, Fe(PF$_3$)$_5$, Co$_2$(PF$_3$)$_8$, Ni (PF$_3$)$_4$, Pd(PF$_3$)$_n$, (F$_3$P)$_3$Fe(PF$_2$)$_2$Fe (PF$_3$)$_3$, Ni(PF$_2$Cl)$_4$, Ni(PF$_3$)$_3$(PH$_3$), Ni(PF$_3$)$_2$(PH$_3$)$_2$.

The method of 'reductive fluorophosphination'[261] of ruthenium(III) and osmium(III) chlorides in the presence of hydrogen was found to yield the hydrido complexes, H$_2$Ru(PF$_3$)$_4$ and H$_2$Os(PF$_3$)$_4$. In a similar manner hydrido complexes of cobalt and its homologues of type HM(PF$_3$)$_4$ could be obtained[262].

An indirect photochemical synthesis for monosubstituted derivatives of chromium or tungsten hexacarbonyls and of cyclopentadienyl manganese tricarbonyl has been described[263]. Using tetrahydrofuran (THF) as a solvent the primary fragments, [M(CO)$_5$] and [C$_5$H$_5$Mn(CO)$_2$], respectively, are stabilised through formation of the complexes (THF)M(CO)$_5$ and C$_5$H$_5$Mn(Co)$_2$(THF). Upon addition of the phosphine ligand the THF molecule is displaced.

Methylsilylcobalt tetracarbonyl, MeH$_2$SiCo(CO)$_4$, reacts with PF$_5$ with formation of, besides MeF$_2$SiCo(CO)$_4$, a series of hydridocarbonyl trifluorophosphine complexes, such as HCo(CO)$_3$(PF$_3$), HCo(CO)$_2$(PF$_3$)$_2$, and HCo(CO)(PF$_3$)$_3$. The essential step in this reaction is the reduction of PF$_5$ by the silane[264]. σ-Bonded cobalt carbonyl derivatives of type RCo(CO)$_4$ (R = H, CF$_3$ C$_2$F$_5$, C$_3$F$_7$) react smoothly with PF$_3$ with formation of all possible members of the series RCo(PF$_3$)$_n$(CO)$_{4-n}$. The substitution reaction of PF_3 with $HCo(CO)_4$, for example, proceeds spontaneously even at $-20\,°C$ [265].

The tetrakis-trifluorophosphine complexes, $Pd(PF_3)_4$ and $Pt(PF_3)_4$ have been obtained in the reaction of PF_3 with compounds of Pd^{II} and Pt^{IV}, respectively, in the presence of copper powder and under pressure[266]. If the addition of copper is omitted, PF_3 may serve as the reducing agent[266, 267], e.g.

$$PdCl_2 + 5PF_3 \rightarrow Pd(PF_3)_4 + PF_3Cl_2 \tag{3.75}$$

$$PtCl_4 + 6PF_3 \rightarrow Pt(PF_3)_4 + 2PF_3Cl_2 \tag{3.76}$$

$$PdCl_2 + 2Cu + 4PF_3 \rightarrow Pd(PF_3)_4 + 2CuCl \tag{3.77}$$

$$PtCl_4 + 4Cu + 4PF_3 \rightarrow Pt(PF_3)_4 + 4CuCl \tag{3.78}$$

In an analogous manner platinum complexes of type PtL_4 ($L = CF_3PF_2$ and $(CF_3)_2PF$) could be obtained.

Nitrosyl manganese tetracarbonyl, $Mn(NO)(CO)_4$, reacts with PF_3 at elevated temperature and under pressure. All possible compounds of the series $Mn(NO)(PF_3)_n(CO)_{4-n}$ (with the exception of $n = 4$) have been obtained; they could be separated by gas chromatography[268].

A simple low-pressure synthesis for PF_3 complexes of rhodium and iridium has been described[269]. Displacement of coordinated olefins in complexes such as $RhCl(cyclo\text{-}C_8H_{14})_2$ or $IrCl(cyclo\text{-}C_8H_{14})_2$ with PF_3 at $25\,°C/1$ atm gives $[RhCl(PF_3)_2]_2$ and $[IrCl(PF_3)_2]_n$, respectively[269]. Based on these two compounds a number of further complexes have been obtained, e.g. in the case of $[RhCl(PF_3)_2]_2$.

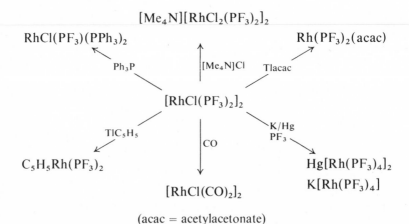

(acac = acetylacetonate)

The ethylene–rhodium complex, $[(C_2H_4)_2RhCl]_2$ may serve as a precursor in the synthesis of a series of complexes of PF_3 and fluorophosphines, as illustrated in the following scheme[270]:

$$RhCl(CCl_3PF_2)_3$$

$$\uparrow \quad CCl_3PF_2$$

$$\underset{F_3P}{\overset{F_3P}{>}}Rh\underset{Cl}{\overset{Cl}{<}}Rh\underset{PF_3}{\overset{PF_3}{<}} \quad \xrightarrow{Ph_3P} \quad \underset{Cl}{\overset{Ph_3P}{>}}Rh\underset{PPh_3}{\overset{PF_3}{<}}$$

$$\nearrow PF_3 \text{ (excess)}$$

$$\underset{\backslash\backslash}{\overset{/\!/}{>}}Rh\underset{Cl}{\overset{Cl}{<}}Rh\underset{\backslash\backslash}{\overset{/\!/}{<}} \quad \xrightarrow{Me_2NPF_2} \quad \underset{Me_2NF_2P}{\overset{Me_2NF_2P}{>}}Rh\underset{Cl}{\overset{Cl}{<}}Rh\underset{PF_2NMe_2}{\overset{PF_2NMe_2}{<}}$$

$$Me_2NPF_2 \Updownarrow \text{vacuum}$$

$$(Me_2NPF_2)_3RhCl$$

A binuclear complex is formed in the reaction of difluoroiodophosphine with dimanganese decacarbonyl[271]:

$$PF_2I + Mn_2(CO)_{10} \rightarrow (CO)_4Mn\underset{I}{\overset{PF_2}{<\!>}}Mn(CO)_4 + 2CO \qquad (3.79)$$

This complex is of particular interest as it constitutes the first example in a new class of binuclear complexes involving two different bridging groups, PF_2 and I.

Phosphorus trihalide complexes of type $(PX_3)M(CO)_5$ (X = Br, I; M = Cr, Mo, W) have been obtained in the reaction of the methylmethoxy-carbene pentacarbonyl complexes of Cr, Mo and W with PBr_3 or PI_3 [272]:

$$(CO)_5MC(OMe)Me + PX_3 \xrightarrow[\text{pentane/hexane}]{\text{hexane or}}$$

$$(CO)_5M(PX_3) + \text{'C(OMe)Me''} \qquad (3.80)$$

3.5.3.2 Physical and chemical properties and reactions of transition metal complexes of phosphorus halides and halophosphines

The fluorophosphine complexes of iron and cobalt, $Fe(CO)_4PF_{3-n}(CF_3)_n$ and $CF_3Co(CO)_3PF_3$, were found to represent interesting examples of 'stereochemically non-rigid' molecules which lead to 'deceptively simple' n.m.r. spectra because of intramolecular ligand exchange[273, 274]. Evidence has been obtained that trifluoromethylfluorophosphine ligands, $(CF_3)_nPF_{3-n}$, have a preference for equatorial sites, methylfluorophosphites, $(MeO)_nPF_{3-n}$, for axial sites within the trigonal-bipyramidal framework of iron penta-carbonyl[274].

According to ^{19}F n.m.r. evidence, a slow, spontaneous re-distribution of ligands takes place in aminofluorophosphine nickel dicarbonyl complexes[275]:

$$2Ni(CO)_2L_2 \rightarrow Ni(CO)_3L + Ni(CO)L_3 \qquad (3.81)$$

(L = $C_5H_{10}NPF_2$ ($C_5H_{10}N$ = piperidine group), Me_2NPF_2, $MePF(NMe_2)$), etc.)

The hydrolysis of monochlorophosphine complexes of platinum of the type, cis-$PtX_2(MR_2Cl)(M'R'_3)$ (X = Cl, Br, I; M = P; M' = P, As; R, R' = alkyl or aryl), in the first stage yields $[PtX_2(MR_2OH)(M'R'_3)]$. Elimination of HX from these complexes gives $[(R'_3M')XPt(R_2MO)_2PtX(M'R'_3)]$ [276], e.g.

$$cis\text{-}[PtCl_2(PR_2Cl)(M'R'_3)] \xrightarrow{OH^-} cis\text{-}[PtCl_2(PR_2OH)(M'R'_3)] \quad (3.82)$$
$$\Updownarrow -HCl$$
$$[(M'R'_3)Pt(Cl)(PR_2O)_2Pt(Cl)(M'R'_3)]$$

Exchange of fluorine by amino groups in PF_3 complexes of zerovalent nickel, for example, has been effected by the reaction of the PF_3 complex with amines, and gives rise to aminofluorophosphine complexes of nickel(0) [277].

The kinetics of the substitution reactions of $Ni(PF_3)_4$ and $Pt(PF_3)_4$ with cyclohexyl isocyanide have been investigated and an S_N1 mechanism has been established[278].

$$M(PF_3)_4 \rightarrow M(PF_3)_3 + PF_3 \quad \text{slow} \qquad (3.83)$$
$$L + M(PF_3)_3 \rightarrow M(PF_3)_3L \quad \text{fast} \qquad (3.84)$$

The activation energies for dissociation of the metal–phosphorus bond in the complexes of Ni, Pd and Pt show the following trend, Ni > Pd < Pt [278].

3.5.3.3   Spectroscopic investigations

Most of the published work in this field is concerned with transition metal complexes of fluorophosphines. Reference is again made to the exhaustive review article by Nixon[71], and only recent developments or those concerning P—X compounds where X ≠ F will be discussed here.

A detailed investigation of the ^{19}F n.m.r. spectra of cis- and trans-$(PF_3)_2$ $Cr(CO)_4$ [279] has shown that the coupling constants cis-$^2J(PMP')$ are greater than trans-$^2J(PMP')$ while for phosphine complexes of second and third row transition metals trans-$^2J(PMP')$ are greater than cis-$^2J(PMP')$ [280].

The values of $^2J(PMP')$ in disubstituted metal carbonyl derivatives of PF_3 may be rationalised by the relatively high s character in the metal–phosphorus bond in PF_3 complexes, compared to the metal–phosphorus bond in complexes of other phosphine ligands[281].

The ^{31}P n.m.r. spectra of a large series of metal carbonyl complexes with PX_3 ligands (X = F, Cl, OMe, etc.) have been studied[282] and have been discussed in terms of σ- and π-contributions to the metal–phosphorus bonding.

The ^{19}F n.m.r. spectra of a series of *fac*-trisubstituted fluorophosphine molybdenum carbonyl complexes of type $L_3Mo(CO)_3$ (L = Et_2NPF_2, Me_2NPF_2, $(CF_3)_2PF$, CH_2ClPF_2, CCl_3PF_2, CF_3PF_2, and PF_3) have been analysed as examples of the $(AX_n)_3$ spin system, and signs of the coupling constants, $^3J_{PF}$, $^1J_{PF}$, and $^2J_{PP}$ have been determined. The values of $^2J_{PP}$ were found to increase with the electronegativity of the groups other than fluorine, bonded to trivalent phosphorus[283].

3.5.3.4 Vibrational spectra

I.R. studies have been conducted on the pentacarbonyl complexes of Cr^0, Mo^0 and W^0 with PBr_3 and PI_3. From the observation of the ν_{CO} band at shortest wavelength the following order of charge transfer, P → metal → CO, has been derived (values for PH_3, PCl_3, and PF_3 are included for comparison)[272],

$$PH_3 > PI_3 > PBr_3 > PCl_3 > PF_3$$

The following values for $\nu_{CO}(A_1)$ in $(CO)_5M(PX_3)$ complexes have been observed ((a) in hexane, (b) in hexadecane, (c) in CS_2).

Table 3.3

Ligand	$(CO)_5Cr(PX_3)$	$(CO)_5Mo(PX_3)$	$(CO)_5W(PX_3)$	Reference
PH_3	2074 (a)	2081 (a)	2083 (a)	284
PI_3	2081 (a)	2089 (a)	2087 (a)	272
PBr_3	2087 (a)	2094 (a)	2093 (a)	272
PCl_3	2089 (a)	2095 (b)	2095 (b)	272, 285
PF_3			2101 (c)	286

The vibrational spectra of $Ni(PF_3)_4$[287], $Ni(PCl_3)_4$[288], $Pd(PF_3)_4$[287] and $Pt(PF_3)_4$[287] have been investigated. The expected T_d symmetry for $Ni(PCl_3)_4$ has been established from polarisation measurements on the Raman spectrum in benzene or carbon tetrachloride solutions[288].

Quantum mechanical calculations on the σ-electron transfer between iron and phosphorus in $(CO)_4Fe(PCl_3)$ have been made[289, 290]. The total charge transfer between phosphorus and iron could be determined from i.r. spectra; if the σ-electron contribution is known, the transfer of π-electron density from iron to phosphorus can be assessed[290].

The magnetic rotatory dispersion (Faraday effect) of a series of nickel(0)–phosphorus halide complexes of the type $Ni[P(XYZ)]_4$ (X, Y, Z = OR, halogen, alkyl, aryl) has been studied, and the magneto-optical contributions of σ- and π-electrons in the Ni—P bond have been determined. The σ-contribution within the series $Ni[P(OR)_{3-n}Cl_n]_4$ has been found to be constant while the π-contribution varies considerably[291].

3.5.3.5   Structural investigations

Electron diffraction studies on Ni(PF$_3$)$_4$ [292] and Pt (PF$_3$)$_4$ [292] have been conducted, and the bond angles and distances shown in Table 3.4 have been observed. The parameters for PF$_3$ and F$_3$P·BH$_3$ are included for comparison.

Table 3.4

Compound	d(P—F)Å	(F—P—F) (degrees)	d(M—P)Å
PF$_3$*	1.569	97.7	—
Ni(PF$_3$)$_4$	1.561	98.4	2.116
Pt(PF$_3$)$_4$	1.546	98.9	2.230
F$_3$P·BH$_3$†	1.538	99.8	1.836

*Morino, Y., Kuchitsu, K. and Moritani, T. (1969). *Inorg. Chem.*, **8**, 867
†Kuczkowski, R. L., and Lide, D. R. (1967). *J. Chem. Phys.*, **46**, 357

An independent electron diffraction study of Ni(PF$_3$)$_4$ has yielded the following parameters, d(Ni—P) 2.099 Å, d(P—F) 1.561 Å, and Ni—P—F 118.4 degrees[293].

3.6   HALIDES OF PENTA-COORDINATE PHOSPHORUS

With rare exceptions, derivatives of phosphorus pentafluoride as the only truly penta-coordinate phosphorus halides are to be covered in this section. Considerable progress has been made during the 1969–1970 period, and both preparative chemists and spectroscopists have contributed towards the synthesis and structural investigations of 5-coordinate phosphorus halides, notably of fluorophosphoranes. The fluorophosphoranes have continued to attract interest, in particular, because of their most important structural feature, i.e. their 'dynamic stereochemistry'. This manifests itself in simplified n.m.r. spectra, and is caused by intramolecular positional exchange of ligands which is rapid, relative to the time-scale of the n.m.r. experiment. A review article by Heller[294] deals with aspects of these phenomena.

3.6.1   Preparation of halophosphoranes

A few further examples of fluorophosphoranes, prepared by the standard methods, have become known, although some anomalies were noted, for example, in the preparation of the t-butylfluorophosphorane (Bu^t)$_n$PF$_{5-n}$ ($n = $ 1–3) [97] and pentafluorophenylfluorophosphorane series (C$_6$F$_5$)$_n$PF$_{5-n}$ ($n = $ 1–3) [98]. t-Butyldichlorophosphine did not undergo the usual redox reaction with SbF$_3$, as most other chlorophosphines[295], and prior oxidation with chlorine, followed by chlorine–fluorine exchange with SbF$_3$, was required[97]. The same observation was made in the case of the reaction of (C$_6$F$_5$)$_n$PCl$_{3-n}$ ($n = $ 1,2) with SbF$_3$. Bis-(t-butyl)chlorophosphine, surprisingly underwent the normal redox reaction with SbF$_3$,

$$3Bu^t_2PCl + 3SbF_3 \rightarrow 3Bu^t_2PF_3 + 2Sb + SbCl_3 \qquad (3.85)$$

The new fluorophosphoranes, $CF_2=CFPF_4$ and $(CF_2=CF)_2PF_3$, were prepared by the oxidative fluorination of the appropriate chlorophosphines, $CF_2=CFPCl_2$ and $(CF_2=CF)_2PCl$ with antimony pentafluoride[328].

Bis(trifluoromethyl)amino-bis-(trifluoromethyl)phosphine was found to react with chlorine at $-50\,°C$ to give a 5-coordinate compound which decomposed on standing with intramolecular fluorination,

$$(CF_3)_2NP(CF_3)_2 + Cl_2 \rightarrow (CF_3)_2NP(CF_3)_2Cl_2 \qquad (3.86)$$

$$\begin{array}{c} F_3C \\ {\searrow} \\ F_2C \end{array}\!\!N\!-\!P(CF_3)_2Cl_2 \rightarrow (CF_3)_2PCl_2F + CF_3N=CF_2 \qquad (3.87)$$

Based on ^{19}F n.m.r. evidence, the structure of $(CF_3)_2NP(CF_3)_2Cl_2$ is suggested to be trigonal-bipyramidal with the two CF_3 groups occupying the axial positions[296].

Further mixed chlorofluorophosphoranes, $CF_3PF_2Cl_2$, $C_3F_7PF_2Cl_2$, and $(CF_3)_2PFCl_2$, were prepared by the reaction of the respective fluorophosphines with chlorine. The chlorofluorophosphoranes could be fluorinated with SbF_3 to give pure fluorophosphoranes, such as $C_3F_7PF_4$ and $(C_3F_7)_2PF_3$[297].

Evidence for an interesting fluorophosphorane, involving 5-coordinate and 3-coordinate phosphorus linked directly to each other, has been obtained in the ^{19}F n.m.r. investigation of the slow decomposition of dimethylfluorophosphine[100] which is represented by

$$3Me_2PF \rightarrow Me_2PF_3 + Me_2PPMe_2 \qquad (3.88)$$

The reaction apparently proceeds via an intermediate,

$$\begin{array}{c} F \\ Me | \\ {\searrow}\!P\!-\!PMe_2 \\ Me | \\ F \end{array}$$

whose formation and gradual disappearance can be followed in the ^{19}F n.m.r. spectra of ageing samples of Me_2PF. It is suggested that the phosphinofluorophosphorane reacts with a further molecule of Me_2PF to give the ultimate products, Me_2PF_3 and Me_2PPMe_2[100].

$$Me_2PF_2PMe_2 + Me_2PF \rightarrow Me_2PF_3 + Me_2PPMe_2 \qquad (3.89)$$

A detailed account has appeared on the synthesis of the hydride derivatives of PF_5, HPF_4 and H_2PF_3 in the reaction of phosphorous acid with hydrogen fluoride[298]. The compounds have already been the subject of detailed structural studies, both by n.m.r. and vibrational spectroscopy.

Considerable effort has been expended on the synthesis and structural investigation of fluorophosphoranes containing heteroatoms, such as nitrogen, oxygen, or sulphur. As a preparative route to such compounds, the reaction of phosphorus pentafluoride and its derivatives with the appropriate element–trimethylsilyl compound has proved useful in many instances.

Small ring compounds, containing 5-coordinate phosphorus as part of a ring system, such as the diazafluorophosphetidines, have continued to attract interest, and reviews have appeared which are concerned with chemical and stereochemical aspects of compounds such as $[F_3PNR']_2$, $[F_2RPNR']_2$, etc.[299-301]. The syntheses of virtually all the pertinent compounds are based on Si—N cleavage reactions, and the following examples are illustrative,

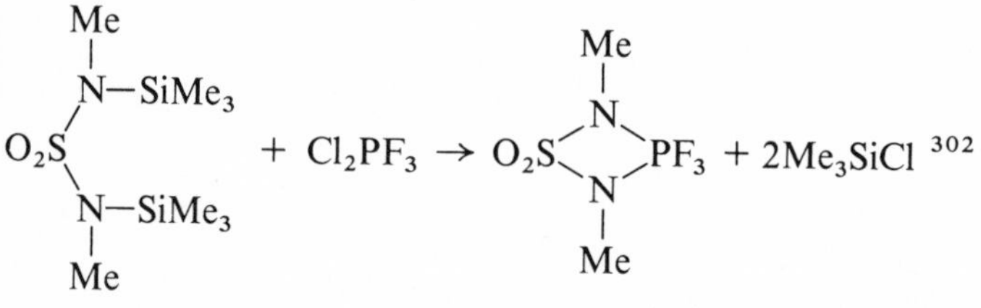

$$
\begin{array}{c}
\text{Me} \\
|\\
\text{N—SiMe}_3 \\
\text{O}_2\text{S} \qquad\qquad + \text{Cl}_2\text{PF}_3 \rightarrow \text{O}_2\text{S} \quad \text{N} \quad \text{PF}_3 + 2\text{Me}_3\text{SiCl} \quad ^{302} \\
\text{N—SiMe}_3 \\
|\\
\text{Me}
\end{array} \qquad (3.90)
$$

(29)   (30)

$$
(29) + (30) \rightarrow \text{O}_2\text{S}\overset{\text{Me}\ \ \text{F}\ \ \text{Me}}{\underset{\text{Me}\ \ \ \ \text{Me}}{\text{N—P—N}}}\text{SO}_2 \qquad (3.91)
$$

(31)

$$
(30) + \begin{array}{c}\text{Me}\\|\\\text{N—SiMe}_3\\ \text{O=C}\\ \text{N—SiMe}_3\\|\\\text{Me}\end{array} \longrightarrow \text{O}_2\text{S}\overset{\text{Me}\ \ \text{F}\ \ \text{Me}}{\underset{\text{Me}\ \ \ \ \text{Me}}{\text{N—P—N}}}\text{C=O} \qquad (3.92)
$$

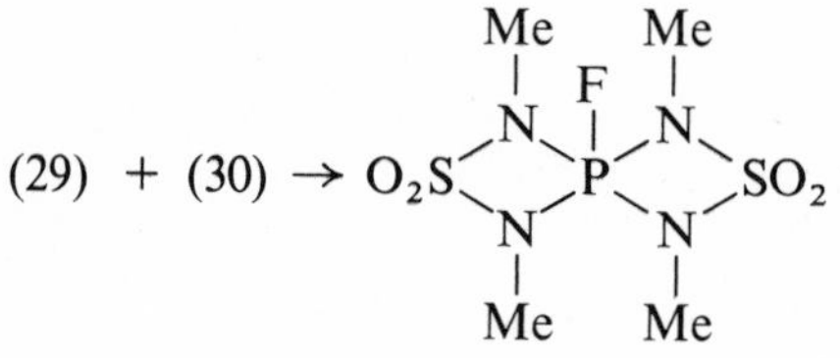

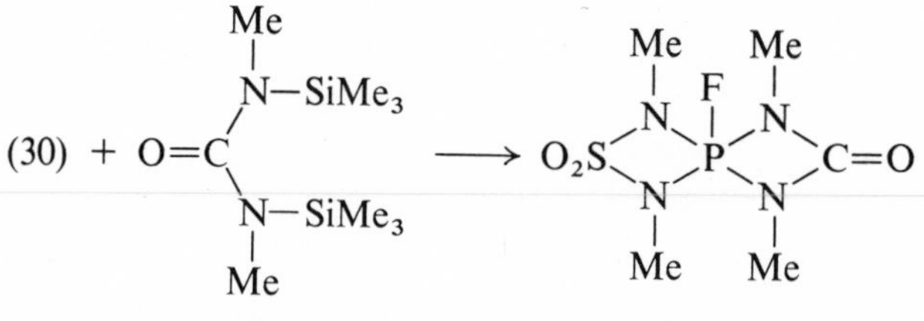

(32)   (33)

Mixed chloro-fluoro-diaza-phosphetidines, $(MeNP)_2Cl_nF_{6-n}$ $(n = 1-5)$, have been obtained in the equilibration of $(MeNPCl_3)_2$ and $(MeNPF_3)_2$ [303]. The compound (34) was found to be formed in the reaction of N,N'-

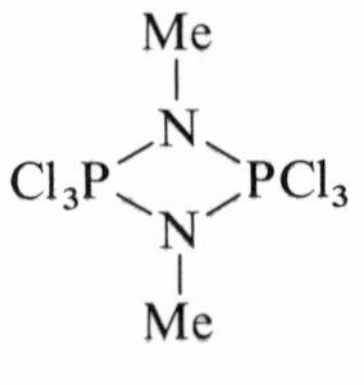

$$
\begin{array}{c}
\text{Me}\\|\\ \text{N}\\
\text{Cl}_3\text{P} \qquad \text{PCl}_3\\
\text{N}\\|\\\text{Me}
\end{array}
\qquad\qquad
\text{O}_2\text{S}\overset{\text{Me}\ \ \text{Cl}\ \ \text{Me}}{\underset{\text{Me}\ \ \ \ \text{Me}}{\text{N—P—N}}}\text{SO}_2
$$

(34) 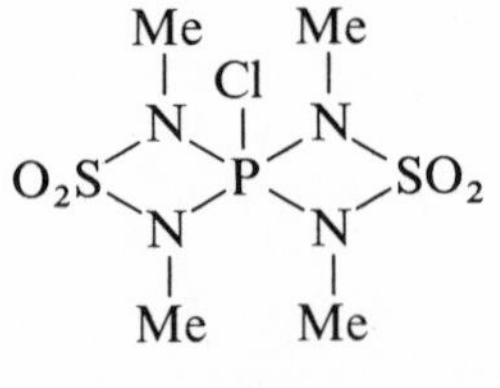   (35)

dimethylsulphamide with PCl_5 [304]. If the same reaction is carried out in the presence of pyridine as a base, the spiro compound (35) is obtained. Chlorine-fluorine exchange in (35), with formation of the monofluorophosphorane (36), is possible with silver fluoride,

$$(35) + AgF \rightarrow \quad (36) \qquad (3.93)$$

Further reaction of (34) with N,N'-dimethylsulphamide gives the tricyclic compound (37):

$$(34) + O_2S(NHMe)_2 \rightarrow \quad (37) \qquad (3.94)$$

An unusual tricyclic chlorophosphorane (38) has been obtained in the reaction of acetohydrazide with PCl_5 in the presence of triethylamine[305]:

(38)

The oxyfluorophosphorane, $MeOPF_4$, is formed in the reaction,

$$PF_5 + P(OMe)_3 \rightarrow F_4POMe + FP(OMe)_2 \qquad (3.95)$$

The compound undergoes further, complicated transformation[306]. Oxy-fluorophosphoranes containing catechol groups bonded to 5-coordinate phosphorus have been obtained in the reaction of PF_5 and of its substitution products with the bis-(trimethylsilyl)ether of catechol[307], e.g.:

$$+ 4Me_3SiF \quad (3.96)$$

$$(3.97)$$

$$(3.98)$$

Stable oxyfluorophosphoranes have also been obtained in the reaction of PF_5, RPF_4 and R_2PF_3 with aryltrimethylsilyl ethers (aryl = C_6H_5 or C_6F_5)[308, 309].

$$RPF_4 + 2ArOSiMe_3 \rightarrow RPF_2(OAr)_2 + 2Me_3SiF \quad (3.99)$$
$$R_2PF_3 + ArOSiMe_3 \rightarrow R_2PF_2OAr + Me_3SiF \quad (3.100)$$

With PF_5 no stable $ArOPF_4$ is formed but, depending on the conditions, the reaction may be directed to give, predominantly, either $[(ArO)_4P][PF_6]$ (isomeric with $(ArO)_2PF_3$) or $(ArO)_3PF_2$.

Fluorophosphoranes containing mercapto groups could also by synthesised via the silyl cleavage reaction, in this case of silylthioethers[310, 311], e.g.:

$$PF_5 + R'SSiMe_3 \rightarrow R'SPF_4 + Me_3SiF \quad (3.101)$$
$$RPF_4 + R'SSiMe_3 \rightarrow RPF_3SR' + Me_3SiF \quad (3.102)$$
$$R_2PF_3 + R'SSiMe_3 \rightarrow R_2PF_2SR' + Me_3SiF \quad (3.103)$$

Formation of $MeSPF_4$, amongst other products, has also been reported in the reaction of $P(SMe)_3$ with PF_5 [311]. The reaction proceeds via a 1:1 adduct. Reaction of PF_5 with $(Me_2N)_2S$ has also been reported to give a 1:1 adduct whose decomposition gives Me_2NPF_4 and other products[311].

The diamino-substituted derivative of PF_5 has been obtained in the ammonolysis of PF_5 [312]:

$$PF_5 \xrightarrow{NH_3} (H_2N)_2PF_3 + NH_4PF_6 + \text{solids} \quad (3.104)$$

3.6.2 Structural investigations on halides of penta-coordinate phosphorus

The study of the structure of the halides of penta-coordinate phosphorus continued to attract the interest of spectroscopists and techniques such as mass spectroscopy, n.q.r. spectroscopy, and, in particular, n.m.r. and vibrational spectroscopy have been employed. Further progress has been made in gaining understanding of the intramolecular exchange processes taking place in fluorophosphoranes, mostly based on n.m.r. work.

3.6.2.1 *Mass spectroscopy*

A large series of fluorophosphoranes, R_nPF_{5-n} (n = 0–3:R = hydrocarbon or dialkylamino group) has been investigated by mass spectroscopy, and characteristic fragmentation patterns have been listed and discussed[313].

3.6.2.2 *Chlorine-35 n.q.r. spectroscopy*

The method has been applied to the study of the compounds $PhPCl_4$ and Ph_2PCl_3 [314]. Only one frequency was observed for both compounds, and the possibility of a covalent structure has been considered[314].

3.6.2.3 *Vibrational spectra*

Phosphorus pentafluoride and some of its mixed halo derivatives have been the subject of study by this technique[127, 315–319]. Further refinements have been made in the description of the bonding situation in these molecules, including changes in the assignments of some bands, determination of force constants, etc.[127, 316–318, 320, 321]. Force constants have also been reported for the series of methylfluorophosphoranes[322], based on published data for Me_nPF_{5-n} (n = 1–3).

The trigonal-bipyramidal structure, so far invariably observed for all fluorophosphoranes[295], has been confirmed by i.r. and Raman spectroscopy for two further compounds, CCl_3PF_4 [323] and $C_6H_5SPF_4$ [324].

According to ^{31}P n.m.r. spectroscopic investigations[325] of Ph_3PCl_2 in polar solvents, the conclusion was made that the compound normally exists in the ionic form, $[Ph_3PCl]Cl$, although there was evidence that the covalent form may exist in nitrobenzene. I.R. spectroscopic evidence has now been obtained which suggests that Ph_3PCl_2 in haloform solutions exists in the form of hydrogen or deuterium-bonded solvates of the non-ionic, dimeric structure shown below.

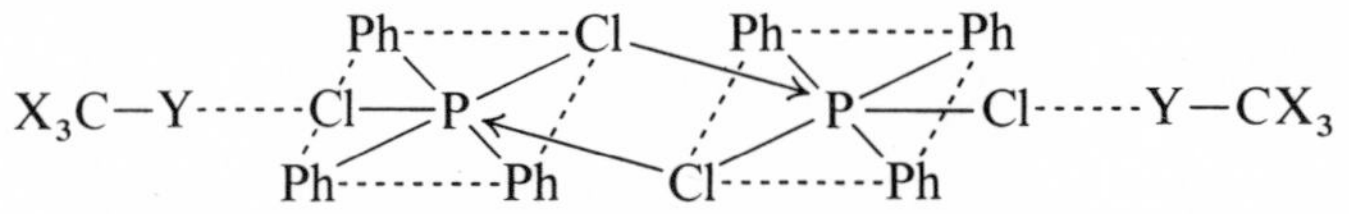

(Y = H or D)

The compound $Ph_3PCl_2 \cdot CHBr_3$, for example, has been obtained in the form of crystals[326].

3.6.2.4   Thermochemical investigation of $(Cl_3PNMe)_2$

A trigonal-bipyramidal structure for $(Cl_3PNMe)_2$ has been well established:

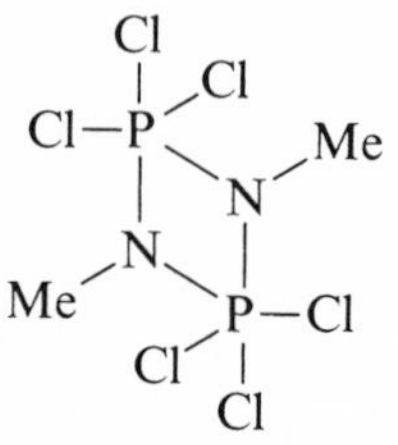

In accord with this structure, different bond energies are to be expected, and were, in fact, observed: 69 kcal mol^{-1} for the axial (weaker) P—N bond, 77.5 kcal mol^{-1} for the equatorial (stronger) P—N bond[327].

3.6.2.5   Nuclear magnetic resonance

The n.m.r. method, especially ^{19}F and ^{31}P n.m.r., has been used extensively, both as a means of characterisation of fluorophosphoranes and in order to study the intramolecular positional exchange processes in such molecules. N.M.R. data have been included in many publications concerned with synthetic aspects of the halophosphoranes[97, 98, 248, 297, 302, 304, 305, 310, 312, 313, 328]. The positive ^{31}P chemical shifts are usually taken as an indication of the presence of 5-coordinate phosphorus, some typical δ_P values are as follows (all relative to 85% H_3PO_4).

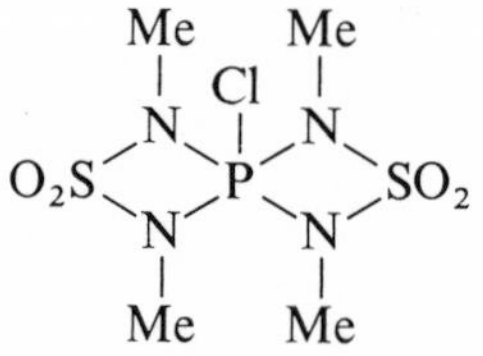

(40)

(41)

$\delta_P + 86$ p.p.m.[304]

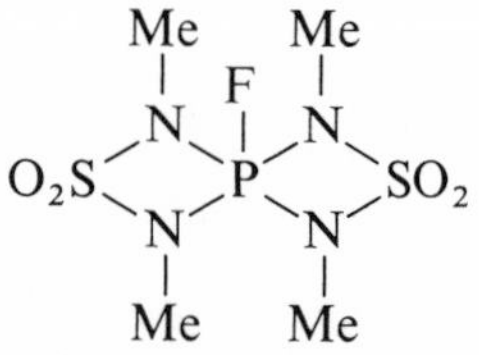

$\delta_P + 84\cdot6$ p.p.m.

$MeSPF_4$ (42) $\delta_P + 34\cdot2$ p.p.m.[310]

$PhPF_3SMe$ (43) $\delta_P + 18.9$ p.p.m.[310]
Ph_2PF_2SMe (44) $\delta_P + 39.0$ p.p.m.[310]

(45) $\delta_P + 52\cdot7$ p.p.m.

(46) $\delta_P + 76\cdot8$ p.p.m.[302]

(47) $\delta_P + 85\cdot0$ p.p.m.

(48) $\delta_P + 67\cdot0$ p.p.m.

PF_5 $\delta_P + 80\cdot3$ p.p.m.[306]
(the previous value was shown
to be in error).

The stereochemistry of fluorophosphoranes involving small ring systems of the above types, and also of the catechol-substituted derivatives, e.g. (49)–(51) is of special interest. Based on a large number of data the trigonal-bipyramid has been established as the normal configuration of fluoro-phosphoranes, and it has been concluded that the axial positions are always occupied by the most electronegative groups, i.e. fluorine. The distinction between axial and equatorial fluorine can be made on the basis of both δ_F and J_{P-F} values, which are characteristically lower for axial than for equatorial fluorine[295, 329]. It has been found that the monofluorophosphorane (49) is exceptional, in that the single fluorine atom occupies an equatorial position within the trigonal-bipyramid,

J_{P-F} 1018 Hz[307]
δ_F (CCl$_3$F) $+ 70\cdot4$ p.p.m.

(49)

These values may be compared to those for two other monofluorophos-phoranes, (47) and (48),

(47): J_{P-F} 1035 Hz[302, 304], $\delta_F + 88.5$ p.p.m.
(48): J_{P-F} 1038 Hz[302], $\delta_F + 102.8$ p.p.m.

where both J_{P-F} and δ_F values also suggest equatorial fluorine. Apparently the steric requirements of the small heteroatom rings are such that the rings can only bridge one axial and one equatorial position each, thus forcing the single fluorine atom into an equatorial position. It has been predicted that these may be cases of 5-coordinate, trigonal bipyramidal systems in which the influence of the electronegativity of substituents such as fluorine may be outweighed by steric factors[329].

As far as the catechol derivatives are concerned, a comparison of the data for (49) with those of the compounds (50) and (51) is informative.

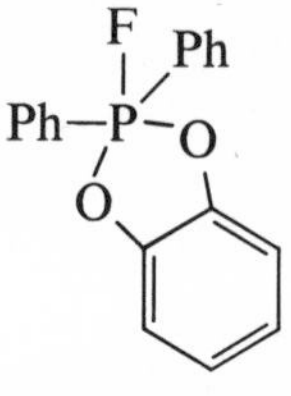

J_{P-F} 797 Hz [307]
δ_F + 31·0 p.p.m.

(50)

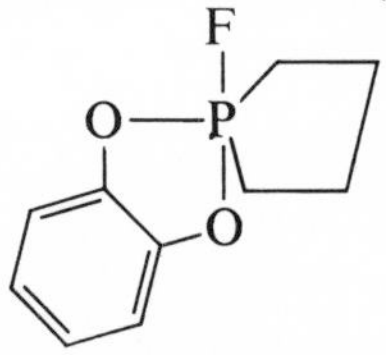

J_{P-F} 829 Hz.[307] δ_F + 21·7 p.p.m.

(51)

Both J_{P-F} and δ_F, in these cases, are indicative of fluorine occupying an axial position within the trigonal-bipyramid around phosphorus. Apparently the carbon ring in (51) is sufficiently flexible to span two equatorial positions.

The temperature-dependence of the ^{19}F n.m.r. spectra of a series of mercapto-substituted fluorophosphoranes[310] has been studied.

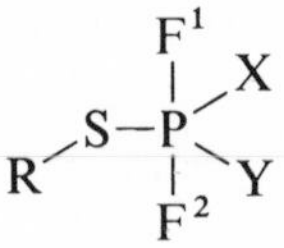

(X = Y = F; X ≠ Y = Me, Ph, or F; R = Me, Et, or Ph)

Evidence was obtained in the compounds $RSPF_4$ for a slowing down not only of the intramolecular fluorine exchange but also of the P—S bond rotation, relative to the n.m.r. time-scale, from the observation of the non-equivalence of F^1 and F^2 at low temperature. In compounds of types RPF_3SR' and R_2PF_2SR' (where no exchange takes place at room temperature), non-equivalence of F^1 and F^2, caused by the slowing down of the P—S bond rotation, has also been observed on cooling[310].

A similar situation was noted for the related nitrogen-containing fluorophosphoranes containing an unsymmetrically substituted amino group[330]:

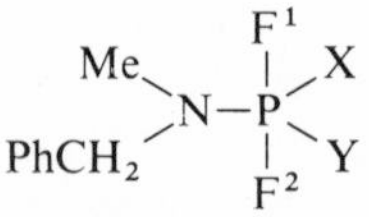

(X = Y = F; X = F, Y = Ph; or X = Y = Ph)

Non-equivalence of the axial fluorine atoms, F^1 and F^2, also became evident in the ^{19}F n.m.r. spectra at low temperature. In the case of the tetrafluorophosphorane, $Me(PhCH_2)NPF_4$ $(X = Y = F)$ both P—N bond rotation and intramolecular exchange were slowed down on cooling.

The variable-temperature ^{19}F n.m.r. spectra of a further type of fluorophosphorane,

$(R^3 = F$ or hydrocarbon group; $R^1 = R^2$ or $R^1 \neq R^2$; R^1, R^2 = hydrocarbon group)

have also been studied[331]. Fast positional exchange between axial and equatorial sites takes place in these compounds and could not be slowed down sufficiently on cooling in any case where $R^1 = R^2$. However, when $R^1 \neq R^2$, magnetic non-equivalence of axial and equatorial fluorine became apparent at $c. -40°C$ [331]. Analysis of the ^{19}F spectra in the temperature region $-20°C$ to $+20°C$ for the compound

gave the activation energy of the exchange process as 18 ± 3 kcal mol^{-1} [331].

An interesting alternative to the mechanism of intramolecular exchange which is thought to account for the simplified n.m.r. spectra of fluorophosphoranes at room temperature has been suggested[332]. Based on a study of the variable-temperature ^1H spectrum, exchange in Me_3PF_2 is proposed to occur via a fluorine-bridged dimer:

A similar intermolecular exchange process seems to take place in Me_2PF_3 [332].

Energy barriers for the intramolecular exchange process in PF_5 and PF_2Cl_3 have been compared for square-pyramidal and tetragonal-pyramidal intermediates, and were found to be similar for both alternatives[321].

Magnitudes and signs of indirect spin coupling constants, relative to $^1J(CH) > 0$, were determined by the heteronuclear INDOR technique in the series Me_nPF_{5-n} $(n = 1, 2, 3)$ and it was concluded that $^1J(PF_a) < 0$, $^1J(PF_e) < 0$, and $^1J(PC) > 0$ [333] (F_a, F_e refer to axial and equatorial fluorine within the trigonal-bipyramidal fluorophosphorane structure).

3.6.2.6 Electron diffraction study of $(F_3PNMe)_2$

Further evidence for the trigonal-bipyramidal structure of the diaza-fluoro-phosphetidine, $(F_3PNMe)_2$, has been obtained through electron diffraction[334]. The known general trends in the bond parameters in fluorophosphoranes, e.g. the invariably greater length of axial bonds (P—N and P—F) compared to equatorial bonds, have been fully confirmed.

3.7 PHOSPHORUS–HALOGEN COMPOUNDS INVOLVING HEXA-COORDINATE PHOSPHORUS

Activity in this area was considerable during the last two years but was confined almost entirely to hexa-coordinate species having phosphorus–fluorine bonds. A very large number of publications cover the use of hexa-fluorophosphate as a counter ion in the precipitation of complex cations involving Main Group elements and, especially, transition metals. No attempt will be made to cover all these reports and reference is made to an earlier review[639].

Hexa-coordinate phosphorus species within the scope of this review may be divided into two classes (a) adducts of 5-coordinate phosphorus halides (i.e. phosphorus pentafluoride) with donor molecules, and (b) hexahalo-phosphates and their substituted derivatives, $[PX_nY_{6-n}]^-$ (where X = halogen; Y = hydrogen, halogen, or organic group; $n = 5, 4, 3$). Considerable effort has been expended recently on the synthesis and spectroscopic (especially n.m.r.) characterisation of such species.

3.7.1 Adducts between phosphorus pentafluoride and Lewis bases

A large number of adducts of type $F_5P\cdot B$ has been prepared, in an effort to obtain novel herbicides, by the combination of PF_5 with the appropriate base in a suitable solvent. B may be, for example, oximes[335-337], carboxylic acid amides[338], or nitriles[339].

A number of adducts of PF_5 with primary amines, $F_5P\cdot NH_2R$ ($F_5P\cdot 2NH_2R$ for R = Me), have been prepared[340]. While the exact nature of the 1:2 adduct between PF_5 and $MeNH_2$ could not be clarified, the 1:1 adducts have been shown by ^{19}F n.m.r. on solutions in acetonitrile to contain the $F_5P\cdot B$ grouping, the coordination of the amine to phosphorus occurring via nitrogen. The adducts, $F_5P\cdot NH_2R$, have been shown to undergo various transformations with formation of amino-substituted fluorophosphoranes, such as $RNHPF_4$ and $(RNH)_2PF_3$ [340]. A further series of 1:1 adducts, $F_5P\cdot B$, has been prepared where B = Me_3P, Pr^n_3P, $(Me_2N)_3P$, $MeP(NMe_2)_2$, and Me_2PNMe_2 [109]. According to spectroscopic evidence (i.r., n.m.r.) all these adducts contain a phosphorus–phosphorus bond. The adducts, formed by the methylaminophosphines, all undergo decomposition upon heating with formation of dimethylaminofluorophosphoranes.

^1H and ^{19}F n.m.r. data have been reported for the 1:1 adducts, $Me_3N\cdot PF_5$ and $Me_2O\cdot PF_5$ [341]. The spectra are consistent with the octahedral structure expected. In the case of $Me_3N \cdot PF_5$ no evidence was found for exchange between free and complexed PF_5, nor between fluorine of the free and the complexed PF_5. By contrast, rapid exchange between free and complexed dimethylether was noted in the ^1H n.m.r. spectrum of $Me_2O \cdot PF_5$ [341].

The adduct, $F_5P \cdot Me_3N$ has also been obtained by the addition of Me_3N to a solution of PF_5 in fluoroform where partial ionisation has been postulated.

$$HCF_3 + PF_5 \rightleftharpoons H^+ + [CF_3PF_5]^- \qquad (3.105)$$

The adduct was obtained in an attempt to isolate the trimethylammonium salt, $[Me_3NH][CF_3PF_5]$ [342].

3.7.2 New modes of formation of the hexafluorophosphate ion

The solution of phosphorus pentafluoride in fluoroform shows electric conductivity and ionisation in accordance with

$$HCF_3 + PF_5 \rightleftharpoons H^+ + [CF_3PF_5]^- \qquad (3.106)$$

has been postulated. Upon addition of trimethylamine (molar ratio Me_3N: $PF_5 = 1:1$) the adduct $Me_3N \cdot PF_5$ is obtained as the major product, besides some $[Me_3NH][PF_6]$. Fluoroform thus serves to transfer the F^- ion, not the CF_3 group[342].

Treatment of PF_5 with nitrogen oxides, such as NO, N_2O_3, NO_2, and N_2O_5, gave rise to formation of $[PF_6]^-$, in the form of salts with the $[NO]^+$ or $[NO_2]^+$ cation, or as a mixture of both nitrosonium and nitronium salts. Indications of the formation of the 5-coordinate anions, $[POF_4]^-$ have also been obtained[343], the evidence being based mainly on i.r. spectra.

A new study has been undertaken to confirm the identity of the complex, $SF_4 \cdot PF_5$. I.R. and Raman spectra served to confirm the ionic structure, $[SF_3][PF_6]$, assignments for PF_3 (isoelectronic with $[SF_3]^+$) and authentic $[PF_6]$ being employed[344].

3.7.3 Novel fluorophosphate anions

The fluoro-hydride anion, $[PF_5H]^-$, has been obtained in the following reactions[345, 346].

$$PF_3 + [HF_2]^- \xrightarrow{\text{room temp.}} [PF_5H]^- \qquad (3.107)$$

$$Me_2NPF_2 + 3[HF_2]^- \xrightarrow[\text{MeCN}]{70\,°C} [Me_2NH_2]^+[PF_5H]^- + 3F^- \quad (3.108)$$

Fluoro-hydride anions containing CF_3 groups bonded to phosphorus have been prepared[346].

$$CF_3PF_2 + [HF_2]^- \rightarrow [CF_3PF_4H]^- \ (cis \text{ and } trans) \qquad (3.109)$$

$$(CF_3)_2PF + [HF_2]^- \rightarrow [(CF_3)_2PF_3H]^- \qquad (3.110)$$

The latter ion is also formed in the dimethylaminolysis of $(CF_3)_2PF$ where $[HF_2]^-$ probably occurs as a reaction intermediate[346].

$$3(CF_3)_2PF + 3Me_2NH \rightarrow 2(CF_3)_2PNMe_2 + [Me_2NH_2][(CF_3)_2PF_3H]$$
(3.111)

The anion $[(CF_3)_2PF_4]^-$ is formed as a by-product in this reaction.

Phosphorus pentafluoride was found to react with Me_3SnCF_3 with formation of a product which, upon treatment with tetraphenylarsonium chloride in dichloromethane, gave a mixture of the tetraphenylarsonium salts, $[Ph_4As][CF_3PF_5]$ and $[Ph_4As][(CF_3)_2PF_4]$ [347]. The following course of reaction has been suggested:

$$Me_3SnCF_3 + PF_5 \rightleftharpoons Me_3SnCF_3 \cdot PF_5 \rightarrow [Me_3Sn][CF_3PF_5] \quad (3.112)$$
$$[Me_3Sn][CF_3PF_5] + Me_3SnCF_3 \rightarrow [Me_3Sn][(CF_3)_2PF_4] + Me_3SnF$$
(3.113)

Several routes towards the synthesis of trifluoromethyl substituted fluorophosphates have been reported[348]. The anions, $[(CF_3)_nPF_{6-n}]$ $(n = 1–3)$ could be obtained by the addition of caesium fluoride to the appropriate fluorophosphorane, $(CF_3)_nPF_{5-n}$. The disubstituted derivative could also be prepared as follows[348].

$$Ag[(CF_3)_2P(:O)O] + 2SF_4 \rightarrow Ag[(CF_3)_2PF_4] + 2SOF_2 \quad (3.114)$$

According to Chan and Willis[348] the reaction of PF_5 with Me_3SnCF_3 which might be expected to yield fluorophosphates, e.g.

$$Me_3SnCF_3 + PF_5 \rightarrow [Me_3Sn][CF_3PF_5]^- \quad (3.115)$$

did not proceed in a clear-cut fashion.

Addition of caesium fluoride to fluorophosphoranes also served as a preparative route towards the synthesis of the pentafluorophosphates $Cs[MePF_5]$, $Cs[PhPF_5]$, $Cs[Me_2NPF_5]$, $Cs[Ph_2PF_4]$, and $Cs[MePhPF_4]$. No stable product was obtained, under similar conditions, from Me_2PF_3 and CsF, but the anion $[Me_2PF_4]^-$ has now been obtained in the form of several stable salts[349], e.g.

$$2Me_2PF_3 + 2Me_3SiN=PR_3$$

$$\downarrow$$

$$\left[Me_2P \begin{matrix} N=PR_3 \\ N=PR_3 \end{matrix} \right]^+ [Me_2PF_4]^- + 2Me_3SiF$$
(3.116)

Further examples of the formation of anions, $[RPF_5]^-$, during the slow rearrangement of certain aminofluorophosphoranes have been observed[248], e.g.

$$2RPF_3NR_2' \rightarrow [RP(F)(NR_2')_2][RPF_5]^-$$
(3.117)
$$(R = Et, Ph; R' = Me)$$

3.7.4 Spectroscopic and structural studies of hexafluorophosphate and its derivatives

Crystal structures of a number of hexafluorophosphates have been determined during the last 20 years[69]. A further structure, that of pyridinium hexafluoro-phosphate (and of the corresponding hexafluoro-arsenate and -antimonate), has been determined. The anions have been shown to be regular octahedra, with an F^1—P—F^2 angle of 89.5 degrees and a P—F bond distance of 1.593 Å [350].

For the monosubstitution products of $[PF_6]^-$, $[RPF_5]^-$ (where R = hydrocarbon group or Me_2N), the ^{19}F and ^{31}P n.m.r. spectra are consistent with an octahedral arrangement, involving a 4:1 environment of fluorine atoms[248, 351].

The highly-positive δ_P values for $[RPF_5]^-$ are indicative of highly-shielded phosphorus.

The identity of the novel hydrido-fluoro anion, $[PF_5H]^-$, has also been established by ^1H and ^{19}F n.m.r. spectroscopy:

The following parameters for a solution of $K[PF_5H]$ in acetonitrile are illustrative[345, 346]:

δ_H	-5.4 p.p.m. (SiMe$_4$ ref.)
δ_F (equatorial)	$+56.4$ p.p.m. (CCl$_3$F ref.)
δ_F (axial)	$+66.2$ p.p.m.
J_{P-F} (equatorial)	817 Hz
J_{P-F} (axial)	729 Hz
J_{F-F}	41 Hz
$J_{F(axial)-H}$	~ 0 Hz
$J_{F(equatorial)-H}$	126–128 Hz
J_{P-H}	955 Hz

Solubility problems prevented measurements of ^{31}P n.m.r. spectra of the caesium salts, $Cs[RPF_5]$ (R = Me, Ph, Me_2N), $Cs[MePhPF_4]$ and $Cs[Ph_2PF_4]$. In some cases, even ^{19}F n.m.r. spectra were difficult to obtain[351]. In the case of $Cs[Ph_2PF_4]$ (solution in acetonitrile) a single ^{19}F resonance (J_{P-F} 947 Hz) was obtained. It is in agreement with a *trans* structure for the anion but in the absence of variable-temperature, ^{19}F spectra, which cannot be obtained because of the poor solubility of the salt even at room temperature, an exchange process within the anion cannot be ruled out.

The interpretation of the ^{19}F n.m.r. spectra of the CF_3-substituted fluorophosphates by a first-order approach has been reported[347, 348], and it was suggested that both $[(CF_3)_2PF_4]^-$ and $[(CF_3)_3PF_3]^-$ have a *trans* configuration[347, 348]. A re-interpretation of the ^{19}F spectrum of $Cs[(CF_3)_2PF_4]$ has been given which confirms the *trans* arrangement of CF_3 groups in the anion. The accurate spectral parameters for solutions of $Cs[(CF_3)_2PF_4]$ in acetonitrile are as follows[352]:

$$^1J_{P-F} \pm 898.0 \pm 2.0 \text{ Hz}$$
$$^2J_{P-C-F} \pm 148.5 \pm 1.0 \text{ Hz}$$
$$^3J_{FCPF} \ 14-3 \pm 0.2 \text{ Hz}$$

Shift difference, PF_4–CF_3, 7.01 p.p.m.

The n.m.r. spectra of the hydrido-fluoro anions, $[CF_3PF_4H]^-$ and $[(CF_3)_2PF_3H]^-$, could be interpreted on a first-order basis, in terms of an octahedral structure. Both *cis* and *trans* isomers were observed for $[CF_3 PF_4H]^-$ in some instances, depending on the method of preparation. The *trans* structure was also most probable for $[(CF_3)_2PF_3H]^{- \ 346}$.

Magnitudes and signs of indirect spin coupling constants, relative to $^1J(CH) > 0$, for the anion $[MePF_5]^-$ were studied by the heteronuclear INDOR technique. It was shown that $^1J(PF_{axial}) < 0$, $^1J(PF_{equat.}) < 0$, and $^1J(PC) > 0$ [333].

According to Raman spectroscopic studies the structure of the ionic isomer of Br_2PF_3 is $[PBr_4][PF_6]$, both in the solid state and in solution in polar solvents[353]. The ion, $[PBr_4]^+$, has also been shown to exist in the compounds PBr_5 and PBr_7 which are to be formulated as $[PBr_4]Br$ and $[PBr_4][Br_3]$ [353].

References

1. Knoll, F. and Bergeshoff, G. (1966). *Monatsh. Chem.*, **97**, 808
2. Dimroth, K. and Hoffmann, R. (1964). *Angew. Chem.*, **76**, 433
3. Burg, A. B. and Mahler, W. (1961). *J. Amer. Chem. Soc.*, **83**, 2388
4. Kemmitt, R. D. W., McRae, V. M., Peacock, R. D. and Wilson, I. L. (1969). *J. Inorg. Nucl. Chem.*, **31**, 3674
5. Morse, K. W. and Parry, R. W. (1967). *J. Amer. Chem. Soc.*, **89**, 172
6. Kan, J. K. S., Morton, J. R. and Bernstein, H. J. (1966). *Can. J. Chem.*, **44**, 1957
7. Fessenden, R. W. and Schuler, R. H. (1966). *J. Chem. Phys.*, **45**, 1845
8. Wei, M. S., Current, J. H. and Gendell, O. (1970). *J. Chem. Phys.*, **52**, 1592
9. Nelson, W., Jackel, G. and Gordy, W. (1970). *J. Chem. Phys.*, **52**, 4572
10. Andrews, L. and Frederick, D. L. (1969). *J. Phys. Chem.*, **73**, 2774
11. Kokoszka, G. F. and Brinckman, F. E. (1968). *Chem. Commun.*, 349
12. Babkina, E. I. and Vereshkinskii, I. V. (1968). *Zh. obshchei Khim.*, **38**, 1772
13. Babkina, E. I. and Vereshkinskii, I. V. (1969). *Khim. Vysok. Energii*, **3**, 450
14. Sandoval, A. A., Moser, H. C. and Kiser, R. W. (1963). *J. Phys. Chem.*, **67**, 124
15. Halmann, M. and Klein, Y. (1964). *J. Chem. Soc.*, 4324; (1964) ASTM Mass Spectrometry *Symposium, Paris*; (1966) in *Advances in Mass Spectrometry*, Institute of Petroleum, vol. 3, p. 276
16. Ritter, H. (1855). *Annalen*, **95**, 210
17. Germann, F. E. E. and Traxler, R. N. (1927). *J. Amer. Chem. Soc.*, **49**, 309
18. Rudolph, R. W., Taylor, R. C. and Parry, R. W. (1966). *J. Amer. Chem. Soc.*, **88**, 3729
19. Rudolph, R. W. and Schiller, H. W. (1968). *J. Amer. Chem. Soc.*, **90**, 3581
20. Solan, D. and Timms, P. L. (1968). *Chem. Commun.*, 1540
21. Besson, A. and Fournier, L. (1910). *Compt. Rend.*, **150**, 102
22. Sandoval, A. A. and Moser, H. C. (1963). *Inorg. Chem.*, **2**, 27

23. Stock, A., Brandt, A. and Fischer, H. (1925). *Chem. Ber.*, **58**, 643
24. Finch, A. (1959). *Can. J. Chem.*, **37**, 1793
25. Teague, J. L. and Sandoval, A. A. (1969). *Radiochim. Acta*, **11**, 57
26. Baudler, M., Gehlen, O., Kipker, K. and Backes, P. (1967). *Z. Naturforsch.*, **22b**, 1354
27. Payne, D. S. (1967). *Topics in Phosphorus Chemistry*, vol. 4, p. 85. (New York: John Wiley & Sons)
28. Wyllie, D., Ritchie, M. and Ludlam, E. B. (1940). *J. Chem. Soc.*, 583
29. Baudler, M. and Wetter, G. (1964). *Z. Anorg. Allg. Chem.*, **329**, 3
30. Basco, N. and Jee, K. K. (1967). *Chem. Commun.*, 1146
31. Jolly, W. L., Lindahl, C. B. and Kapp, R. W. (1962). *Inorg. Chem.*, **1**, 958
32. Levchenko, E. S., Sheinkmann, I. E. and Kirsanov, A. V. (1959). *Zh. Obshchei Khim.*, **29**, 1474
33. Feshchenko, N. G. and Kirsanov, A. V. (1960). *Zh. Obshchei Khim.*, **30**, 3041
34. Leung, Y. C. and Waser, J. (1956). *J. Phys. Chem.*, **60**, 539
35. Baudler, M. and Fricke, G. (1963). *Z. Anorg. Allg. Chem.*, **320**, 11
36. Garrett, A. G. and Urry, G. (1963). *Inorg. Chem.*, **2**, 400
37. Lindahl, C. B. and Jolly, W. L. (1964). *Inorg. Chem.*, **3**, 1634
38. Auger, V. and Billy, M. (1904). *Compt. Rend.*, **139**, 597
39. Burg, A. B., Brendel, G., Caron, A. P., Juvinall, G. L., Mahler, W., Moedritzer, K. and Slota, P. J. (1957). *W.A.D.C. Tech. Report, 56*, Part II, (1962). *J. Inorg. Nucl. Chem.*, **24**, 319
40. Feshchenko, N. G., Alekseeva, T. J. and Kirsanov, A. V. (1963). *Zh. Obshchei Khim.*, **33**, 1013
41. Levchenko, E. S., Piven, Yu. V. and Kirsanov, A. V. (1960). *Zh. Obshchei Khim.*, **30**, 1976
42. Kirsanov, A. V. and Feshchenko, N. G. (1963). *USSR Patent* 149, 776/7; *Chem. Abs.*, **58**, 11401
43. Feshchenko, N. G., Iridionova, L. F., Korol, O. I., Kirsanov, A. V. (1970). *Zh. Obshchei Khim.*, **40**, 773
44. Gautier, A. (1874). *Compt. Rend.*, **78**, 286
45. Linke, K. H. (1964). *Z. Naturforsch.*, **19b**, 534
46. Feher, F., Hirschfeld, D. and Linke, K. H. (1962). *Acta Crystallogr.*, **15**, 1182
47. Tarible, M. (1901). *Compt. Rend.*, **132**, 204
48. Baudler, M. and Fricke, G. (1963). *Z. Anorg. Allg. Chem.*, **319**, 211
49. Baudler, M. and Meugel, M. (1970). *Z. Anorg. Allg. Chem.*, **374**, 159
50. Kolitowska, J. H. (1937). *Roczniki Chem.*, **17**, 616
51. Kolitowska, J. H. (1935). *Roczniki Chem.*, **15**, 29
52. Falius, H. (1963). *Z. Anorg. Allg. Chem.*, **326**, 79
53. Cowley, A. H. and Cohen, S. T. (1964). *Inorg. Chem.*, **3**, 780
54. Finch, A., Gardner, P. J. and Gupta, K. K. S. (1969). *J. Chem. Soc. A*, 2958
55. Burg, A. B. and Mahler, W. (1958). *J. Amer. Chem. Soc.*, **80**, 2334
56. Hayter, R. G. (1964). *Inorg. Chem.*, **3**, 711
57. Issleib, K. and Schwager, G. (1961). *Z. Anorg. Allg. Chem.*, **310**, 43; (1961) **311**, 83
58. Feshchenko, N. G., Kovaleva, T. V. and Kirsanov, A. V. (1969). *Zh. Obshchei Khim.*, **39**, 2188
59. Johnson, F. A. and Rudolph, R. W. (1967). *J. Chem. Phys.*, **47**, 5449
60. Cowley, A. H., White, W. D. and Damasco, M. C. (1969). *J. Amer. Chem. Soc.*, **91**, 1922
61. Finch, A., Hameed, A., Gardner, P. J. and Paul, N. (1969). *Chem. Commun.*, 391
62. Shanmugasundaram, G. and Nagarajan, G. (1968). *Monatsh. Chem.*, **100**, 789
63. Frankiss, S. G. and Miller, F. A. (1965). *Spectrochim. Acta*, **21**, 1235
64. Moeller, T. and Huheey, J. E. (1962). *J. Inorg. Nucl. Chem.*, **24**, 315
65. Cowley, A. H. and Cohen, S. T. (1965). *Inorg. Chem.*, **4**, 1200
65a. Nagarajan, G. and Sivaprakasam, R. (1970). *Acta Phys.*, **28**, 367
66. Frankiss, S. G., Miller, F. A., Stammreich, H. and Sans, Th. T. (1966). *Chem. Commun.*, 318
67. Kolditz, L. (1965). *Advan. Inorg. Chem. Radiochem.*, **7**, 5
68. Payne, D. S. (1967). *Topics in Phosphorus Chemistry*, vol. 4, p. 86, (New York: John Wiley & Sons)
69. Schmutzler, R. (1965). *Advan. Fluorine Chem.*, **5**, 31
70. Drozd, G. I. (1970). *Uspekhi Khim.*, **39**, 3

71. Nixon, J. F. (1970). *Advan. Inorg. Chem. Radiochem.*, **13**, 363
72. Nixon, J. F. (1968). *J. Chem. Soc. A*, 2689
73. Cavell, R. G. (1964). *J. Chem. Soc.*, 1992
74. Morse, J. G., Cohn, K., Rudolph, R. W. and Parry, R. W. (1967). *Inorg. Synth.*, **10**, 147
75. Müller, A., Niecke, E. and Glemser, O. (1967). *Z. Anorg. Allg. Chem.*, **350**, 256
76. Müller, A., Niecke, E., Krebs, B. and Glemser, O. (1968). *Z. Naturforsch.*, **23b**, 588
77. Delwaulle, M. L. and Bridoux, M. (1959). *Compt. Rend.*, **248**, 1342
78. Müller, A., Glemser, O. and Niecke, E. (1966). *Z. Naturforsch.*, **21b**, 732
79. Burg, A. B. and Slota, P. J. (1958). *J. Amer. Chem. Soc.*, **80**, 1107
80. Houben-Wehl (1964). *Methoden der organischen Chemie*, Bd. XII/2, p. 45 ff, (Stuttgart: Organische Phosphorverbindungen)
81. Nöth, H. and Vetter, H. J. (1963). *Chem. Ber.*, **96**, 1109
82. Nöth, H. and Vetter, H. J. (1961). *Chem. Ber.*, **94**, 1505
83. Gerrard, W. (1940). *J. Chem. Soc.*, 1464
84. Anschütz, L., Broeker, W. and Ohnheiser, A. (1944). *Ber. Deut. Chem. Ges.*, **77**, 439
85. Jefferson, R., Nixon, J. F. and Painter, T. M. (1969). *Chem. Commun.*, 622; (1969) 1263
86. van Dyke, C. H. (1968). *J. Inorg. Nucl. Chem.*, **30**, 81
87. Barlow, C. G. and Nixon, J. F. (1966). *J. Chem. Soc. A*, 228
88. Goya, A. E., Rosario, M. D. and Gilje, J. W. (1969). *Inorg. Chem.*, **8**, 725
89. Whigan, D. B., Gilje, J. W. and Goya, A. E. (1970). *Inorg. Chem.*, **9**, 1279
90. Roesky, H. W. (1969). *Inorg. Nucl. Chem. Lett.*, **5**, 891
91. Boranov, Yu. I., Filipov, O. F., Varshavskii, S. L. and Kabachnik, M. I. (1968). *Doklady Akad. Nauk SSSR*, **182**, 337
92. USSR Patent 210, 155
93. Smirnov, E. A., Zinovev, Yu. M., Petrunin, V. A. (1968). *Zh. Obshchei Khim.*, **38**, 1551
94. Levin, Ya. A., Galeev, V. S., Trutneva, E. K. (1967). *Zh. Obshchei Khim.*, **37**, 1872
95. Sommer, K. (1970). *Z. Anorg. Allg. Chem.*, **379**, 56
96. Sommer, K. (1970). *Z. Anorg. Chem.*, **376**, 37
97. Fild, M. and Schmutzler, R. (1970). *J. Chem. Soc. A*, 2359
98. Fild, M. and Schmutzler, R. (1969). *J. Chem. Soc. A*, 840
99. Ang, H. G. and Schmutzler, R. (1969). *J. Chem. Soc. A*, 702
100. Seel, F., Rudolph, K. and Gombler, W. (1967). *Angew. Chem.*, **79**, 686; Seel, F. and Rudolph, K. (1968). *Z. Anorg. Allg. Chem.*, **363**, 233; Seel, F., Rudolph, K. and Budeuz, R. (1965). *Z. Anorg. Allg. Chem.*, **341**, 196
101. Rudolph, R. W. and Parry, R. W. (1965). *Inorg. Chem.*, **4**, 1339
102. Schmidt, M. and Schröder, H. H. J., *Z. Anorg. Allg. Chem.*, in press
103. Schmidt, M. and Schröder, H. H. J. (1970). *Angew Chem.*, **82**, 808
104. Feshchenko, N. G., Melnichuk, E. A. and Kirsanov, A. V. (1969). *Zh. Obshchei Khim.*, **39**, 2139
105. Feshchenko, N. G., Kraleva, T. V. and Kirsanov, A. V. (1969). *Zh. Obshchei Khim.*, **39**, 2184
106. Brown, D. H., Crosbie, K. D., Fraser, G. W. and Sharp, D. W. A. (1969). *J. Chem. Soc. A*, 872
107. Krokhina, S. S., Pyrkin, R. I., Levin, Ya. A. and Ivanov, B. E. (1968). *Izv. Akad. Nauk SSSR, Ser. Khim.*, **6**, 1420
108. Holmes, R. R. and Wagner, R. P. (1963). *Inorg. Chem.*, **2**, 384
109. Brown, D. H., Crosbie, K. D., Fraser, G. W. and Sharp, D. W. A. (1969). *J. Chem. Soc. A*, 551
110. Spangenberg, S. F. and Sisler, H. H. (1969). *Inorg. Chem.*, **8**, 1006
111. Feshchenko, N. G., Gorbatenko, Zh. K. and Kirsanov, A. V. (1969). *Zhur. Obshchei Khim.*, **39**, 2596
112. Schmitz-Du Mont, O. and Kleiber, H. (1969). *Z. Anorg. Allg. Chem.*, **371**, 115
113. Horman, J. S. and Sharp, D. W. A. (1970). *J. Chem. Soc. A*, 1935
114. Ross, J. A. and Martz, M. D. (1969). *J. Org. Chem.*, **34**, 399
115. Ponomarev, V. V., Golubtsov, S. A., Andrianov, K. A. and Kondrashova, G. N. (1969) *Izv. Akad. Nauk SSSR, Ser. Khim.*, **7**, 1545
116. Sollott, G. P. and Peterson, W. R. (1969). *J. Organometal. Chem.*, **19**, 143
117. Brooks, R. and Bunton, C. A. (1970). *J. Org. Chem.*, **35**, 2642
118. Joshifuji, M., Fujishima, J., Okazaki, R. and Inamoto, V. (1970). *Chem. Ind. (London)*, 625
119. Lustig, M. (1968). *Inorg. Chem.*, **7**, 2054

120. Flaskerud, G. G., Pullen, K. E. and Shreeve, J. M. (1969). *Inorg. Chem.*, **8**, 728
121. Hagen, A. P. and MacDiarmid, A. G. (1970). *Inorg. Nucl. Chem. Lett.*, **6**, 413
122. Chaigneau, M. and Santarromana, M. (1971). *Compt. Rend.*, **272**, 70
123. Chaigneau, M. and Santarromana, M. (1969). *Compt. Rend.*, **286C**, 825
124. Chaigneau, M. and Santarromana, M. (1969). *Compt. Rend.*, **269C**, 1643
125. Smirnov, E. A., Kugazev, B. A., Petrunin, A. V. and Zinovev, Yu. M. (1969). *Zh. Neorg. Khim.*, **14**, 847
126. Clune, A. E. and Cohn, K. (1968). *Inorg. Chem.*, **10**, 2067
127. Ramaswamy, K. and Krishna Rao, B. (1969). *Z. Phys. Chem. (Leipzig)*, **241**, 18
128. Cyvin, S. J., Cyvin, B. N. and Müller, A. (1969). *J. Mol. Struct.*, **4**, 341
129. Pillai, M. G. K. and Pillai, P. P. (1968). *Indian J. Pure Appl. Phys.*, **6**, 404
130. Ramaswamy, K. and Krishna Rao, B. (1969). *Z. Phys. Chem. (Leipzig)*, **240**, 127
131. Ramaswamy, K. and Krishna Rao, B. (1969). *Z. Phys. Chem. (Leipzig)*, **242**, 155
132. Kim, S. T. and Overend, J. (1969). *J. Phys. Chem.*, **73**, 406
133. Stenzenberger, H. and Schindlbaur, H. (1970). *Spectrochim. Acta*, **26A**, 1707
134. Green, J. H. S. and Kynaston, W. (1969). *Spectrochim. Acta*, **25A**, 1677
135. Fild, M. and Holmes, R. R. (1971). *Spectrochim. Acta*, **27A**, 1525, 1537
136. Latscha, H. P., Hormuth, P. B. and Vollmer, V. (1969). *Z. Naturforsch.*, **24b**, 1237
137. Lucken, E. A. C. and Williams, D. F. (1969). *Mol. Phys.*, **16**, 17
138. de Ketelaere, R., Muylle, E., Vanermen, V., Claeys, E. and van der Kelen, G. P. (1969). *Bull. Soc. Chim. Belges*, **78**, 219
139. Imbery, D. and Friebolin, H. (1968). *Z. Naturforsch.*, **23b**, 759
140. Cowley, A. H., Dewar, M. J. S. and Jackson, W. R. (1968). *J. Amer. Chem. Soc.*, **90**, 4185
141. Goldwhite, H. and Rowsell, D. G. (1969). *Chem. Commun.*, 713
142. Ionov, S. P. and Ionova, G. V. (1969). *Zh. fiz. Khim.*, **43**, 825
143. Semin, G. K., Babushkina, T. A. and Svergun, V. I. (1969). *Uch. Zap. Mosk. Pedagog. Inst.*, **222**, 78
144. Hirsta, E. and Morino, Y. (1970). *J. Mol. Spectrosc.*, **33**, 460
145. Morino, Y., Kuchitzu, K. and Moritani, T. (1969). *Inorg. Chem.*, **8**, 867
146. Gillespie, R. J. (1963). *J. Chem. Soc.*, 4672
147. Gillespie, R. J. (1967). *Angew. Chem. Int. Edn. Engl.*, **6**, 819
148. Brown, R. D. and Peel, J. B. (1968). *Aust. J. Chem.*, **21**, 2605
149. Hillier, J. H. and Saunders, V. R. (1970). *Trans. Faraday Soc.*, **66**, 2401
150. Bassett, P. J., Lloyd, D. R., Hillier, I. H. and Saunders, V. R. (1970). *Chem. Phys. Lett.*, **6**, 253
151. Vilkov, L. V., Khaikin, L. S. and Evdokinov, V. V. (1966). *Dokl. Akad. Nauk SSSR*, **168**, 810
152. Morris, E. D. and Nordman, C. E. (1969). *Inorg. Chem.*, **8**, 1673
153. Schmidpeter, A. and Weingand, C. (1969). *Angew. Chem. Int. Edn. Engl.*, **8**, 615
154. Becke-Goehring, M. (1968). *Fortschr. Chem. Forsch.*, **10**, 207
155. Schmidpeter, A., Weingand, C. and Hafner-Roll, E. (1969). *Z. Naturforsch.*, **24b**, 799
156. Lustig, M. (1969). *Inorg. Chem.*, **8**, 443
157. Roesky, H. W. and Grimm, L. F. (1969). *Chem. Ber.*, **102**, 2319
158. Roesky, H. W. and Grimm, L. F. (1970). *Chem. Ber.*, **103**, 3114
159. Roesky, H. W. and Grimm, L. F. (1970). *Chem. Ber.*, **103**, 1664
160. Roesky, H. W. and Böwing, W. G. (1970). *Chem. Ber.*, **103**, 2281
161. Beerman, M. (1971). *Adv. Inorg. Chem. Radiochem.*, in press
162. Roesky, H. W., Holtschneider, G. and Giere, H. H. (1970). *Z. Naturforsch.*, **25b**, 252
163. Roesky, H. W. (1969). *Z. Anorg. Allg. Chem.*, **367**, 151
164. Roesky, H. W. and Grimm, L. F. (1970). *Angew. Chem. Int. Edn. Engl.*, **9**, 244
165. Moran, E. F. and Reider, D. P., U.S. *Pat.*, 3,467,704/1969
166. Kireev, V. V., Kolesnikov, G. S. and Raigorodskii, I. M. (1969). *Uspekhi Khim.*, **38**, 1504
167. Becke-Goehring, M. and Jung, D. (1970). *Z. Anorg. Allg. Chem.*, **372**, 233
168. Bodnarchuk, N. D. and Gavrilenko, B. B. (1969). *Zh. Obshchei Khim.*, **39**, 1961
169. Clipsham, R., Hart, R. M. and Whitehead, M. A. (1969). *Inorg. Chem.*, **8**, 2431
170. Stokes, H. N. (1895). *J. Amer. Chem. Soc.*, **17**, 275
171. Lehr, W. (1967). *Z. Anorg. Allg. Chem.*, **350**, 18
172. Lehr, W. (1969). *Z. Anorg. Allg. Chem.*, **371**, 225
173. Roesky, H. W. and Niecke, E. (1968). *Inorg. Nucl. Chem. Lett.*, **4**, 463
174. Glemser, O., Niecke, E. and Thamm, H. (1970). *Z. Naturforsch.*, **25b**, 754

175. Glemser, O., Niecke, E. and Roesky, H. W. (1969). *Chem. Commun.*, 282
176. Chivers, T. and Paddock, N. L. (1969). *Chem. Commun.*, 337
177. Steger, E. and Klemm, D. (1967). *J. Inorg. Nucl. Chem.*, **29,** 1812
178. Chapman, A. C., Paine, D. H., Searle, H. T., Smith, D. R. and White, R. F. M. (1961). *J. Chem. Soc.*, 1768
179. Niecke, E., Glemser, O. and Roesky, H. W. (1969). *Z. Naturforsch.*, **24b,** 1187
180. Green, B. and Sowerby, D. B. (1970). *J. Chem. Soc. A*, 987
181. Green, B. and Sowerby, D. B. (1969). *Chem. Commun.*, 628
182. Green, B. and Sowerby, D. B. (1969). *Inorg. Nucl. Chem. Lett.*, **5,** 989
183. Das, S. K., Feakins, D., Last, W. A., Nabi, S. N., Ray, S. K., Shaw, R. A. and Smith, B. C. (1970). *J. Chem. Soc. A*, 616
184. Sowerby, D. B. and Audrieth, L. F. (1961). *Chem. Ber.*, **94,** 2670
185. Lehr, W. (1969). *Naturwissenschaften,* **56,** 214
186. Lehr, W. and Pietschmann, J. (1970). *Chem-Ztg., Chem. App.*, **94,** 362
187. Douglas, W. M., Cooke, M., Lustig, M. and Ruff, J. K. (1970). *Inorg. Nucl. Chem. Lett.*, **6,** 409
188. Biddlestone, M. and Shaw, R. A. (1969). *J. Chem. Soc. A*, 178
189. Biddlestone, M., Shaw, R. A. and Taylor, D. (1969). *Chem. Commun.*, 320
190. Biddlestone, M. and Shaw, R. A. (1968). *Chem. Commun.*, 407
191. Cotson, S. and Hodd, K. A. (1969). *J. Inorg. Nucl. Chem.*, **31,** 245
192. Coxon, G. E. and Sowerby, D. B. (1969). *J. Chem. Soc. A*, 3012
193. Chivers, T. and Paddock, N. L. (1969). *J. Chem. Soc. A*, 1687
194. Shokol, V. A., Kisilenko, A. A. and Derkach, G. I. (1969). *Zh. Obshchei Khim.*, **39,** 874
195. Khomenko, D. P., Kozlov, E. S. and Dyadyusha, G. G. (1970). *Spectroscopy Lett.*, **3,** 129
196. Khomenko, D. P., Dyadyusha, G. G. and Kozlov, E. S. (1970). *Zh. Strukt. Khim.*, **11,** 660
197. Ionin, B. I. (1968). *Zh. Obshchei Khim.*, **38,** 1659
198. Lund, L. G., Paddock, N. L., Proctor, J. E. and Searle, H. T. (1960). *J. Chem. Soc.*, 2542
199. Schlueter, A. W. and Jacobson, R. A. (1968). *J. Chem. Soc. A*, 2317
200. Lanoux, S. (1971). *J. Inorg. Nucl. Chem.*, **33,** 279
201. Latscha, H. P. (1968). *Z. Anorg. Allg. Chem.*, **362,** 7
202. Kaplansky, M. and Whitehead, M. A. (1967). *Can. J. Chem.*, **45,** 1669
203. Adamson, G. W. and Bart, J. C. J. (1969). *Chem. Commun.*, 1036
204. Cox, J. W. and Corey, E. R. (1967). *Chem. Commun.*, 123
205. Bullen, G. J., Mallinson, P. R. and Burr, A. H. (1969). *Chem. Commun.*, 691
206. Davies, M. J. and Paul, J. W. (1969). *Acta Crystallogr. Suppl.*, **25A,** S116
207. Falius, H. (1970). *Angew. Chem.* **82,** 702; (1970). *Angew. Chem. Int. Edn. Engl.*, **9,** 733
208. Falius, H. (1969). *Z. Anorg. Allg. Chem.*, **365,** 51
209. Charlton, T. L. and Cavell, R. G. (1969). *Inorg. Chem.*, **8,** 281
210. Fühler, K. (1959). *Dissertation.* Technische Hochschule, Stuttgart
211. Bühler, K. (1969). *Z. Naturforsch.*, **24b,** 1484
212. Falius, H. (1968). *Angew. Chem.*, **80,** 616
213. Tebbe, F. N. and Muetterties, E. L. (1970). *Inorg. Chem.*, **9,** 629
214. Houk, L. W. and Lustig, M. (1970). *Inorg. Chem.*, **9,** 2462
215. Colburn, C. B., Hill, W. E. and Sharp, D. W. A. (1970). *J. Chem. Soc. A*, 2221
216. Charlton, T. L. and Cavell, R. G. (1970). *Inorg. Chem.*, **9,** 379
217. Keat, R. (1970). *J. Chem. Soc. A*, 2732
218. Centofanti, L. F. and Parry, R. W. (1970). *Inorg. Chem.*, **9,** 744
219. Roesky, H. W. (1969). *Z. Naturforsch.*, **24b,** 818
220. Roesky, H. W. (1968). *Chem. Ber.*, **101,** 636
221. Lindner, E. and Beer, H. (1970). *Chem. Ber.*, **103,** 2802
222. Lindner, E. and Schless, H. (1960). *Chem. Ber.*, **99,** 3331
223. Lindner, E. (1970). *Angew. Chem.*, **82,** 143; (1970) *Angew. Chem. Int. Edn., Engl.*, **9,** 114
224. Paul, R. C., Khurana, H., Vasisht, S. K. and Chadher, S. L. (1969). *J. Indian Chem. Soc.*, **46,** 915
225. Glemser, O., Biermann, U. and von Halasz, S. P. (1969). *Inorg. Nucl. Chem. Lett.*, **5,** 501
226. Glemser, O., Biermann, U. and von Halasz, S. P. (1969). *Inorg. Nucl. Chem. Lett.*, **5,** 643
227. Biermann, U. and Glemser, O. (1969). *Chem. Ber.*, **102,** 3342
228. Gutmann, V. and Imhof, J. (1970). *Monatsh. Chem.*, **101,** 1
229. Suvorov, A. V. and Shubaev, V. L. (1970). *Zh. Neorg. Khim.*, **15,** 1181
230. Kidd, R. G. and Truax, D. R. (1969). *Chem. Commun.*, 160

231. Oliver, J. G. and Worall, I. J. (1970). *J. Chem. Soc. A*, 848
232. van der Veer, W. and Jellinek, F. (1970). *Rec. Trav. Chim.*, **89**, 833
233. Gutmann, V. (1967). *Halogen Chemistry*, vol. 2, p. 399, (New York: Academic Press)
234. Paul, C. R. and Vasisht, S. K. (1966). *J. Indian Chem. Soc.*, **43**, 141
235. Selig, H. and Aminadov, N. (1970). *Inorg. Nucl. Chem. Lett.*, **6**, 595
236. Hägele, G. and Kuchen, W. (1970). *Chem. Ber.*, **103**, 2885
237. Kosfeld, R., Hägele, G. and Kuchen, W. (1968). *Angew. Chem. Int. Edn.*, **7**, 814
238. Kuznetsov, N. T. and Khimchuk, G. S. (1969). *Zh. Neorg. Khim.*, **14**, 2705
239. Roesky, H. W. (1970). *J. Inorg. Nucl. Chem.*, **32**, 1845
240. Binder, H. and Fluck, E. (1969). *Z. Anorg. Allg. Chem.*, **365**, 170
241. Shlyapochnikov, V. A., Strukov, O. G., Dubov, S. S. and Shitov, L. N. (1963). *Zh. Neorg. Khim.*, **14**, 2913
242. Cyvin, S. J., Vizi, B., Muller, A. and Krebs, B. (1969). *J. Mol. Struct.*, **3**, 173
243. Shokol, V. A., Kisilenko, A. A. and Derkach, G. I. (1969). *Zh. Obshchei Khim.*, **39**, 1485
244. Nyquist, R. A. and Muelder, W. W. (1968). *J. Mol. Struct.*, **2**, 465
245. Durig, J. R. and Clark, J. W. (1970). *J. Chem. Phys.*, **50**, 107
246. Hill, W. E., Sharp, D. W. A. and Colburn, C. B. (1969). *J. Chem. Phys.*, **50**, 612
247. Cavell, R. G., Charlton, T. L. and Pinkerton, A. A. (1969). *Chem. Commun.*, 424
248. Reddy, G. S. and Schmutzler, R. (1970). *Z. Naturforsch.*, **25b**, 1199
249. Mitchell, K. A. R. (1968). *Can. J. Chem.*, **46**, 3499
250. Olie, K. and Mijlhoff, F. C. (1969). *Acta Crystallogr.*, **B25**, 974
251. Vilkov, L. V., Khaikin, L. S., Vasilev, A. F. and Tulyakova, T. F. (1968). *Zh. Strukt. Khim.*, **9**, 1071
252. Kruck, T. (1967). *Angew. Chem.*, **79**, 27
253. Burg, A. B. and Street, G. B. (1966). *Inorg. Chem.*, **5**, 1532
254. Nixon, J. F. (1967). *J. Chem. Soc. A*, 1136
255. Reddy, G. S. and Schmutzler, R. (1967). *Inorg. Chem.*, **6**, 823
256. Clark, R., Hargader, J. P., Haas, H. and Sheline, R. K. (1968). *Inorg. Chem.*, **7**, 673
257. Barlow, C. G., Nixon, J. F. and Webster, M. (1968). *J. Chem. Soc. A*, 2216; Barlow, C. G. Nixon, J. F. and Swain, J. R. (1969). *J. Chem. Soc. A*, 1082
258. Johnson, T. R. and Nixon, J. F. (1969). *J. Chem. Soc. A*, 2518
259. Timms, P. L. (1969). *Chem. Commun.*, 1033
260. Timms, P. L. (1970). *J. Chem. Soc. A*, 2526
261. Kruck, T. and Prasch, A. (1969). *Z. Anorg. Allg. Chem.*, **371**, 1
262. Kruck, T., Lang, W., Derner, N. and Stadler, M. (1968). *Chem. Ber.*, **101**, 3816
263. Strohmeier, W. and Müller, F. J. (1969). *Chem. Ber.*, **102**, 3608
264. Gondal, S. K., MacDiarmid, A. G., Saalfeld, F. E. and McDowel, M. V. (1969). *Inorg. Nucl. Chem. Lett*, **5**, 351
265. Udovich, C. A. and Clark, R. J. (1969). *Inorg. Chem.*, **8**, 938
266. Kruck, T. and Baur, K. (1969). *Z. Anorg. Allg. Chem.*, **364**, 192
267. Nixon, J. F. and Sexton, M. D. (1970). *J. Chem. Soc. A*, 321
268. Udovich, C. A. and Clark, R. J. (1970). *J. Organometal. Chem.*, **25**, 199
269. Bennett, M. A. and Patmore, D. J. (1969). *Chem. Commun.*, 1510
270. Clement, D. A., Nixon, J. F. and Sexton, M. D. (1969). *Chem. Commun.*, 1509
271. Grobe, J. and Kober, F. (1969). *Z. Naturforsch.*, **24b**, 1660
272. Fischer, E. O. and Knauss, L. (1969). *Chem. Ber.*, **102**, 223
273. Udovich, C. A. and Clark, R. J. (1969). *J. Amer. Chem. Soc.*, **91**, 526
274. Udovich, C. A., Clark, R. J. and Haas, H. (1969). *Inorg. Chem.*, **8**, 1066
275. Nixon, J. F., Murray, M. and Schmutzler, R. (1970). *Z. Naturforsch.*, **25b**, 110
276. Chatt, J. and Heaton, B. T. (1968). *J. Chem. Soc. A*, 2745
277. Kruck, T., Hoefler, M., Jung, H. and Blume, H. (1969). *Angew. Chem. Int. Ed. Engl.*, **8**, 522
278. Johnston, R. D., Basolo, F. and Pearson, R. G. (1971). *Inorg. Chem.*, **10**, 247
279. Johnson, T. R., Lynden-Bell, R. M. and Nixon, J. F. (1970). *J. Organometal. Chem.*, **21**, P15
280. Jenkins, J. M. and Shaw, B. L. (1963). *Proc. Chem. Soc.*, 279
281. Ogilvie, F., Clark, R. J. and Verkade, J. G. (1969). *Inorg. Chem.*, **8**, 1904
282. Mathieu, R., Lenzi, M. and Poilblanc, R. (1970). *Inorg. Chem.*, **8**, 2030
283. Harris, R. K., Woplin, J. R. and Schmutzler, R. (1971). *Ber. Bunsengesellschaft Phys. Chem.*, **75**, 134

284. Fischer, E. O., Louis, E., Bathelt, W., Moser, E. and Müller, J. (1968). *J. Organometal. Chem.*, **14,** P9; Fischer, E. O., Louis, E. and Schneider, R. J. J. (1968). *Angew. Chem.*, **80,** 122
285. Poilblanc, R. and Bigorgne, M. (1962). *Bull. Soc. Chim.*, **29,** 1301
286. Strohmeier, W. and Müller, F. J. (1967). *Chem. Ber.*, **100,** 2812
287. Edwards, H. G. M. and Woodward, L. A. (1970). *Spectrochim. Acta*, **26A,** 897
288. Edwards, H. G. M. and Woodward, L. A. (1970). *Spectrochim. Acta*, **26A,** 1077
289. Kahn, O. (1969). *J. Chim. Phys.*, **66,** 1869
290. Kahn, O. and Bigorgne, M. (1969). *J. Chim. Phys.*, **66,** 874
291. Cassoux, P. and Labarre, J. F. (1969). *J. Chim. Phys.*, **66,** 1420
292. Marriott, J. C., Salthouse, J. A., Ware, M. J. and Freeman, J. M. (1970). *Chem. Commun.*, 595
293. Almeninngen, A., Andersen, B. and Astrup, E. E. (1970). *Acta Chem. Scand.*, **24,** 1579
294. Heller, J. (1969). *Chimia*, **23,** 351
295. Schmutzler, R. (1967). *Halogen Chemistry*, vol. 2, p. 31, Ed. by Gutmann, V. (London: Academic Press)
296. Ang, H. G. (1969). *J. Inorg. Nucl. Chem.*, **31,** 3311
297. Nixon, J. F. (1969). *J. Inorg. Nucl. Chem.*, **31,** 1615
298. Blaser, B. and Worms, K. H. (1968). *Z. Anorg. Chem.*, **361,** 15
299. Haiduc, I. (1970). *The Chemistry of Inorganic Ring Systems*, part 2, p. 787. (New York: Wiley–Interscience)
300. Grapov, A. F., Mel'nikov, N. N. and Razvodovskaya, L. V. (1970). *Uspekhi Khim.*, **39,** 39
301. Becke-Goehring, M. (1970). *Chem.-Ztg.*, *Chem. App.*, **94,** 179
302. Becke-Goehring, M. and Weber, H. (1969). *Z. Anorg. Allg. Chem.*, **365,** 185
303. Utvary, K. and Czysch, W. (1969). *Monatsh. Chem.*, **100,** 681
304. Becke-Goehring, M. and Wald, H. J. (1969). *Z. Anorg. Allg. Chem.*, **371,** 88
305. Ebeling, J. and Schmidpeter, A. (1969). *Angew. Chem.*, **81,** 707
306. Maier, L. and Schmutzler, R. (1969). *Chem. Commun.*, 961
307. Doak, G. O. and Schmutzler, R. (1970). *Chem. Commun.*, 476
308. Peake, S. C. and Schmutzler, R. (1968). *Chem. Commun.*, 665
309. Fild, M., Hewson, M. J. C., Peake, S. C. and Schmutzler, R. (1971). *Inorg. Chem.*, **10,** in press
310. Peake, S. C. and Schmutzler, R. (1970). *J. Chem. Soc. A*, 1049
311. Brown, D. H., Crosbie, K. D., Darragh, J. I., Ross, D. S. and Sharp, D. W. A. (1970). *J. Chem. Soc. A*, 914
312. Lustig, M. and Roesky, H. W. (1970). *Inorg. Chem.*, **9,** 1289
313. Blazer, T. A., Schmutzler, R. and Gregor, I. K. (1969). *Z. Naturforsch*, **24b,** 1081
314. Kaplansky, M., Clipsham, R. and Whitehead, M. A. (1969). *J. Chem. Soc. A*, 584
315. Wyatt, R., Roberts, J. T., Wentz, R. E. and Wilt, P. M. (1969). *J. Chem. Phys.*, **50,** 2552
316. Selig, H., Halloway, J. H., Tyson, J. and Claasen, H. H. (1970). *J. Chem. Phys.*, **53,** 2559
317. Lockett, P., Fowler, W. and Wilt, P. M. (1970). *J. Chem. Phys.*, **53,** 452
318. Levin, I. W. (1970). *J. Mol. Spectrosc.*, **33,** 61
319. Deiters, R. M. and Holmes, R. R. (1968). *J. Chem. Phys.*, **48,** 4796
320. Beattie, I. R., Livingston, K. M. S. and Reynolds, D. J. (1969). *J. Chem. Phys.*, **51,** 4269
321. Holmes, R. R., Deiters, R. M. and Golen, J. A. (1969). *Inorg. Chem.*, **8,** 2612
322. Ramaswamy, K. and Krishna Rao, B. (1969). *Z. Phys. Chem.*, **242,** 215
323. Holmes, R. R. and Fild, M. (1970). *J. Chem. Phys.*, **53,** 4161
324. Norbury, A. H., Peake, S. C. and Schmutzler, R. (1971). *Spectrochim. Acta*, **27A,** 151
325. Wiley, G. A. and Stine, W. R. (1967). *Tetrahedron Lett.*, 2321
326. Arzoumanidis, G. G. (1969). *Chem. Commun.*, 217
327. Fleig, H. and Becke-Goehring, M. (1970). *Z. Anorg. Allg. Chem.*, **376,** 215
328. Ramaswamy, K. and Krishna Rao, B. (1969). *Z. Phys. Chem.*, **242,** 18
329a. Muetterties, E. L., Mahler, W. and Schmutzler, R. (1963). *Inorg. Chem.*, **2,** 613;
329b. Muetterties, E. L., Mahler, W., Packer, K. J. and Schmutzler, R. (1964). *Inorg. Chem.*, **3,** 1298
330. Peake, S. C. and Schmutzler, R. (1969). *Colloques Internationaux de Centre National de la Recherche Scientifique* (C.N.R.S.), No. 182, 99
331. Dunmur, R. E., Murray, M., Schmutzler, R. and Gagnaire, D. (1970). *Z. Naturforsch.*, **25b,** 903
332. Furtsch, T. A., Dierdorf, D. S. and Cowley, A. H. (1970). *J. Amer. Chem. Soc.*, **93,** 5759

333. Dreeskamp, H., Schumann, C. and Schmutzler, R. (1970). *Chem. Commun.*, 671
334. Almenningen, A., Andersen, B. and Astrup, E. E. (1969). *Acta Chem. Scand.*, **23,** 2179
335. Brokke, M. E., Lukes, G. E. and Arneklev, D. R. *U.S. Pat.*, 3 375 277/1968
336. Brokke, M. E., Lukes, G. E. and Arneklev, D. R. *U.S. Pat.*, 3 449 374/1969
337. Brokke, M. E., Lukes, G. E. and Arneklev, D. R. *U.S. Pat.*, 3 455 674/1969
338. Brokke, M. E., Lukes, G. E. and Arneklev, D. R. *U.S. Pat.*, 3 435 043/1969
339. Brokke, M. E., Lukes, G. E. and Arneklev, D. R. *U.S. Pat.*, 3 438 983/1969
340. Harman, J. S. and Sharp, D. W. A. (1970). *J. Chem. Soc. A,* 1138
341. Lunazzi, L. and Brownstein, S. (1969). *J. Magnetic Resonance,* **1,** 119
342. Jander, J. and Börner, D. (1969). *Annalen,* **726,** 13
343. Peacock, R. D. and Wilson, I. L. (1969). *J. Chem. Soc. A,* 2030
344. Azeem, M., Brownstein, M. and Gillespie, R. J. (1969). *Can. J. Chem.,* **47,** 4159
345. Nixon, J. F. and Swain, J. R. (1969). *Inorg. Nucl. Chem. Lett.,* **5,** 295
346. Nixon, J. F. and Swain, J. R. (1970). *J. Chem. Soc. A,* 2075
347. Jander, J., Börner, D. and Engelhardt, U. (1969). *Annalen,* **726,** 19
348. Chan, S. S. and Willis, C. J. (1968). *Can. J. Chem.,* **46,** 1237
349. Stadelmann, W., Stelzer, O. and Schmutzler, R., unpublished work
350. Copeland, R. F., Conner, S. H. and Meyers, E. A. (1966). *J. Phys. Chem.,* **70,** 1288
351. Peake, S. C., Hewson, M. J. C. and Schmutzler, R. (1970). *J. Chem. Soc. A,* 2364
352. Bishop, E. O., Carey, R. P., Nixon, J. F. and Swain, J. R. (1970). *J. Chem. Soc. A,* 1074
353. Dhamelincourt, P. and Crunelle-Cras, M. (1970). *Compt. Rend.,* **271B,** 124

4
Phosphonitriles

D. B. SOWERBY
University of Nottingham

4.1   INTRODUCTION

Phosphonitriles or phosphazenes belong to a class of phosphorus–nitrogen compounds which contain within their structure a ring consisting of equal numbers of alternate phosphorus and nitrogen atoms. Each phosphorus atom carries two substituents and the formal valence requirements of the ring atoms are completed by π-bonding within the ring. The best known compounds are the trimeric and tetrameric chlorides, $P_3N_3Cl_6$ (1) and $P_4N_4Cl_8$ (2), but compounds with larger rings are known.

(1)   (2)

The definition chosen is more restricted than that sometimes used but during the past years considerable advances have been made in all aspects of the chemistry of phosphorus–nitrogen compounds that such a restriction is necessary.

Earlier work has been summarised in reviews by Paddock[1], Shaw and co-workers[2,3], and Schmulbach[4] and in books by Haiduc[5], Allcock[6], and Pantel and Becke-Goehring[7]. General preparative aspects have been reviewed[8], and *Inorganic Synthesis* contain details for the preparation of alkoxy and aryloxy[9], mercapto[10], fluoro[11,12] phenylbromo[13] and phenyl-fluoro[14] derivatives. The present review covers the literature of the past 5 years up to March 1971.

4.2   PREPARATION OF PHOSPHONITRILES

4.2.1   From phosphorus halides

4.2.1.1   *Chlorides*

The original preparation of chlorophosphonitriles:

$$nPCl_5 + nNH_4Cl \rightarrow (PNCl_2)_n + 4nHCl \qquad (4.1)$$

involved the direct interaction between phosphorus pentachloride and ammonium chloride. This has been re-investigated to show that the highest yields result when mole ratios of 1:4 are used and the ammonium chloride is finely divided[15]. A new approach which can give up to 65% yields of petrol-soluble cyclic compounds is to carry out the reaction in the presence of four moles of pyridine which acts as a hydrogen chloride acceptor[16]. The main advantage is that reaction is complete within minutes rather than requiring several hours as when carried out conventionally in solvents such 1,1,2,2-tetrachloroethane or chlorobenzene.

Various methods have, however, been used to decrease reaction times when a solvent is used.

(a) By use of finely-divided ammonium chloride[17]. This, as prepared by the gas phase reaction between ammonia and hydrogen chloride, had a mean diameter of 13 μm.

(b) By addition of various metal chlorides which behave as catalysts[18, 19]. Anhydrous Co^{II}, Mn^{II}, Cu^{II}, Ti^{IV}, Al^{III}, Sn^{IV} and Zn^{II} chlorides have been used and in some cases the reaction time can be cut to a third that of a non-catalysed reaction. Zinc chloride is apparently particularly useful and gives an increased yield[20]. The mechanism for catalysis is not known with certainty but the active species are those which can coordinate to nitrogen atoms. The presence of metal halides may also promote the formation of cationic species which are precursors of the cyclic phosphonitriles. For example, in a reaction with zinc chloride, species of the type $[Cl_3P(NPCl_2)_nCl]^+$ $ZnCl_3^-$ with $n = 1-10$ have been prepared[21]. The distribution of products between linear and cyclic compounds can also be affected and recently $MoCl_5$ has been shown to promote the formation of the trimer[22].

(c) By addition of metals[23, 24]. These are probably converted to chlorides during the reaction but some have been shown to accelerate the loss of hydrogen chloride (Zn, Co, Al, Cu and Fe) while others retard the reaction (Ni, Mg, Ti, Mn and Sn). As noted above, the product distribution can also be changed.

(d) By addition of phosphoryl chloride[25, 26]. In addition to reducing the time required for complete reaction to approximately 3 h, this catalyst also increases the yield of cyclic compounds from $\approx 60\%$ in uncatalysed reactions to $\approx 90\%$. Hydrated metal salts, $MgSO_4 \cdot H_2O$ and $CuCl_2 \cdot 2H_2O$, are also effective in reducing reaction times, but in these cases the proportion of linear products is much higher.

The nature of the linear products and the products that can be formed in reactions with metal halides has been elucidated by examining the reaction between phosphorus pentachloride and the trimeric and tetrameric chloro-phosphonitriles[27]. In each case, ^{31}P n.m.r. spectroscopy showed that a single compound was obtained, i.e. $[Cl(Cl_2P=N)_nPCl_3]^+PCl_6^-$ where $n = 3$ or 4 respectively, which on treatment with boron or aluminium trichlorides exchanged the hexachlorophosphate anion for either BCl_4^- or $AlCl_4^-$. The products were heat-stable liquids. A recent patent[28] has shown that petrol-insoluble linear polymers can be converted to petrol-soluble species by treatment with further ammonium chloride.

Although ammonium chloride is usually the source of the nitrogen in the phosphorus pentachloride reaction, other compounds have been used.

Good yields of cyclic products can be obtained by use of gaseous ammonia if the rate of addition is controlled[29], and mercury amido-chloride, $Hg(NH_2)Cl$, has been shown by g.l.c. and ^{31}P n.m.r. spectroscopy to give the compounds $(PNCl_2)_{3-8}$ [30]. The active phosphorus species in the PCl_5—NH_4Cl reaction has been identified as the PCl_4^+ ion[31] by showing that reactivity increased along the series $PCl_6^- < PCl_5 < py \cdot PCl_5 < PCl_4^+$.

The ammonolysis of phosphorus pentachloride is an exceptionally complex process depending markedly on the conditions of the reaction and new products, e.g. $P(NH_2)_4Cl$ [32] and $[(H_2N)_3P{=}N{-}P(NH_2)_3]^+Cl^-$ [33], continue to be isolated. However, the mechanism postulated for the reactions leading to chlorophosphonitriles[34] has received strong support from a careful examination of the rate of hydrogen chloride evolution and the change in conductivity and ^{31}P n.m.r. spectra as the reaction proceeds[35]. There are two well-defined steps. Initially, $[Cl_3P{=}N{-}PCl_3]^+PCl_6^-$ is formed and a maximum in the hydrogen chloride evolution curve can be correlated with the reaction:

$$3PCl_5 + NH_4Cl \rightarrow P_3NCl_{12} + 4HCl \qquad (4.2)$$

In the early stages, ^{31}P n.m.r. spectroscopy shows that two compounds containing one type of phosphorus atom are present. These may be PCl_4NH_2 and PCl_3NH and reactions (4.3)–(4.6), which lead to the overall reaction (4.2), have been suggested. The second stage involves chain growth and cyclisation.

$$PCl_5 + NH_4Cl \rightarrow PCl_4NH_2 + 2HCl \qquad (4.3)$$
$$PCl_4NH_2 \rightarrow PCl_3NH + HCl \qquad (4.4)$$
$$PCl_3NH + PCl_5 \rightarrow PCl_3NPCl_4 + HCl \qquad (4.5)$$
$$PCl_3NPCl_4 + PCl_5 \rightarrow [PCl_3{=}N{-}PCl_3][PCl_6] \qquad (4.6)$$

The former probably involves attack of PCl_3NH, arising from reaction with further ammonium chloride, on the cation $P_2NCl_6^+$ (4.7)

$$Cl_3P{=}N{-}PCl_3^+ + PCl_3NH \rightarrow Cl_3P{=}N{-}PCl_2{=}N{-}PCl_3^+ + HCl \ (4.7)$$

and, when the concentration has increased, on $P_3N_2Cl_8^+$ also

$$Cl_3P{=}N{-}PCl_2{=}N{-}PCl_3^+ + PCl_3NH \rightarrow$$
$$Cl_3P{=}N{-}PCl_2{=}N{-}PCl_2{=}N{-}PCl_3^+ + HCl \qquad (4.8)$$

Eventually the solution contains only $P_2NCl_6^+$, $P_3N_2Cl_8^+$ and $P_4N_3Cl_{10}^+$ and further chain growth arises by these units combining as shown in equations (4.9) and (4.10), probably via imine groups at the ends of chains. Chain growth can be followed by ^{31}P n.m.r. and these reactions are correlated with a second maximum in the rate of hydrogen chloride evolution and with the maximum in the conductivity curve.

$$Cl_3P{-}(NPCl_2)_n{-}NPCl_3^+Cl^- + NH_4Cl \rightarrow$$
$$Cl_3P{-}(NPCl_2)_n{-}NPCl_2{=}NH + 3HCl \qquad (4.9)$$

$$Cl_3P{-}(NPCl_2)_n{-}NPCl_2{=}NH + Cl_3P{-}(NPCl_2)_m{-}NPCl_3^+ \rightarrow$$
$$Cl_3P{-}(NPCl_2)_{n+m}{-}NPCl_3^+ + HCl \qquad (4.10)$$

Trimeric chlorophosphonitrile probably arises by elimination of PCl_4^+

from $P_4N_3Cl_{10}^+$ by a mechanism similar to that postulated by Kobayashi[36], and similar processes would give rise to the other cyclic species.

4.2.1.2 Bromides

Bromophosphonitriles, $(PNBr_2)_n$, can be prepared by similar reactions between phosphorus pentabromide and ammonium bromide but, if tetrachloroethane is the solvent, the products are contaminated by small amounts of species containing one or two chlorine atoms[37]. This can be avoided if 1,2-dibromoethane is used. As for the chloride system, the major products are the trimer and tetramer ($n = 3$, 4) but small amounts of pentamer[38] and hexamer[37] can also be isolated. The reaction requires long periods of time for completion and, as yet, no effective catalysts have been reported.

4.2.1.3 Bromide chlorides

Mixed bromochlorotriphosphonitriles were obtained initially by Rice and co-workers[39] from reactions between phosphorus pentachloride and ammonium bromide or phosphorus tribromide, bromine and ammonium bromide. Attempts to separate pure compounds by fractional crystallisation from the resulting mixtures were not successful, but by g.l.c. compounds with the stoichiometry, $P_3N_3Cl_nBr_{6-n}$ where $n = 0-6$, were shown to be present[40, 41]. Infrared and ^{31}P n.m.r. spectra indicated that the compounds separated were mixtures of the geminal and non-geminal isomers[42, 43]. These arise due to different arrangements of the halogen atoms and three compounds (3), (4) and (5) are possible for the stoichiometry $P_3N_3Br_2Cl_4$. Compound (3) is a geminal isomer and (4) and (5) are respectively the *cis* and *trans* non-geminal isomers.

(3)   (4)   (5)

The pure non-geminal compounds can, however, be separated by preparative g.l.c. from the mixture which results from the chlorination of molten hexabromotriphosphonitrile with mercury(II) chloride[43].

4.2.1.4 Phenyl chlorides and bromides

Trimeric and tetrameric phosphonitriles, non-geminally substituted by chlorine and phenyl groups, result when phenyltetrachlorophosphorane, $PhPCl_4$, reacts either in the melt or in chlorobenzene solution with ammonium chloride. Two isomeric trimers were previously isolated[44] and Grushkin and co-workers[45] have isolated three of the four possible non-geminal tetramers. The compounds were identified by ^{1}H n.m.r. spectroscopy of amine derivatives and dielectric constant measurements etc., and are shown in formulae (6), (7) and (8); compound (9) was not observed. In the formulae, the corners of the square represent the phosphorus atoms of the P_4N_4 ring and the disposition above (*cis*) or below (*trans*) the ring can be shown.

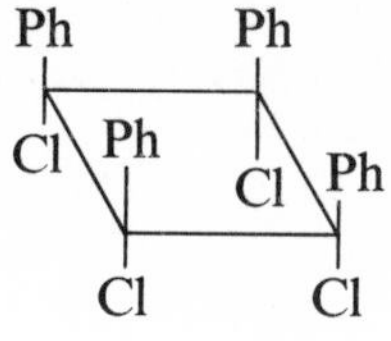

(6) 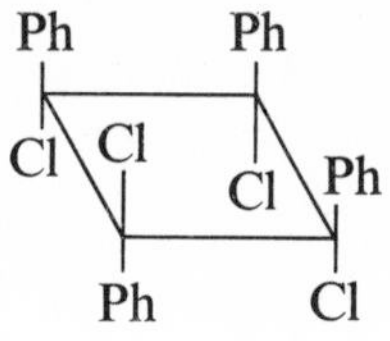 (7)

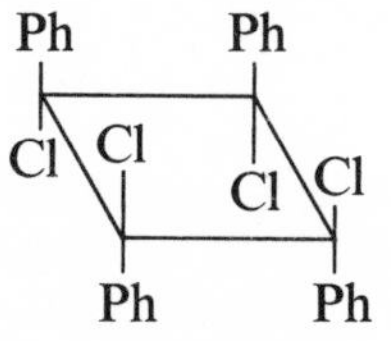

(8) 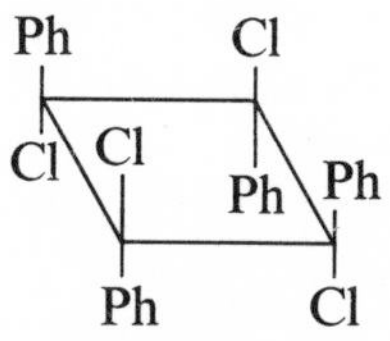 (9)

Phenyldibromophosphine, bromine and ammonium bromide give the *cis* and *trans* non-geminal trimers, $Ph_3Br_3P_3N_3$, which can be interconverted by heating solutions in acetonitrile (*trans* → *cis*) or bromobenzene (*cis* → *trans*)[46].

4.2.1.5 Fluoroalkyls

Bis(heptafluoropropyl)trichlorophosphorane reacts[47] with ammonium chloride in refluxing 1,1,2,2-tetrachloroethane to give the trimer, $P_3N_3(C_3F_7)_6$, and tetramer, $P_4N_4(C_3F_7)_8$.

There has been little work on the isolation of possible intermediates from these reactions with substituted halophosphoranes, but the reaction mechanisms are expected to be very similar to that discussed above for phosphorus pentachloride itself.

4.2.2 From iminodiphosphoranes

The linear compound, $[Ph_2P(NH_2)\!\!=\!\!N\!\!-\!\!P(NH_2)Ph_2]Cl$, first prepared by Bezmann and Smalley[48], is a strong electrolyte in acetonitrile[49] and a variety of substituted triphosphonitriles can be prepared by elimination of hydrogen chloride with a chlorophosphorane (equation (4.11)). With dimethyltrichlorophosphorane, ring closure occurs and the hydrochloride of 1,1-dimethyl-3,3,5,5-tetraphenyltriphosphonitrile (10, R = R' = Me) can be isolated[50].

$$[Ph_2P(NH_2)=N-P(NH_2)Ph_2]Cl + RR'PCl_3 \longrightarrow 4\,HCl +$$

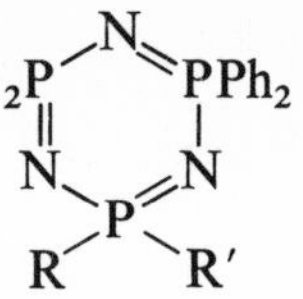

(10) (4.11)

A similar method has been used to prepare the first optically active phosphonitrile[51]. Phenyltolylchlorophosphine on treatment with chlorine and ammonia yields racemic $[Ph(CH_3 \cdot C_6H_4)(NH_2)P=N-P(NH_2)(CH_3 \cdot C_6H_4)Ph]^+Cl^-$ which can be resolved via the α-bromo-d-camphor-π-sulphonate. On treatment with phosphorus pentachloride in benzene, the d-form gives compound (11) with a d-configuration ($[\alpha_{5461}^{32}] = +4.2°$ in methanol).

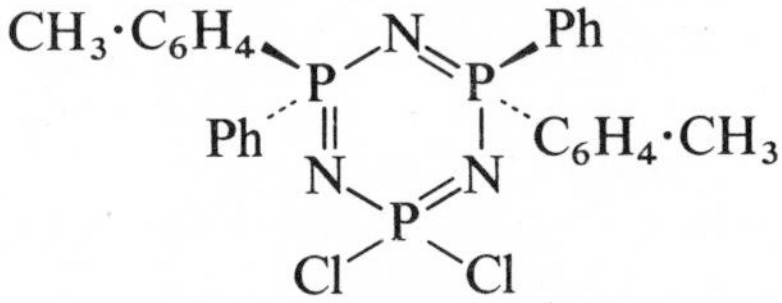

(11)

Ring closure reactions with the iminodiphosphorane (12) can also lead to phosphonitriles (Equation (4.12)) but a tautomeric proton shift, (13a) → (13b), is required[52].

$$Ph_2(NH_2)P=N-P(=NH)Ph_2 + RP(OPh)_2 \qquad\qquad (4.12)$$

R = OPh, Me ,Et

(12)

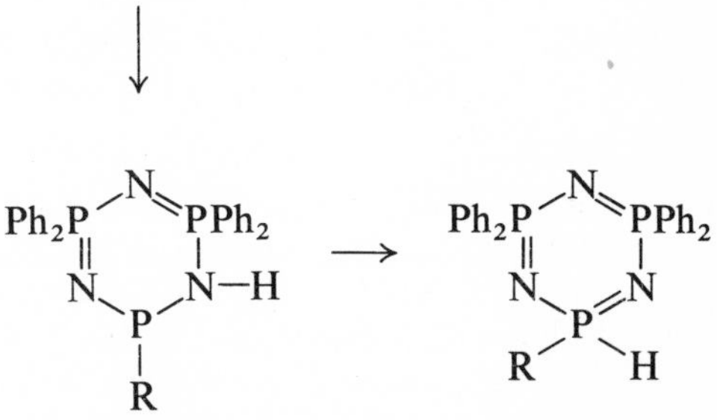

(13a)   (13b)

A similar reaction takes place between (12) and thiophosphoric anhydrides, $(RPS_2)_2$, in which hydrogen sulphide is eliminated but the tautomer which involve protonation of a ring nitrogen (14a) is formed[53]. Subsequent treatment with methyl iodide however gives the substituted phosphonitrile (14c).

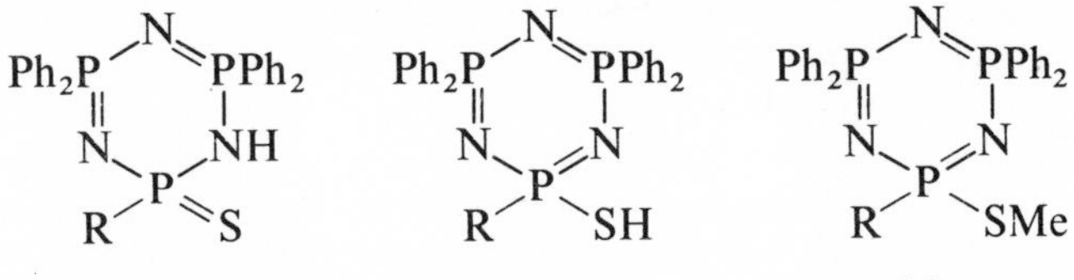

(14a)   (14b)   (14c)

4.2.3 From phosphorus azides

Thermal decomposition of phosphorus(III) azides leads to elimination of nitrogen and the probable generation of nitrene-like intermediates which stabilise by polymerisation to phosphonitriles. As the reactions are generally difficult to control and often lead to high yields of polymeric compounds, this approach has not been exploited. A non-geminal hexaphenyldichloro-tetraphosphonitrile, $P_4N_4Ph_6Cl_2$, has however been obtained[54, 55] by treatment of an equimolar mixture of phenyldichlorophosphine and diphenyl-chlorophosphine with sodium azide and then heating the mixture to 300 °C.

The halogen atom in diphenylchlorophosphine can be exchanged on treatment with either triphenyl- or trimethyl-azidosilane and the resulting azido-intermediate decomposes to give a good yield of hexaphenyltri-phosphonitrile[56]. Similarly, the products with phenyldichlorophosphine were mixed phenyl-azido-chloro-phosphonitriles.

The low-temperature photolysis of a mixture of phosphorus trichloride and hydrazoic acid[57] did not lead to loss of nitrogen with formation of Cl_3PNH, one of the postulated intermediates in the PCl_5–NH_4Cl reaction. Instead, hydrogen chloride was also lost (Equation (4.13)) to give a polymeric product with the empirical formula $P_5N_8Cl_9$.

$$5PCl_3 + 6HN_3 \rightarrow P_5N_8Cl_9 + 6HCl + 5N_2 \qquad (4.13)$$

4.2.4 Other methods

Thermal decomposition of dimethyldiaminophosphonium chloride[58] gives the polymeric dimethylphosphonitrile (4.14)

$$[Me_2P(NH_2)_2]Cl \rightarrow (Me_2PN)_n + NH_3 + HCl \qquad (4.14)$$

while in the presence of phosphorus pentachloride[59], 1,1-dimethyltetra-chlorotriphosphonitrile can be obtained in 26–59% yield. Hexakis(per-fluorophenyl)triphosphonitrile and the high polymer can be obtained[60] by chlorination of amidobis(perfluorophenyl)phosphine followed by dehydro-chlorination with triethylamine (4.15).

$$(C_6F_5)_2PCl \xrightarrow{NH_3} (C_6F_5)_2PNH_2 \xrightarrow{Cl_2} (C_6F_5)_2P(NH_2)Cl_2 \xrightarrow[-HCl]{Et_3N}$$
$$(C_6F_5)_6P_3N_3 \qquad (4.15)$$

4.2.5 Preparation of related compounds

A large number of phosphonitrilic-type compounds in which phosphorus atoms are replaced by other non-metals are now known but here it is possible to mention only a selection. Boron atoms can be incorporated by reactions (4.16) [61] and (4.17) [62] while an antimony compound[63] results from reaction (4.18).

$[Ph_2P(NH_2)=N-P(NH_2)Ph_2]Cl + RBCl_2$ $\qquad$ (4.16)

R = Ph or Cl

$$\left[\begin{array}{c} \overset{+}{N} \\ Ph_2P \overset{\diagup}{\cdots} PPh_2 \\ | \qquad | \\ H-N \qquad N-H \\ \diagdown \underset{|}{B} \diagup \\ R \end{array} \right] Cl^- + 2\,HCl$$

$[Cl_3P=N-PCl_3]Cl + 2MeNH_3Cl + BCl_3$ $\qquad$ (4.17)

$$\begin{array}{c} \overset{+}{N} \\ Ph_2P \overset{\diagup}{\cdots} PPh_2 \\ | \qquad | \\ Me-N \qquad N-Me \\ \diagdown \overset{-}{B} \diagup \\ Cl \quad Cl \end{array} \quad + 6\,HCl$$

$[Ph_2P(NH_2)=N-P(NH_2)Ph_2]Cl + SbCl_5$ $\qquad$ (4.18)

$$\left[\begin{array}{c} \overset{+}{N} \\ Ph_2P \overset{\diagup}{\cdots} PPh_2 \\ | \qquad | \\ H-N \qquad N-H \\ \diagdown \underset{Cl_3}{Sb} \diagup \end{array} \right] Cl^- + 2\,HCl$$

The properties of cyclic compounds containing two phosphorus atoms and one carbon atom (15) have been recently examined in detail by Schmidpeter and his co-workers[64].

$$\begin{array}{cc} \begin{array}{c} N \\ X_2P \diagup \diagdown PX_2 \\ \| \qquad \| \\ N \qquad N \\ \diagdown C \diagup \\ | \\ Y \end{array} & \begin{array}{c} N \\ X_2P \diagup \diagdown C-Y \\ \| \qquad \| \\ N \qquad N \\ \diagdown C \diagup \\ | \\ Y \end{array} \\ (15) & (16) \end{array}$$

The 'parent' compound [(15), X = Cl, Y = Ph][65] results from ammonolysis of a substituted amidine hexachloroantimonate (4.19).

$$[PhC(NPCl_3)_2]SbCl_6 + 2NH_4Cl \rightarrow \underset{(15)}{PhC(NPCl_2)_2N} + 4HCl + NH_4SbCl_6 \qquad (4.19)$$

Two carbon atoms may be incorporated in place of phosphorus [(16), X = Y = Cl] by treating sodium dicyanimide with phosphorus pentachloride[66].

An arsenic analogue of the phosphonitriles, $As_4N_4Ph_8$, was prepared by thermal decomposition of diphenylazidoarsine[67], but attempts to extend this method by use of alkylazidoarsines gave diarsines rather than the analogous arsenic–nitrogen compounds[68]. The octaphenyl derivative mentioned above has also been prepared by ammonolysis of diphenylarsenic trichloride[69] and, under different conditions, the analogous trimer, $As_3N_3Ph_6$, has been isolated[70].

4.3 SUBSTITUTION REACTIONS

A large number of reactions have been carried out in which halogen atoms, particularly in the trimeric chloride, are replaced by other atoms or groups. Sub-division in this section is in terms of the new kind of bonds that are formed; reactions leading to phosphorus–fluorine, –nitrogen, –oxygen, –carbon and –sulphur bonds are considered separately.

Substitution reactions in general result from nucleophilic attack on phosphorus and two limiting paths, geminal and non-geminal, can be envisaged. In the former, substitution of the first halogen by an X group promotes further reaction at the same phosphorus atom, while in the latter, further reaction is directed to an unsubstituted phosphorus atom. The possibility of *cis–trans* isomerism also arises when a non-geminal substitution scheme is followed. In some cases, the electronic effect of the substituent readily explains the path adopted, e.g. the strong electron-withdrawing effect of fluorine promotes a geminal scheme while electron donation from, say, dimethylamine groups leads to non-geminal replacement; however, in other cases this simple approach is by no means completely satisfactory. Although a large number of reactions involve substitution in the trimeric chloride, partially-substituted compounds are also being examined to assess the steric and electronic consequences of the first substituents.

Recently, Allcock and co-workers[71] have examined a series of exchange reactions between organic nucleophiles and organo-substituted triphosphonitriles. Among reactions examined are displacements of alkoxy, aryloxy, and anilino groups by amide, phenoxide and ethoxide ions. Organometallic reagents such as Grignard and organolithium compounds are also effective but lead to ring cleavage as well as substitution.

Reactions with tetramers are much more complex and there is still relatively little work in this area. Figure 4.1 shows the inter-relationship between compounds of different degrees of substitution; the phosphorus atoms are represented by the corners of the square and lines above and below the ring represent the relative orientation of the substituents. As for the trimer, the positions of the substituents can be designated by numbering the phosphorus atoms (1,3,5,7) and the arrangement with respect to the ring by *cis* and *trans*. With penta- and higher phosphonitriles, the systems become more complex and so far only mono or completely-substituted products have been isolated.

4.3.1 By fluorine

The reaction of anionic fluorinating agents with the trimeric chloride has been re-examined and information on the corresponding tetramer is now available[72]. Reactions with sodium fluoride in nitromethane give the mixed chloride fluorides, $P_3N_3F_nCl_{6-n}$ with $n = 1–3$, while in nitrobenzene the more highly-fluorinated products result. Similar reactions with the tetramer are very slow but with potassium fluorosulphite, particularly after the first fluorine has been substituted, reaction is rapid and the resulting chloride

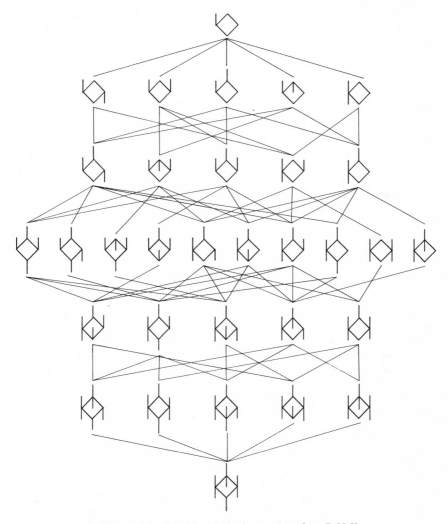

Figure 4.1   Possible substitution products from $P_4N_4X_8$

fluorides, $P_4N_4F_nCl_{8-n}$ must be removed immediately from the reaction system. The individual compounds can be separated by fractional distillation. In both cases replacement of the chlorine atoms is geminal; the unambiguous structures for the trimeric compounds is confirmed by ^{19}F n.m.r. spectroscopy. The tetramer reaction is more complicated but n.m.r. indicated that substitution of the third chlorine atom occurred predominantly at the phosphorus atom nearest to that already substituted, i.e. the trisubstituted compound is the 1,1,3-derivative (17) rather than the 1,1,5-isomer (18).

(17)   (18)

Non-geminally substituted chloride fluorides cannot be prepared by the direct fluorination reaction but can be obtained indirectly by treatment of non-geminally substituted dimethylaminofluorophosphonitriles with hydrogen chloride. Amine groups are removed and non-geminal isomers in the trimeric[73], $P_3N_3Cl_4F_2$ (19), and $P_3N_3Cl_3F_3$ (20), and tetrameric series[74], $P_4N_4Cl_3F_5$ (21) result

(19)   (20)   (21)

The position of fluorination in the *cis* and *trans* non-geminal 1,3-bis-dimethylaminotetrachlorotriphosphonitriles (22) depends on the fluorinating agent[75]. With the potassium fluorosulphite (4.20), attack is at the more reactive PCl_2 group rather than at the $PCl(NMe_2)$ group, which is less susceptible to nucleophilic attack owing to π-bonding between the phosphorus and amine-nitrogen atom. The products are the *cis*- and *trans*-5,5-difluorides (23) which can be separated by g.l.c.

Antimony trifluoride, on the other hand, reacts at the amine-substituted phosphorus atoms to give the isomeric 1,3-difluorides (24). To account for this specificity, a mechanism involving coordination of the antimony trifluoride to the most basic ring nitrogen atom, i.e. that flanked by the phosphorus atoms carrying amine substituents, has been proposed. The completely fluorinated isomers (25) result when the 1,3-difluorides react with potassium fluorosulphite.

Fluorination of hexabromotriphosphonitrile[76], with either silver fluoride in benzene or potassium fluorosulphite in nitrobenzene gives mixed bromide fluorides, $P_3N_3F_{6-n}Br_n$. Little is yet known about these compounds but

(22)   (23)   (4.20)

$$\downarrow SbF_3$$

(24)   (25)

from g.l.c. and i.r. spectroscopy geminal structures are proposed for compounds with $n = 2–4$.

4.3.2   By nitrogen

4.3.2.1   Reactions with the trimeric chloride

Reactions with ammonia and amines lead to phosphorus–nitrogen exocyclic bond formation and constitute the largest class of reactions. The diamino compound, $P_3N_3Cl_4(NH_2)_2$, isolated from the reaction of the trimeric chloride and ammonia is now known to have the geminal structure (26)[77,78]. The ^{31}P n.m.r. spectrum of the compound itself is suggestive of this structure but conclusive proof rests on the spectrum of the bistrichlorophosphazo derivative (27) obtained by reaction with phosphorus pentachloride (4.21).

(26)   (27)   (4.21)

No monoamino compound, $P_3N_3Cl_5(NH_2)$, has been isolated from the direct reaction with ammonia, but it can be prepared[78] from the diamide (26) by passing hydrogen chloride through a refluxing dioxan solution.

In the reaction with ammonia, the specific formation of a *gem*-diamino compound implies that the monoamide reacts more rapidly than the unsubstituted chloride and promotes reaction at the $PCl(NH_2)$ group. The proton abstraction mechanism (4.22) suggested is similar to that advanced by Shaw and his co-workers[79] for the reaction with t-butylamine which gives geminal substitution products similarly.

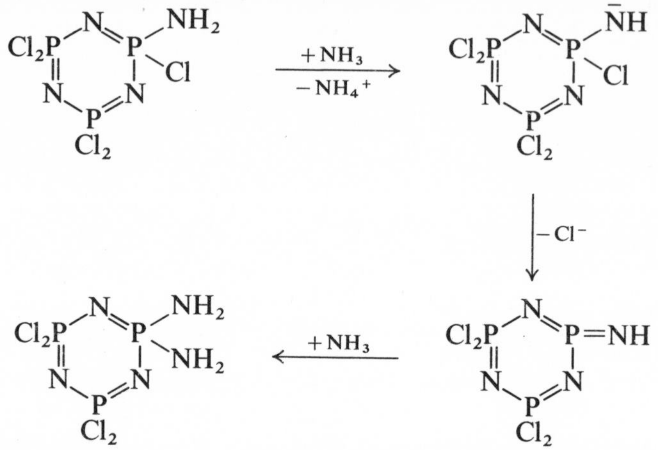

$$(4.22)$$

Thermal analysis of hexa-aminotriphosphonitrile[80], $P_3N_3(NH_2)_6$, shows endothermic effects at 92 and 245 °C. The former is correlated with the loss of half a mole of ammonia (4.23), while the second effect indicates the formation of highly polymeric phospham, $(PN_2H)_n$.

$$2P_3N_3(NH_2)_6 \rightarrow NH_3 + (NH_2)_5P_3N_3—NH—P_3N_3(NH_2)_5 \quad (4.23)$$

Primary amine reactions have been extensively studied and, in addition to a monosubstituted compound, the trimeric chloride with methylamine gives three disubstitution products[81]. The product formed in the largest amount, m.p. 103 °C, has the lowest dipole moment and is assigned the *trans*-non-geminal structure (28). The *cis* isomer (29) constitutes the smallest amount (m.p. 149 °C) while the geminal compound (30), also formed in relatively small amounts, has the highest dipole moment. Only two compounds were isolated in a previous examination of this system[82].

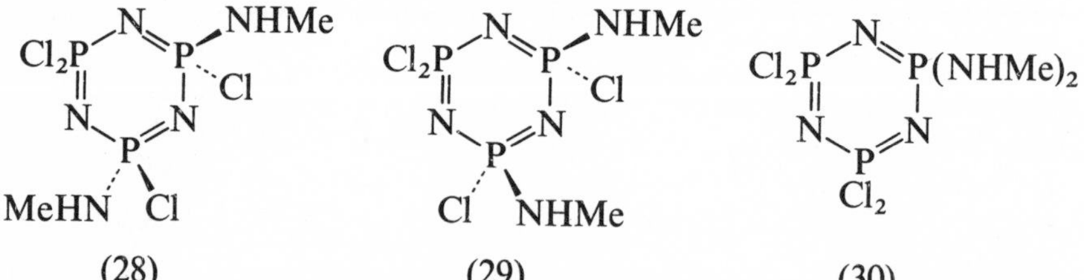

(28)   (29)   (30)

The system with isopropylamine gives mono-, di-, tetra- and hexa-substitution products[83]. The disubstituted compound is a non-geminal isomer but it is not known whether it has a *cis* or *trans* configuration. Treatment with dimethylamine gives the hydrochloride of *trans*-non-geminal $P_3N_3(NMe_2)_4(NHPr^i)_2$ but isomerisation may have occurred during the reaction. The tetrasubstituted compound, $P_3N_3Cl_2(NHPr^i)_4$, with a melting point of 126 °C, is highly basic and also occurs in the original reaction product as the hydrochloride, $P_3N_3Cl_2(NHPr^i)_4 \cdot HCl$. Treatment with dimethylamine gives a product which from ^{1}H n.m.r. spectroscopy contains geminal NMe_2 groups and thus a geminal structure (31) is suggested for the original compound.

$$Cl_2P \overset{N}{\underset{N}{\diagup}} P(NHPr^i)_2$$
$$\underset{\underset{(NHPr^i)_2}{P}}{\| \quad \|}$$

(31)

$$(Pr^iHN)_2P \overset{N}{\underset{N}{\diagup}} \overset{NHPr^i}{\underset{Cl}{P}}$$
$$Pr^iHN \quad Cl$$

(32)

The less basic tetrasubstituted product is the non-geminal *trans* isomer (32). No trisubstituted derivative was identified but a compound with this stoichiometry was prepared in a later investigation[84]. A potentiometric titration with perchloric acid in nitrobenzene solution however suggested that the product was a 1:1 adduct of the geminal di- and tetra-substituted compounds. This was confirmed by separation of the components by either column chromatography or crystallisation of the hydrochlorides. A similar adduct between the two known tetrasubstituted compounds, (31) and (32) was isolated when eight moles of isopropylamine were used.

t-Butylamine-substituted triphosphonitriles, $P_3N_3Cl_n(NHBu^t)_{6-n}$ are known for $n = 0,2,4$ and 5 [79]. In each case only one isomer is obtained, and ^{1}H n.m.r. spectra of the dimethylamine derivatives indicate that the compounds in which $n = 2$ and 4 have geminal structures.

Various monosubstituted compounds have been reported, e.g. with ethylamine[85], butylamine[85], hexylamine[86], and amino esters[87], while the hexasubstituted octylamine derivative is also known[86]. Di- and tetra-substituted products have been prepared with the ethylester of glycine[88] and, by further substitution with ethyleneimine, both have been shown to have geminal structures.

Mono- and hexa- allylamine derivatives are known[89], but a patent[90] claims the preparation of compounds of all degrees of substitution from both the trimeric chloride and bromide. For di-, tri- and tetra-substitution both geminal and non-geminal isomers are claimed.

The first aniline derivative, $P_3N_3Cl_4(NHPh)_2$, was isolated from toluene solution and shown to have a geminal structure by ^{31}P n.m.r. spectroscopy[91]. A re-examination[92] showed that mixtures containing six products of the type $P_3N_3Cl_{6-n}(NHPh)_n$, where $n = 1,2$(two isomers),3,4 and 6, were usually obtained and that substitution followed a predominantly geminal path.

The overall view of reactions with primary amines is that a non-geminal path is predominantly followed when the alkyl group is small, but with greater branching, substitution by both non-geminal and geminal routes occurs until, with the highly branched t-butylamine, the latter is exclusively followed. With aniline, a relatively weak nucleophile, geminal substitution is followed probably because electron supply to the phosphorus is not sufficiently great to force reaction at a second phosphorus atom. The proposed mechanism for geminal substitution has already been briefly discussed

(4.22) and involves initially a base-catalysed proton-abstraction mechanism[79].

Although the nature of a substituent already present largely determines the position of further substitution, it is not always the case. This is shown[93] by the reactions in (4.24) and (4.25).

$$
P_3N_3Cl_5(NHBu^t) \quad
\begin{cases}
\xrightarrow{\;Bu^tNH_2\;} & P_3N_3Cl_4(NHBu^t)_2 \;(gem) \quad (34) \\
\xrightarrow{\;EtNH_2\;} & P_3N_3Cl_4(NHBu^t)(NHEt) \;(non\text{-}gem) \quad (35)
\end{cases}
\qquad (4.24)
$$

(33)

$$
P_3N_3Cl_5(NHEt) \quad
\begin{cases}
\xrightarrow{\;EtNH_2\;} & P_3N_3Cl_4(NHEt)_2 \;(non\text{-}gem) \quad (37) \\
\xrightarrow{\;Bu^tNH_2\;} & P_3N_3Cl_4(NHEt)(NHBu^t) \;(gem) \quad (38)
\end{cases}
\qquad (4.25)
$$

(36)

Ethylamine gives a non-geminal disubstituted compound (37) but with t-butylamine the geminal product (34) is formed. The mixed disubstituted products (35) and (38) however have, respectively, non-geminal and geminal arrangements showing that the strength of the attacking nucleophile is the determining factor.

Probably the most intensively examined system is that between hexachlorotriphosphonitrile and dimethylamine[94-96]. As expected for reaction with a strong nucleophile, a non-geminal path is followed at low temperatures, but it is clear that the actual experimental conditions can greatly affect the course of the reaction. A higher temperature and more rapid addition of the amine, for example, promotes the formation of the geminal bis- and tris-compounds. Until recently, only three of the complete series of possible geminal and non-geminal substitution products had not been isolated. Of these, the *trans* non-*gem* tetrakis-compound (39) has now been separated[97] and the pentakis-compound (40) has been identified as a component in the mixture obtained by reacting the trimeric chloride with both eight and ten moles of dimethylamine[97].

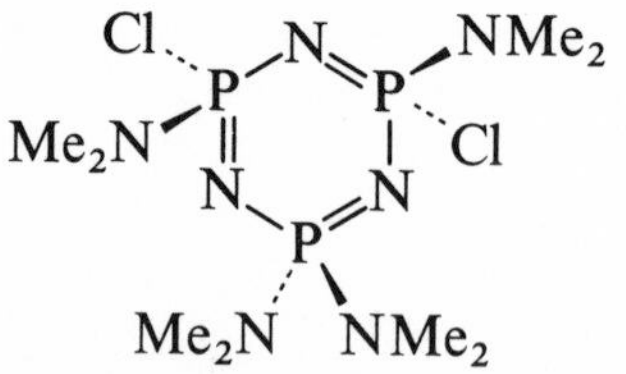

(39)

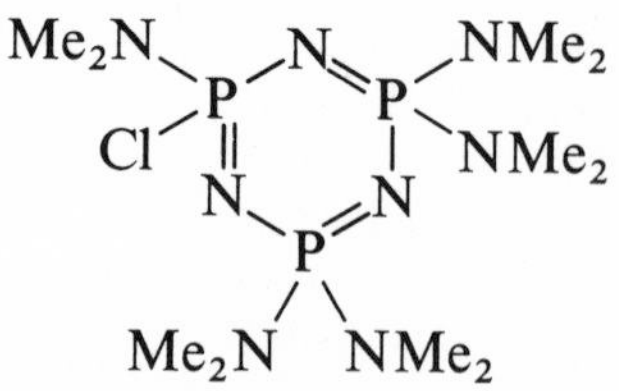

(40)

The detailed structural assignments of the products of the dimethylamine reaction follow from dipole moments[95] and g.l.c. measurements[97] but ^1H n.m.r. spectroscopy is of the greatest value. Each amine environment gives rise to a doublet due to spin coupling with the nearest phosphorus atom, although 'virtual coupling' often complicates the spectra. Using these data it is possible to show that in both di- and tri-substitution the *trans*-non-geminal isomer is formed in greater amounts than the *cis*, while in the latter the geminal trisdimethylamino compound is also produced. The greater yield of *trans* isomers is not considered to be a result of steric effects, but to arise through an electronic effect[98]. A '*cis*-effect' is postulated in which electrons are transferred most readily between groups that are *cis* to each other. If the monodimethylamino derivative reacts further, the primary effect of the substituent is to direct reaction towards an unsubstituted phosphorus atom. Then, a chlorine atom *trans* to the amino group already present is substituted preferentially as the *cis* chlorines are relatively de-activated by accepting negative charge from the *cis* amine group. For a similar reason, the *trans*-trisdimethylamino derivative is expected to be formed in greater amounts than the *cis* derivative which in turn will be greater than that for the geminal isomer.

The corresponding diethylamine system has been less carefully examined but claims for a tetrakis-compound formulated as the geminal isomer and the pentakis-derivative, $P_3N_3Cl(NEt_2)_5$, have been made[99].

Pyrrolidine, $HN(CH_2)_4$, gives all degrees of substitution with two isomers; the *cis*- and *trans*-non-geminal compounds, being isolated for di- and tetra-substitution[100]. Three trisubstituted compounds are known which implies that the geminal compound is also formed. As is the case with many highly-substituted compounds, the high basicity allows isolation as hydrochlorides. With 2-phenyl pyrrolidine[101], one to four chlorine atoms can be replaced in the hexachloride but in no case is there evidence for the presence of more than one isomer. Derivatives containing chlorine and both pyrrolidine and aziridine residues are known[102].

There are discrepancies between two reports on the reaction with piperidine. Keat and Shaw[98] report that the results fit into the same basic pattern as for dimethylamine and seven products can be isolated. These include the mono- and hexakis-compounds and a tetrakis-isomer shown to have a *cis*-configuration. The bis-derivatives formed are the *cis* and *trans* non-geminal isomers while substitution of three chlorine atoms leads to the geminal and the *trans* tris compounds. A later report by Kropacheva and Mukhina[103] agrees with the formation of the mono isomer, two bis-isomers and the hexakis-derivatives but they isolated only a single tris-compound melting at the same temperature as the *trans* isomer above. Of the two tetrakis-derivatives isolated, the lower melting isomer is probably analogous to Keat and Shaw's *cis* isomer while that melting at 113–115 °C may be the *trans* compound. Interestingly, a good yield of the pentakis-compound is reported[103], while in the earlier investigation[98] no evidence could be found for this compound.

Kinetic investigations of the reaction between the trimeric chloride and both piperidine and diethylamine show that replacement of the first chloride is fast compared with subsequent reactions[104]. For both amines in toluene solution a mixed second- and third-order rate law is followed; in addition the piperidine reaction is catalysed by tri-n-butylamine which suggests that the third order term can be correlated with base catalysis. Similar reactions with aniline and a series of substituted anilines[105] were second order in the initial stages indicating that reactions proceeded by an S_N2-type mechanism. Activation energies are given and the reaction velocities parallel the basicity of the substituting nucleophile.

With morpholine[106], the *cis*- and *trans*-dimorpholino-tetrachlorides are formed as well as pairs of tri- and tetra-substituted derivatives. One of the trimorpholides is a geminal compound but both the latter are non-geminal isomers. Structure assignments follow from the preparation of mixed morpholine-ethyleneimine compounds.

The absence of geminal tetrakis-compounds is not unexpected but the absence of penta-amine derivatives is a little surprising. Reasons for this behaviour have been advanced[94, 107, 108]. It seems that the observation may have arisen due to difficulties in detection and isolation of pentakis-compounds rather than in their non-formation. In some cases only small amounts may be formed[97, 100], but this does not seem to be so with ethyleneimine[109] and morpholine[106].

Although certain isomers are formed either in small amounts or not at all in direct reactions between amines and the trimeric chloride, it is possible to synthesize some of them by isomerisation reactions. As expected, inter-conversion of geminally and non-geminally substituted compounds does not seem possible, but *cis–trans* isomerisation of non-geminal compounds can often be achieved by treatment with aluminium chloride[110], substituted ammonium chlorides[111], or hydrogen chloride or bromide[112]. The hexakis-dimethylamino compound initially gives a hydrochloride with anhydrous hydrogen chloride but, on prolonged treatment, deaminolysis occurs[112] and the *cis*-tetrakis-, $[P_3N_3Cl_2(NMe_2)_4]$ and the *cis*- and *trans*-tris- compounds $[P_3N_3Cl_3(NMe_2)_3]$ have been identified.

Reactions with *o*-phenylenediamine[113] in the presence of triethylamine give mono- (41), bis- and tris- (42) compounds (42).

(41)

(42)

A bisphenylhydrazine derivative, $P_3N_3Cl_4(NHNHPh)_2$ with a *trans*-non-geminal structure, has also been prepared[114] which gives the *trans*-tetrakis-dimethylamino derivative (43) on treatment with an excess of the amine. No isomerisation occurs in this reaction but the *cis* isomer (44) and its hydrochloride can be prepared from $cis\text{-}P_3N_3Cl_2(NMe_2)_4$ and phenyl-hydrazine.

$$\text{(43)} \qquad\qquad\qquad \text{(44)}$$

Although the trimeric chloride cleaves trimethylamine to give dimethyl-amino derivatives[115], interaction with pyridine leads to species best repre-sented as $[P_3N_3Cl_{6-n}(py)_n]^{n+}nCl^-$ [116]. These phosphonitrile-pyridinium salts result from partial heterolysis of P—Cl bonds and stabilisation of the resultant cations by pyridine coordination. Formation of such species explains the enhanced rates of hydrolysis and alcoholysis of $P_3N_3Cl_6$ in pyridine solutions and also the formation of 4-substituted pyridines from reactions in the presence of N,N-dialkylanilines[116, 117]. The hexapyridinium salt, $[P_3N_3(py)_6]Cl_6$, is an unstable hygroscopic compound[117], which from i.r. spectroscopy does not contain phosphorus–chlorine bonds. More stable derivatives can be formed by adding metal chlorides, or in some cases metals, and products such as $[P_3N_3(py)_6]Cl_3(AlCl_4)_3$ and $[P_3N_3(py)_6](CuCl_4)_3$ are known[118].

4.3.2.2 Reactions with the higher chlorides

As mentioned earlier (see Figure 4.1), substitution reactions with the tetra-meric chloride are considerably more complex than those for the trimer, and there is relatively little work reported on the preparation of partially-aminolysed compounds. With ammonia, the completely-substituted com-pound, $P_4N_4(NH_2)_8$, is known[119] and recently di- and tetra-substituted compounds have been re-examined[120]. The ^{31}P n.m.r. spectra of the com-pounds and their trichloro-phosphazo derivatives show that the disub-stituted compound has a 1,5-structure (45) while the tetra-substituted compound is the 1,1,5,5-isomer (46).

$$\text{(45)} \qquad\qquad\qquad \text{(46)}$$

Mono- and di-substituted primary amine derivatives[121], e.g. $P_4N_4Cl_7$ (NHR) where R = Me, Et, Pr^n and Pr^i, and $P_4N_4Cl_6(NHR)_2$ where R = Pr^n and Pr^i, are known but the structures for the latter are unknown.

In a preliminary communication, Lehr[122] has reported the preparation of the mono- and all five of the possible isomeric bis-dimethylamino-tetraphosphonitriles [(47)–(51)].

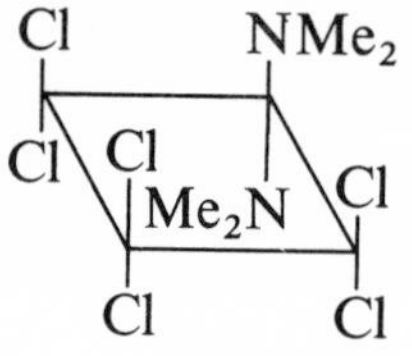

(47) 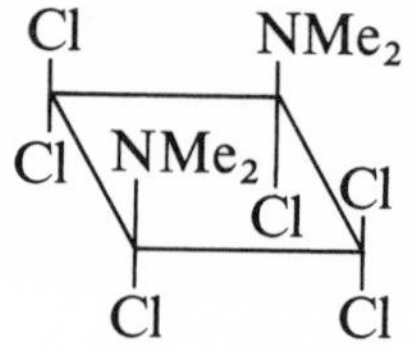 (48) 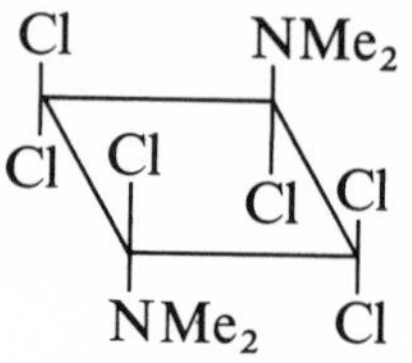 (49)

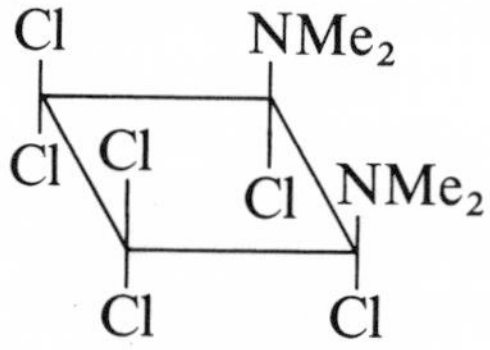

(50) 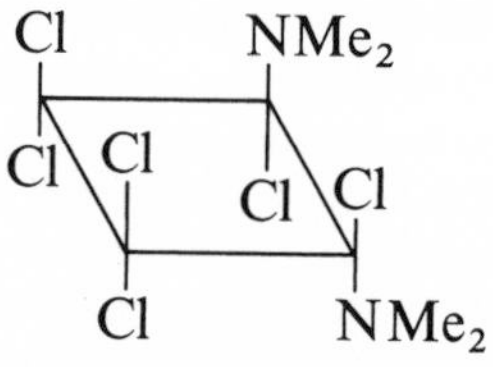 (51)

The *trans*-1,5-isomer (49) is formed in the greatest amount and *cis–trans* isomerism occurs when the non-geminal compounds are heated with pyridine hydrochloride. Further substitution gives a mixture of tris-compounds containing four of the possible isomers.

Provisional structures[123] have been advanced on the basis of ^{1}H and [31]P n.m.r. spectroscopy and basicity measurements for various dimethyl-amine derivatives, $P_4N_4Cl_n(NMe_2)_{8-n}$, obtained in an earlier investigation[124]. Compounds were obtained which represent all degrees of substitution with the exception of seven and the structures of the isolated compounds were related to those of the *trans*-1,5-bis- (49) and one or both of the 1,1,5-tris- (52) or 1-*cis*-3-*trans*-5-tris- (53) derivatives.

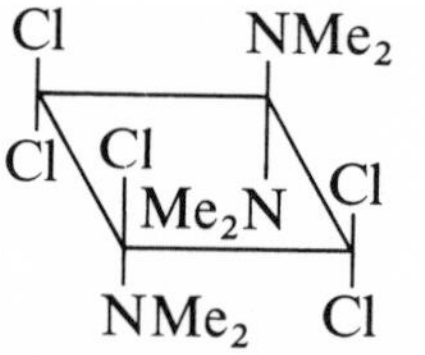

(52) 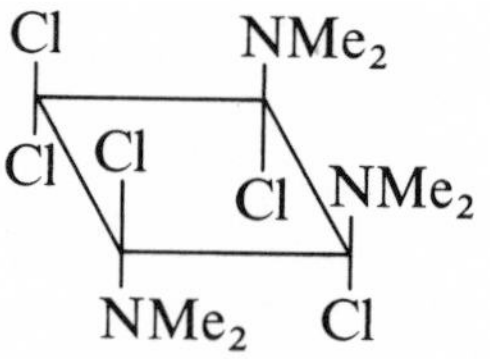 (53)

Stahlberg and Steger[125] have also examined the dimethylamine reaction and extended this to the tetrameric bromide. Complex reaction products are always obtained whose composition can be determined by thin-layer chromatography and infrared spectroscopy. The formation of four isomeric bis-dimethylamino compounds, $P_4N_4Cl_6(NMe_2)_2$, is noted and attention is drawn to the greater possibility of stabilising different ring conformations in tetramer products than in those from trimers.

With a large excess of N-methylcyclohexylamine the tetrameric chloride gave a high yield of a single tetrasubstitution product[126], shown by n.m.r. spectroscopy to be the 1-cis-3-trans-5-trans-7-tetrakis-compound (54).

(54)

Berlin and his co-workers[126] postulate that amination of tetramers takes place successively at phosphorus atoms 1,5,3 and 7; each substitution takes place *trans* to the previous one. This is the order expected if the 'cis-effect' operates but the results with dimethylamine indicate the system is more complex and that geminal isomers are more readily formed than in corresponding reactions with the trimer.

With the exception of $P_5N_5Cl_9(NMe_2)$, only completely-substituted products have been reported[127] for reactions between dimethylamine and the higher phosphonitriles, $(PNCl_2)_{5-8}$.

4.3.2.3 Reactions with the trimeric bromides

Steger and his co-workers[128-130] have reported the formation of dimethylamine derivatives containing one to three amine groups and showing strong similarities with the chloride system. The bromide reactions give greater yields of the *cis*-bis-amine and *gem*-tris-amine derivatives compared with the analogous chloride reactions but the *cis*-tris derivative was not observed. Infrared spectra of initial reaction products give evidence for tetrakis-isomers but these were not isolated.

4.3.2.4 Reactions with the fluorides

Monosubstitution of hexafluorotriphosphonitrile occurs with ammonia[131], methylamine[132] and dimethylamine[132] while Chivers and Paddock[133] have prepared series of compounds, $P_nN_nF_{2n-x}(NMe_2)_x$ where $n = 3-6$ and $x = 1-2$, by reactions with either dimethylamine or dimethylaminotrimethylsilane. Disubstitution is non-geminal and the compound $P_4N_4F_6$ $(NMe_2)_2$, for example, has a 1,5-structure. Glemser and co-workers[134] have prepared the non-geminal bis monomethylamino- and dimethylamino-compounds, $P_3N_3F_4(NRR')_2$, while the tetrameric fluoride reacts with ammonia at $-80\,°C$ to give $P_4N_4F_7NH_2$ [135]. Similar monosubstituted derivatives of the pentameric and hexameric fluorides have been prepared[136].

The amine groups can be replaced by either chlorine or bromine by reaction with the appropriate hydrogen halide and a series of monochlorides or bromides, $P_nN_nF_{2n-1}Cl$ or $P_nN_nF_{2n-1}Br$ where $n = 3$–6, have been isolated[133]. The dichlorides and bromides, $P_3N_3F_4Cl_2$ [134], $P_3N_3F_4Br_2$ [134] and $P_4N_4F_6Cl_2$ [133] have non-geminal structures and this provides an alternative to the preparative route discussed earlier[73, 74].

4.3.2.5   Reactions of amine substituted compounds

The preparation of a geminal bistrichlorophosphazo derivative (27) from $P_3N_3Cl_4(NH_2)_2$ (26) and phosphorus pentachloride aided in the assignment of geminal structure to (26) [77, 78]. Mixed phenylhalogenophosphoranes can be used in place of phosphorus pentachloride but their reactivity depends on the number of halogen atoms and whether the chloride or bromide is used[137]. With diphenyltrichloro phosphorane reaction occurs at both amine groups, but only one amine group is attacked by the less reactive triphenyldibromophosphorane and compound (55) is obtained. The corresponding dichlorophosphorane similarly gives a monophosphazo compound but the second amine group is replaced by chlorine giving compound (56). A bistriphenylphosphazo compound (57) can however be prepared from Ph_3PCl_2 and the monoamide (55).

(55)   (56)   (57)

The pentachloride (56) can be arylated in a Friedel–Crafts reaction; the chlorine geminal to the phosphazo group is replaced.

A Grignard reaction on the bistrichlorophosphazo derivative (27) readily gave compound (57) but no further chlorine atoms could be replaced[138]. The pentachloro triphenylphosphazo-compound (56) similarly resulted from $P_3N_3Cl_5(N{=}PCl_3)$ but here further phenylation, as in the Friedel–Crafts reaction above, occurred on prolonged reflux at the chlorine atom geminal to the phosphazo group.

A spiro compound (58), $P_5N_6Cl_8Me$, has been isolated from a ring closure reaction (4.26) on $P_3N_3Cl_4(N{=}PCl_3)_2$ (27) and the structure confirmed by ^{31}P and ^1H n.m.r. spectra[139].

Monoamidofluorophosphonitriles also react with phosphorus pentachloride and trichlorophosphazo derivatives of the trimer[131], tetramer[135], pentamer[136], and hexamer[136] are known. A trifluorophosphazo compound, $P_3N_3F_5N{=}PF_3$, can be prepared by use of PF_3Cl_2 instead of phosphorus pentachloride[140] and reactions of the monoamide with sulphur tetrafluoride and thionyl chloride give $P_3N_3F_5NSF_2$ and $P_3N_3F_5NSO$ respectively[141]. The latter on heating in the presence of pyridine loses sulphur dioxide and a sulphur-bridged dimer, $P_3N_3F_5N{\cdot}S{\cdot}NP_3N_3F_5$, is formed.

$$P_3N_3Cl_4(N{=}PCl_3)_2 + (Me_3Si)_2NMe$$

$$\downarrow$$

$$2Me_3SiCl$$

$$+$$

(4.26)

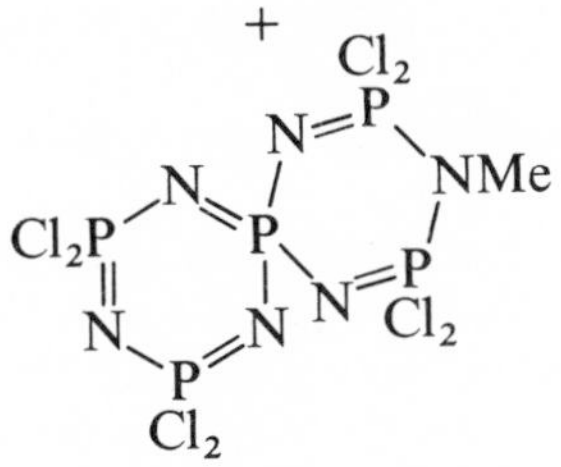

(58)

Thionyl chloride also reacts with the monoamide of the tetramer[135] to give $P_4N_4F_7NSO$.

Roesky and co-workers[136] have shown that the phosphazo side-chain can be extended by reaction alternately with hexamethyldisilazane (4.27, 4.29) and phosphorus pentachloride (4.28, 4.30).

$$P_3N_3F_5N{=}PCl_3 + (Me_3Si)_2NH \rightarrow$$
$$P_3N_3F_5N{=}PCl_2{-}NH{-}SiMe_3 + Me_3SiCl \quad (4.27)$$
$$P_3N_3F_5N{=}PCl_2{-}NH{-}SiMe_3 + PCl_5 \rightarrow$$
$$P_3N_3F_5N{=}PCl_2{-}N{=}PCl_3 + HCl + Me_3SiCl \quad (4.28)$$
$$P_3N_3F_5N{=}PCl_2{-}N{=}PCl_3 + (Me_3Si)_2NH \rightarrow$$
$$P_3N_3F_5N{=}PCl_2{-}N{=}PCl_2{-}NH{-}SiMe_3 + Me_3SiCl \quad (4.29)$$
$$P_3N_3F_5N{=}PCl_2{-}N{=}PCl_2{-}NH{-}SiMe_3 + PCl_5 \rightarrow$$
$$P_3N_3F_5N{=}PCl_2{-}N{=}PCl_2{-}N{=}PCl_3 + HCl + Me_3SiCl \quad (4.30)$$

4.3.3 By oxygen

4.3.3.1 Reactions with halides

Fitzsimmons and Shaw[142] reported complete substitution of the chlorine atoms in both the trimeric and tetrameric compounds with a variety of alcohols and phenols. In addition, a monochloropentaphenoxide, P_3N_3Cl $(OPh)_5$, was readily isolated which contrasts with the difficulty experienced in isolation of the analogous amine-substituted derivatives. Kinetic investigations show that the reaction between hexachlorotriphosphonitrile and sodium methoxide or butoxide in the corresponding alcohol is first order in both phosphonitrile and alkoxide ion but zero order in alcohol[143]. This suggests a similar S_N2 mechanism to that postulated for aminolysis. With increasing substitution by butoxy groups there is a strong decrease in reaction rate and an increase in activation energy. The reaction between the hexachloride and methanol in the presence of pyridine is first order in phosphonitrile but second order in alcohol[144] and the data are interpreted as showing that the rates of substitution of two chlorine atoms in $P_3N_3Cl_6$ are very similar. New chlorofluoroalkoxytriphosphonitriles can be obtained from reorganisation (4.31) of a mixture of the hexakis-compound, P_3N_3 $(OCH_2R_f)_6$ and the tris-compound, $P_3N_3Cl_3(OCH_2R_f)_3$, where $R_f = C_2F_5$ or C_3F_7.

$$P_3N_3(OCH_2R_f)_6 + P_3N_3Cl_3(OCH_2R_f)_3 \rightarrow$$
$$P_3N_3Cl_2(OCH_2R_f)_4 + P_3N_3Cl(OCH_2R_f)_5 \quad (4.31)$$

Spiro derivatives can result from alkanediols[146] and compounds have been isolated from reactants which include ethyleneglycol, 1,3-propanediol and 2,4-butanediol. With 1,2-propanediol, two isomeric forms of the completely-substituted compound were obtained. With longer carbon chains, e.g. 1,6-hexanediol, polymers were obtained as each hydroxy-group substituted a chlorine atom in a different phosphonitrile[147].

All degrees of substitution of hexachlorotriphosphonitrile have been achieved with sodium phenoxide and sodium p-bromophenoxide[148]. The reactions both follow a non-geminal path but the separation and purification of products is difficult. Mixed phenoxy-dimethylamino derivatives, prepared either by reaction of chlorophenoxy compounds with the amine or by phenolysis of chlorodimethylamino derivatives of known structures, show that overwhelmingly a non-geminal reaction path is followed[149, 150]. Cis–trans isomers are formed for the di-, tri- and tetra-substituted derivatives.

Similar results are obtained from ^1H n.m.r. spectroscopy and the adsorption behaviour of mixed amido-phenoxy-triphosphonitriles[151]. From the reaction of a mixture of chlorophenoxy compounds with ammonia the following were isolated: cis- and trans-$P_3N_3(OPh)_4(NH_2)_2$ and the cis-(59), trans- (60) and gem- (61) forms of $P_3N_3(OPh)_3(NH_2)_3$, providing evidence that the geminal path is followed but to a small extent only.

(59)   (60)   (61)

There also appears to be a slight preference in the non-geminal path for the formation of cis over trans isomers.

The reaction with six moles of sodium p-nitrophenoxide in toluene however, led to only small amounts of $P_3N_3Cl(OC_6H_4NO_2)_5$ and $P_3N_3Cl_3(OC_6H_4NO_2)_3$, and attempts to prepare other compounds in the series by this method were unsuccessful[152]. The completely-substituted compound could however be obtained by refluxing the trimeric chloride in xylene with potassium hydroxide and a large excess of p-nitrophenol. In the presence of Raney nickel the compound was reduced smoothly to the amino derivative, $P_3N_3(OC_6H_4NH_2)_6$, and urethanes were obtained from further treatment with alcohols[153].

No partially-substituted compounds could be isolated when the hexachloride reacted with catechol in the presence of a tertiary amine;

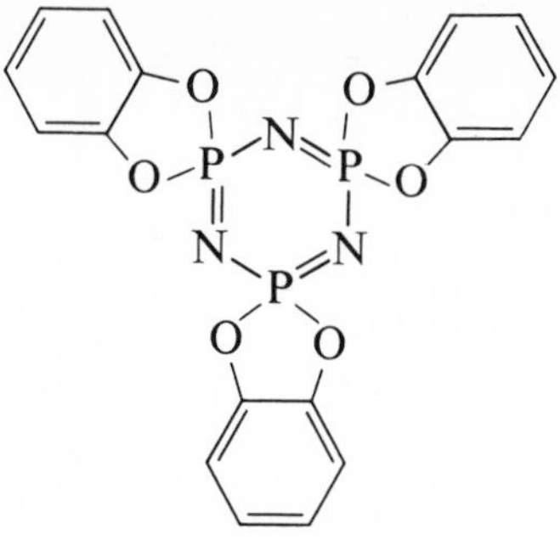

(62)

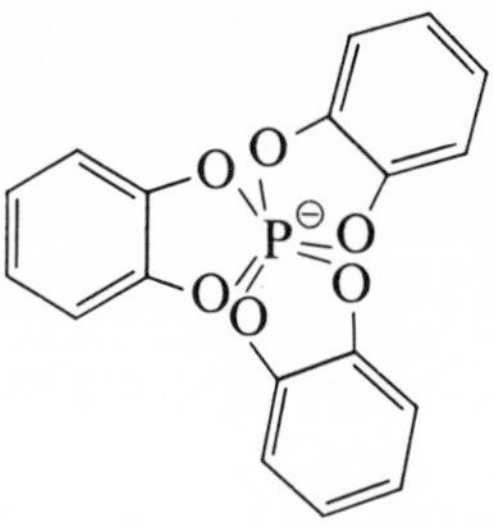

(63)

$Et_3NH^{\oplus}$

only the trisubstituted spiro derivative (62) and a decomposition product (63) were obtained[154]. Similar spiro derivatives result from reactions with 2,3-dihydroxynaphthalene and 2,2′-dihydroxydiphenyl[113]. The dioxynaphthyl derivative shows the same property of including solvent molecules as the catechol derivative[155].

Polymers containing P_3N_3 groups and residues from diols are obtained from reactions with 2,2-bis-(p-hydroxyphenyl)propane[156] resorcinol[156] and hydroquinone[156–158], while with 4,4′-dihydroxydiphenyl, chlorine-free compounds with the formulae $P_3N_3(OC_6H_4 \cdot C_6H_4OH)_6$ and $P_3N_3(OC_6H_4 \cdot C_6H_4O)_3$ can be obtained[159]. o-Aminophenol, however, causes ring cleavage with the trimeric chloride and the phosphorane (64) is isolated[160]. The same compound results from $P_3N_3F_6$, $P_3N_3Br_6$, $P_4N_4Cl_8$ and $(PNCl_2)_n$, and also from spiro-derivatives such as the tris-catechol compound (62).

A large number of liquid trimeric and tetrameric derivatives containing fluoroalkoxide groups and either substituted phenol or aniline residues have been prepared[161, 162]. Acetoxime with the trimeric chloride in the presence of triethylamine gives the monosubstituted product[163], $P_3N_3Cl_5(ON{=}CMe_2)$, which with dimethylamine at 0 °C gives a tetrakisdimethylamide (65). The remaining chlorine can be replaced by various alkoxide groups.

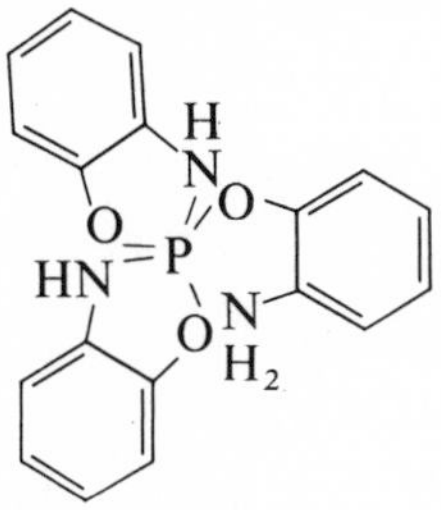

(64)

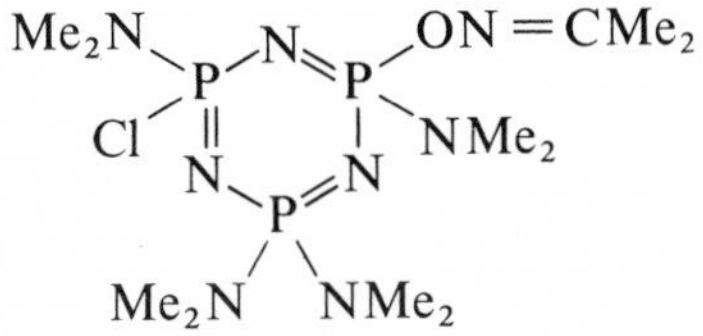

(65)

p-Quinone dioxime[164] reacts with $P_3N_3Cl_6$ to give three products with the empirical formula $PNC_{12}H_{10}N_4O_4$ but different physical properties, while with 2,6-dimethylol-p-cresol[165, 166], a variety of polymeric species result, some of which are soluble in organic solvents.

Potassium O,O-diethylthionophosphate, $KOP(S)(OEt)_2$, in butanol can replace two chlorines in the hexachloride giving $P_3N_3Cl_4[OP(S)(OEt)_2]_2$ or a completely-substituted product can be prepared[167]. In addition a thio derivative, $P_3N_3[SP(S)(OEt)_2]_6$, is known and both the oxygen and sulphur derivatives of the tetramer have been prepared.

Dimethyl formamide[168, 169] reacts with both the trimeric and tetrameric chlorides to form adducts similar to those discussed earlier for pyridine. Compound (66), for example, can be isolated from the trimer and with aluminium or copper chlorides the anion can be converted into $AlCl_4^-$ or $CuCl_4^{2-}$.

$$\left[\begin{array}{c} (Me_2NCHO)_2P \overset{N}{\underset{\parallel}{\diagup}} \overset{N}{\diagdown} P(OCHNMe_2)_2 \\ N \diagdown \underset{P}{} \diagup N \\ Cl_2 \end{array} \right] Cl_4$$

(66)

As with amine reactions, there is little information on the behaviour of the higher compounds with alcohols and phenols. Completely-substituted methoxy compounds, $P_nN_n(OMe)_{2n}$ where $n = 5-8$, have been prepared from the respective chloride and sodium methoxide in benzene solution[127, 170]; the corresponding 2,2,2-trifluoroethoxides and phenoxides are also known[127].

Compounds containing both amine and alkoxide groups have been prepared. For example, reaction of 1,1-diamino-tetrachlorotriphospho-nitrile, $P_3N_3Cl_4(NH_2)_2$, with sodium fluoroalkoxides gave compounds with the formula $P_3N_3(OR)_4(NH_2)_2$, where $R = CF_3 \cdot CH_2$, $CHF_2 \cdot CF_2 \cdot CH_2$, $C_2F_5 \cdot CH_2$ and $C_3F_7 \cdot CH_2$. The derivatives, which will have structures based on (67), lose ammonia on heating giving phospham-type polymers and react with fluorinated α,ω-glycols to eliminate ammonia and give elastomers.

$$(RO)_2P \overset{N}{\underset{\parallel}{\diagup}} \overset{N}{\diagdown} P(NH_2)_2$$
$$N \diagdown \underset{P}{} \diagup N$$
$$(OR)_2$$

(67)

Similar chlorine-free compounds result from the action of sodium fluoro-alkoxides on the gem-bisaniline compound, $P_3N_3Cl_4(NHPh)_2$ [91]. All the chlorine atoms in tris- and tetrakis- methylamino and dimethylamino chlorotriphosphonitriles can similarly be substituted using a solution of the sodium salt (NaOR where R = Me, Et, Pr, or Am) in the appropriate alcohol[172, 173]. With allyl alcohol in the presence of pyridine, the non-geminally substituted compounds, $P_3N_3(NMe_2)_3(OCH_2CH=CH_2)_3$ and $P_3N_3(NMe_2)_2(OCH_2CH=CH_2)_4$ can be prepared from the corresponding chlorodimethylamino compounds[174].

4.3.3.2 Reactions of oxygen substituted compounds

Certain alkoxy-phosphonitriles rearrange on heating to form N-alkyl species (phosphazanes) as shown in equation (4.32). In the trimeric series, this occurs when R is methyl or benzyl and for tetramers when R is methyl[175].

$$(4.32)$$

Aryloxy and fluorolakoxy compounds do not rearrange nor apparently do the methoxides of the higher phosphonitriles, $[PN(OMe)_2]_{5-8}$ [170]. In the trimeric and tetrameric series alkyl iodides catalyse the change[176] and the alkyl group of the catalyst can be incorporated in the rearrangement product, e.g. when the trimeric ethoxide is heated with i-propyl iodide (4.33) ethyl iodide is eliminated and i-propyl groups are substituted at the ring nitrogen atoms.

$$P_3N_3(OEt)_6 \ + \ 3Pr^iI \ \longrightarrow$$

$$+ \ 3EtI$$

$$(4.33)$$

On the other hand, the rearrangement of $P_4N_4(OPr^i)_8$ is not catalysed by i-propyl iodide but with methyl iodide the N-methyl compound (68) results.

$$(68)$$

Heating the trimeric ethoxide, $P_3N_3(OEt)_6$, with either titanium(IV) or zirconium(IV) chloride[177] initially gives a polymeric structure based on (69) which is converted to the completely-rearranged structure (70) on boiling.

Butyl chloride is eliminated from $P_3N_3(OBu)_6$ on heating with mono-chlorosilanes such as $Me_2PhSiCl$ to give products with the general formula $P_3N_3(OBu)_n(OSiR_3)_{6-n}$ [178]. Although purification is difficult, both $P_3N_3(OBu)_5(OSiMe_2Ph)$ and $P_3N_3(OSiMe_2Ph)_6$ have been isolated from the hexabutoxide and there is no infrared evidence for rearrangement to N-

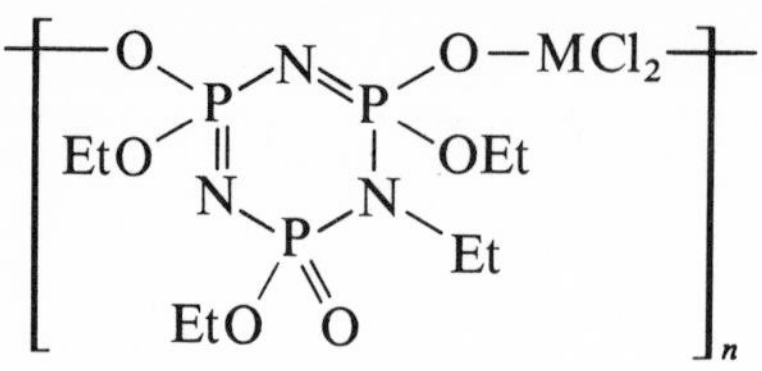

(69)

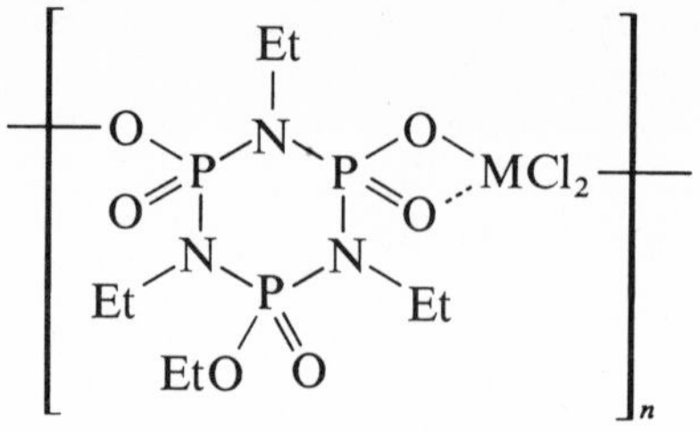

(70)

substituted species. The monobutoxide, $P_3N_3Cl_5(OBu)$, similarly gives $P_3N_3Cl_5(OSiMe_2Ph)$ which hydrolyses in air to $P_3N_3Cl_5(OH)$; a second hydroxy species, $P_3N_3(OBu)_5OH$, results when acetone solutions of $P_3N_3(OBu)_5(OSiMe_2Ph)$ are hydrolysed.

Both trimeric and tetrameric ethoxyphosphonitriles undergo ring cleavage on treatment with benzoyl chloride[179], to give ethyl chloride, triphenyl-s-triazine (by trimerisation of benzonitrile) and ethyl phosphenate. A similar reaction takes place when substituted sodium or ammonium benzoates and the hexachloride are heated to $\approx 250\,°C$ [180].

The base-catalysed hydrolysis of alkoxy- and aryloxy-phosphonitriles[181, 182] shows that both the hydrolytic stability and mechanism of hydrolysis are markedly dependent on the nature of the exocyclic groups. In water–methanol solutions of sodium hydroxide the trifluoroethoxy groups are removed from $P_3N_3(OCH_2CF_3)_6$ in a stepwise, non-geminal fashion and after acidification the intermediates (71) and (72) can be isolated.

(71)

(72)

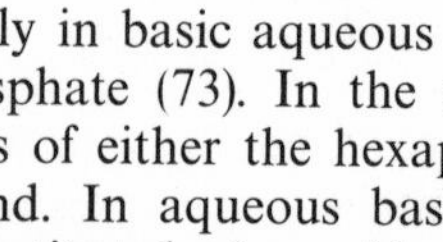

(73)

This behaviour should be contrasted with that of the tris-catechol derivative which hydrolyses very rapidly in basic aqueous dioxane to ammonia and sodium 1,2-dioxyphenylphosphate (73). In the same circumstances there was no detectable hydrolysis of either the hexaphenoxide or the tris-2,2′-dihydroxydiphenyl compound. In aqueous basic diglyme, on the other hand, one phenoxide or substituted phenoxide group could be removed from the compounds $P_3N_3(OC_6H_4X)_6$, where $X = $ 4-nitro, 4-methyl, 2-nitro, or hydrogen. As the ease of reaction paralleled the increase in acidity of the resulting phenol, the difference is related to electronic rather than steric effects.

4.3.4 By carbon

4.3.4.1 Reactions with halides

Friedel–Crafts phenylation of the trimeric chloride follows a geminal substitution path and a re-examination[183] has shown that greatly improved yields can be obtained when either aluminium or a tertiary base is added to the mixture or when the reaction is repeated with fresh aluminium chloride. The three approaches are designed to reduce the effect of the liberated hydrogen chloride which is thought to retard the reaction by protonation of the phosphonitrile. Phenylation of the *trans* non-geminal triphenyl, $P_3N_3Cl_3Ph_3$, in the presence of aluminium chloride gives the isomeric *cis*- and *trans*-tetraphenyl, $P_3N_3Cl_2Ph_4$, and the pentaphenyl derivatives, but no reaction occurs when the catalyst is either iron(III) or tin(IV) chloride[184]. *Cis–trans* isomerisation of the non-geminal $P_3N_3Cl_3Ph_3$ is catalysed by both aluminium and iron(III) chlorides.

With the tetrameric chloride a Friedel–Crafts reaction[185] in the presence of triethylamine gave a derivative of the trimer (74), identical with that produced by phenylation of trichlorophosphazo derivatives of the trimer[137, 138],

$$Cl_2P \overset{N}{\underset{N}{\overset{\displaystyle \nwarrow}{\underset{P}{\parallel}}}} \overset{N=PPh_3}{\underset{P}{\overset{\displaystyle \diagup}{\underset{\displaystyle Cl_2}{\underset{N}{\diagdown}}}}} Ph$$

(74)

and a product of unknown structure which with dimethylamine gave $P_4N_4Cl_2Ph_5(NMe_2)$.

Grignard reactions with both the trimeric and tetrameric chlorides have been re-examined. In a variety of solvents, phenylmagnesium bromide gives only a small yield of the hexaphenyl, $P_3N_3Ph_6$; the bulk of the product consist of magnesium-containing linear phosphorus–nitrogen species[186].

$$[Ph_3P{=}N{-}PPh_2{=}N{-}PPh_2{-}NH_2]^+ClO_4^-$$
(75)

A metal-free perchlorate (75) can be isolated after treatment with ammonium salts or liquid ammonia, but this is also only a minor product. As no partially-phenylated trimeric compounds were isolated, ring cleavage is thought to occur early in the reaction, probably after the introduction of the first phenyl group; the linear compounds so produced then react rapidly with the Grignard reagent.

Two different products have been isolated when diphenylmagnesium in dioxane reacts with the hexachloride. The compound[187] formed in the larger amount has the formula $P_6N_6Ph_7Cl_5$ and, as hydrolysis yields two moles of triphenylphosphine oxide, one mole of phenylphosphonic acid and three moles of phosphoric acid, is given the structure (76). The second product, $P_6N_6Ph_2Cl_8$, consists of two trimeric units linked by a P—P bond (77)[188].

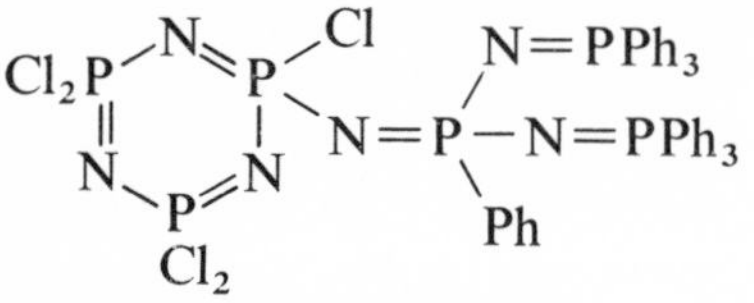

(76)

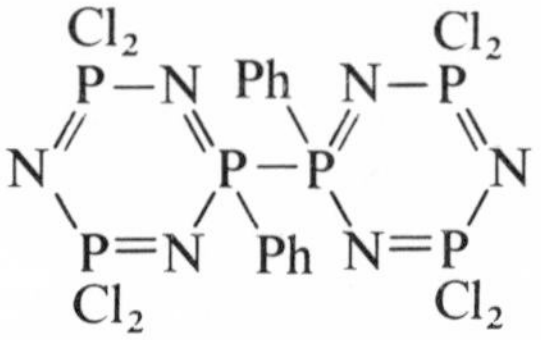

(77)

The reaction of the tetrameric chloride[189] with phenylmagnesium bromide is slower than that for the trimer and, in addition to the octaphenyl, $P_4N_4Ph_8$, yields two major products each with the formula $P_4N_4Cl_4Ph_4$. One is the geminal 1,3-tetraphenyl (78) while the second is the ring contraction product (74) discussed above.

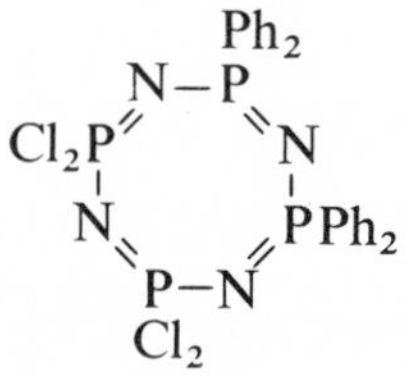

(78)

The relative yields of the two products depends on the basicity of the solvent, with higher basicity favouring ring contraction. Linear compounds are also obtained and, as for the trimer, it is thought that ring opening occurs early in the reaction.

In contrast to the chloride reaction, the trimeric fluoride reacts smoothly in tetrahydrofuran with phenylmagnesium bromide to give good yields of the mono- and gem-di- substituted compounds[190]. This should be contrasted with the corresponding reaction with phenyl-lithium[191], where disubstitution gives a small amount of the geminal compound but mainly the cis and trans non-geminal isomers. Structures can be assigned unambiguously to the cis- and trans-$P_3N_3F_4Ph_2$ compounds by ^{31}P and ^{19}F n.m.r. spectroscopy in conjunction with dipole moment measurements. It is interesting that here the yield of the cis-diphenyl is greater than that for the trans (compared with the reverse order for substitution with secondary amines); a modification of the 'cis-effect' is proposed. This states that there is labilisation of the phosphorus–halogen bond which is cis to the exocyclic group which forms the better π-bond to phosphorus. Thus in a monoaminochloride, the trans-disubstituted compound would predominate but in the monophenyl-fluoride the better π-bonding ability of fluorine would lead to a greater proportion of the cis-diphenylated product.

Other phenylated fluorotriphosphonitriles[192] have been obtained from the hexafluoride by a combination of reactions with phenyl-lithium, which promotes non-geminal substitution, and Friedel–Crafts reactions which give geminal products. The reactions are summarised in (4.34) and (4.35).

$$P_3N_3F_6 \xrightarrow[\text{AlCl}_3]{\text{PhH}} P_3N_3F_4Ph_2 \xrightarrow{\text{PhLi}} P_3N_3F_3Ph_3 \qquad (4.34)$$
$$\qquad\qquad \text{(1,1-isomer)} \qquad \text{(1,1,3-isomer)}$$

$$P_3N_3F_6 \xrightarrow{\text{PhLi}} P_3N_3F_4Ph_2 \xrightarrow[\text{AlCl}_3]{\text{PhH}} P_3N_3F_2Ph_4 \qquad (4.35)$$
$$\qquad\quad \text{(mainly} \qquad\qquad \text{(1,1,3,3-isomer)}$$
$$\qquad\quad \text{1,3-isomers)}$$

In contrast to phenyl-lithium, substitution by methyl-lithium[193] follows a geminal path. For the trimeric fluoride, compounds containing only one or two methyl groups can be isolated due to competition with an addition reaction but for the tetramer both di- and tetra-methyl derivatives are produced. N.M.R. spectra indicate that the latter, contrary perhaps to expectation, is the 1,1,5,5-isomer. One or two fluorines in the trimer and tetramer can be replaced by butyl-lithium[194] but the orientation of the groups is not known. Attempts to produce more highly-substituted compounds give only degradation products. A series of mono(pentafluorophenyl) compounds, $P_nN_n(C_6F_5)F_{2n-1}$ where $n = 3$–8 have been prepared and ^{19}F n.m.r. spectra indicate that the P_3N_3 system withdraws electrons strongly from the pentafluorophenyl ring[195].

4.3.4.2 Reactions of carbon-substituted compounds

Replacement of the chlorine atoms in the geminal-diphenylated chloro-triphosphonitrile, $P_3N_3Cl_4Ph_2$, by ammonia gives both a geminal bisamino compound (79) and the hydrochloride of the tetrakis-derivative[196]. With aniline a similar geminal replacement scheme is followed and mono-, di-, and tetra-substituted compounds can be isolated. Although similar stoichio-metries result from reactions with dimethylamine, ^{1}H n.m.r. spectra show that chlorine replacement in this case follows the non-geminal path.

Chloropentaphenyltriphosphonitrile is hydrolysed by moist pyridine to $P_3N_3Ph_5(OH)$ which has an infrared spectrum which indicates a hydroxy-(80) rather than an oxo- (81) form[197].

$$\text{(79)} \qquad\qquad \text{(80)} \qquad\qquad \text{(81)}$$

The compound is a weak acid and can be converted into the oxygen-bridged compound, $P_3N_3Ph_5$—O—$P_3N_3Ph_5$, on treatment with $P_3N_3Ph_5Cl$ in the presence of pyridine. Kinetic measurements[198] support a base-catalysed hydrolysis mechanism for $P_3N_3Ph_5Cl$. The monochloride can be fluo-rinated to $P_3N_3Ph_5F$ with sodium or ammonium fluoride and reaction with silver perchlorate in pyridine gives a pyridinium perchlorate, $P_3N_3Ph_5$ $(C_5H_5N)ClO_4$.

[In contrast to the pentaphenyl compound above, the hydrolysis product of the chloropentaphenoxide[199], $P_3N_3(OPh)_5Cl$, is considered to have the oxo-structure analogous to (81)].

Reactions of diphenyltetrachlorophosphonitrile with sodium or potassium[200] phenoxides give two products, $P_3N_3Ph_2Cl_2(OPh)_2$, which, by ^1H n.m.r. spectroscopy of their dimethylamine derivatives, were shown to have *cis* and *trans* non-geminal structures. Completely-substituted alkoxy-derivatives of both $P_3N_3Ph_2Cl_4$ and $P_3N_3Ph_4Cl_2$ can be prepared by refluxing with the alkoxide in the parent alcohol[199], however the reaction rate decreases rapidly with increasing phenylation of the ring and several days are required for completion with $P_3N_3Ph_4Cl_2$. In addition to normal alcoholysis products, partially hydrolysed species were sometimes isolated and as infrared and n.m.r. spectra indicate the presence of NH groups, the compounds are considered to have structures based on (82) and (83).

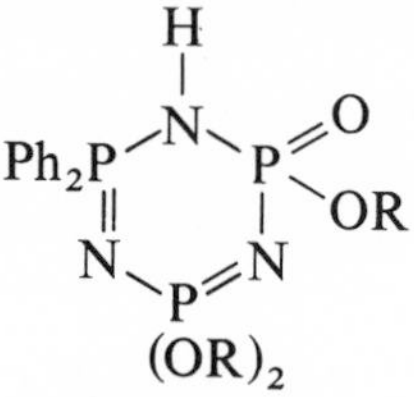

(82) 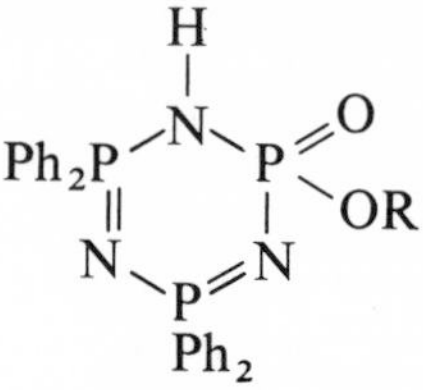   (83)

Compound (83) (R = ethyl) was also obtained by treatment of $P_3N_3Ph_4$ $(OEt)_2$ with hydrogen chloride, in a reaction similar to the alkyl-halide-catalysed rearrangement of alkoxy-phosphonitriles discussed earlier.

The chlorodimethylaminotriphosphonitriles, $P_3N_3Cl_{6-n}(NMe_2)_n$ where $n = 2,3,4$ and 6, react with trimethyloxonium fluoroborate to give mono-methyl derivatives, together with a dimethylated product, $[P_3N_3(NMe_2)_6 Me_2][BF_4]_2$, for the hexakis[201]. In all cases the compounds give trimethyl-amine on hydrolysis and ^1H n.m.r. spectra confirm that alkylation occurs at an amine group. This result is interesting as the most basic centres are considered to be ring-nitrogen atoms and reaction might be expected to occur there. With the isopropylamine derivative, *gem*-$P_3N_3Cl_2(NHPr^i)_4$, alkylation was at a ring nitrogen atom and further, with phosphonitriles containing both isopropylamine and dimethylamine groups, alkylation took place first at a ring-nitrogen but further reaction gave the dimethylated derivative (84). Only one site is available in hexaphenyltriphosphonitrile and alkylation occurs at a ring nitrogen atom.

In the tetrameric series, the non-geminal hexaphenyl compound[202] (85) gives a variety of amine derivatives and with aqueous pyridine, a dihydroxide results.

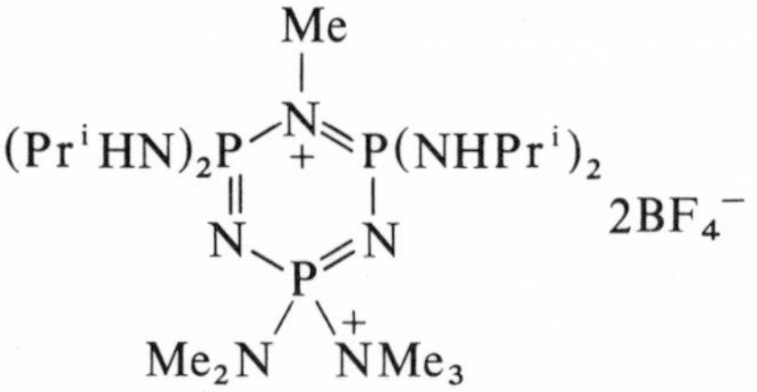

(84) 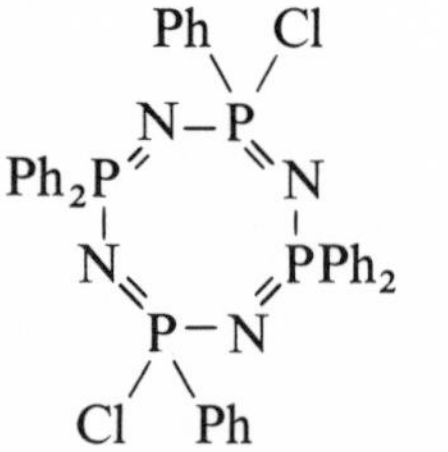   (85)

Treatment of the latter with phosphorus pentachloride regenerates the dichloride but if thionyl chloride is used both the original dichloride and a low melting isomer of unknown structure are formed. The dihydroxide is converted to a dibromide with phosphorus pentabromide and the original chloride can be fluorinated in acetonitrile with caesium fluoride. Reactions with alcohols and diols have also been carried out[55].

Cis- and *trans*-diazides can be obtained from the dichloride (85) and lithium azide[203], the lower melting compound with the higher dipole moment (3.7 D) is assigned the *cis* structure. When the *trans* compound was heated with triphenylphosphine, two moles of nitrogen were evolved and the bistriphenylphosphazo derivative, $P_4N_4Ph_6(N{=}PPh_3)_2$, was produced.

4.3.5 By sulphur

The chlorine atoms in the hexachloride are not substituted by direct reaction with a thiol and, although reaction occurs in the presence of a tertiary base such as pyridine, pure products could not be isolated. The sodium salts are better reagents but the degree of replacement depends markedly on the solvent[204]. Derivatives containing two, four or six alkylthiolate groups are formed predominantly but trisubstituted compounds, $P_3N_3Cl_3(SR)_3$, also result. With sodium benzenethiolate only the bis- and hexakis-products, $P_3N_3Cl_4(SPh)_2$ and $P_3N_3(SPh)_6$, were isolated. N.M.R. spectroscopy proves that geminal substitution takes place.

The trimeric chloride is completely substituted by thiourea giving P_3N_3 $[SC(NH)NH_2]_6$, while three moles of dithio-oxamide or six moles of thio-semicarbazide give compounds (86) and (87) respectively[205].

(86)   (87)

With toluene-3,4-dithiol, tris(1-methylphenyl-3,4-dithio)triphosphonitrile could be prepared[113].

In contrast to the chloride reactions mentioned above, compounds representing all degrees of substitution were isolated from the reaction between hexafluorotriphosphonitrile and sodium ethylthiolate[206]. ^{19}F and ^{31}P n.m.r. spectra indicated however that the reaction gave geminal isomers as in the case of the chloride reaction.

4.3.6   Miscellaneous reactions

Adducts containing a molecule of hydrogen halide have been isolated during the preparation of aminophosphonitriles and interaction is considered to be with the most basic ring-nitrogen atom. Several hydrochlorides have been mentioned in the section on aminolysis. The hexakis-dimethylamino-compound also forms a monohydrobromide and, although there is a large amount of decomposition with hydrogen iodide, small quantities of P_3N_3 $(NMe_2)_6 \cdot HI$ and the hydrogen tri-iodide can be obtained[112]. Solid dihydro-chlorides have been reported for $P_3N_3(NHR)_6$, where R is propyl or butyl, while a trihydrochloride can be isolated for the amyl derivative[207].

The reaction between hexachlorotriphosphonitrile and aluminium bromide in carbon disulphide is exothermic and a 1:1 addition compound, $P_3N_3Cl_6 \cdot AlBr_3$, is precipitated[208]. The corresponding hexabromide is more basic and $P_3N_3Br_6 \cdot 2AlBr_3$ is formed in addition to the 1:1 adduct; further aluminium bromide could not be complexed even when a large excess was used. The adducts are insoluble in non-donor solvents and are decomposed by donors. Infrared spectra can best be interpreted in terms of coordination of aluminium bromide to ring-nitrogen atoms.

The fluorophosphonitriles, $P_nN_nF_{2n}$ where $n = 3-6$, form addition compounds with two moles of antimony pentafluoride and, in addition, the pentamer forms a 1:1 adduct with arsenic pentafluoride[209]. It seems unlikely in view of the very low basicities of the fluorophosphonitriles that coordination is via the ring-nitrogen atoms and the most likely alternative is that the antimony fluoride molecules coordinate to exocyclic fluorines to form bridged structures.

The trimeric fluoride forms a 1:1 addition compound with caesium fluoride in acetonitrile[210], which is a 1:1 electrolyte in nitromethane. Infrared and ^{19}F n.m.r. spectroscopy do not however distinguish between the alternative formulations as the salt of a cyclic (88) or linear (89) anion.

$$F_3P{=}N{-}PF_2{=}N{-}PF_2{-}N^-$$
(89)

(88)

Hexamethyltriphosphonitrile forms solid 1:1 adducts with both tin(IV) and titanium(IV) chlorides[211], while the corresponding tetramer reacts with molybdenum hexacarbonyl[212] to eliminate two moles of carbon monoxide and forming $P_4N_4Me_8 \cdot Mo(CO)_4$. With the methiodide, $P_4N_4Me_8 \cdot MeI$, the compound $P_4N_4Me_9^+[Mo(CO)_5I]^-$ is produced. Crystal structures have been determined for certain compounds between phosphonitriles and metal chlorides; details are included in Section 4.5.1.

A tetrachloroethane adduct with hexaphenyltriphosphonitrile, $P_3N_3Ph_6 \cdot 3C_2H_2Cl_4$, can be obtained by slow crystallisation from tetrachloroethane but other phenyl-substituted phosphonitriles did not behave similarly[213].

The low values of ΔH and ΔG for the decomposition reaction indicate clathrate formation. Similarly the 1:1 acetonitrile adduct with dichloro-tetraphenyltriphosphonitrile is almost certainly a clathrate[214].

Solid-phase interactions between a variety of phosphonitriles and hexa-methylbenzene indicate that phosphonitriles can also show weak acceptor properties[215].

Isothiocyanato-phosphonitriles are stable compounds but attempts to isolate isocyanates by treatment of the trimeric bromide with silver cyanate in nitromethane were not successful[216]. Infrared spectra showed that bromine atoms were replaced but the products polymerised. With $NaB_{10}H_{13}$ the trimeric bromide gives mixed bromophosphonitriles of the type $P_3N_3Br_{6-n}$ $(B_{10}H_{13})_n$ where $n = 1,2,3,6$ [217, 218], while $P_3N_3(B_{12}H_{12})_3$ is obtained from $Na_2B_{12}H_{12}$ in tetrahydrofuran[219].

Lithium aluminium hydride did not react with hexachlorotriphospho-nitrile at $0\,^{\circ}C$, but at $20\,^{\circ}C$ phosphine was evolved to leave a chlorine-free residue probably consisting of a complex mixture of polymeric phosphorus hydrides[220]. A vanadium-containing phosphonitrile (90) results from reaction between $VOCl_3$ and 1,1,5,5-tetrakis(trichlorophosphazo)tetra-chlorotetraphosphonitrile[221].

$$
\begin{array}{c}
Cl_2 \\
N-P \\
Cl_2P \diagup \quad \diagdown N \\
\big| \qquad\qquad \big| \\
N \diagdown \quad \diagup P(N{=}VCl_3)_2 \\
P-N \\
Cl_2
\end{array}
$$

(90)

4.4 POLYMER FORMATION

Polymeric species based on phosphonitriles result from the following approaches. (a) Thermal polymerisation of a halide, in particular the trimeric chloride. (b) Replacement of the halogen atoms in such polymerised species by methods similar to those described in the preceding section. (c) Formation of polymers by reaction between phosphonitriles and difunctional reagents. (d) Polymerisation due to unsaturation in exocyclic groups. These approaches fall into two classes, i.e. those in which polymerisation depends on ring opening and the formation of P—N chains, (a) and (b), and the latter two where the phosphonitrile ring is maintained and polymerisation depends on linking these together.

Thermal polymerisation of hexachlorotriphosphonitrile depends markedly on the purity of the compound and with rigorously purified material occurs only slowly at $300\,^{\circ}C$ [222]. The reaction involves an addition chain reaction but the nature of the active centre is considered uncertain. Addition of dry oxygen greatly accelerates the process and polymerisation of material purified only by a single vacuum sublimation takes place rapidly at $270\,^{\circ}C$.

A second investigation of the bulk polymerisation of the hexachloride[223] indicates that second-order kinetics are followed with an energy of activation of 57 kcal mol^{-1}. First-order kinetics are followed in the presence of benzoic acid. A mechanism is suggested for the uncatalysed reaction in which the propagation step (4.36) is the result of nucleophilic attack on a molecule of the trimer by the linear species (91) which is in equilibrium with the cyclic trimer at elevated temperatures.

$$Cl_2P^+\!=\!N\!-\!PCl_2\!=\!N\!-\!PCl_2\!=\!N^- + P_3N_3Cl_6 \rightarrow$$

$$(91) \qquad Cl_2P^+\!=\!N\!-\!(PNCl_2)_4\!-\!PCl_2\!=\!N^- \qquad (4.36)$$

Radiation-induced polymerisation of the trimeric chloride and bromide takes place at temperatures much below those required for thermal polymerisation[224] and the rate of polymerisation is markedly decreased for solid solutions of the two halides.

Short linear polymers may be stabilised by thermal polymerisation in the presence of phosphorus pentachloride or hydrogen chloride[225] but it is possible to obtain chlorophosphonitrile species with up to 15 000 repeating units which are soluble in benzene[226]. Polymerisation for longer periods gives insoluble species which are elastomeric and insoluble due to cross-linking between P—N chains.

Direct formation of polyphosphonitriles by thermal methods occurs only when the exocyclic groups are halides or isothiocyanate. The nature of the exocyclic groups is clearly of great importance and their effect on inter-conversion between cyclic oligomers and the polymer have been investigated[227].

Polymeric phosphonitriles containing groups other than halogen can be prepared by methods similar to those discussed in Section 4.3 by use of the benzene-soluble polymeric chloride as starting material. The chlorine atoms were completely replaced by aniline and ethylamine but with methylamine or ammonia both substitution and cross-linking occurred[228] to give partially substituted compounds insoluble in organic solvents. The extent of substitution by secondary amines is governed by their steric properties, e.g. dimethylamine and piperidine give complete substitution while methyl-phenylamine and diphenylamine will not substitute even one chlorine per phosphorus atom. On the other hand, diethylamine[229] in tetrahydrofuran replaced half the chlorine atoms non-geminally in polychlorophospho-nitrile and species containing two different amino-residues can be prepared by subsequent reaction with ammonia, methylamine, ethylamine, etc.

Similar reactions between $(PNCl_2)_n$ and the sodium salts of a variety of alkoxides give chlorine-free polymers[226, 230, 231] but, as found for the amine-substituted compounds, the temperature at which depolymerisation sets in is markedly lowered from that in the polymeric chloride[226]. Unlike the chloride, however, many of the products have high hydrolytic stability and stability towards acids and organic solvents. Allen and co-workers[231] have shown that aryloxy-substituted polymers have higher transition temperatures and greater thermal stability than alkoxy-derivatives and the co-polymer containing equal numbers of trifluoroethoxy and heptafluorobutoxy groups[232] is an elastomer which is inert to water, acids and organic solvents. A completely-phenylated polymer results from reaction with phenyl-lithium[233]. Polyphosphonitriles on thermal decomposition usually give a mixture of cyclic compounds and shorter chain polymers, complicated in some cases by elimination of volatile decomposition products from side chains and the formation of black residues[234, 235].

Examples have already been given of the formation of polymeric compounds by use of a difunctional reagents[155–158, 164, 165]. Others involve reactions between hexamidotriphosphonitrile and either diacylchlorides[236] or di-isocyanates[237] and between the non-geminal dichloride, $P_4N_4Ph_6Cl_2$, and dihydroxy compounds[238]. Boron-containing polymers can be obtained by treatment of the trimeric chloride or hexabutoxide with boron diols or triols[239].

The patent literature contains many examples on the formation of further polymeric species and specialised reviews such as those by Saito[240] and Horn[241] cover aspects of this behaviour.

4.5 PROPERTIES

4.5.1 Structures

4.5.1.1 Trimeric compounds

Corbridge[242] has reviewed earlier structural data. It is reported in two papers[243, 244] that the P_3N_3 ring in hexabromotriphosphonitrile has a slight chair conformation similar to that found for the corresponding chloride, and that within experimental error the P—N bond lengths (1.576 Å) are equal[244]. The phosphorus ring angles (N—P—N) are 118.5 degrees but two different angles at nitrogen, 119.3 and 122.4 degrees, are observed. This difference and the chair conformation result from intermolecular forces.

The hexaisothiocyanate, $P_3N_3(NCS)_6$, also contains a non-planar ring system[245] but here one nitrogen atom deviates by 0.15 Å from the plane described by the other ring atoms. The ring bonds vary from 1.52 to 1.64 Å with a mean value of 1.58 Å and, as there is no chemical reason for bond inequality in this uniformly-substituted phosphonitrile, the discrepancy probably results from large thermal motions. The mean ring angle is 121.0 degrees though the variation is quite large. The P—N bond to the isothiocyanate group is short, 1.64 Å, which implies partial donation of the nitrogen lone-pair into a phosphorus orbital; this multiple bonding is supported by the large average P—N—C angle (152 degrees).

In the hexaphenyl derivative[246] the ring system has an irregular, slight chair, conformation but the P—N bonds are statistically equal in length (1.597 Å). The mean ring angles are 117.8 degrees for N—P—N and 122.1 degrees for P—N—P. Hexaphenoxytriphosphonitrile[247] provides a further example of a non-planar ring system with two of the nitrogen atoms displaced in opposite directions from the plane of the other four atoms. Thus of all the uniformly-substituted trimeric compounds examined, a planar ring system has been observed only for the fluoride. In the phenoxide the ring angles show variations which conform to C_2 symmetry but there is no difference in length between chemically equivalent bonds and the mean P—N distance is 1.575 Å. These latest data confirm the trend that shorter ring bond lengths are associated with the higher electronegativity of the exocyclic groups, and that with increasing electronegativity there is a decrease in the P—N—P and exocyclic P angles.

The benzene and bromobenzene inclusion compounds[248, 249] of the triscatechol derivative of the trimer, tris(o-phenylenedioxy)triphosphonitrile, have a unit cell containing two molecules of the phosphonitrile and one guest molecule. The host molecules stack in such a way as to form continuous channels along the c-axis with a diameter of 4.5 Å and guest molecules are arranged randomly in these channels. The structure is stabilised by host–guest contacts: removal of the guest molecules causes collapse of the lattice. The phosphonitrile ring is planar with bond distances of 1.575 Å. An interesting feature of the structure is the small (97 degrees) exocyclic O—P—O angle.

In the diphenyl analogue, tris-(2,2'-dioxydiphenyl)triphosphonitrile[250], the ring bonds average 1.572 Å and the ring has a slightly distorted boat conformation. The exocyclic angle (O—P—O) is approximately 103 degrees which indicates the absence of strain in this structure; in contrast to the situation above, the greater bulkiness of the side-groups does not allow the formation of channels and thus inclusion compounds are not formed.

Interesting results emerge from structures where the ring system carries different substituents. In the geminally-substituted compound, $P_3N_3Ph_2Cl_4$, the ring has a slight chair conformation and there are three different bond distances as shown below (92)[251].

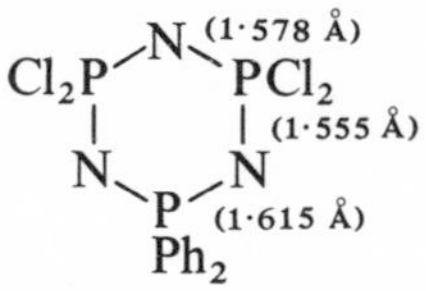

Cl₂P—N(1·578 Å)—PCl₂
| | (1·555 Å)
N—P—N
Ph₂ (1·615 Å)

(92)

F₂P—N(1·558 Å)—PF₂
| | (1·539 Å)
N—P—N
Ph₂ (1·618 Å)

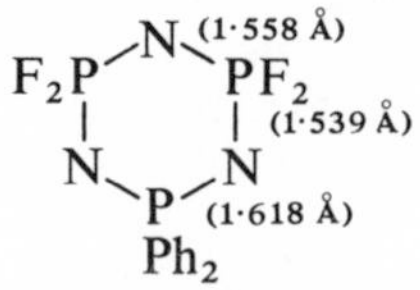

(93)

This variation provides strong support for the theoretical picture of bonding in these compounds. In the segment Ph_2PNPCl_2, the greater electronegativity of chlorine will lead to a greater degree of multiple bonding in the N—P(Cl₂) bond than in the N—P(Ph₂) bond and in agreement the former has the shortest bond distance. The situation with the corresponding fluoride[252], $P_3N_3Ph_2F_4$, is very similar with respect to variation in bond distances (93), but the effect is more pronounced due to the greater electronegativity of fluorine. This leads to better orbital overlap and greater delocalisation of the lone pair on the nitrogen atom between the phenyl and fluorine-substituted phosphorus atoms. The ring conformation is however markedly different from that in $P_3N_3Ph_2Cl_4$. The phenyl-substituted phosphorus atom and the nitrogen atom opposite are displaced from the plane of the other four atoms, to give an overall boat conformation.

In the case of the tetraphenyl compound[253], $P_3N_3Ph_4Cl_2$, the ring has a slight boat conformation but there are again three different P—N bond lengths (94).

The results for $P_3N_3Cl_5F$ [254] indicate that the fluorine atom is randomly distributed over the halogen atom positions. But, in spite of the disorder in the molecule, the bond lengths were shown to be equal (1.563 Å) and the ring system is close to planarity.

Direct evidence for protonation at a ring-nitrogen atom in a hydrogen chloride adduct comes from the structure of the hydrochloride of the geminally-substituted tetra-i-propylamine derivative, $P_3N_3Cl_2(NHPr^i)_4$ [255]. The compound is ionic and consists of chloride ions and $[P_3N_3Cl_2(NHPr^i)_4H]^+$ cations. On basicity arguments (see Section 4.5.5) protonation would be expected at the nitrogen flanked by the amine-substituted phosphorus atoms and this is confirmed by the structure (95).

$$\underset{\text{(94)}}{}\qquad\qquad\underset{\text{(95)}}{}$$

The ring has a slight boat conformation and there are three different P—N bond lengths. The two shorter bonds fall in the expected range for phosphonitriles and the variation is related to electronegativity differences of the substituents as discussed previously, while bonds to the protonated nitrogen are very long in agreement with interruption of the π-bonding system. The exocyclic phosphorus–nitrogen(amine) bonds are very short (1.609 Å), implying that the positive charge at the protonated nitrogen promotes a large degree of multiple bonding due to delocalisation of the amine lone-pairs. The configuration at the protonated nitrogen is planar and the ring angle (P—N—P) is large (132 degrees).

In the thermal rearrangement product of hexamethoxytriphosphonitrile which contains N-methyl groups [see equation (4.32) R = Me], the ring system has a twisted-boat conformation and the average bond length is 1.66 Å [256]. This is still considerably shorter than a single P—N bond (1.77 Å) suggesting that, even though π-bonding is not formally required, some occurs due to delocalisation of nitrogen lone-pairs into phosphorus orbitals. The degree of multiple bonding is, as expected, much less than in the phosphonitriles. The configuration around the ring-nitrogen atoms is very close to planar.

Finally an electron-diffraction examination of the trimeric chloride [257] indicates slight puckering of the ring to a chair conformation with bond lengths of 1.585 Å.

4.5.1.2 Tetrameric compounds

Two different conformers of the tetrameric chloride are known. The structure of the metastable K-form was determined earlier [258] and showed that the P_4N_4 ring had a tub (S_4) conformation. In the stable T-form [259], the ring is in a chair conformation with approximately C_{2h} symmetry. The ring bond distances are equal (1.559 Å) but slightly shorter than those in the K-form, while the chair arrangement requires larger angles at nitrogen

(134, 138 degrees) than found for the metastable form (131 degrees). The ring angles at phosphorus average 120.5 degrees.

A third type of ring conformation is found in $P_4N_4(OMe)_8$ [260]. Here the phosphorus atoms are nearly co-planar with nitrogen atoms displaced alternatively above and below this plane to give a 'saddle' arrangement which approaches D_{2d} symmetry.

Two centrosymmetric compounds, $P_4N_4Cl_4Ph_4$ and $P_4N_4(NHMe)_4Ph_4$, have chair-type ring conformations[261] similar to that for the T-form of the tetrameric chloride. The arrangement of substituents to give a centre of symmetry is *cis-cis-trans-trans*. As in the tetrameric chloride structure, there are two significantly different P—N—P ring angles; for the chlorophenyl compound these are 132 and 139 degrees and for the amino derivative, 125 and 131 degrees.

The P_4N_4 ring in the *cis* form of $P_4N_4Cl_4Ph_4$ has a crown conformation[262] which is flattened due to two opposite phosphorus atoms being disposed closer to the plane of the four nitrogens than the other phosphorus atoms. The symmetry approximates to C_{2v}. The nitrogen angles are all large, ranging between 133.2 and 141.8 degrees and it appears that this ring conformation arises because of steric effects due to the phenyl groups being all in *cis* positions.

In the geminally-substituted tetramer, 1,1,5,5-tetramethyl-3,3,7,7,-tetra-fluorotetraphosphonitrile, there are two different ring bond lengths[263] Those adjacent to the methyl-substituted phosphorus atoms are 1.59 Å while those including the fluorine-substituted atoms are 1.53 Å. This arrangement is associated with the different electronegativities of fluorine and methyl groups and the greater degree of delocalisation of nitrogen lone-pairs into orbitals on the fluorine-substituted phosphorus atoms. The ring adopts a 'saddle' conformation similar to that for the octamethoxide.

Four distinct phosphorus–nitrogen bond lengths are observed with the corresponding geminal dimethyl compound[264], $P_4N_4Me_2F_6$, and the variation (96) can be readily rationalised in terms of different degrees of π-bonding. Again, the ring adopts a saddle conformation.

$$
\begin{array}{c}
\text{Me}_2 \\
\text{N}-\text{P} \quad (1\cdot584\text{ Å}) \\
\text{F}_2\text{P} \qquad \text{N} \\
| \qquad\qquad | \quad (1\cdot470\text{ Å}) \\
\text{N} \qquad\quad \text{PF}_2 \\
\text{P}-\text{N} \quad (1\cdot532\text{ Å}) \\
\text{F}_2 \,(1\cdot487\text{ Å})
\end{array}
$$

(96)

Protonation of the tetrameric ring, as in $[P_4N_4Me_8H]CuCl_3$ and $[P_4N_4Me_8H]_2CoCl_4$, leads to a similar variation in bond lengths. In the copper compound[265], the proton and the $CuCl_3$ group are covalently bonded to opposite nitrogen atoms, to give rise to the four distinct pairs of ring bonds as shown in (97). The ring-nitrogen angles also show considerable variation but those carrying the $CuCl_3$ and H groups are close to 120 degrees; this indicates that the proton and metal are attached via nitrogen lone pairs. The ring adopts a 'tub' conformation similar to that of the parent methyl compound.

The structure of the cobalt compound[266] is different and consists of tetrahedral $CoCl_4^{2-}$ ions and two protonated $P_4N_4Me_8$ rings. One ring is in the 'tub' conformation similar to that in the copper compound but in the other ring the conformation is more nearly a 'saddle' type. In each ring however there are four pairs of ring bonds arranged as in (98). Variations both here and in the copper compound result from modification of the π-system by protonation.

4.5.1.3 Higher homologues

Data are beginning to accumulate on the structures of phosphonitriles containing larger ring systems. The P_5N_5 ring system in the pentameric chloride[267] is approximately planar but it deviates from D_{5h} symmetry as two nitrogen atoms are puckered towards the ring centre. The phosphorus–nitrogen lengths range between 1.49 and 1.55 Å (average, 1.52 Å) but there is no implication of alternating double bonds. The ring angles at phosphorus (118.4 degrees) are similar to those in other phosphonitriles but the nitrogen angles vary from 133.4 to 159.0 degrees (average 148.6 degrees). This large average value correlates with the shortness of the P—N bonds and implies a large degree of π_s-bonding.

For the hexameric dimethylamino compound[268], $P_6N_6(NMe_2)_{12}$, the twelve-membered ring adopts a highly puckered conformation, comparable

(97)

(98)

in many respects to the 'tub' conformation in the P_4N_4 series. The ring bonds (1.56 Å) are equal in length while shortening of the P—N (amine) bonds to 1.67 Å and the near planarity of the amine nitrogens both imply multiple bonding. A striking feature of this structure is the very large angle at the ring nitrogen atoms (147.5 degrees). Such large angles are usually associated, as in the case of the tetrameric fluoride, with electronic effects which promote delocalisation of the nitrogen lone-pair of electrons but this cannot be so when the exocyclic groups are amines. Evidence is presented that the large angles result from steric interactions between the bulky amino-groups.

The compound, $P_6N_6(NMe_2)_{12}$, can behave as a macrocyclic ligand[269] and, when treated with a mixture of copper-(I) and -(II) chlorides, gives $[P_6N_6(NMe_2)_{12}CuCl]^+CuCl_2^-$. The anion is linear and in the cation copper(II) is coordinated to four of the ring-nitrogen atoms and a chlorine atom in a distorted square-pyramidal arrangement. This leads to two types of ring bonds; those which involve a coordinated nitrogen are longer

(1.62 Å) than the others (1.55 Å). The structure of the ring tightens on coordination and both the angles at phosphorus (107.5 degrees) and nitrogen (133.6 degrees) are reduced from their values in the uncomplexed hexamer.

Nitrilononachlorahexaphosphonitrile, $P_6N_7Cl_9$, which is obtained in small quantities from the reaction between phosphorus pentachloride and ammonium chloride, contains a condensed ring structure[270] (99) with molecular symmetry. C_{3v}.

(99)

Although the molecule as a whole is non-planar, the central nitrogen and the three attached phosphorus atoms are almost co-planar. Surprisingly, however, these P—N bonds are the longest known for any phosphonitrile (1.723 Å), and it seems most likely that the non-planarity of the whole molecule is a result of σ-bonding requirements.

The only octameric structure determined is that of the methoxy-compound, $P_8N_8(OMe)_{16}$ [271]. The molecule is centrosymmetric and the ring consists of two approximately planar and parallel six atom segments joined by a step formed from the remaining two phosphorus and nitrogen atoms. The ring bonds are equal in length (average 1.561 Å) within experimental error and the angles at nitrogen (136.7 degrees) are larger than in the corresponding tetramer (132 degrees).

4.5.1.4 Polymers

X-ray patterns have been given for oriented fibres of polyphosphonitriles containing trifluoroethoxy, chlorophenoxy and phenylphenoxy groups[226, 231]. The chain repeat-distances all fall between 4.8 and 4.9 Å which indicates that the substituents fit into the polymer chain without causing undue steric hindrance. Polymerised fluorotriphosphonitrile[272] shows two different diffraction patterns depending on the temperature. The conformer formed between 25 and −36 °C has a repeating distance of 6.49 Å, while in the region of −56 °C the repeating distance is 4.92 Å. The latter has been correlated with the presence of a cis–trans planar conformation.

4.5.2 Bonding

Irrespective of ring size, bonding in phosphonitriles can be considered in terms of (a) a σ-bonded skeleton which arises from approximately sp³-hybrid orbitals on each phosphorus atom and approximately sp^2-hybrids on the ring nitrogen atoms, (b) a π-system which results from combination between the remaining p orbital on each nitrogen and a d orbital on phosphorus; this is designated as the π_a-(antisymmetric) system, (c) a second π-system which arises from combination between filled sp^2-hybrid orbitals (formally containing a lone-pair) on nitrogen and an empty d orbital on phosphorus; this system is designated as the π_s (symmetric) system, (d) π-interaction between unshared electron pairs on exocyclic groups and suitable d orbitals at phosphorus.

These systems do not operate independently and it is clear that any changes in electron density at phosphorus can have a marked effect on the relative magnitudes of the π_a and π_s systems. This in turn affects the ring conformation, particularly for oligomers higher than trimers, and the whole problem is potentially of great complexity.

The general use of d orbitals in bonding has been considered by Mitchell[273] and calculations for the phosphonitriles show that of the five available orbitals, the d_{xz} orbital is mainly involved in π_a bonding and the $d_{x^2-y^2}$ in the π_s system[274]. A large amount of support for the basic correctness of this approach has come from the results of crystal-structure determinations discussed in the previous section.

From symmetry considerations, the π_a-system is expected to be hetero-morphic while the π_s-system will be homomorphic and, although it is difficult to separate the two systems completely, it should be possible to observe an alternation in certain properties with change in ring size due to the presence of the π_s-system. The most recent evidence is from the measure-ment of ionisation potentials[275]. For the fluorophosphonitriles, $(PNF_2)_n$, the values alternate with those for odd values of n being greater than those where n is even. There is a similar situation for the chloride series and for other exocyclic groups; the ionisation potential of the tetramer is always lower than that for the trimer. This behaviour can be correlated with ionisation of an electron from the π_s-system. Photoelectron spectroscopy of the fluorides gives further ionisation potentials which can be correlated with electron loss from the highest level of the π_a-system, from lone-pairs on fluorine and from phosphorus–fluorine σ-bonds. The pattern of levels agree with those obtained from Hückel molecular orbital calculations and bear out previous conclusions that the higher π-system is homomorphic and can be correlated with the π_s-system.

4.5.3 Vibrational spectroscopy

Infrared spectroscopy is widely used and most papers dealing with pre-parative aspects contain some discussion of the absorption spectra but in this section only papers concerned primarily with more detailed inter-pretation of infrared and Raman spectra are considered.

Infrared and Raman data for hexabromotriphosphonitrile[276] indicate that in solution the molecule has D_{3h} symmetry but that deviations toward C_{3v} symmetry occur in the solid. Heat capacity, heat content, free energy and entropy data have been calculated for the trimeric fluoride, chloride and bromide[277]. The spectrum of the hexaisothiocyanate, $P_3N_3(NCS)_6$, is best interpreted on the basis of D_{3h} symmetry[278].

Detailed analyses have been given of the infrared and Raman spectra of the geminal chlorofluorotriphosphonitriles[279], $P_3N_3Cl_nF_{6-n}$, and of the non-geminal bromide chlorides[43, 280]. The dimethylamine-substituted trimeric chlorides and bromides have also been examined[281] and an important correlation has been shown between phosphorus–nitrogen (amine) absorptions in the 700 cm^{-1} region and the presence of geminal or non-geminal amino groups. Infrared spectra of trimeric compounds substituted with alkoxy, amino and thiolate groups have also been discussed[282].

Spectra of oriented samples of $P_3N_3Cl_6$ [283], $P_3N_3Br_6$ [283], and $P_3N_3Cl_5$ Br [284] with polarised infrared radiation generally confirm previous assignments for these molecules.

The tetrameric chloride[276, 285, 286] and bromide[276, 285] in solution give spectra that agree with D_{2d} symmetry, while the solid-phase spectra indicate lowering of symmetry to S_4. Polarised infrared spectral data have also been obtained for both compounds[283] and have been interpreted on the basis of ·S_4 and C_i site symmetries. The most recent examination of the vibrational spectrum of the tetrameric chloride[287] confirms the D_{2d} molecular symmetry in solution and for the melt and indicates that in the gas phase the symmetry is D_{4h}. Spectra were observed for both the K- and T-forms of the solid and can be interpreted on the basis of the molecular symmetry being S_4 and C_{2h} respectively.

The infrared and Raman spectra[288] of solutions of $P_5N_5Br_{10}$ indicate that the molecular symmetry is close to D_{5h} but the coincidence of certain infrared and Raman frequencies indicates distortion.

A helical structure with C_2 symmetry is suggested for polychlorophosphonitrile[289] from analysis of infrared data on the material as a stretched fibre, a thin film or as a rubber swollen by carbon tetrachloride.

4.5.4   N.M.R. spectroscopy

As with infrared spectroscopy, n.m.r. spectroscopy has been widely used in assigning structures to substituted phosphonitriles. ^1H n.m.r. has been particularly useful for suitable derivatives and details of the spectra can be found in many preparative papers.

Proton resonances of amino-substituted phosphonitriles[290] are generally not first order but a doublet for each different amine environment is usually observed. In some cases, the amine signals are not sufficiently well separated and use can be made of a benzene-induced solvent shift which moves signals due to cis-amine groups upfield to a greater extent than those in trans positions[291]. The spectra are generally complicated because of the complexity of the spin system and there is a build-up of intensity between the doublet components which becomes more important in dimethylamine derivatives as the ring size increases[127]. Another complicating factor is that spin–spin coupling can occur between chemically equivalent phosphorus atoms, e.g. in triphosphonitriles which carry ethoxy and phenyl substituents[292]. These effects have been considered theoretically[293, 294] and recently Harris and his co-workers[295, 296] have concentrated on a description of the $(AX_n)_3$ system found in cis-trisubstituted triphosphonitriles.

Phosphorus resonances are clearly very important and such data have been tabulated[297]. Spin coupling between phosphorus atoms in these and similar compounds has been reviewed[298]. Recent phosphorus data are concerned with trimeric chloride bromides[42, 43], a series of monosubstituted triphosphonitriles[299] and amine derivatives[300]. Chemical shift and coupling constant data together with the results of dynamic polarisation experiments are also available for a variety of substituted tri- and tetra-phosphonitriles[301]. Similar dynamic nuclear polarisation experiments have been carried out with the chlorophosphonitriles, $(PNCl_2)_{3-7}$ [302].

4.5.5 Basicity measurements

In a series of papers[303-309], Feakins, Shaw and co-workers have assessed the basicity of a large number of substituted phosphonitriles. The experimental technique[303] involves a potentiometric titration with perchloric acid in nitrobenzene solution and the results are expressed in terms of pK_a' values. These values are related by a constant to the acid dissociation constants (pK_a) in aqueous solution. Protonation is considered to occur in all cases at ring-nitrogen atoms[304] so that all the results are directly comparable. In some cases the titration yields two pK_a' values $[pK_{a(1)}'$ and $pK_{a(2)}']$.

A selection of the results for homogeneously-substituted trimeric and tetrameric phosphonitriles is shown in Table 4.1. With aliphatic amines as substituents the compounds are all strong bases and diprotonation can also be detected in each case. This base strengthening[304], compared with the corresponding chlorides, arises because of increased electron release to phosphorus from the amine groups with consequent greater localisation of the lone pair on the ring-nitrogen atoms. A similar release of electrons from methoxy and ethoxy groups leads also to base strengthening which is much less for the corresponding sulphur derivatives in agreement with smaller electron release from sulphur[305]. The variation within each series is related to the inductive effect of the alkyl groups but in the case of the amino compounds electron release in each case is very large and a saturation effect comes into play. The high basicity of the ethyl derivatives is in line with the known electron releasing power of this group.

In all cases where second pK_a' values are observed, these are higher for the tetramer than for the analogous trimer. This can probably be associated with steric effects which are expected to be much lower for the larger ring system.

The pK_a' values for compounds carrying different ring substituents depend both on the nature of the substituents and also their arrangement. In general, base strengthening groups in a geminal position will lead to a higher basicity than with the corresponding non-geminal arrangement, e.g. pK_a' for gem-$P_3N_3Cl_3(NMe_2)_3$ is -4.4, while for the non-geminal isomers, values of -5.5 and -5.4 were obtained[306]. It is not possible in general, to detect significant differences between cis and trans non-geminal isomers.

By a consideration of the electronic effects of substituents, it is possible in trimeric compounds to assess the extent of base strengthening at the nitrogen atoms adjacent (α) and opposite (γ) the given substituent. In all cases the effect at the α atoms is greater than at γ. Thus in the example above, protonation is expected at the nitrogen atom between the amine-substituted atoms [indicated by an asterisk in (100)] as this is subject to the greatest

Table 4.1   pK'$_a$ values for homogeneously substituted tri- and tetra-phosphonitriles

X	$P_3N_3X_6$		$P_4N_4X_8$		Reference
	$pK'_{a(1)}$	$pK'_{a(2)}$	$pK'_{a(1)}$	$pK'_{a(2)}$	
NHMe	8.8	−2.0	8.2	3.4	304
NHEt	8.2	−1.3	8.1	3.8	304
NHPr^i	8.4	−1.5	8.1	2.2	304
NMe$_2$	7.6	−3.3	8.3	0.6	304
NEt$_2$	8.5	−3.9	8.3	−0.9	304
Et	6.4		7.6	0.2	305
Ph	1.5		2.2	−5.8	305
OMe	−1.9		−1.0		305
OEt	−0.2		0.6		305
OPh	−5.8		−6.0		305
SEt	−2.8				305
SPh	−4.8				305
Cl	< −6.0		< −6.0		305

base strengthening, i.e. 3α, due to the three amines on adjacent atoms. Base strengthening occurs at the other nitrogen atoms as shown but the effects are smaller. For the non-geminal isomers (101), all nitrogen atoms are of equal basicity but the strengthening is lower than for the unique nitrogen in the geminal compound.

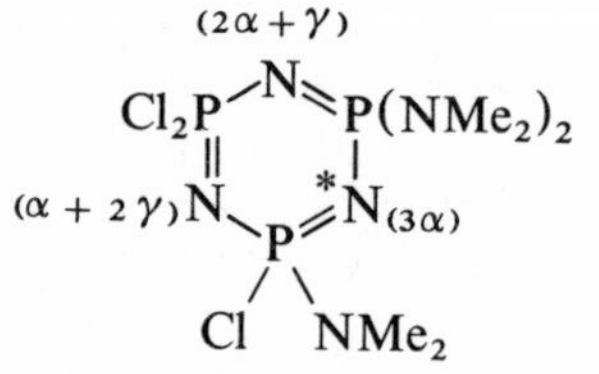

(100)

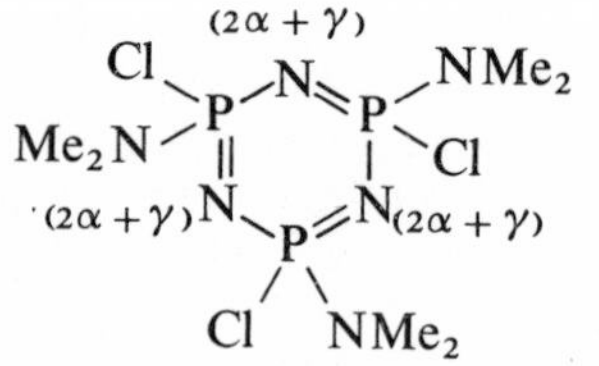

(101)

From a detailed analysis of the pK'$_a$ values for a large number of derivatives containing different ring substituents[307–309], it is possible to evaluate a series of substituent constants (α_R and γ_R). To a first approximation these are additive and some representative values are included in Table 4.2.

4.5.5   Mass spectrometry and other measurements

Fragmentation patterns have been reported for the bromides[310], $(PNBr_2)_{3-6}$, chlorides[311, 312], $(PNCl_2)_{3-8}$, and fluorides[313], $(PNF_2)_{3-16}$. In all cases loss of halogen atoms from the original ring structure is an important process in which even-electron species are formed predominantly. This leads to an alternation in the relative intensities of ions with the number of attached halogen atoms and, as doubly charged species are fairly abundant, the alternation is also observed in this series. With trimeric compounds linear ions, $P_nN_{n-1}X_m^+$, are also formed but they are relatively unimportant when compared with the cyclic ions. For all higher phosphonitriles well-defined series of cyclic ions based on all smaller rings can be detected but it is clear that there are preferred modes of ring fragmentation for hexamers and higher homologues.

The behaviour of the trimeric and tetrameric isothiocyanates, [PN $(NCS)_2]_{3,4}$, is very similar to the halides but processes involving fragmentation of the thiocyanate group, i.e. loss of S and CS, are also observed[314].

Table 4.2   Substituent constants for $P_3N_3X_6$ [308, 309]

R	NHMe	NHEt	NMe$_2$	NEt$_2$
α_R	5.8	5.8	5.6	5.5
γ_R	3.1	3.6	2.8	3.1
R	OMe	OPh	SEt	SPh
α_R	3.6	3.1	3.6	3.0
γ_R	1.8	1.3	1.8	1.5

Fragmentation patterns have been reported for mixed trimeric chloride bromides, $P_3N_3Cl_nBr_{6-n}$ where $n = 1-5$ [315]. The cyclic and linear ions observed are similar to those found for the parent halides, but bromine atoms are lost preferentially. Ionisation potential measurements show that there is a large decrease (from 10.27 to 9.83 eV) on substitution of one chlorine atom in $P_3N_3Cl_6$ by bromine, but that in subsequent substitutions the change is much less. This can be rationalised in terms of the lower electron-withdrawing power of bromine and the consequent greater localisation of the lone-pair of electrons on the adjacent nitrogen atom.

The ^{35}Cl n.q.r. of the K- and T-forms of the tetramer[316, 317] show two and four lines respectively in agreement with molecular symmetries assigned from x-ray data, and there is a transition between the two forms at 63 °C.

Vapour pressure data have been obtained by the Knudsen effusion method for trimeric[318] and tetrameric[318] chlorides, the trimeric[319] and tetrameric[319] bromides and the hexameric chloride[319]. The molar polarisation, molar refraction and Kerr constant have been evaluated for the trimeric chloride[320]. Dipole moments for pyrrolidine, piperidine, morpholine and ethyleneimine substituted chlorotriphosphonitriles[321] have been measured and correlated with structure. E.S.R. spectra have been recorded of radical anions generated from the phenyl-substituted tri- and tetra-phosphonitriles[322].

References

1. Paddock, N. L. (1964). *Quart. Rev. Chem. Soc.,* **18,** 168
2. Shaw, R. A., Fitzsimmons, B. W. and Smith. B. C. (1962). *Chem. Rev.,* **62,** 247
3. Keat, R. and Shaw, R. A. (1970). *Organophosphorus Chemistry,* Vol. 1. (London: Chemical Society)
4. Schmulbach, C. D. (1962). *Progr. in Inorg. Chem.,* **4,** 275
5. Haiduc, I. (1970). *The Chemistry of Inorganic Ring Systems,* (New York: Wiley-Interscience)
6. Allcock, H. R. (1967). *Heteroatom Ring Systems and Polymers,* (New York: Academic Press)
7. Pantel, S. and Becke-Goehring, M. (1969). *Sechs- und Achtgliedrige Ringsysteme in der Phosphor-Stickstoff Chemie.* (Berlin: Springer-Verlag)
8. Shaw, R. A., Keat, R. and Hewlett, C. (1965). *Preparative Inorganic Reactions,* **2,** 1
9. Fitzsimmons, B. W. and Shaw, R. A. (1966). *Inorg. Syn.,* **8,** 77
10. Carroll, A. P. and Shaw, R. A. (1966). *Inorg. Syn.,* **8,** 84
11. Schmutzler, R. (1967). *Inorg. Syn.,* **9,** 76
12. Moeller, T. and Tsang, F. (1967). *Inorg. Syn.,* **9,** 78
13. Nannelli, P., Chu, S-K., Manhas, B. and Moeller, T. (1968). *Inorg. Syn.,* **11,** 201
14. Allen, C. W. and Moeller, T. (1970). *Inorg. Syn.,* **12,** 293
15. Zhivukhin, S. M., Tolstoguzov, V. B., Kireev, V. V. and Kuznetsova, K. G. (1965). *Russ. J. Inorg. Chem.,* **10,** 178
16. Zhivukhin, S. M., Kireev, V. V., Popilin, V. P. and Kolesnikov, G. S. (1970). *Russ. J. Inorg. Chem.,* **15,** 630
17. Kobayashi, E. (1966). *Kogyo Kagaku Zasshi,* **69,** 618 (*Chem. Abstr.,* 1968, **69,** 108185)
18. Paddock, N. L. and Searle, H. T. (1969). *U.S. Patent* 3 407 047 (*Chem. Abstr.,* 1969, **70,** 39397)
19. Zhivukhin, S. M., Kireev, V. V., Kolesnikov, G., Popilin, V. P. and Filippov, E. A. (1969). *Russ. J. Inorg. Chem.,* **14,** 548
20. Paddock, N. L. and Searle, H. T. (1969). *U.S. Patent* 3 462 247 (*Chem. Abstr.,* 1969, **71,** 82031)
21. Nichols, G. M. (1969). *U.S. Patent* 3 449 091 (*Chem. Abstr.,* 1969, **71,** 31905)
22. Jenkins, R. W. and Lanoux, S. (1970). *J. Inorg. Nucl. Chem.,* **32,** 2453
23. Kobayashi, E. (1967). *Kogyo Kagaku Zasshi,* **70,** 628 (*Chem. Abstr.,* 1968, **68,** 56122)
24. Zhivukhin, S. M., Kireev, V. V., Kolesnikov, G. S. and Popilin, V. P., *USSR Patent* 242, 165, (*Chem. Abstr.,* 1969, **71,** 72474)
25. Emsley, J. and Udy, P. B. (1967). *Chem. Commun.,* 633
26. Emsley, J. and Udy, P. B. (1971). *J. Chem. Soc. A,* 768
27. Moran, E. F. (1968). *J. Inorg. Nucl. Chem.,* **30,** 1405
28. Ura, M. and Ogihara, K. (1967). *Japanese Patent* 14 693 (*Chem. Abstr.,* 1968, **68,** 97, 204)
29. Wunsch, G., Schiedermaier, R., Kiener, V., Fluck, E. and Heckmann, G. (1970). *Chem. Ztg.,* **94,** 832
30. Horn, H-G. and Becke-Goehring, M. (1969). *Naturwissenschaften.,* **56,** 137
31. Lehr, W. and Schwarz, M. (1968). *Z. Anorg. Allg. Chem.,* **363,** 43
32. Schmidpeter, A. and Weingard, C. (1969). *Angew. Chem. Int. Ed. Engl.,* **8,** 615
33. Becke-Goehring, M. and Scharf, B. (1967). *Z. Anorg. Allg. Chem.,* **353,** 320
34. Becke-Goehring, M. (1968). *Fortschr. Chem. Forsch.,* **10,** 207
35. Emsley, J. and Udy, P. B. (1970). *J. Chem. Soc. A,* 3025
36. Kobayashi, E. (1966). *Nippon Kagaku Zasshi,* **87,** 135, (*Chem. Abstr.,* 1966, **65,** 11744)
37. Coxon, G. E. and Sowerby, D. B. (1967). *J. Chem. Soc. A,* 1566
38. Coxon, G. E., Sowerby, D. B. and Tranter, G. C. (1965). *J. Chem. Soc.,* 5697
39. Rice, R. G., Daasch, L. W., Holden, J. R. and Kohn, E. J. (1958). *J. Inorg. Nucl. Chem.,* **5,** 190
40. Coxon, G. E., Palmer, T. F. and Sowerby, D. B. (1966). *Inorg. Nucl. Chem. Lett.,* **2,** 215
41. Rotzche, H., Stahlberg, R. and Steger, E. (1966). *J. Inorg. Nucl. Chem.,* **28,** 687
42. Engelhardt, G., Steger, E. and Stahlberg, R. (1966). *Z. Naturforsch.,* **21b,** 1231
43. Coxon, G. E. and Sowerby, D. B. (1967). *Inorg. Chim. Acta,* **1,** 381
44. Grushkin, B., Sanchez, M. G. and Rice, R. G. (1964). *Inorg. Chem.,* **3,** 623
45. Grushkin, B., Berlin, A. J., McClanahan, J. L. and Rice, R. G. (1966). *Inorg. Chem.,* **5,** 172

46. Manhas, B. S., Chu, S-K. and Moeller, T. (1968). *J. Inorg. Nucl. Chem.*, **30**, 322
47. Prons, V. N., Grinblam, M. P. and Klebanskii, A. L. (1970). *Zh. Obshch. Khim.*, **40**, 2127
48. Bezman, I. I. and Smalley, J. H. (1960). *Chem. Ind. (London)*, 839
49. Ahmed, I. Y. and Schmulbach, C. D. (1967). *J. Phys. Chem.*, **71**, 2358
50. Bermann, M. and Utvary, K. (1969). *J. Inorg. Nucl. Chem.*, **31**, 271
51. Schmulbach, C. D., Derderian, C., Zeck, O. and Sahuri, S. (1971). *Inorg. Chem.*, **10**, 195
52. Schmidpeter, A. and Ebeling, J. (1968). *Angew. Chem. Int. Ed. Engl.*, **7**, 209
53. Schmidpeter, A. and Weingand, C. (1969). *Z. Naturforsch.*, **24b**, 177
54. Herring, D. L. and Douglas, C. M. (1965). *Inorg. Chem.*, **4**, 1012
55. Herring, D. L. and Douglas, C. M. (1969). *U.S. Patent* 3 454 634 (*Chem. Abstr.*, 1969, **71**, 81522)
56. Kratzer, R. H. and Paciorek, K. L. (1965). *Inorg. Chem.*, **4**, 1767
57. Bunting, R. K. and Schmulbach, C. D. (1966). *Inorg. Chem.*, **5**, 533
58. Sisler, H. H., Frazier, S. E., Rice, R. G. and Sanchez, M. G. (1966). *Inorg. Chem.*, **5**, 326
59. Frazier, S. E. and Sisler, H. H. (1966). *Inorg. Chem.*, **5**, 925
60. Magnelli, D. D., Tesi, G., Lowe, J. V. and McQuistion, W. E. (1966). *Inorg. Chem.*, **5**, 457
61. Sherif, F. G. and Schmulbach, C. D. (1966). *Inorg. Chem.*, **5**, 322
62. Becke-Goehring, M. and Müller, H-J. (1968). *Z. Anorg. Allg. Chem.*, **362**, 51
63. Schmulbach, C. D. and Derderian, C. (1970). *J. Inorg. Nucl. Chem.*, **32**, 3397
64. Schmidpeter, A. and Schindler, N. (1970). *Z. Anorg. Allg. Chem.*, **372**, 214
65. Schmidpeter, A. and Schindler, N. (1968). *Z. Anorg. Allg. Chem.*, **362**, 281
66. Becke-Goehring, M. and Jung, D. (1970). *Z. Anorg. Allg. Chem.*, **372**, 233
67. Reichle, W. T. (1962). *Tetrahedron Lett.*, 61
68. Revitt, D. M. and Sowerby, D. B. (1969). *Inorg. Nucl. Chem. Lett.*, **5**, 459
69. Haque, R. U. and B. ud Din. (1966). *Pakistan J. Sci. Ind. Res.*, **9**, 121
70. Sisler, H. H. and Stratton, C. (1966). *Inorg. Chem.*, **5**, 2003
71. Allcock, H. R., Kugel, R. L. and Walsh, E. J. (1970). *Chem. Commun.*, 1283
72. Emsley, J. and Paddock, N. L. (1968). *J. Chem. Soc. A*, 2590
73. Green, B. and Sowerby, D. B. (1969). *Chem. Commun.*, 628
74. Green, B. and Sowerby, D. B. (1969). *Inorg. Nucl. Chem. Lett.*, **5**, 989
75. Green, B. and Sowerby, D. B. (1970). *J. Chem. Soc. A*, 987
76. Steger, E. and Klemm, D. (1967). *J. Inorg. Nucl. Chem.*, **29**, 1812
77. Lehr, W. (1967). *Z. Anorg. Allg. Chem.*, **350**, 18
78. Feistel, G. R. and Moeller, T. (1967). *J. Inorg. Nucl. Chem.*, **29**, 2731
79. Das, S. K., Keat, R., Shaw, R. A. and Smith, B. C. (1965). *J. Chem. Soc.*, 5032
80. Azhikina, Yu V., Koroleva, M. Ya, Maslennikov, B. M. and Kulikova, L. Ya (1968). *Inorg. Mater.*, **4**, 1492
81. Lehr, W. (1967). *Z. Anorg. Allg. Chem.*, **352**, 27
82. Ford, C. T., Dickson, F. E. and Bezman, I. I. (1965). *Inorg. Chem.*, **4**, 890
83. Das, S. K., Keat, R., Shaw, R. A. and Smith B. C. (1966). *J. Chem. Soc. A*, 1677
84. Das, S. K., Feakins, D., Last, W. A., Nabi, S. N., Ray, S. K., Shaw, R. A. and Smith, B. C. (1970). *J. Chem. Soc. A*, 616
85. Koopman, H. and Daams, J. (1967). *U.S. Patent* 3 351 456 (*Chem. Abstr.*, 1968, **68**, 12462)
86. Bogeat, G. and Cauquis, G. (1966). *Bull. Soc. Chim. Fr.*, 2735
87. Kropacheva, A. A. and Kashnikova, N. M. (1967). *Khim. Org. Soedin. Fosfora, Akad. Nauk SSSR, Otd. Obshch. Tekh. Khim.*, 186 (*Chem. Abstr.*, 1968, **69**, 10066)
88. Kropacheva, A. A. and Kashnikova, N. M. (1965). *J. Gen. Chem. USSR*, **35**, 2219
89. Allcock, H. R., Forgione, P. S. and Valan, K. J. (1965). *J. Org. Chem.*, **30**, 947
90. Allcock, H. R. and Thomas, W. M. (1967). *U.S. Patent* 3 329 663 (*Chem. Abstr.*, 1967, **67**, 64883)
91. Lederle, H., Ottmann, G. and Kober, E. (1966). *Inorg. Chem.*, **5**, 1818
92. Desai, V. B., Shaw, R. A. and Smith, B. C. (1970). *J. Chem. Soc. A*, 2023
93. Keat, R. and Shaw, R. A. (1968). *Angew. Chem. Int. Ed. Engl.*, **7**, 212
94. Keat, R. and Shaw, R. A. (1965). *J. Chem. Soc.*, 2215
95. Koopman, H., Spruit, F. J., Van Deursen, F. and Bakker, J. (1965). *Rec. Trav. Chim.*, **84**, 341
96. Ford, C. T., Dickson, F. E. and Bezman, I. I. (1967). *Inorg. Chem.*, **6**, 1594
97. Green, B. and Sowerby, D. B. (1971). *J. Inorg. Nucl. Chem.*, in the press
98. Keat, R. and Shaw, R. A. (1966). *J. Chem. Soc.*, 908

99. Joffre, S. P. (1967). *U.S. Patent* 3 311 622 (*Chem. Abstr.*, 1967, **67**, 64450)
100. Kropacheva, A. A. and Kashnikova, N. M. (1965). *J. Gen. Chem. USSR*, **35**, 1978
101. Kropacheva, A. A. and Kashnikova, N. M. (1967). *Khim. Org. Soedin Fosfora, Akad. Nauk. SSSR, Otd. Obshch. Tekh. Khim.*, 188. (*Chem. Abstr.*, 1968, **69**, 10321)
102. Kropacheva, A. A. and Kashnikova, N. M. (1968). *J. Gen. Chem. USSR.*, **38**, 135
103. Kropacheva, A. A. and Mukhina, L. E. (1969). *Khim. Geterotsikl. Soedin*, 162. (*Chem. Abstr.*, 1969, **70**, 114966)
104. Capon, B., Hills, K. and Shaw, R. A. (1965). *J. Chem. Soc.*, 4059
105. Yokoyama, M. and Cho, H. (1964). *Kogakuin Daigaku Kenkyu Hokoku*, **15**, 22. (*Chem. Abstr.*, 1968, **68**, 72771)
106. Mukhina, L. E. and Kropacheva, A. A. (1968). *J. Gen. Chem. USSR*, **38**, 314
107. Hills, K. and Shaw, R. A. (1964). *J. Chem. Soc.*, 130
108. Feakins, D., Last, W. A. and Shaw, R. A. (1964). *J. Chem. Soc.*, 4464
109. Kobayashi, Y., Chasin, L. A. and Clapp, L. B. (1963). *Inorg. Chem.*, **2**, 212
110. Keat, R., Shaw, R. A. and Stratton, C. (1965). *J. Chem. Soc.*, 2223
111. Keat, R. and Shaw, R. A. (1965). *J. Chem. Soc.*, 4067
112. Nabi, S. N., Shaw, R. A. and Stratton, C. (1969). *Chem. Ind. (London)*, 166
113. Allcock, H. R. and Kugel, R. L. (1966). *Inorg. Chem.*, **5**, 1016
114. Pitina, M. R., Negrebetskii, V. V. and Shvetsov-Shilovskii, N. I. (1969). *Zh. Obshch. Khim.*, **39**, 1216. (*Chem. Abstr.*, 1969, **71**, 70572)
115. Burg, A. B. and Caron, A. (1959). *J. Amer. Chem. Soc.*, **81**, 836
116. Stepanov, B. I. and Migachev, G. I. (1966). *J. Gen. Chem. USSR*, **36**, 1454
117. Stepanov, B. I. and Migachev, G. I. (1965). *J. Gen. Chem. USSR*, **35**, 2245
118. Migachev, G. I. and Stepanov, B. I. (1966). *Russ. J. Inorg. Chem.*, **11**, 929
119. Sowerby, D. B. and Audrieth, L. F. (1961). *Chem. Ber.*, **94**, 2670
120. Lehr, W. and Pietschmann, J. (1970). *Chem. Ztg.*, **94**, 362
121. Mattogno, G. and Monaci, A. (1965). *Ricerca Sci.*, **8**, 1139
122. Lehr, W. (1969). *Naturwissenschaften*, **56**, 214
123. Desai, V. B., Shaw, R. A., Smith, B. C. and Taylor, D. (1969). *Chem. Ind. (London)* 1177
124. Ray, S. K., Shaw, R. A. and Smith, B. C. (1963). *J. Chem. Soc.*, 3236
125. Stahlberg, R. and Steger, E. (1968). *J. Inorg. Nucl. Chem.*, **30**, 737
126. Berlin, A. J., Grushkin, B. and Moffett, L. R. (1968). *Inorg. Chem.*, **7**, 589
127. Allen, G., Oldfield, D. J., Paddock, N. L., Rallo, F., Serreqi, J. and Todd, S. M. (1965). *Chem. Ind (London)*, 1032
128. Stahlberg, R. and Steger, E. (1966). *J. Inorg. Nucl. Chem.*, **28**, 684
129. Engelhardt, G., Steger, E. and Stahlberg, R. (1966). *Z. Naturforsch.*, **21b**, 586
130. Stahlberg, R. and Steger, E. (1967). *J. Inorg. Nucl. Chem.*, **29**, 961
131. Roesky, H. W. and Niecke, E. (1968). *Inorg. Nucl. Chem. Lett.*, **4**, 463
132. Glemser, O., Niecke, E. and Roesky, H. W. (1969). *Chem. Commun.*, 282
133. Chivers, T., Oakley, R. T. and Paddock, N. L. (1970). *J. Chem. Soc. A*, 2324
134. Glemser, O., Niecke, E. and Thamm, H. (1970). *Z. Naturforsch.*, **25b**, 754
135. Roesky, H. W. and Grosse-Böwing, W. (1970). *Inorg. Nucl. Chem. Lett.*, **6**, 781
136. Roesky, H. W., Grosse-Böwing, W. and Niecke, E. (1971). *Chem. Ber.*, **104**, 653
137. Keat, R., Miller, M. C. and Shaw, R. A. (1967). *J. Chem. Soc. A*, 1404
138. Feldt, M. K. and Moeller, T. (1968). *J. Inorg. Nucl. Chem.*, **30**, 2351
139. Lehr, W. (1969). *Z. Anorg. Allg. Chem.*, **371**, 225
140. Roesky, H. W. and Grosse-Böwing, W. (1969). *Z. Naturforsch.*, **24b**, 1250
141. Niecke, E., Glemser, O. and Thamm, H. (1970). *Chem. Ber.*, **103**, 2864
142. Fitzsimmons, B. W. and Shaw, R. A. (1964). *J. Chem. Soc.*, 1735
143. Sorokin, M. F. and Latov, V. A. (1966). *Kinetika Kataliz.*, **7**, 42. (*Chem. Abstr.*, 1966, **64**, 19348)
144. Zhivukhin, S. M., Tolstoguzoc, V. B. and Lukashevski, Z. (1965). *Russ. J. Inorg. Chem.*, **10**, 901
145. Prons, V. N., Grinblam, M. P., Klebanskii, A. L. and Nikolaev, G. A. (1970). *Zh. Obshch. Khim.*, **40**, 2128
146. Pornin, R. (1966). *Bull. Soc. Chim. Fr.*, 2861
147. Matuszko, A. J. and Chang, M. S. (1966). *J. Org. Chem.*, **31**, 2004
148. Dell, D., Fitzsimmons, B. W. and Shaw, R. A. (1965). *J. Chem. Soc.*, 4070
149. Ford, C. T., Dickson, F. E. and Bezman, I. I. (1965). *Inorg. Chem.*, **4**, 419

150. Dell, D., Fitzsimmons, B. W., Keat, R. and Shaw, R. A. (1966). *J. Chem. Soc. A,* 1680
151. McBee, E. T., Okuhara, K. and Morton, C. J. (1966). *Inorg. Chem.,* **15,** 450
152. Kober, E., Lederle, H. and Ottmann, G. (1966). *Inorg. Chem.,* **5,** 2239
153. Ottmann, G., Lederle, H., Hooks, H. and Kober, E. (1967). *Inorg. Chem.,* **6,** 394
154. Allcock, H. R. (1964). *J. Amer. Chem. Soc.,* **86,** 2591
155. Allcock, H. R. and Siegel, L. A. (1964). *J. Amer. Chem. Soc.,* **86,** 5140
156. Zhivukhin, S. M., Kireev, V. V. and Zelenetskii, A. N. (1966). *Zh. Prikl. Khim.,* **39,** 234. (*Chem. Abstr.,* 1966, **64,** 12579)
157. Kajiwara, M. and Saito, H. (1970). *Kogyo Kagaku Zasshi,* **73,** 1947
158. Okuhashi, T. and Watanabe, Y. (1970). *Kogyo Kagaku Zasshi,* **73,** 1164. (*Chem. Abstr.,* 1970, **73,** 110187)
159. Kajiwara, M. and Saito, H. (1970). *Kogyo Kagaku Zasshi,* **73,** 1954. (*Chem. Abstr.,* 1971, **74,** 76916)
160. Allcock, H. R. and Kugel, R. L. (1969). *J. Amer. Chem. Soc.,* **91,** 5452
161. Lederle, H., Kober, E. and Ottmann, E. (1966). *J. Chem. Eng. Data.,* **11,** 221
162. Ottmann, G., Lederle, H. and Kober, E. H. (1966). *Ind. Eng. Chem., Prod. Res. Development,* **5,** 202
163. Pitina, M. R., Ivanova, T. M. and Shvetsov-Shilovskii, N. I. (1967). *Zh. Obshch. Khim.,* **37,** 2076 (*Chem. Abstr.,* 1968, **68,** 48942)
164. Kajiwara, M. and Saito, H. (1969). *Kogyo Kagaku Zasshi,* **72,** 2566. (*Chem. Abstr.,* 1970, **72,** 111899)
165. Kajiwara, M. and Saito, H. (1967). *Kogyo Kagaku Zasshi,* **70,** 1101. (*Chem. Abstr.,* 1968, **68,** 96452)
166. Kajiwara, M. and Saito, H. (1968). *Kogyo Kagaku Zasshi,* **71,** 1470. (*Chem. Abstr.,* 1969, **70,** 20451)
167. Nakanishi, M., Inamasu, S. and Mukai, T. (1966). *Japanese Patent* 21, 779. (*Chem. Abstr.,* 1967, **66,** 75669)
168. Stepanov, B. I. and Migachev, G. I. (1966). *Zh. Vses. Khim. Obshchestva im. D.I. Mendeleeva,* **11,** 472. (*Chem. Abstr.,* 1966, **65,** 18493)
169. Stepanov, B. I. and Migachev, G. I. (1968). *Zh. Obshch. Khim.,* **38,** 194. (*Chem. Abstr.,* 1968, **69,** 35624)
170. Rallo, R. (1965). *Ricera Sci.,* **8,** 1134
171. Lenton, M. V. and Lewis, B. (1966). *J. Chem. Soc. A,* 665
172. Pitina, M. R. and Shvetsov-Shilovskii, N. I. (1966). *J. Gen. Chem. USSR,* **36,** 517
173. Pitina, M. R. and Shvetsov-Shilovskii, N. I. (1967). *Khim. Org. Soedin Fosfora, Akad. Nauk SSSR, Otd. Obshch. Tekh. Khim.,* 159. (*Chem. Abstr.,* 1968, **69,** 43333)
174. Joel, D. and Wende, A. (1966). *Plaste Kaut.,* **13,** 643. (*Chem. Abstr.,* 1967, **66,** 28343)
175. Fitzsimmons, B. W., Hewlett, C. and Shaw, R. A. (1964). *J. Chem. Soc.,* 4459
176. Fitzsimmons, B. W., Hewlett, C. and Shaw, R. A. (1965). *J. Chem. Soc.,* 7432
177. Buslaev, Yu A., Levin, B. V. and Rumyantseva, Z. G., Petrosyants, S. P. and Mironova, V. V. (1969). *Russ. J. Inorg. Chem.,* **14,** 1711
178. Belykh, S. I., Zhivukhin, S. M., Kireev, V. V. and Kolesnikov, G. S. (1969). *Russ. J. Inorg. Chem.,* **14,** 668
179. Fitzsimmons, B. W., Hewlett, C. and Shaw, R. A. (1965). *J. Chem. Soc.,* 4799
180. Sonobe, K. (1967). *Kogyo Kagaku Zasshi,* **70,** 2305 (*Chem. Abstr.,* 1968, **68,** 87002)
181. Allcock, H. R. and Walsh, E. J. (1969). *J. Amer. Chem. Soc.,* **91,** 3102
182. Allcock, H. R. and Walsh, E. J. (1970). *Chem. Commun.,* 580
183. McBee, E. T., Okuhara, K. and Morton, C. J. (1965). *Inorg. Chem.,* **4,** 1672
184. Grushkin, B., Sanchez, M. G., Ernest, M. V., McClanahan, J. L., Ashby, G. E. and Rice, R. G. (1965). *Inorg. Chem.,* **4,** 1538
185. Desai, V. B., Shaw, R. A. and Smith, B. C. (1968). *Angew. Chem. Int. Ed. Engl.,* **7,** 887
186. Biddlestone, M. B. and Shaw, R. A. (1969). *J. Chem. Soc. A,* 178
187. Biddlestone, M., Shaw, R. A. and Taylor, D. (1969). *Chem. Commun.,* 320
188. Biddlestone, M. and Shaw, R. A. (1968). *Chem. Commun.,* 407
189. Biddlestone, M. and Shaw, R. A. (1970). *J. Chem. Soc. A,* 1750
190. Allen, C. W. (1970). *Chem. Commun.,* 152
191. Allen, C. W. and Moeller, T. (1968). *Inorg. Chem.,* **7,** 2177
192. Allen, C. W., Tsang, F. Y. and Moeller, T. (1968). *Inorg. Chem.,* **7,** 2183
193. Paddock, N. L., Ranganathan, T. N. and Todd, S. M. (1971). *Can. J. Chem.,* **49,** 164
194. Moeller, T., Failli, A. and Tsang, F. Y. (1965). *Inorg. Nucl. Chem. Lett.,* **1,** 49

195. Chivers, T. and Paddock, N. L. (1968). *Chem. Commun.*, 704
196. Desai, V. B., Shaw, R. A. and Smith, B. C. (1969). *J. Chem. Soc. A*, 1977
197. Schmulbach, C. D. and Miller, V. R. (1966). *Inorg. Chem.*, **5**, 1621
198. Schmulbach, C. D. and Miller, V. R. (1968). *Inorg. Chem.*, **7**, 2189
199. Fitzsimmons, B. W., Hewlett, C., Hills, K. and Shaw, R. A. (1967). *J. Chem. Soc. A*, 679
200. Ford, C. T., Barr, J. M., Dickson, F. E. and Bezman, I. I. (1966). *Inorg. Chem.*, **5**, 351
201. Rapko, J. N. and Feistel, G. (1970). *Inorg. Chem.*, **9**, 1401
202. Bilbo, A. J., Grieve, C. M., Herring, D. L. and Salzbrunn, D. E. (1968). *Inorg. Chem.*, **7**, 2670
203. Sharts, C. M., Bilbo, A. J. and Gentrey, D. R. (1966). *Inorg. Chem.*, **5**, 2140
204. Carroll, A. P. and Shaw, R. A. (1966). *J. Chem. Soc. A*, 914
205. Janik, B., Zeszutko, V. and Pelczar, T. (1966). *J. Gen. Chem. USSR*, **36**, 1451
206. Niecke, E., Glemser, O. and Roesky, H. W. (1969). *Z. Naturforsch.*, **24b**, 1187
207. Denny, K. and Lanoux, S. (1969). *J. Inorg. Nucl. Chem.*, **31**, 1531
208. Coxon, G. E. and Sowerby, D. B. (1969). *J. Chem. Soc. A*, 3012
209. Chivers, T. and Paddock, N. L. (1969). *J. Chem. Soc. A*, 1687
210. Douglas, W. M., Cooke, M., Lustig, M. and Ruff, J. K. (1970). *Inorg. Nucl. Chem. Lett.*, **6**, 409
211. Lappert, M. F. and Srivastava, G. (1966). *J. Chem. Soc. A*, 210
212. Dyson, J. and Paddock, N. L. (1966). *Chem. Commun.*, 191
213. Whitaker, R. D., Barreiro, A. J., Furman, P. A., Guida, W. C. and Stallings, E. S. (1968). *J. Inorg. Nucl. Chem.*, **30**, 2921
214. Whitaker, R. D. and Guida, W. C. (1969). *J. Inorg. Nucl. Chem.*, **31**, 875
215. Das, S. K., Shaw, R. A., Smith, B. C. and Thakur, C. P. (1966). *Chem. Commun.*, 33
216. Steger, E. and Bachmann, G. (1970). *Z. Chem.*, **10**, 306
217. Kuznetsov, N. T. (1966). *Inorg. Mater.*, **2**, 1953
218. Kuznetsov, N. T. and Klimchuk, G. S. (1967). *Inorg. Mater.*, **3**, 525
219. Kuznetsov, N. T. and Klimchuk, G. S. (1970). *Russ. J. Inorg. Chem.*, **15**, 1495
220. Levin, B. V., Rumyantseva, Z. G. and Mironova, V. V. (1967). *Inorg. Mater.*, **3**, 523
221. Slawisch, A. and Pietschmann, J. (1970). *Z. Naturforsch.*, **25b**, 321
222. Colclough, R. O. and Gee, G. (1968). *J. Polymer Sci. C*, **16**, 3639
223. MacCallum, J. R. and Werninck, A. (1967). *J. Polymer Sci. A-1*, **5**, 3061
224. Cordischi, D., Delle Site, A., Mele, A. and Porta, P. (1966). *J. Macromol. Chem.*, **1**, 219
225. Allen, G., Lewis, C. J. and Todd, S. M. (1970). *Polymer*, **11**, 31
226. Allcock, H. R., Kugel, R. L. and Valan, K. J. (1966). *Inorg. Chem.*, **5**, 1709
227. Allcock, H. R. (1966). *Inorg. Chem.*, **5**, 1320
228. Allcock, H. R. and Kugel, R. L. (1966). *Inorg. Chem.*, **5**, 1716
229. Allcock, H. R. and Mack, D. P. (1970). *Chem. Commun.*, 685
230. Mirhej, M. E. and Henderson, J. F. (1966). *J. Macromol. Chem.*, **1**, 187
231. Allen, G., Lewis, C. J. and Todd, S. M. (1970). *Polymer*, **11**, 44
232. Rose, S. H. (1968). *J. Polymer Sci. B.*, **6**, 837
233. MacCallum, J. R. and Tanner, J. (1968). *J. Polymer Sci. A-1*, **6**, 3163
234. MacCallum, J. R. and Tanner, J. (1970). *J. Macromol. Sci. Chem.*, **4**, 481
235. MacCallum, J. R. and Wernick, A. R. S. (1971). *J. Macromol. Sci. Chem.*, **5**, 651
236. Yokoyama, M. and Konya, S. (1967). *Kogyo Kagaku Zasshi*, **70**, 1453. (*Chem. Abstr.*, 1968, **68**, 69431)
237. Yokoyama, M. and Konya, S. (1969). *Kogyo Kagaku Zasshi*, **72**, 376. (*Chem. Abstr.*, 1969, **70**, 97520)
238. Bilbo, A. J. Douglas, C. M., Fetter, N. R. and Herring, D. L. (1968). *J. Polymer Sci. A-1*, **6**, 1671
239. Deryabin, A. V., Zhivukhin, S. M., Kireev, V. V. and Kolesnikov, G. S. (1968). *Plast. Massy.*, **3**, 29. (*Chem. Abstr.*, 1968, **69**, 3200)
240. Saito, H. (1968). *Kobunshi*, **17**, 391. (*Chem. Abstr.*, 1968, **69**, 59570)
241. Horn, H-G. (1969). *Chem. Ztg.*, **93**, 241
242. Corbridge, D. E. C. (1966). *Topics in Phosphorus Chemistry*, **3**, 57
243. Giglio, E. and Puliti, R. (1967). *Acta. Crystallogr.*, **22**, 304
244. Zoer, H. and Wagner, A. J. (1970). *Acta Crystallogr.*, **B26**, 1812
245. Faught, J. B., Moeller, T. and Paul, I. C. (1970). *Inorg. Chem.*, **9**, 1656
246. Ahmed, F. A., Singh, P. and Barnes, W. H. (1969). *Acta Crystallogr.*, **B25**, 316
247. Marsh, W. C. and Trotter, J. (1971). *J. Chem. Soc. A*, 169

248. Siegel, L. A. and Van den Hende, J. H. (1967). *J. Chem. Soc. A*, 817
249. Smith, G. W. and Wood, D. (1966). *Nature (London)*, **210**, 520
250. Allcock, H. R., Stein, M. T. and Stanko, J. A. (1970). *Chem. Commun.*, 944
251. Mani, N. V., Ahmed, F. R. and Barnes, W. H. (1965). *Acta Crystallogr.*, **19**, 693
252. Allen, C. W., Faught, J. B., Moeller, T. and Paul, I. C. (1969). *Inorg. Chem.*, **8**, 1719
253. Mani, N. V., Ahmed, F. R. and Barnes, W. H. (1966). *Acta Crystallogr.*, **21**, 375
254. Olthof, R. (1969). *Acta Crystallogr.*, **B25**, 2040
255. Mani, N. V. and Wagner, A. J. (1971). *Acta Crystallogr.*, **B27**, 51
256. Ansell, G. B. and Bullen, G. J. (1968). *J. Chem. Soc. A*, 3026
257. Davies, M. I. and Paul, J. W. (1969). *Acta Crystallogr.*, **A25**, S116
258. Hazekamp, R., Migchelson, T. and Vos, A. (1962). *Acta Crystallogr.*, **15**, 539
259. Wagner, A. J. and Vos, A. (1968). *Acta Crystallogr.*, **B24**, 707
260. Ansell, G. B. and Bullen, G. J. (1966). *Chem. Commun.*, 430
261. Bullen, G. J., Mallinson, P. R. and Burr, A. H. (1969). *Chem. Commun.*, 691
262. Bullen, G. J. and Tucker, P. A. (1970). *Chem. Commun.*, 1185
263. Marsh, W. C. and Trotter, J. (1971). *J. Chem. Soc. A*, 569
264. Marsh, W. C. and Trotter, J. (1971). *J. Chem. Soc. A*, 573
265. Trotter, J. and Whitlow, S. H. (1970). *J. Chem. Soc. A*, 455
266. Trotter, J. and Whitlow, S. H. (1970). *J. Chem. Soc. A*, 460
267. Schlueter, A. W. and Jacobson, R. A. (1968). *J. Chem. Soc. A*, 2317
268. Wagner, A. J. and Vos, A. (1968). *Acta Crystallogr.*, **B24**, 1423
269. Marsh, W. C., Paddock, N. L., Stewart, C. J. and Trotter, J. (1970). *Chem. Commun.*, 1190
270. Harrison, W., Oakley, R. T., Paddock, N. L. and Trotter, J. (1971). *Chem. Commun.*, 357
271. Paddock, N. L., Trotter, J. and Whitlow, S. H. (1968). *J. Chem. Soc. A*, 2227
272. Allcock, H. R., Konopski, G. F., Kugel, R. L. and Stroh, E. G. (1970). *Chem. Commun.*, 985
273. Mitchell, K. A. R. (1969). *Chem. Rev.*, **69**, 157
274. Mitchell, K. A. R. (1968). *J. Chem. Soc. A*, 2683
275. Branton, G. R., Brion, C. E., Frost, D. C., Mitchell, K. A. R. and Paddock, N. L. (1970). *J. Chem. Soc. A*, 151
276. Manley, T. R. and Williams, D. A. (1967). *Spectrochim. Acta*, **23A**, 149
277. Manley, T. R. and Williams, D. A. (1967). *Spectrochim. Acta*, **23A**, 1221
278. Stahlberg, R. and Steger, E. (1967). *Spectrochim. Acta*, **23A**, 2185
279. Emsley, J. (1970). *J. Chem. Soc. A*, 109
280. Stahlberg, R. and Steger, E. (1967). *Spectrochim. Acta*, **23A**, 2057
281. Stahlberg, R. and Steger, E. (1967). *Spectrochim. Acta*, **23A**, 2005
282. Kotova, G. G., Zimina, K. I., Myannik, E. I., Sher, V. V. and Sanin, P. I. (1968). *Neftekhimiya*, **8**, 289. (*Chem. Abstr.*, 1968, **69**, 23322)
283. Stahlberg, U. and Steger, E. (1967). *Spectrochim. Acta*, **23A**, 627
284. Stahlberg, U., Stahlberg, R. and Steger, E. (1967). *Spectrochim. Acta*, **23A**, 2691
285. Manley, T. R. and Williams, D. A. (1968). *Spectrochim. Acta*, **24A**, 1661
286. Griffith, W. P. and Rutt, K. J. (1968). *J. Chem. Soc. A*, 2331
287. Hisatsune, I. C. (1969). *Spectrochim. Acta*, **25A**, 301.
288. Coxon, G. E. and Sowerby, D. B. (1968). *Spectrochim. Acta*, **24A**, 2145
289. Manley, T. R. and Williams, D. A. (1969). *Polymer*, **10**, 307
290. Keat, R., Ray, S. K. and Shaw, R. A. (1965). *J. Chem. Soc.*, 7193
291. Keat, R. and Shaw, R. A. (1968). *J. Chem. Soc. A*, 703
292. Hewlett, C. and Shaw, R. A. (1966). *J. Chem. Soc. A*, 56
293. Harris, R. K. (1966). *Inorg. Chem.*, **5**, 701
294. Finer, E. G. (1967). *J. Mol. Spectrosc.*, **23**, 104
295. Finer, E. G. and Harris, R. K. (1969). *J. Chem. Soc. A*, 1972
296. Finer, E. G., Harris, R. K., Bond, M. K., Keat, R. and Shaw, R. A. (1970). *J. Mol. Spectrosc.*, **33**, 72
297. Mark, V., Dungan, C. H., Crutchfield, M. M. and Van Wazer, J. R. (1967). *Topics in Phosphorus Chemistry*, **5**, 227
298. Finer, E. G. and Harris, R. K. (1971). *Progr. N.M.R. Spectroscopy*, **6**, 61
299. Heatley, F. and Todd, S. M. (1966). *J. Chem. Soc. A*, 1152
300. Latscha. H. P. (1968). *Z. Anorg. Allg. Chem.*, **362**, 7
301. Dwek, R. A., Richards, R. E., Taylor, D. and Shaw, R. A. (1970). *J. Chem. Soc. A*, 244

302. Dwek, R. A., Paddock, N. L., Poindexter, E. H. and Potenza, J. A. (1969). *J. Amer. Chem. Soc.*, **91**, 5436
303. Feakins, D., Last, W. A. and Shaw, R. A. (1964). *J. Chem. Soc.*, 2387
304. Feakins, D., Last, W. A. and Shaw, R. A. (1964). *J. Chem. Soc.*, 4464
305. Feakins, D., Last, W. A., Neemuchwala, N. and Shaw, R. A. (1965). *J. Chem. Soc. A*, 2804
306. Feakins, D., Last, W. A., Nabi, S. N. and Shaw, R. A. (1966). *J. Chem. Soc. A*, 1831
307. Feakins, D., Nabi, S. N., Shaw, R. A. and Watson, P. (1968). *J. Chem. Soc. A*, 10
308. Feakins, D., Last, W. A., Nabi, S. N., Shaw, R. A. and Watson, P. (1969). *J. Chem. Soc. A*, 196
309. Feakins, D., Shaw, R. A., Watson, P. and Nabi, S. N. (1969). *J. Chem. Soc. A*, 2468
310. Coxon, G. E., Palmer, T. F. and Sowerby, D. B. (1967). *J. Chem. Soc. A*, 1568
311. Brion, C. E. and Paddock, N. L. (1968). *J. Chem. Soc. A*, 388
312. Schmulbach, C. D., Cook, A. G. and Miller, V. R. (1968). *Inorg. Chem.*, **7**, 2463
313. Brion, C. E. and Paddock, N. L. (1968). *J. Chem. Soc. A*, 392
314. Wagner, A. J. and Moeller, T. (1971). *J. Chem. Soc. A*, 596
315. Coxon, G. E., Palmer, T. F. and Sowerby, D. B. (1969). *J. Chem. Soc. A*, 358
316. Dixon, M., Jenkins, H. D. B., Smith, J. A. S. and Tong, D. A. (1967). *Trans. Faraday Soc.*, **63**, 2852
317. Kaplansky, M. and Whitehead, M. A. (1967). *Can. J. Chem.*, **45**, 1669
318. Cotson, S. and Hodd, K. A. (1965). *J. Inorg. Nucl. Chem.*, **27**, 335
319. Cotson, S. and Hodd, K. A. (1969). *J. Inorg. Nucl. Chem.*, **31**, 245
320. Aroney, M. J., LeFevre, R. J. W., Murthy, D. S. N., Peacock, G. J. and Saxby, J. D. (1966). *J. Chem. Soc. B*, 657
321. Kokoreva, I. Y., Syrkin, Ya K., Kropacheva, A. A., Kashnikova, N. M. and Mukhina, L. (1966). *Dokl. Akad. Nauk SSSR*, **166**, 155. (*Chem. Abstr.*, 1966, **64**, 12522)
322. Allcock, H. R. and Birdsall, W. J. (1969). *J. Amer. Chem. Soc.*, **91**, 7541

5
Binary and Ternary Systems Involving Sulphur

T. K. WIEWIOROWSKI

Freeport Sulphur Company, Louisiana

5.1 INTRODUCTION

This chapter presents a review of chemical systems involving elemental sulphur. Much information is available in the chemical literature on the solubility of sulphur in various organic and inorganic solvents. Such data are compiled elsewhere[1,2] and will not be included in this review. Emphasis is placed here on systems in which molten sulphur serves as the solvent. Only in the case of the sulphur-hydrogen sulphide system is the discussion extended to cover a wider range of temperature and pressure, because of the special interest of this system to the sulphur chemist.

As the result of the chemical reactivity of elemental sulphur, there are few sulphur-containing systems in which the components are inert with respect to each other. Nevertheless. the systems described in the first section of this chapter represent cases in which component interaction is either weak or sufficiently slow for the systems to be regarded as binary or ternary mixtures.

Perhaps more interesting from a chemical standpoint are systems in which the second component enters into reversible chemical reactions with molten sulphur. Several such systems are discussed in this chapter. The equilibria in such cases usually entail the formation of heteroatomic polymeric entities. Polymerisation, of course, is well known to occur in pure molten sulphur. The second component, however, may affect this polymerisation process by serving as a polymer termination agent, as in the case of hydrogen sulphide or iodine; by changing the nature of the monomer, as in the case of selenium; or by providing sites for polymer branching, as in the case of arsenic.

5.2 BINARY AND TERNARY SYSTEMS

5.2.1 Sulphur–carbon disulphide

Carbon disulphide is undoubtedly the most widely known solvent for elemental sulphur. It is perhaps fitting to begin this review of sulphur-containing systems with a discussion of the sulphur–carbon disulphide system.

The solubility of sulphur in carbon disulphide has been studied extensively and much information is available on this subject in the early chemical literature. More recently, a study was reported on the solubility of carbon disulphide in molten sulphur[3]. The experimental data were obtained by direct gravimetric measurements and by infrared absorption studies. Both sets of solubility data are incorporated into the sulphur–carbon disulphide phase diagram shown in Figure 5.1. The dashed curves represent the solubility of carbon disulphide in molten sulphur, while curve ABDE is a plot of the solubility of monoclinic sulphur in carbon disulphide. The part of this curve between points D and E represents a set of metastable conditions. Point E (118.9 °C) is the melting point of monoclinic sulphur. The melt is initially composed of octatomic sulphur rings, (S_8^R), but the formation of at least one other molecular species reduces the freezing point of sulphur to 114.5 °C (Point G). Thus, on the phase diagram, curve ABDG separates the liquid phase (above the line) from liquid and solid (below the line) at equilibrium conditions.

Line HFD reflects the effect of carbon disulphide on the temperature of the conversion between orthorhombic and monoclinic sulphur. In the absence of carbon disulphide this transition occurs at 95.5 °C (Point H). As carbon disulphide is introduced into the system, this transition temperature decreases along line HFD.

While curve DE represents a set of metastable conditions attained when monoclinic sulphur (pure S_8^R) is gradually added to the system, line GF reflects another set of metastable conditions which are observed if carbon disulphide is added to a system which consists initially of pure molten sulphur equilibrated with respect to its molecular composition at its freezing temperature (Point G). The existence of metastable conditions on both sides of the equilibrium curve DE is a reflection of the low reaction rates in this system at temperatures below 115 °C. The kinetic aspects of molecular equilibration in molten sulphur have been studied[4] and the reactions are reported to be quite slow.

Carbon disulphide should be regarded as the solvent in the S—CS₂ system along curve AB. At Point B of the phase diagram the atmospheric elevated boiling point of carbon disulphide saturated with sulphur is reached.

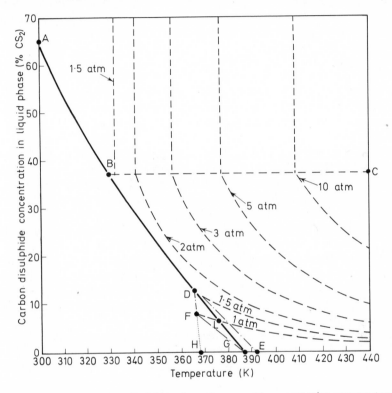

Figure 5.1 Sulphur–carbon disulphide phase diagram with concentration-pressure-temperature relationships superimposed. (From Touro and Wiewiorowski[3], by courtesy of the American Chemical Society).

The area above line BC represents a liquid phase in which carbon disulphide serves as the solvent; directly below this line a mixed solvent prevails. The mixed solvent concept may also be applied to the solubility curve between Points B and D, since in this temperature range sulphur by itself is a solid and carbon disulphide is a gas at atmospheric pressure.

It has been observed that at Point B (the elevated boiling point of carbon disulphide) the molar ratio of carbon disulphide to sulphur is 2 to 1, while at Point D, where sulphur becomes the solvent, this ratio is 1 to 2. This may be interpreted as an indication that intermolecular bonding between carbon disulphide and sulphur may be taking place in the system. The heat of solution for carbon disulphide in molten sulphur is -6420 cal mol^{-1}, indicating strong solute–solvent interaction.

5.2.2 Sulphur–oxygen

Relatively little information is available in the literature on this system. The solubility of oxygen in molten sulphur has been studied over a narrow temperature range between 125 and 150 °C [5]. Undoubtedly, the scarcity of information on the chemical behaviour of oxygen dissolved in molten sulphur must be a consequence of the unusual experimental problems encountered in the study of this system.

It has been clearly demonstrated that a slow liquid phase reaction occurs in molten sulphur between dissolved oxygen and the solvent. It is, therefore, virtually impossible to obtain a solution of molten sulphur saturated with the dissolved gas. Also, the solubility of oxygen in molten sulphur is very low, which further complicates its experimental determination.

The following experimental approach was employed in measuring the solubility of oxygen in molten sulphur. A thermostatically controlled agitated reaction vessel was filled with molten sulphur. The sulphur was contacted for one hour with a vigorous stream of air introduced into the melt through a fine glass frit. It was shown that under these conditions the rate at which oxygen was being introduced into this system was much greater than the rate at which it reacted with the solvent. Consequently, this procedure resulted in nearly saturating the sulphur with oxygen at 1 atm of air over-pressure, or 0.21 atm of partial oxygen pressure. The flow of air was then discontinued and nitrogen was passed through the same fine frit to scrub the dissolved oxygen from the sulphur. The oxygen content of the effluent gas stream was monitored over a period of several hours by means of gas chromatography.

In experiments of this type, the logarithm of oxygen concentration exhibits a linear decrease with time in accordance with a first-order kinetic equation:

$$\frac{-dC_{O_2}^g}{dt} = KC_{O_2}^s$$

where $C_{O_2}^g$ is the oxygen concentration in the effluent gas, $C_{O_2}^s$ is the concentration of oxygen dissolved in molten sulphur, t is the time and K is the observed overall rate constant.

The overall rate constant reflects two simultaneously occurring phenomena – the physical stripping of oxygen from the system and the reaction between dissolved oxygen and molten sulphur, both of which are governed by first-order kinetics. The contributions of each of these processes to the overall experimentally observed rate constant have been determined. This, in turn, allowed for the interpretation of experimentally obtained data in terms of both the kinetics governing the reaction

$$O_2 \xrightarrow[k_r]{\text{in sulphur}} SO_2$$

and the solubility of oxygen in molten sulphur.

In the investigated temperature range the solubility of oxygen in molten sulphur is reported to be in the order of $3 \times 10^{-4}\%$ by weight at 0.21 atm of partial oxygen overpressure. The reaction between molten sulphur and dissolved oxygen is first-order overall and with respect to dissolved oxygen. The reaction rate constant is strongly temperature dependent, increasing from 1.6×10^{-3} to 13.5×10^{-3} min^{-1} in a temperature range from 125.5 to 150.5 °C. The effect of temperature on this rate constant indicates an activation energy of 28 800 cal which presumably reflects the dissociation energy of a terminal sulphur–sulphur bond in a sulphur chain upon attack by a molecule of dissolved oxygen. This value is lower than the dissociation energy of a sulphur–sulphur bond in a polysulphide system in which oxygen does not participate.

5.2.3   Sulphur–sulphur dioxide

In contrast to the system discussed in the preceding section, a solution of sulphur dioxide in molten sulphur is chemically stable and lends itself readily to investigation by somewhat more conventional methods. Touro et al.[7] employed a heated dual light path infrared cell to measure vapour–liquid equilibria in this system. The cell was so designed as to facilitate the measurement of sulphur dioxide concentrations in the vapour and in the molten sulphur for an equilibrated vapour–liquid sample. Of the two light paths, one was fixed at about 50 mm and the other adjustable from about 1–15 mm. The cell was normally operated while half filled with molten

Table 5.1   Equilibrium sulphur dioxide concentration in molten sulphur as a function of temperature and partial SO$_2$ pressure

Temperature (°C)	Partial pressure of SO$_2$ (atm)	SO$_2$ concentration (g SO$_2$/100 g S)
121	1.0	0.052
125	1.0	0.051
125	0.27	0.014
125	0.13	0.007
134	1.0	0.048
150	1.0	0.044
150	0.31	0.014
150	0.20	0.010

sulphur. The two light paths were positioned in the cell in such a manner that either one could be used for liquid phase or vapour phase measurement of SO$_2$ concentration. In most cases, however, infrared absorption measurements were conducted while the cell itself was placed in the spectrophotometer so that the fixed path was aligned for vapour measurements and the shorter path was aligned for dissolved sulphur dioxide measurements. The cell was equipped with an SO$_2$ injection system and a small magnetic stirring bar. Reference cells filled with molten sulphur, similar in design to those reported in an earlier paper[7], were used to minimise interferences.

The paper by Touro et al.[6] includes information on the effect of temperature and partial pressure of SO_2 on the solubility of sulphur dioxide in molten sulphur. Experimental evidence is provided showing that the system obeys Henry's law at partial pressures up to 1 atm of SO_2. Furthermore, the data indicate that the solubility of sulphur dioxide in molten sulphur decreases with increasing temperature. A summary of the available solubility data is presented in Table 5.1.

The following van't Hoff equation expresses the solubility of sulphur dioxide (C_{SO_2}) at 1 atm in molten sulphur as a function of absolute temperature.

$$\ln C_{SO_2} = 931/T - 5.32$$

where C_{SO_2} is expressed by grammes of dissolved SO_2 per 100 g of sulphur. Since Henry's law was found to be operative, the following more general expression also applies:

$$\ln \frac{C_{SO_2}}{p} = 931/T - 5.32$$

where p is the partial pressure of sulphur dioxide expressed in atmospheres. The heat of solution for sulphur dioxide in molten sulphur is $-1850\,\mathrm{cal\,mol^{-1}}$.

5.2.4   Sulphur–naphthalene–n–octadecane

Phase equilibria in this system were studied at 113 °C using infrared spectrophotometry as the analytical tool[8]. An equilateral triangular phase diagram is shown in Figure 5.2. The equilibrium curve separates two-phase compositions (within the curve) from single-phase compositions (outside the curve). The general shape of this curve indicates that the investigated system constitutes an example of most common type of ternary system, one in which two of the component pairs (sulphur–naphthalene and naphthalene–octadecane) are completely miscible and the third component pair (sulphur–octadecane) is partially miscible.

Since sulphur is miscible with only one of the components, one would expect that in a ternary system the sulphur would show a pronounced selectivity for one hydrocarbon over the other. The ternary phase diagram confirms this reasoning. Consider a binary mixture consisting of naphthalene and octadecane, with a composition defined by Point R. If sulphur is gradually added to this mixture, its composition will change along line R–C. Assume that the addition of sulphur is discontinued at Point M (on line R–C) and that this ternary system is allowed to equilibrate. Since Point M is within the equilibrium curve, the system will separate into two phases, a sulphur-rich phase with a composition defined by Point O and an octadecane-rich phase with composition defined by Point N.

Point P on the phase diagram shown in Figure 5.2 is the so-called 'plait' point, that is, the point corresponding to a tie line of zero length, where the two liquid phases merge into a single liquid phase.

It is of particular significance that line N–O, as well as other tie lines defining this ternary system, does not point toward the sulphur apex. As a

result, the sulphur-rich phase is characterised by a relatively high naphthalene content. The ratio of naphthalene to octadecane is obviously higher
in the sulphur-rich phase (Point O) than it is either in the overall ternary
system (Point M) or in the octadecane-rich phase (Point N).

Thus, molten sulphur was demonstrated to exhibit a considerable degree
of selectivity toward naphthalene in this ternary system. This finding was

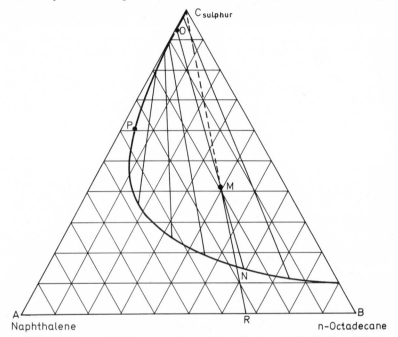

Figure 5.2   Sulphur-naphthalene-n-octadecane phase diagram at 113 °C. (From Wiewiorowski and Slaten[8], by courtesy of the American Chemical Society).

contrary to results reported in previous literature[9], where tie lines in similar
phase diagrams are pointing directly toward the sulphur apex, indicating
a complete lack of solvent selectivity by molten sulphur.

5.3   SYSTEMS INVOLVING THE FORMATION OF HETEROATOMIC POLYMERIC ENTITIES

5.3.1   Sulphur–hydrogen sulphide

This system has been investigated more extensively than any other sulphur-
containing system. Phase equilibria have been studied in a temperature
range below the melting point of sulphur, where the system may be regarded
as a solution of sulphur in liquid hydrogen sulphide, at temperatures above
120 °C, where the chemical equilibria are described in terms of phenomena
observed in the liquid sulphur phase, and at high pressures, where the solubility of sulphur in gaseous hydrogen sulphide occurs.

Data on the solubility of sulphur in liquid hydrogen sulphide have been reported by B. Meyer and J. J. Smith[10], who studied phase equilibria in this system at temperatures ranging from -80 to $+80\ ^\circ C$. The experimental approach employed by these investigators involved sealing known amounts of sulphur and liquid hydrogen sulphide in Pyrex tubes and observing temperatures of dissolution and precipitation during repeated heating and cooling cycles. Table 5.2 summarises the solubility data for sulphur in liquid hydrogen sulphide.

Table 5.2 Solubility of sulphur in liquid hydrogen sulphide as a function of temperature [10]

Temperature (°C)	Solubility (g S/100 g H_2S)
-60	0.14
-40	0.23
-20	0.33
0	0.49
20	0.67
40	0.87
60	1.1
80	1.3

In addition to providing these solubility data, Meyer and Smith[10] made two very significant observations concerning the sulphur-hydrogen sulphide system. First, they concluded that in the temperature range investigated, these two components do not react with each other. Their research yielded evidence indicating that hydrogen polysulphides do not form spontaneously.

Table 5.3 Solubility of sulphur in H_2S gas at 121 °C as a function of pressure

Pressure (atm)	Solubility (g S/100 g H_2S)
68	0.144
136	0.405
204	1.45
272	3.84
340	6.51
408	10.88

A semi-logarithmic plot of solubility versus the reciprocal of absolute temperature yields a straight line, as expected for a non-reactive solution. The chemical stability in this system is also verified by the fact that the experimentally observed dissolution and precipitation temperatures remain unchanged after numerous repeated heating and cooling cycles. Secondly, Meyer and Smith observed that sulphur in equilibrium with hydrogen sulphide melts slightly below the critical temperature of H_2S. Consequently, liquid sulphur and liquid hydrogen sulphide co-exist as two immiscible liquid phases over a range of several degrees near 100 °C. The mutual solubility of the two components under these conditions has not been determined.

Wieland[11] investigated the solubility of sulphur in hydrogen sulphide at temperatures of 66–121 °C and pressures ranging from 68 to 408 atm. It is of interest to note that sulphur has substantial solubility in hydrogen sulphide gas.

The effect of pressure on sulphur solubility in H_2S gas at 121 °C is exemplified by the data shown in Table 5.3. It is clearly evident that increasing pressure favours the solubility of sulphur in this system.

The solubility shown in Table 5.3 is of an order of magnitude comparable with that which characterises the solubility of sulphur in liquid hydrogen sulphide between − 60 and + 80 °C (see Table 5.2). It has not been established whether the solubilisation of sulphur in hydrogen sulphide gas at these high pressures leads to the formation of hydrogen polysulphides in this system. Further research is required to resolve this question. Wieland cites several examples of other systems in which the solubility of a solid in a gas was observed; the sulphur–hydrogen sulphide system is not unique in this respect.

At temperatures above the melting point of sulphur and ambient pressures the sulphur–hydrogen sulphide system is described in terms of phenomena occurring in the liquid sulphur phase. Fanelli[12] measured the solubility of hydrogen sulphide in molten sulphur over a wide temperature range and reported that changes in this solubility with temperature follow a very unusual pattern. While normally the solubility of a gas in a liquid decreases with increasing temperature, in the system under consideration the reverse holds true within a sizeable temperature interval. Between 120 and 180 °C the solubility of hydrogen sulphide in molten sulphur rises sharply; it then proceeds through a broad maximum and finally decreases at temperatures exceeding 385 °C. Fanelli postulated that this unusual thermal behaviour is due to the formation of hydrogen polysulphides in molten sulphur. More recently[13] this reasoning was verified, both by direct experimental evidence as well as from a theoretical standpoint.

Direct experimental confirmation of the formation of hydrogen polysulphides in molten sulphur was obtained by infrared spectroscopy. An electrically heated infrared cell specially designed for handling samples of molten sulphur[7] was employed in these studies. Molten sulphur was saturated with hydrogen sulphide by bubbling the gas through the liquid contained in the cell. An infrared spectrum obtained on the resulting solution of hydrogen sulphide exhibits two bands in the S—H stretching region. One, at $2570 \, cm^{-1}$, is assigned to S—H stretching in hydrogen sulphide (H_2S) and the other, at $2498 \, cm^{-1}$, is due to S—H stretching in hydrogen polysulphides (H_2S_x).

Measurements of the intensity of these two bands at various temperatures enables one to break Fanelli's solubility curve down into its two components. Figure 5.3 provides a graphic illustration of the contribution of H_2S and H_2S_x to the total solubility of hydrogen sulphide in molten sulphur. It is clearly evident that the two compounds exhibit a different temperature

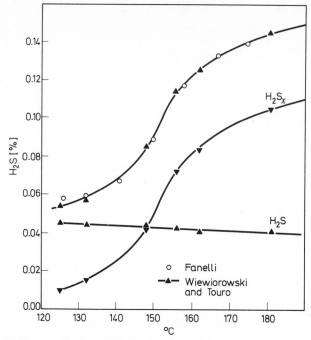

Figure 5.3 The contributions of H_2S and H_2S_x to the total hydrogen sulphide solubility in molten sulphur at 760 mmHg of H_2S overpressure. (From Wiewiorowski and Touro[13], by courtesy of the American Chemical Society).

dependency. The solubility of hydrogen monosulphide, C, decreases slowly with rising temperature in accordance with the van't Hoff equation:

$$\ln C_2/C_1 = -\Delta H/R\,(1/T_2 - 1/T_1)$$

where ΔH, the heat of solution of this component in molten sulphur, is equal to -818 cal mol^{-1}. Thus, the hydrogen monosulphide follows a 'normal', thermodynamically predictable solubility pattern in molten sulphur. The unusual total hydrogen sulphide solubility was shown to be a consequence of hydrogen polysulphide formation in the system.

It should be noted that, while the hydrogen monosulphide concentrations shown in Figure 5.3 reflect actual solubilities at 760 mmHg of H_2S overpressure, the hydrogen polysulphide curve represents concentrations at equilibrium with the corresponding hydrogen monosulphide levels. Presumably, the solubility of hydrogen polysulphides in molten sulphur is quite high, but concentrations larger than those indicated by the curve would tend to decompose into hydrogen monosulphide and sulphur, unless the pressure of H_2S was maintained at levels above 1 atm.

The formation of hydrogen polysulphides in molten sulphur has been thermodynamically interpreted in terms of the chemical equilibria expected to occur in the system[13]. Details of the thermodynamic treatment of this system will not be presented here, since this subject was reviewed previously[14]. It will suffice to mention that this interpretation is based on the assumption that the system is governed by three types of chemical equilibria:

$$\text{Initiation:} \quad S_8 \rightleftharpoons S_8^*$$
$$\text{ring} \quad \text{chain}$$
$$\text{Propagation:} \, S_8 + S_8^* \rightleftharpoons S_{2 \times 8}^*$$
$$S_8 + S_{16}^* \rightleftharpoons S_{3 \times 8}^*$$
$$S_8 + S_8^*(n-1) \rightleftharpoons S_{8n}^*$$
$$\text{Termination:} \, S_{8n}^* + H_2S \overset{K_4}{\rightleftharpoons} H_2S_{8n+1}$$

A mathematical analysis of these simultaneous equilibria leads to equations describing the total concentration of hydrogen polysulphides and the number average chain length of the polysulphides, P', in the $S—H_2S—H_2S_x$ system as a function of H_2S concentration and temperature. Some typical results of such calculations are condensed in Table 5.4.

Table 5.4   Hydrogen sulphide concentrations (H_2S and H_2S_x), termination equilibrium constant, K_4, and polysulphide chain length, P', in molten sulphur as a function of temperature at 760 mmHg of H_2S over-pressure

Temp. °C	H_2S Concentration (% by weight)	H_2S_x Concentration (% by weight) expressed as H_2S	K_4	P' (in S_8 units)
127	0.0450	0.011	1.70×10^{11}	3.3
137	0.0439	0.018	8.18×10^{10}	4.8
147	0.0428	0.036	3.09×10^{10}	8.1
157	0.0418	0.072	1.69×10^{10}	12.4
177	0.0402	0.101	2.80×10^{9}	22.9
197	0.0386	0.118	5.66×10^{8}	31.7
217	0.0360	0.139	4.45×10^{7}	42.5
307	0.0327	0.154	2.19×10^{6}	53.0

As evident from these data, theory predicts that hydrogen polysulphide concentrations should increase with rising temperature in this system. The calculated values are in excellent agreement with the experimentally obtained data shown in Figure 5.3. From the effect of temperature on the values of K_4 shown in Table 5.4, the enthalpy and entropy changes associated with the chain termination reaction defined by this equilibrium constant are calculated to be $-30\,900$ cal mol^{-1} and -25.7 cal mol^{-1}, respectively.

The number average chain lengths, P', shown in Table 5.4 differ markedly from those calculated to exist in molten sulphur in the absence of hydrogen sulphide. For instance, in pure molten sulphur the average length of a polymeric sulphur chain in the order of 114 000 S_8 units at 177 °C.

The high values of K_4 throughout the temperature range of interest indicate that the termination equilibrium is effectively shifted to the right in the presence of appreciable hydrogen sulphide in the system. That is, the polymeric component of the melt consists primarily of hydrogen polysulphides, while there is very little polymeric sulphur in the system. Since the hydrogen polysulphides are considerably shorter than the sulphur polymers encountered in the pure element, one would expect significant differences between the viscosity of pure sulphur and of the sulphur–hydrogen sulphide system. These differences are, of course, well documented in the chemical literature[15] with data illustrating the viscosity-reducing effect of hydrogen sulphide upon molten sulphur, in particular at temperatures above 159 °C.

In addition to the infrared evidence for the formation of hydrogen poly-sulphides in this system, more recent n.m.r. measurements[16] provided further insight into the chemistry of the sulphur–hydrogen sulphide system. N.M.R. spectra enable one to distinguish between the lower homologues of the hydrogen polysulphide series, from H_2S_2 to H_2S_5. Higher polysulphides have not been resolved – they yield a single H_2S_x band. It has been established that when sulphur is saturated with hydrogen sulphide, two n.m.r. bands appear; one assigned to H_2S, the other to H_2S_x, where $x \geqslant 6$. N.M.R. data indicate that hydrogen polysulphides, H_2S_2 to H_2S_5, are notably absent in the sulphur–hydrogen polysulphide system. This is in agreement with values calculated on the basis of thermodynamic considerations. As indicated in Table 5.4, the number-average chain length of hydrogen polysulphides is about 3.3 S_8 units at 127 °C and increases with rising temperature.

5.3.2 Sulphur–selenium

Ward[17] investigated the molecular composition of sulphur–selenium and sulphur–arsenic mixtures by Raman spectroscopy using a continuous-wave He—Ne laser as the excitation source. His work on the sulphur–selenium system covers crystalline as well as liquid compositions. The Raman spectra offer evidence for the existence of ring molecules defined by the formula S_xSe_{8-x}, where $4 \leqslant x \leqslant 8$. As might be expected, the abundance of octa-atomic rings with a high selenium content increases with rising total selenium content in the mixture.

The fact that selenium enters into the octa-atomic ring structures has pronounced effects on the polymerisation phenomena in the sulphur–selenium system. The addition of selenium to sulphur markedly lowers the poly-merisation threshold temperature of molten sulphur. Tobolsky and Owen[18] developed a theory of copolymerisation applicable to the sulphur–selenium system. Their model assumed the monomer species to be S_8 and Se_8 rings capable of copolymerising into mixed sulphur–selenium chains. The Tobolsky-Owen treatment facilitates the calculation of threshold temperatures at which this copolymerisation commences. A refinement of this treatment was reported by Ward and Meyers[19], who took into account that selenium not only copolymerises into the system but also enters the sulphur ring mono-mers substitutionally to yield the mixed sulphur–selenium molecules, S_xSe_{8-x}, as mentioned above. The enthalpy of propagation of such monomer species is lower than that of pure sulphur because of the lower bond energy of a sulphur–selenium or selenium–selenium linkage. Polymerisation thresh-old temperatures, T_ϕ, calculated using this model are in good agreement with experimentally (calorimetrically) determined data. Thus, while the T_ϕ of pure sulphur is 159 °C, the polymerisation of $S_{0.9}Se_{0.1}$ sets in at about 110 °C.

It must be recognised that the relative abundance and availability for polymerisation of such mixed monomers is relatively large as compared to the fraction of either elemental component in the system. This point was quantitatively elaborated upon in a more recent paper by Ward[20]. Using relationships originally derived by Tobolsky and Owen[18] and assuming that the mixed sulphur–selenium monomer has a formula Se_3S_5, Ward calculated the molecular composition of sulphur–selenium alloys at their melting point. For example, an alloy containing 95 atoms of selenium and 5 atoms of sulphur per 100 atoms of the melt will at equilibrium assume the following composition: 2.7% by weight of Se_3S_5, 40% Se_8 and 57.3% Se_pS_g (polymer). Thus, sulphur representing an atomic fraction of 5% in this melt, is present in 60% (by weight) of the molecular species. This composition is assumed to be retained upon rapid cooling of the melt to room temperature.

It must be pointed out that certain assumptions had to be made by Ward to facilitate and simplify the calculations. As already mentioned, the monomer was assumed to have the formula Se_3S_5. Although Raman evidence suggests that this is a serious over-simplification, the molecular equilibrium calculated to be reached upon polymerisation is apparently quite insensitive to the exact composition of the mixed monomer. Calculations based on Se_4S_4 or Se_2S_6 as the mixed monomeric species yield results very similar to those obtained using Se_3S_5 in the model and Ward's assumption is, therefore, entirely reasonable. Another assumption which had to be made in order to facilitate the calculations regards the entropy and enthalpy of the propagation reaction for mixed monomer. Ward assigned to Se_3S_5 the propagation enthalpy of Se_8 and the propagation entropy S_8. While sulphur is the major atomic component of the mixed ring, bond scission during propagation will most likely involve a S—Se or a Se—Se linkage. Therefore, both assumptions appear justified, in view of the lack of direct data on the propagation process for Se_3S_5.

The thermodynamic treatment of polymerisation phenomena in the S—Se system also yields information on the chain length of the polymers. Ward has shown that the number-average chain length of the polymer phase in the Se—S system increases as sulphur is added to the system.

The literature cited in this section provides excellent experimental data, obtained by Raman spectroscopy, on the molecular composition which sulphur–selenium melts attain upon equilibration. However, the kinetic aspects of the formation of mixed sulphur–selenium rings and polymers have not yet been reported, leaving open to investigation an important area of sulphur chemistry.

5.3.3 Sulphur-arsenic

The system composed of sulphur and arsenic differs in several respects from the sulphur–selenium system. Raman spectroscopy was employed by Ward and Myers[17, 19] to follow polymerisation phenomena in this system. It is reported that essentially all arsenic atoms in a sulphur-arsenic melt are associated with polymeric species. In contrast to selenium, arsenic does not enter into the octa-atomic sulphur monomer. Furthermore, trivalent arsenic facilities the formation of branched polymers, a phenomenon not observed in pure molten sulphur or in the sulphur–selenium system.

Ward and Myers[19] point out that, although arsenic is not part of the monomeric entity at equilibrium conditions in a sulphur–arsenic melt or glass, the arsenic-containing polymer does not act as an inert diluent. While the addition of inert diluents to molten sulphur is known to raise its polymerisation temperature, arsenic addition to molten sulphur lowers the polymerisation temperature. As little as 2 atomic % of arsenic decreases the polymerisation temperature of sulphur from 159 ° to about 122 °C, a value which shows little variation with further addition of arsenic up to 15.%. This suggests that above the polymerisation temperature the arsenic–sulphur polymer reacts with monomeric S_8 according to the propagation reaction,

$$S_8 + AsS_n \rightleftharpoons AsS_{n+8}$$

The qualitative description of the sulphur–arsenic system is supplemented[19] by quantitative information regarding the sulphur monomer and polymer fractions in S—As mixtures as a function of temperature above the polymerisation threshold temperature. Information is lacking, however, on the relative concentrations of S—As polymer and S_8 monomer at equilibrium below the transition temperature.

5.3.4 Sulphur-iodine

Chemical equilibria in this system were investigated recently by Harris[21] who conducted spectrophotometric measurements of liquid sulphur and of dilute iodine solutions in liquid sulphur as a function of temperature and wavelength in the visible region. The experimental evidence suggests that near 120 °C essentially all dissolved iodine is present as I_2, but as the temperature is raised the concentration of free I_2 decreases and sulphur–iodine species absorbing at shorter wavelengths appear.

The chemical equilibria occurring in the sulphur–iodine system include, in addition to the previously described initiation and propagation equilibria, a chain termination reaction defined as follows:

$$I_2 + S_{8_n}^* \overset{K_2}{\rightleftharpoons} IS_{8_n}I$$

Spectrophotometric data obtained by Harris on this system are condensed in Table 5.5.

Employing these data and a model of the molecular composition of liquid sulphur at equilibrium described in an earlier paper[22], Harris derived an expression defining the equilibrium constant, K_2, as a function of absolute temperature:

$$K_2 = \exp\left(-14.93 + 16\,020/T\right) \ 1/\text{mol}$$

The chemical equilibria involving iodine in solutions of iodine in molten sulphur of course are not limited to the termination reaction given above, but may also be expressed as follows:

$$I_2 + S^*_{x+y} \rightleftharpoons IS_xI + S_y$$
$$I_2 + S^*_x + S^*_y \rightleftharpoons IS_x\cdot + IS_y\cdot$$
$$I_2 + S^*_{x+y} \rightleftharpoons IS_x\cdot + IS_y\cdot$$

In all cases, iodine atoms assume the terminal position in a polymeric sulphur

Table 5.5 **Experimentally determined concentration of bound iodine in molten sulphur as a function of temperature**

Total $I_2 = 14.9 \times 10^{-3}$ mol l^{-1}		Total $I_2 = 23.1 \times 10^{-3}$ mol l^{-1}		Total $I_2 = 36.3 \times 10^{-3}$ mol l^{-1}	
T (°C)	Bound iodine (10^{-3} mol l^{-1})	T (°C)	Bound iodine (10^{-3} mol l^{-1})	T (°C)	Bound iodine (10^{-3} mol l^{-1})
120.6	0.0	122.4	0.0	120.7	0.0
140.5	1.4	140.8	2.1	130.9	1.8
151.4	3.3	152.6	6.7	140.9	4.4
161.7	7.7	162.8	13.2	156.4	15.1
172.1	9.6	171.9	15.0	169.4	20.9
181.3	9.5	181.5	15.7	181.5	22.9

chain. In this respect the behaviour of iodine in molten sulphur closely resembles that of the hydrogen sulphide–sulphur system.

5.4 CONCLUDING REMARKS

The discussion of sulphur–containing systems presented in this chapter may be regarded as a summary of information currently available on this subject. An effort has been made to present a critical evaluation of relevant and useful data and also to outline areas where further research is most obviously in order.

Note that nearly all cited papers concerning chemical equilibria in molten sulphur were published within the last decade. There is a continuing interest in the chemistry of molten sulphur, in its unique solvent properties and its behaviour in multicomponent systems.

References

1. Linke, W. F. (1965). *Solubilities of Inorganic and Metal-Organic Compounds,* (Washington, D.C.: A.C.S.)
2. Tuller, W. N. (1954). *The Sulphur Data Book,* (New York: McGraw-Hill)
3. Touro, F. J. and Wiewiorowski, T. K. (1966). *J. Phys. Chem.,* **70,** 3534
4. Wiewiorowski, T. K., Parthasarthy, A. and Slaten, B. L. (1968). *J. Phys. Chem.,* **72,** 1890
5. Wiewiorowski, T. K. and Slaten, B. L. (1967). *J. Phys. Chem.,* **71,** 3014
6. Touro, F. J. and Wiewiorowski, T. K. (1966). *J. Phys. Chem.,* **70,** 3531
7. Wiewiorowski, T. K., Matson, R. F. and Hodges, C. T. (1965). *Anal. Chem.,* **37,** 1080

8. Wiewiorowski, T. K. and Slaten, B. L. (1968). *J. Chem. Eng. Data,* **13,** 38
9. Francis, A. W. (1966). *J. Chem. Eng. Data,* **11,** 557
10. Meyer, B. and Smith, J. J. (1969). 'The Chemistry of Hydrogen Sulphide as a Solvent', *4th Ann. Sulphur Symposium, Calgary, Alberta*
11. Wieland, D. R. (1958). 'The Solubility of Elemental Sulphur in Methane, Carbon Dioxide and Hydrogen Sulphide Gas', *Dissert., Agricultural and Mechanical College of Texas*
12. Fanelli, R. (1949). *Ind. Eng. Chem.,* **41,** 2031
13. Wiewiorowski, T. K. and Touro, F. J. (1966). *J. Phys. Chem.,* **70,** 234
14. Tobolsky, A. V., Editor. (1968). *The Chemistry of Sulphides,* 9, (New York: Interscience)
15. Rubero, P. A. (1964). *J. Chem. Eng. Data,* **9,** 481
16. Hyne, J. B., Muller, E. and Wiewiorowski, T. K. (1966). *J. Phys., Chem.,* **70,** 3733
17. Ward, A. T. (1968). *J. Phys. Chem.,* **72,** 4133
18. Tobolsky, A. V. and Owen, G. D. T. (1962). *J. Polym. Sci.,* **59,** 329
19. Ward, A. T. and Myers, M. B. (1969). *J. Phys. Chem.,* **73,** 1374
20. Ward, A. T. (1970). *J. Phys. Chem.,* **74,** 4110
21. Harris, R. E., National ACS Meeting, Los Angeles, 1971, *Phys. Chem.,* 183
22. Harris, R. E. (1970). *J. Phys. Chem.,* **74,** 3102

6
Polonium

K. W. BAGNALL
University of Manchester

6.1 INTRODUCTION

Although several isotopes of polonium occur in nature as members of the thorium ($^{216}Po, ^{212}Po$), actinium ($^{215}Po, ^{211}Po$) and radium ($^{218}Po, ^{214}Po$ and ^{210}Po) decay chains, only one of these naturally occurring isotopes,

^{210}Po, has a half-life sufficiently long to permit studies of its chemistry with weighable (milligramme) amounts. It is formed in the chain by the β-decay of ^{210}Bi ($T_{\frac{1}{2}}$ 5.0 days) but, owing to the difficulty of working up sufficient radium or radon to obtain a worth-while yield of ^{210}Po, the normal production method is to irradiate natural bismuth (^{209}Bi) with neutrons to obtain the parent, ^{210}Bi. The isotope ^{210}Po has a half-life of only 138.4 days, which means that it is intensely radioactive, the specific α-activity being 4.5 Ci mg^{-1}, equivalent to 10^{13} α-disintegrations per minute per milligramme. Two other, longer-lived isotopes can be synthesised, ^{209}Po ($T_{\frac{1}{2}}$ about 42 years) and ^{208}Po ($T_{\frac{1}{2}}$ 2.9 years) by cyclotron bombardment of bismuth with protons (^{209}Bi(p,2n)^{208}Po) or deuterons (^{209}Bi(d,2n)^{209}Po and ^{209}Bi(d,3n)^{208}Po, depending on the energy of the deuterons). However, neither of these isotopes is as easy, or as cheap, to prepare as ^{210}Po, and nearly all of the published work on the chemistry of the element has been performed using this last isotope.

Almost all of the basic research on the chemistry of weighable amounts of polonium was published between 1950 and 1960. Since then little new work has been reported on the fundamental chemistry of the element, although a number of papers have appeared in which trace level studies of the solution chemistry of polonium are described, and there have been several papers concerning separation procedures aimed at isolating gramme quantities of ^{210}Po from neutron irradiated bismuth. Quantities ot this magnitude are required for use as heat sources in thermo-electric generators in earth satellites, where the high heat output resulting from the radioactive decay (about 140 W/g of ^{210}Po) offsets the short half-life when space missions of limited duration are envisaged.

The chemistry of polonium, the last member of Group VIb of the Periodic Table, exhibits a number of the characteristics to be expected of its position in a Group in which the lighter elements are purely non-metallic in character, in accord with their outer electronic configurations, ns^2 np^4, which are approaching the configurations of the next highest noble gas. In Group VIb the elements exhibit a transition from the non-metal character of sulphur to the semi-metallic behaviour of selenium and tellurium and, finally, elementary polonium has the physical properties of a metal. Associated with this change there is a clearly marked tendency for the divalent state to become increasingly more stable with respect to oxidation or disproportionation as the atomic number increases down the group. This trend is particularly well-marked in the case of polonium, for polonium(II) compounds are appreciably more stable in the above respects than the analogous compounds of the lower homologues. At the same time, the highest oxidation state ($+6$) of polonium is much more difficult to attain than is the case with tellurium or selenium, the overall behaviour being very similar to that of other end members of the main groups, such as lead and bismuth.

Earlier reviews[1,2] have covered much of the information given in the following sections, in which the present state of knowledge of the chemistry of polonium is briefly summarised; a more recent review[3] of the stereochemistry of polonium and the other Group VIb elements is also available.

6.2 SEPARATION PROCEDURES

The use of solvent extraction procedures in analysis for polonium has recently been reviewed[4], but because of the radiolytic damage to organic material which results when large quantities of the α-active polonium-210 are present, separation processes which involve organic materials are more commonly used for the isolation of trace level amounts of this isotope. Tri-n-butylphosphate[5] (TBP), amines[6] and carbinols[7] are useful extractants for low-levels of polonium, while on the larger scale the extraction from aqueous nitric acid into a solution of tricaprylammonium nitrate in xylene[8] or into dibutyl carbitol[9] appears to be a reasonable proposition. Ion exchange methods[10] suffer from the same disadvantages as solvent extraction for large scale separations, for which methods that involve precipitation from aqueous solution have proved to be of general utility.

The earlier methods of this kind involved the precipitation of elementary polonium, formed by reduction with stannous chloride, for example, either with or without tellurium present as a carrier. Other workers favoured the spontaneous deposition of polonium (i.e., electrochemical replacement) onto a less noble metal, such as silver, with the deposition conditions arranged so as to ensure that bismuth and other impurities did not plate out[11]. Because polonium metal is more volatile than silver, the separation from the latter is easily achieved by vacuum sublimation. In one process the polonium present in irradiated bismuth is concentrated by a preliminary deposition onto powdered bismuth which is agitated by means of a stream of inert gas[12] and in another the polonium is selectively adsorbed onto active carbon, from which it is desorbed by oxidation of the powdered product with concentrated nitric acid[18].

The volatility of elementary polonium also makes it possible to separate the element from neutron irradiated bismuth metal at high temperatures by distillation. The procedure has the advantage that the residual bismuth metal can be returned to the nuclear reactor for further irradiation without any chemical treatment. Research on the applicability of the method has shown that polonium does not form an ideal solution in bismuth[13], but the concentration of ^{210}Po to greater than 95 wt.% from bismuth containing 25–27 p.p.m. of polonium can be achieved[14]. During the distillation from molten bismuth metal at 650–750 °C, it has been shown[15] that the surface layer becomes depleted in polonium (and any other volatile species present). Molecular distillation[16] seems to be a quite practicable separation method and some concentration of the polonium present in irradiated metallic bismuth can also be achieved by zone melting techniques[17].

In an alternative separation procedure, a suitable carrier (selenium, tellurium, silver, gold or mercury) is added to the molten irradiated bismuth, which is then cooled and used as the anode in a cell in which the bismuth is electrodeposited onto a suitable cathode[19]. This procedure leaves a precipitate of polonium and the carrier element, and the deposited bismuth can be re-irradiated.

The sublimation of polonium from irradiated bismuth oxide has also been studied[20] as a possible separation procedure, but the method does not seem to have been adopted.

The best method of isolating polonium from neutron irradiated bismuth appears to be one in which the molten material is contacted with molten sodium hydroxide at 400–500 °C in an inert atmosphere, for example nitrogen or argon[21]. About 97% of the ^{210}Po present is transferred to the sodium hydroxide phase in a single extraction with as little as 25 wt.% of the alkali. The polonium is recovered from the alkali phase by dissolving it in aqueous nitric acid and then extracting it from this into a solution of a butylphosphonate in kerosine[21]. It is probable that a mixture of sodium polonide and polonite is formed in the alkali phase.

Weighable (>mg) amounts of polonium are conveniently purified by precipitation of polonium monosulphide (p. 192) and vacuum thermal decomposition of this compound, followed by vacuum sublimation of polonium metal[22]. This procedure gives a good separation from lead and most light element impurities, but not from bismuth.

6.3 POLONIUM METAL

6.3.1 Chemical and physical properties

The metal is normally prepared by thermal decomposition of the monosulphide[22] or dioxide[23] in a vacuum and by reduction with suitable reagents in aqueous solution[1]. It is also precipitated by the action of anhydrous liquid ammonia, or concentrated aqueous ammonia, and even primary or secondary aliphatic amines, on polonium compounds, a reaction which is ascribed to atomic hydrogen, formed from ammonia or the amines under the influence of the α-radiation[24]. The metal precipitates in these reactions as a grey–black powder but the sublimed metal has a bright, silvery appearance.

Two crystal modifications of the element are known, the low temperature α-form (simple cubic) and the high temperature β-form (simple rhombohedral). Both of these are metallic in character, the temperature coefficient of resistivity being positive[25], in contrast to the behaviour of the lower homologues. There is no evidence for any ring or long chain polonium species analogous to those found, for example, for sulphur and selenium. The melting point[25] (254 °C) and boiling point[26] (962 °C) are much lower than those of tellurium and are comparable to the values for other elements of same Period, thallium, lead and bismuth. Some physical properties of the element are given in Table 6.1.

The α → β phase transition occurs at about 36 °C, but the high temperature β-form is normal at room temperature because of the internal heating effect of the α-radiation; both forms co-exist[27] between 18 °C and 54 °C. Vapour pressure data[26] and Hartree-Fock calculations[28] for the lowest terms of the ground state have been published.

The metal reacts with air or oxygen to form the dioxide, the reaction being slow at room temperature and rapid at 250 °C. It dissolves in mineral acids, initially yielding a pink solution of polonium(II) which then rapidly oxidises to yellow polonium(IV); it also dissolves in fused alkali (p. 190) and forms the tetrahalides with halogens. Like its lower homologues, polonium metal yields brightly coloured products with sulphur trioxide, selenium trioxide and the corresponding concentrated acids. These red solids were originally[29] formulated as $PoXO_3(X=S, Se)$ but they are probably much more complex

Table 6.1   Some physical properties of polonium metal*

Density, g/cm³	9.196(α), 9.398(β)
Atomic radius, Å	1.64
First ionisation potential, kJ	810
$\Delta H_{vap.}$, kJ mol^{-1}	102.8
Specific electrical resistivity, $\mu\Omega$ cm (0 °C)	42(α), 44(β)
Temperature coefficient of resistivity	0.0046(α), 0.0070(β)
Linear coefficient of expansion, cm cm^{-1} deg^{-1}	2.35×10^{-5}

* Data from references 1, 2 and 45.

and analogous to the polymeric selenium[30] compound. The polonium compounds decompose to polonium monoxide on keeping and dissolve completely in water to yield a solution containing polonium(II), unlike their selenium and tellurium analogues which yield a precipitate of the element with water.

6.3.2   Polonides

Polonium combines with most metals and several non-metals at temperatures between 400 °C and 1000 °C to form polonides which are essentially derived from the hydride, H_2Po. This last has never been isolated in weighable amounts although there is evidence for its formation in tracer quantities

Table 6.2   Crystallographic data for some metal selenides, tellurides and polonides*

	Symmetry and structure type	Lattice parameters, Å					
		X = Se		X = Te		X = Po	
		a_0	c_0	a_0	c_0	a_0	c_0
CaX	Cubic, NaCl	5.91	—	6.345	—	6.514	—
CdX	Hexagonal, ZnO	4.30	7.02	—	—	—	—
	Cubic, ZnS	—	—	6.464	—	6.665	—
HgX	Cubic, ZnS	6.07	—	6.434	—	—	—
	Cubic, NaCl	—	—	—	—	6.250	—
PtX₂	Hexagonal, Cd(OH)₂	3.724	5.062	4.010	5.201	4.104	5.606

* Structural data are from references 69 (Se, Te), 32 and 68 (Po).

(and in very poor yield) by the action of dilute hydrochloric acid on magnesium foil on to which polonium has been plated[31].

Most of the metal polonides are isostructural with the analogous selenides and tellurides (Table 6.2); the lanthanide compounds, prepared by heating the lanthanide metal or its hydride with polonium metal[33-35] at high temperatures are of interest as suitable materials for thermo-electric power

generation in space satellites because of their high thermal stability. They
vaporise incongruently and decompose at the appearance temperature of
polonium, the decomposition products being observed by time-of-flight
mass spectrometry[36].

Experiments on the volatility of trace level amounts of polonium in carbon
monoxide suggest that carbonyl polonide, possibly COPo, should exist, but
although there is some reaction, leading to the formation of a volatile species,
when carbon monoxide is passed over a heated silver foil plated with a
weighable amount of elementary polonium[37], the compound could not be
isolated.

6.4　BIVALENT POLONIUM

Solutions of polonium(II) in dilute acid are easily obtained by reduction of
polonium(IV) with reagents such as sulphur dioxide at room temperature or
by dissolving any of the known polonium(II) compounds in acid; the
$PoCl_{6(aq)}^{2-}/PoCl_{4(aq)}^{2-}$ reduction potential in M-hydrochloric acid has been
measured[38], the value found being approximately 0.72 V, assuming that the
ionic species were as shown above. Solutions of bivalent polonium-210 in
aqueous media oxidise to polonium(IV) in 8 min in the absence of reducing
agents. This results from the formation of oxidising species under the action
of the α-radiation of the polonium-210, and the time required for the oxidation
is therefore independent of the polonium concentration. The longer lived
polonium isotopes should require proportionately longer times for this
oxidation process.

The only compounds recorded for this oxidation state are the monoxide,
monosulphide, dichloride and dibromide; a polonium(II) fluoride, a blue-
grey solid, is formed when a suspension of the white polonium(IV) fluoride
in dilute hydrofluoric acid is treated with sulphur dioxide[37], but its composition is unknown.

The monoxide, a black solid, is formed[29] in the spontaneous decomposition (probably α-radiation induced) of the sulphoxide 'PoSO₃' or
selenoxide 'PoSeO₃' (p. 191). Addition of alkali to a freshly-prepared solution
of polonium(II) yields a dark brown precipitate which is presumably the
corresponding hydroxide or hydrated oxide[39]. Both products oxidise rapidly
to polonium(IV) species in water (oxidation by α-radiolysis products) or in
contact with air. The monosulphide, also a black solid, is precipitated by
hydrogen sulphide from solutions of any polonium compound in dilute
mineral acid and is also obtained by treating polonium(IV) hydroxide with
yellow ammonium sulphide, in which it is insoluble. Its solubility product is
about 5.5×10^{-29} and the compound decomposes to the elements at 275 °C
in a vacuum[22].

The dark ruby-red dichloride[39] and the purple-brown dibromide[40] are
best prepared by reduction of the solid tetrahalides with sulphur dioxide
(PoCl₂) or hydrogen sulphide (PoBr₂) at 25 °C. Reduction of polonium
tetrabromide with sulphur dioxide appears never to go to completion, in

contrast to the reaction with the tetrachloride. Reaction of polonium tetra-iodide with hydrogen sulphide yields polonium metal[41]. The dichloride is also obtained by reduction of the solid tetrachloride with carbon monoxide (150 °C), hydrogen sulphide (150 °C) or hydrogen (200 °C), but prolonged heating in an atmosphere of these gases results in reduction to polonium metal.

Both of the dihalides are hygroscopic and are moderately volatile; the dichloride sublimes, with some decomposition, at 190 °C in nitrogen and the dibromide at 110°C/30 μm, also with some decomposition. The latter melts and appears to disproportionate at 270–280 °C in nitrogen. The crystal symmetry of the dichloride appears to be orthorhombic, but the observed unit cell would contain only a single molecule of the compound so that it is very probable[39] that the true symmetry is monoclinic or triclinic with an angle, or angles, close to 90 degrees.

Very little is known about the chemistry of these compounds. A brown ammine seems to be formed when the dichloride is heated in dry ammonia gas at 200 °C, and in hydrobromic acid solution solvent extraction studies with tracer levels of polonium(II) have provided evidence for the existence of $PoBr_3^-$ (or $[PoBr_3(H_2O)]^-$) and $PoBr_4^{2-}$ ions, the latter predominating in more concentrated hydrobromic acid solutions[42]. Similar experimental work indicates that analogous iodocomplex[43] and chlorocomplex[44] ions are formed in the appropriate halogen acids. However, weighable amounts of polonium tetraiodide are not reduced in aqueous media[41], so that the polonium(II) iodocomplex results may be erroneous.

6.5   TRIVALENT POLONIUM

Many of the earlier studies of the chemical behaviour of trace level polonium, summarised in a previous review by the author[45], erroneously attributed the results to the presence of polonium(III) species in solution. Although these experiments most probably involved bivalent polonium, electrochemical studies with weighable amounts of polonium suggest that polonium(III) may have a transient existence as an intermediate in the oxidation of polonium(II) to polonium(IV) in aqueous solution[38]. No compounds of this oxidation state have ever been isolated.

6.6   QUADRIVALENT POLONIUM

Polonium(IV) is the normal oxidation state for the element in the absence of reducing agents, but it is questionable whether this would be so for stable isotopes of polonium if any such existed. The Po^{IV}/Po^{II} reduction potential is given in Section 6.4 and the $Po_{aq}^{4+}/Po_{(s)}$ potential (E_0^H), determined[38] in M-nitric acid, is $+0.76$ V, close to that of silver.

The solution chemistry of this oxidation state has been studied extensively, most of the reported work being concerned with trace level amounts of ^{210}Po. There is abundant evidence of complex formation in aqueous solution, mostly obtained by solvent extraction techniques, and in one recent publication[46] stability constants were reported for polonium(IV) complexes with thenoyltrifluoroacetone(TTA), ethylenediamine tetra-acetic acid (EDTA),

Table 6.3 Stability constants of some polonium (IV) chelate complexes [46]

Thenoyltrifluoroacetone, TTA	$[Po(OH)_2(TTA)]^+$	4×10^6
	$[Po(OH)_2(TTA)_2]$	1.3×10^{13}
Oxalic acid, H_2Ox	$[Po(OH)_2(Ox)_2]^{2-}$	6×10^7
Tartaric acid, H_4Tart	$[Po(OH)_2(Tart)]^{2-}$	2×10^7
Citric acid, H_3Cit	$[Po(OH)_2(HCit)_2]^{2-}$	1.1×10^9
Nitrilotriacetic acid, H_3NTA	$[Po(OH)_2(HNTA)_2]^{2-}$	1.5×10^8
Ethylenediaminetetra-acetic acid, H_4EDTA	$[Po(OH)_2(HEDTA)]^-$	1.0×10^8

citric, oxalic and tartaric acids (Table 6.3). Earlier solvent extraction results have been summarised in previous reviews[1, 45] and are not repeated here.

6.6.1 Polonium dioxide

This oxide is appreciably more basic than the tellurium compound forming, for example, compounds such as the disulphate, $Po(SO_4)_2$, for which tellurium analogues are unknown. The hydrated compound, obtained as a pale yellow, flocculent precipitate when dilute alkali is added to a solution of quadrivalent polonium, shows only a slight acidic character, $K_c(= [PoO_3^{2-}]/[OH^-]^2)$ being[47] 8.2×10^{-5} at 22 °C. Fusion of the dioxide with potassium hydroxide in air, or with potassium nitrate, yields a colourless melt which probably contains the polonite[37], K_2PoO_3.

Polonium dioxide is prepared[48] from the elements at 250 °C; it exists in two crystal modifications. The yellow, low temperature form[23, 48] possesses face-centred cubic symmetry and, like uranium dioxide, is an oxide of variable stoichiometry, as shown by the variation in the length of the unit cell edge[23]. The red, high temperature form[23, 48] is of tetragonal symmetry, a result which is explicable on ionic radius considerations, for the ratio of the ionic radii Po^{4+}/O^{2-} is close to 0.73, the lower limit for cubic coordination[48].

At high temperatures the oxide exhibits other colour changes, becoming chocolate brown at 885 °C in oxygen (1 atm) and it sublimes at this temperature. The compound decomposes to the elements at 500 °C in a vacuum[23], the metallic polonium so formed being identified by x-ray crystallography. However, in more recent work[20] on the sublimation of polonium dioxide from a mixture with bismuth oxide, it appears that the vapour pressure behaviour in argon at 780 °C is consistent with transport of the dioxide in the argon gas stream, an observation which seems to be at variance with the vacuum decomposition results.

6.6.2 Tetrahalides

It is uncertain whether the white solid, obtained by treating polonium(IV) hydroxide or tetrachloride with dilute hydrofluoric acid, is the tetrafluoride or a basic compound[37]. Its solubility increases markedly with increasing hydrofluoric acid concentration, presumably because of the formation of fluorocomplex species[49].

Polonium tetrachloride is a yellow solid which probably possesses monoclinic symmetry[50], whereas the bright red tetrabromide appears to crystallise with face-centred cubic symmetry[40]. The tetrachloride is best prepared from the elements[35, 50] at 200 °C; complete conversion of polonium metal to the tetrabromide occurs most rapidly by heating the metal in a stream of nitrogen saturated with bromine vapour between 200 °C and 250 °C, although reaction of the metal with bromine vapour (200 mm pressure) is reasonably satisfactory[40, 51]. The salmon-pink dichlorodibromide, $PoCl_2Br_2$, is prepared by reaction of the dichloride with bromine vapour at room temperature and there is some evidence for the formation of the dichlorodi-iodide and dibromodi-iodide in the reaction of the dichloride and dibromide with iodine in carbon tetrachloride, although neither of the dihalides reacts with iodine vapour[40, 41].

Polonium tetraiodide is precipitated as a black solid (cf., TeI_4) when dilute (0.1 M) hydriodic acid is added to a solution of a quadrivalent polonium compound, but is best prepared from the elements at 40 °C at low (~ 1 mm) pressure of iodine. It sublimes in nitrogen, with some decomposition to the metal, at 200 °C and it is possible that the primary reaction on heating is decomposition to an unstable lower iodide, perhaps PoI_2, which subsequently disproportionates. The tetraiodide is much less soluble in ethanol, acetone and other ketones than are the tetrachloride or tetrabromide. Unlike the latter it is not appreciably hygroscopic[41]. All three tetrahalides are rather easily hydrolysed to white solids of uncertain composition which are probably oxohalide species.

Solvent extraction experiments with trace level polonium have provided evidence for the existence of anionic halocomplexes in halogen acid solution. Pentahalospecies, PoX_5^- (X = Cl[44], Br[42], I[43]) exist only at low acid concentrations and the corresponding hexahaloanions predominate at high acidity. The iodide complexing results are in reasonable agreement with an earlier[41] study of the solubility of polonium tetraiodide in aqueous hydriodic acid, from which the equilibrium constants for

$$PoI_4 + I^- \rightleftharpoons PoI_5^- \, ([HI] < 0.02 \text{ N})$$

and

$$PoI_4 + 2I^- \rightleftharpoons PoI_6^{2-} \, ([HI] > 0.02 \text{ N})$$

were found to be 6.7×10^{-5} and 5.9×10^{-3} respectively at 22 °C.

The colour of solutions of the tetraiodide in dilute hydriodic acid changes with temperature, the solutions being green at 0 °C and red-brown above 20 °C; the solubility of the tetraiodide in this medium is also a function of temperature, a minimum being observed at 30 °C. These results may indicate changes in the hydration number of the anions concerned[41]. It is quite probable that the $[PoX_5]^-$ species is actually $[PoX_5(H_2O)]^-$.

The visible spectrum[52] of polonium(IV) in 12.2 M hydrochloric acid shows a well defined maximum at 418 nm; at this acid concentration the $PoCl_6^{2-}$ ion should be the major species present. The greenish-yellow alkali metal (K,Rb,Cs), ammonium and tetramethylammonium hexachloropolonates(IV) have been isolated from aqueous acid solution; the caesium salt is precipitated from concentrated hydrochloric acid solution[39] and the others crystallise when the solutions are evaporated to small bulk[53]. The corresponding hexa-bromo-(Cs,NH$_4$) and hexaiodo-polonates(IV) (Cs) are also known. The former are dark red[40] and the latter black[41]. All of these compounds possess face-centred cubic symmetry ($Fm\ 3m - O_h^5$) and the anion is octahedral.

Very few complexes of the tetrahalides with neutral donor ligands have been prepared in weighable amounts; a study of the solubility of polonium tetrachloride in tri-n-butyl phosphate (TBP) suggests that the complex [PoCl$_4$·2TBP] is formed[5] and there is evidence for the formation of ammines of uncertain composition when polonium tetrachloride[39] or tetrabromide[40] is exposed to gaseous ammonia.

In contrast to the tetrahalides, the polonium(IV) pseudohalides have been scarcely investigated. There is no record of any work on polonium(IV) thiocyanates and there is only one report of an attempted preparation of a cyanide[47]. In this work a white, crystalline solid was obtained when polonium(IV) hydroxide or the tetrachloride was treated with aqueous hydrocyanic acid. The product could be a basic chloride, formed by hydrolysis, for it is almost insoluble (0.089 mg of ^{210}Po/l) in the 0.05 M acid. However, the solubility increases somewhat with increasing hydrocyanic acid concentration.

6.6.3   Oxo-acid compounds

6.6.3.1   Sulphate, selenate and chromate

The polonium(IV) sulphate, selenate and chromate systems clearly demonstrate the increased basic character of polonium dioxide as compared with the tellurium compound. The white, hydrated disulphate, $Po(SO_4)_2 \cdot nH_2O$, is formed when polonium(IV) hydroxide or tetrachloride is treated with sulphuric acid (> 0.25 M), while with more dilute acid (0.05–0.25 M) the white basic compound, $2PoO_2 \cdot SO_3$, is obtained[29], analogous to tellurium(IV) sulphate. The disulphate is less soluble in dilute sulphuric acid than the basic salt, and a study of the solubility of the latter indicates that it is metastable with respect to the disulphate in 0.05–0.25 M sulphuric acid. The hydrated disulphate loses water at 100 °C, or when washed with anhydrous ether, yielding the deep purple anhydrous compound. Both sulphates decompose to the dioxide at 550 °C, the basic salt exhibiting a reversible colour change (to yellow) at 250 °C.

The white basic selenate, which also becomes yellow at 250 °C, is obtained in the same way as the sulphates, using 0.007–2.5 M selenic acid; its solubility in selenic acid is lower than that of the basic sulphate at low acidity, but increases by two orders of magnitude between 0.025 M and 2.5 M selenic acid[29].

An orange yellow chromate, probably $Po(CrO_4)_2$, has been prepared in the same way as the sulphates and selenate, using aqueous 1.0 M chromium trioxide[37]. It is readily hydrolysed to a dark brown solid of composition close to $2PoO_2 \cdot CrO_3$.

6.6.3.2 Nitrates

Reaction of polonium tetrachloride or dioxide with liquid dinitrogen tetroxide yields a white product which seems to be the solvated tetranitrate. This loses dinitrogen tetroxide on standing and the product decomposes rapidly in a vacuum to yield[54] a basic salt in which the ratio NO_3^- : Po is 3:2. This may have the structure (A), below. (A) is also obtained

$$(A) \xrightarrow{100\,°C} (B)$$

(A)   (B)

by treating polonium tetrachloride or polonium(IV) hydroxide with 0.5 M nitric acid for about 12 h, the insoluble product then being vacuum dried at room temperature. The compound also results from reaction of the metal with gaseous dinitrogen tetroxide over a period of several days, although the metal does not appear to react with liquid dinitrogen tetroxide or its solution in ethyl acetate. (A) decomposes at 100 °C to a yellowish white solid (B) in which the ratio NO_3^-/Po is 1:2 (cf., tellurium(IV) nitrate). This decomposes to polonium dioxide[54] at 130 °C. Both (A) and (B) are probably derived from an unknown basic nitrate, $PoO(NO_3)_2$.

The solubility of these species in aqueous nitric acid is quite low, but increases with increasing acid concentration[55] and ion-exchange experiments show that at high nitric acid concentrations there is a marked increase in the concentration of anionic polonium species[54].

6.3.3 Other compounds

A white iodate, possibly $Po(IO_3)_4$, is precipitated from nitric acid solutions of polonium(IV) on addition of a nitric acid solution of iodic acid; its solubility decreases with increasing concentration of the latter[37]. An insoluble phosphate, a white, gelatinous material of composition $2PoO_2 \cdot H_3PO_4$, is obtained by treating polonium tetrachloride with 1 M diammonium phosphate, or polonium(IV) hydroxide with 2 M phosphoric acid[37]. Apart from analyses and solubility data, nothing is known about these compounds.

Several salts of carboxylic acids have been reported, and solubility data for the acetate, formate, oxalate and tartrate are available; except for the formate, it appears that these acids yield anionic polonium(IV) complex species[47]. The EDTA complex is more stable in alkaline than in acid solution[37], and complexing with this reagent can be used in the separation of trace level polonium from radium D,E and F mixtures[56]. The main difficulty with work on such compounds is that the intense α-radiation of ^{210}Po causes almost complete decomposition of the compound within a few hours, leaving free carbon, polonium dioxide, and organic degradation products, so that research in this area with weighable amounts of the element will only be practicable with the longer lived polonium isotopes.

6.7   SEXIVALENT POLONIUM

A volatile compound, believed to be polonium hexafluoride, is obtained when polonium (the longer lived isotope, ^{208}Po) deposited onto platinum is heated with fluorine[57]. The product appears to be stable for a limited time in the vapour phase, while radiation decomposition is very rapid in the solid, leading to the formation of an involatile species which is probably polonium tetrafluoride.

The trioxide, PoO_3, is believed to be formed on the trace scale by anodic deposition of polonium from acid solution[58], and an alkali soluble species is formed when polonium(IV) chromate is left in contact with aqueous chromium trioxide for some time. This species could be a polonium(VI)–chromium(VI) complex acid, for polonium(IV) hydroxide is precipitated from its alkaline solution on the addition of hydrogen peroxide[37]. Fusion of polonium dioxide with a mixture of potassium chlorate and hydroxide yields a bluish solid which is more soluble in water than the corresponding polonium(IV) compound. This product may well contain some polonate(VI).

Trace level solvent extraction studies in which the polonium in the aqueous phase has been treated with chlorine at 60–70 °C have been interpreted[59, 60, 70] on the basis of oxidation of polonium(IV) to polonium(VI). These observations are not in agreement with experiments using weighable amounts of the element[37]. Although this may be because of the intense α-radiation in the latter case, the trace level results should be treated with reserve. The solvent extraction data, ascribed to the formation of $PoO_2Cl_3^-$ and $PoO_2Cl_4^{2-}$ ions, could not be explained on the basis of the quadrivalent species, $PoCl_5^-$ and $PoCl_6^{2-}$.

6.8   ACETYLACETONE, DITHIOCARBAMATE, DITHIZONE AND OXINE COMPLEXES

The extraction of weighable amounts of polonium[37] from aqueous hydrochloric acid with mono- or β-diketones has confirmed earlier trace level studies with these ligands. Evaporation of the methyl isobutyl ketone or acetylacetone extracts yields yellow oils which give yellow, crystalline solids on recrystallisation from petroleum ether (60–80 °C), and the same products are obtained by treating solid polonium tetrachloride or tetrabromide with the ketone. Two moles of the halogen acid are liberated in the latter reaction, which yields a yellow dichloro- and an orange dibromo-polonium complex species respectively. The acetylacetone products are believed[37] to be the cyclic species(C), formed by reaction at the terminal methyl groups of the diketone.

(X = Cl, Br)

(C)

Analogous species are known for tellurium(IV). The compound (C) loses halogen when shaken with aqueous potassium hydroxide, yielding a volatile purple-violet solid in both cases. This is also obtained by treating polonium dichloride or dibromide with acetylacetone. The monoketone products probably involve two molecules of the ketone and the primary product may be of the form $X_2Po(CH_2COR)_2$.

Evidence for the formation of volatile diethyldithiocarbamate[61] (dtc), dithizone[62] (dz) and oxine[61] (ox) complexes has been obtained by trace-level ion-exchange and solvent extraction studies. The dtc and ox complexes appear to be 1:1 species, whereas the dithizone complex may be $PoOdz_2$ or $Podz_2$, for the oxidation state of the polonium is uncertain. The last has also been isolated in weighable amounts as a red solid[5]. Other volatile complexes, based on tracer level extraction work, have also been reported[61].

6.9 ORGANOMETALLIC COMPOUNDS

Tracer scale studies have shown that di-α-naphthyl[63], di-p-anisyl[64] and some other[67] polonides are formed by treating polonium tetrachloride, mixed with tellurium tetrachloride, with the appropriate Grignard reagent; the resulting mixture of polonide and telluride was subsequently separated chromatographically. Diphenyl polonide, together with diphenyl polonium dichloride, appears to be formed in the β-decay of bismuth (^{210}Bi) triphenyl[65, 66]. The diaryl polonium dichlorides are more usually obtained on the trace scale from triarylbismuth (^{210}Bi) dichloride[65, 66] or by treating the corresponding polonide with halogen; the difluoride is said to be formed from tritolylbismuth difluoride. The analogous triaryl polonium halides are also reported to exist[63, 64].

The only report of the formation of weighable amounts of an organo-polonium species concerns dimethyl polonide. The very large losses of polonium during the electrodeposition of the element from acetic acid solution were probably due to the formation of this compound by reaction of polonium metal with methyl radicals, formed by the α-bombardment of the acetic acid[47]. It is unlikely that any preparative studies of organo-polonium species will ever be undertaken because of the radiation damage problems raised by the relatively high α-activity of even the longest-lived polonium isotope.

References

1. Bagnall, K. W. (1962). *Advances in Inorganic and Radiochemistry*, **4**, 198. (New York: Academic Press)
2. Bagnall, K. W. (1966). *The Chemistry of Selenium, Tellurium and Polonium*, (Amsterdam: Elsevier)
3. Geary, W. J. (1968). *Inorganic Sulphur Chemistry*, 49
4. Bishop, C. T. (1970). *U.S.A.E.C. unclassified document* MLM-1721
5. Bagnall, K. W. and Robertson, D. S. (1957). *J. Chem. Soc.*, 509
6. Moore, F. L. (1957). *Anal. Chem.*, **29**, 1660
7. Moore, F. L. (1960). *Anal. Chem.*, **32**, 1048
8. Barbano, P. G. and Rigali-Camen, L. (1967). *Radiochim. Acta*, **8**, 214
9. Schulz, W. W. and Richardson, G. L. (1968). *Ind. Eng. Chem., Process Des. Develop.*, **7**, 149
10. Koch, H. and Schmidt, H. (1963). *Kernenergie*, **6**, 39
11. Pauly, J. (1960). *Bull. Soc. Chim. France*, 2022
12. Chong, C. H. H. and Prisc, M. D. (1970). *Nuclear Applic. Technol.*, **9**, 667
13. Endebrock, R. W. and Engle, P. M. (1953). *U.S.A.E.C. unclassified document* AECD-4146
14. Jennings, A. S., Proctor, J. F. and Fernandez, LeV. P. (1966). *U.S.A.E.C. unclassified document* DP-1066 (Vol. 2)
15. Bradley, R. F. and Webster, D. S. (1969). *Nuclear Sci. Eng.*, **35**, 159
16. Love, C. M. (1970). *U.S.A.E.C. unclassified document* MLM-1661
17. Konovalov, E. E., Peizulaev, Sh. I. and Emel'yanov, V. P. (1967). *Radiokhimiya*, **9**, 215
18. Commissariat à l'Energie Atomique. (1966). *Neth. Patent Appl.* 6 603 452
19. Baltisberger, R. J. (1970). *U.S. Patent* 3 491 003
20. Hasty, R. A. (1967). *J. Inorg. Nucl. Chem.*, **29**, 2679
21. Schulz, W. W., Schiefelbein, G. F. and Bruns, L. E. (1968). *U.S. Patent* 3 463 739; (1969). *Ind. Eng. Chem., Process Des. Develop.*, **8**, 508
22. Bagnall, K. W. and Robertson, D. S. (1957). *J. Chem. Soc.*, 1044
23. Bagnall, K. W. and D'Eye, R. W. M. (1954). *J. Chem. Soc.*, 4295
24. Bagnall, K. W., Robinson, P. S. and Stewart, M. A. A. (1958). *J. Chem. Soc.*, 3426
25. Maxwell, C. R. (1949). *J. Chem. Phys.*, **17**, 1288
26. Brooks, L. S. (1955). *J. Amer. Chem. Soc.*, **77**, 3211
27. Goode, J. M. (1957). *J. Chem. Phys.*, **26**, 1269
28. Fischer, C. F. (1968). *Can. J. Phys.*, **46**, 2336
29. Bagnall, K. W. and Freeman, J. H. (1956). *J. Chem. Soc.*, 4579
30. Barr, J., Gillespie, R., Kapoor, R. and Malhotra, K. C. (1968). *Can. J. Chem.*, **46**, 149
31. Paneth, F. and Johannsen, A. (1922). *Chem. Ber.*, **55**, 2622
32. Wittemann, W. G., Giorgi, A. L. and Vier, D. T. (1960). *J. Phys. Chem.*, **64**, 434
33. Kershner, C. J., DeSando, R. J., Heidelberg, R. F. and Steinmeyer, R. H. (1966). *J. Inorg. Nucl. Chem.*, **28**, 1581
34. Kershner, C. J. and DeSando, R. J. (1970). *J. Inorg. Nucl. Chem.*, **32**, 2911
35. Jarrett, J. H. and Brite, D. W. (1969). *U.S.A.E.C. unclassified document* BNWL-1113
36. Steiger, R. P. and Miles, J. C. (1970). *J. Inorg. Nucl. Chem.*, **32**, 3469
37. Bagnall, K. W., Freeman, J. H., Robertson, D. S., Robinson, P. S. and Stewart, M. A. A. (1958). *U.K.A.E.A. unclassified report* C/R 2566
38. Bagnall, K. W. and Freeman, J. H. (1956). *J. Chem. Soc.*, 2770
39. Bagnall, K. W., D'Eye, R. W. M. and Freeman, J. H. (1955). *J. Chem. Soc.*, 2320
40. Bagnall, K. W., D'Eye, R. W. M. and Freeman, J. H. (1955). *J. Chem. Soc.*, 3959
41. Bagnall, K. W., D'Eye, R. W. M. and Freeman, J. H. (1956). *J. Chem. Soc.*, 3385
42. Iofa, B. Z. and Yushchenko, A. S. (1969). *Radiokhimiya*, **11**, 677
43. Iofa, B. Z. and Yushchenko, A. S. (1970). *Radiokhimiya*, **12**, 65
44. Yushchenko, A. S. and Iofa, B. Z. (1966). *Radiokhimiya*, **8**, 621
45. Bagnall, K. W. (1957). *The Chemistry of the Rare Radioelements*, (London: Butterworths)
46. Koch, H. and Falkenberg, W. D. (1966). *Solvent Extraction Chemistry, Proc. Int. Conf., Goteborg*, 26

47. Bagnall, K. W. and Freeman, J. H. (1957). *J. Chem. Soc.*, 2161
48. Martin, A. W. (1954). *J. Phys. Chem.*, **58**, 911
49. Moyer, H. V. (1956). *'Polonium'*, *U.S.A.E.C. unclassified document* TD–5221, 48
50. Joy, E. F. (1947). *U.S.A.E.C. unclassified document* M–4123 *(declassified 1955)*
51. Joy, E. F. (1954). *Chem. Eng. News*, **32**, 3848
52. Hunt, D. J. (1954). *U.S.A.E.C. unclassified document* MLM–979
53. Staritzky, E. (1951). *U.S.A.E.C. unclassified document* LA–1286
54. Bagnall, K. W., Robertson, D. S. and Stewart, M. A. A. (1958). *J. Chem. Soc.*, 3633
55. Orban, E. (1954). *U.S.A.E.C. unclassified document* MLM–973
56. Vebersik, V. and Horova, K. (1958). *Z. Anal. Chem.*, **162**, 401
57. Weinstock, B. and Chernick, C. L. (1960). *J. Amer. Chem. Soc.*, **82**, 4116
58. Haissinsky, M. and Cottin, M. (1947). *Compt. Rend.*, **224**, 467
59. Iofa, B. Z. and Yuschenko, A. S. (1969). *Radiokhimiya*, **11**, 668
60. Yushchenko, A. S. and Iofa, B. Z. (1969). *U.S.S.R. Patent* 230 116 ((1970). *Chem. Abstr.*, **72**, 22966h)
61. Kimura, K. and Ishimori, T. (1958). *Proc. 2nd Int. Conf. on Peaceful Uses of Atomic Energy*, **28**, 151
62. Kimura, K. and Mabuchi, H. (1955). *Bull. Chem. Soc. Japan*, **28**, 535
63. Nefedov, V. D., Toropova, M. A., Zhuravlev, V. E. and Levchenko, A. V. (1965). *Radiokhimiya*, **7**, 203
64. Nefedov, V. D., Zhuravlev, V. E., Toropova, M. A., Grachev, L. N. and Levchenko, A. V. (1965). *Radiokhimiya*, **7**, 245
65. Murin, A. N., Nefedov, V. D., Zaitsev, V. M. and Grachev, S. A. (1960). *Dokl. Akad. Nauk SSSR*, **133**, 123
66. Nefedov, V. D., Vobetsky, M. and Borak, I. (1965). *Radiokhimiya*, **7**, 627, 628
67. Zhuravlev, V. E. and Antipina, N. F. (1967). *Radiokhimiya*, **9**, 726
68. Goode, J. M. (1952). *U.S.A.E.C. unclassified document* MLM–677
69. Wyckoff, R. W. G. (1960). *Crystal Structures*, (New York: Interscience)
70. Starik, I. Ye. and Ampelogova, N. I. (1961). *Radiokhimiya*, **3**, 261

7
Arsenic

R. G. CAVELL and A. R. SANGER
University of Alberta

7.1   INTRODUCTION

For the first article of the series we have surveyed the literature from 1967 to 1970 inclusive. Because of space limitations we have also made use of other review articles where convenient, and appropriate references will be found in the text. Our object has been to survey the inorganic chemistry of arsenic, and to this end we have excluded from consideration complexes of the transition metals with organoarsines, and we have also excluded organo-metallic arsenic chemistry, both of which are to be reviewed elsewhere. Some overlap is however unavoidable and some organoarsenic chemistry will be discussed, particularly in the Sections on sulphide and halide chemistry. The literature is essentially complete to the end of 1970 and a few relevant 1971 references have been included. Appropriate reviews of a general character[1], arsenic halogen chemistry[2], organometallic arsenic chemistry[3], arsine complexes of transition metals[4] and fluorophenyl arsenic chemistry[5] should be referred to for a coverage of the literature prior to 1966–1967.

7.2   ELEMENTAL ARSENIC

The crystal structure[6], the low temperature (0.5–4.2 K) specific heat[7] and the temperature-dependent magnetic susceptibility[8] of elemental arsenic have been reported. Gaseous arsenic at 485 °C possesses a regular tetrahedral structure with an As—As bond length (by electron diffraction)[9] of 2.435 ± 0.004 Å and a stretching force constant (from the thermal vibrational amplitudes) of 1.5 ± 0.3 mdyn Å^{-1} [9]. At 900 °C, gaseous arsenic shows Raman bands due to As_4, and a weak line at 421 cm^{-1}, the intensity of which increases with temperature, is assigned to the As_2 species[10]. The observed Raman stretch agrees well with the value calculated from the extensively investigated electronic spectrum of As_2, which continues to excite much interest as is illustrated by the number of recent studies which have been published[11-15]. The ultraviolet emission spectrum of atomic arsenic has also been investigated[16] as well as the e.p.r. spectrum[17] of the $^4S_{\frac{3}{2}}$ state of ^{75}As. The hyperfine coupling (66.2 Hz), quadrupole coupling (0.13 Hz) and g_J (1.9983) constants of the latter have been given[17]. Thermodynamic functions for As_2, As_4 and AsF_n ($n = 1$–5) have been calculated[18].

7.3 SMALL MOLECULES OF TRANSIENT EXISTENCE

The emission spectrum of the AsO molecule has been extensively analysed[16, 19–21], but only one recent report on the spectra of AsS and AsS^+ has appeared[22]. Excited As atoms ($^2D_0$, $^2P_0$) and As_2 molecules are produced upon flash photolysis of $AsCl_3$ or $AsBr_3$ [23]. Electronic spectra of AsH, AsH_2 and AsD_2 species, generated by flash photolysis of AsH_3 or AsD_3 have been analysed[25, 26]. Irradiation (γ rays) of AsH_3 in low-temperature Xe or Kr matrices[27] afforded the AsH_2 radical; however, hyperfine couplings were not observed. Microwave discharge through AsF_3 vapour gave a spectrum which has been assigned to AsF [29]. Flash photolysis of $AsCl_3$ alone gave arsenic atoms[24] or AsCl which rapidly decomposed[23], and flashing $AsCl_3$ in the presence of O_2 gave AsO [24]. Flash photolysis of AsH_3 or $AsCl_3$ in the presence of C_2N_2 in a N_2 atmosphere gave the AsCN molecule[28]. To date, little chemistry has been done with these species beyond identification, but in view of the recent interest in high-temperature reactions and laser- and discharge-excited reactions, it seems appropriate to include these species in a chemical review since future studies will probably utilise these results.

7.4 ARSENIC HYDRIDES

7.4.1 Synthesis

Reduction of As_2O_3 by $LiAlH_4$ at 148–170 °C gave AsH_3 [30]. Hydrolysis of annealed mixtures of silicides, germanides, phosphides, arsenides or nitrides gave As_3H_5, H_2PAsH_2, H_2NAsH_2, $HAs(NH_2)_2$, $As(NH_2)_3$, and H_3GeAsH_2, among the volatile products identified by mass spectrometry[31]. Shaking an excess of HI and Hg with $(CF_3)_nAsI_{3-n}$ is an efficient way of preparing $(CF_3)_nAsH_{3-n}$ [32].

7.4.2 Spectra and structure

The infrared spectra (4000–250 cm^{-1}) of AsH_3 in the vapour and solid states, and the Raman spectrum in the liquid state, have been reported[33], and the high-resolution infrared spectrum has also recently been analysed[34]. The potential functions, including a lone pair–X interaction for AsX_3 (X = H, halogens), have been described[35], and amplitudes[36] and frequencies[37, 38] of vibrations for AsX_3 (X = H,D,T) have been calculated.

The n.m.r. spectra of AsH_3 [39, 40], $RAsH_2$ [39], and AsH_2^- [41] have been reported, and the variation of chemical shift of AsH_3 with temperature and concentration described[40]. Rapid exchange of protons of AsH_3 or NH_4^+ with AsH_2^- occurs[41].

Newer, and more accurate, values of rotational constants for isotopically-substituted AsH_3 have been obtained from the microwave spectrum[42]; however, the ground-state structure could not be refined because only one stable isotope of arsenic is available. The older parameters ($< HAsH = 91$ degrees 50 minutes, $d(As-H) = 1.512$ Å) are consistent with the new results.

Theoretical studies have been concerned with the effect of the non-bonding electrons on the barrier to rotation of CH_3AsH_2 [43] and with the evaluation of LCAO—MO—SCF calculations on AsH_3 [44].

7.4.3   Chemical reactions of As—H bonds

In ether, H/D exchange (monitored by i.r.) between Me_2AsD and Et_2AsH yields a value of $K_{eq} = 1.2$ (± 0.1) for this equilibrium, with an initial rate given by $k = 4.95 (\pm 0.25) \times 10^{-5} \, l \, mol^{-1} \, s^{-1}$ (35 °C) [45]. Rapid H/D exchange between Me_2AsH and CD_3OD or $(CF_3)_2AsD$ occurs[45].

Insertion of PhNCO into As—H bonds gives the ureidoderivatives[46]. Insertion of $(CF_3)_2CO$ into AsH_3 affords $H_2AsC(CF_3)_2OH$ rather than the As—O bonded compound[47], and ketene inserts at the C—C bond to yield $R_2AsCOCH_3$ [48]. Insertions of the unsaturated hydrocarbons $Ph_2AsC=CH$, $PhC=CH$ [49], and fluorocarbons $CF_2=CFCF_3$ [50] and $CF_3C\equiv CCF_3$ [51] into As—H bonds have been observed, and kinetic studies on the latter reaction indicate intermolecular hydrogen transfer[52]. In contrast, arsination of the cyclobutene analogue occurs (equation (7.1))[50], and this reaction has been shown to be generally applicable[53].

$$MeAsRH + \begin{matrix} CF_2-(CF_2)_n \\ | \qquad | \\ XC=CX^1 \end{matrix} \longrightarrow \begin{matrix} CF_2-(CF_2)_n \\ | \qquad | \\ XC=CAsMeR \end{matrix} + HX^1$$

(7.1)

$$(X = X^1 = Cl, F; n = 1,2)$$

Carbene insertion into $(CF_3)_2AsH$ on reaction with $(CF_3)_2CN_2$ yields $(CF_3)_2AsC(CF_3)_2H$ with elimination of N_2 [54]. With CF_3COCl, Me_2AsH gave $Me_2AsCOCF_3$ in low yield[55], and this product when treated with $(CF_3CO)_2O$ gave Me_2AsCF_3 and $CF_3CO_2AsMe_2$ with elimination of CO [55].

Though Cl_2 or Br_2 rapidly reacted with $MeAsH_2$ to precipitate $(MeAs)_n$, I_2 reacted at room temperature to give $MeAsHI$ and $MeAsI_2$ [56]. The former decomposed to $(MeAs)_n$ in less than 200 days[56].

Cyclisation occurred on the condensation of $PhAsHCH_2CH_2NH_2$ with aldehydes and ketones (equation (7.2))[57],

$$PhAsHCH_2CH_2NH + RR^1CO \longrightarrow \begin{matrix} CH_2 \diagdown NH \\ | \qquad \diagdown CRR^1 \\ CH_2 \diagup AsPh \end{matrix} + H_2O$$

$$(R,R^1 = H, alkyl, Ph, (CH_2)_n, \quad n = 4, 5)$$

(7.2)

and $(C_6H_{11}N)_2C$ did not insert into the AsH bond in $C_6H_5AsH_2$ but reacted according to equation (7.3) [58].

$$PhAsH_2 + (C_6H_{11}C)_2N \rightarrow C_6H_{11}C{=}NHCHC_6H_{11} + (PhAs)_n \quad (7.3)$$

Alkali metal salts of arsines have been prepared by the reaction of primary or secondary arsines with alkyl-lithium[280].

7.5 ARSENIC HALIDES

7.5.1 Trivalent arsenic halides

7.5.1.1 Structures and spectra

Most of the recent work has been directed towards spectroscopic and structural studies of these compounds. Electron diffraction (e.d.) studies of AsF_3 [59, 60] $AsCl_3$ [59] and AsI_3 [62] structures have been reported, as well as a crystal structure of $AsBr_3$ [61]. The bond data for the former two have been obtained from a combination of e.d. and microwave results[59]. New vibrational data have been given for AsF_3 in gas and liquid phases[63-66], including high

Table 7.1   Bond lengths and bond angles of arsenic trihalides (AsX_3)

X	(AsX) *length* (Å)	<XAsX *angle*	*Reference*
F	1.7080 (± 0.004)	95° 58′ ($\pm 17'$)	59
F	1.706 (± 0.002)	96.2 (± 0.2)°	60
Cl	2.1621 (± 0.0009)	98° 38′ ($\pm 22'$)	59
Br	2.36 (± 0.01)	97.7 (± 0.03)°	61
I	2.557 (± 0.005)	100.2 (± 0.4)°	62

resolution studies[63] with extensive analyses of band contours[65]. Analyses of Coriolis and centrifugal distortion constants[65, 67-69] have been undertaken in order to obtain the force constants of the molecules, but assumed bond parameters in some cases differ from electron diffraction results[59]. Microwave spectra have also been extensively used for the purpose[59, 70] of obtaining bond parameters and force constants. Raman spectra of liquid $AsCl_3$ and $AsCl_3$ in various solvents have been reported[71]. In the latter case, the protonic solvents (e.g. ROH,H_2O) show complex formation, and there was no evidence for chloride loss until complete precipitation of As_4O_6 occurred[71]. Ether did not interact[71]. Spectra of $AsBr_3$ (liquid and solid)[71, 72] and AsI_3 (solid and solution)[73-76] have also been reported. A large number of force constant and vibrational amplitude calculations have been reported for the simple AsX_3 halides (X = F,Cl,Br,I)[59, 63, 70, 77-81], the mixed halides $AsX_{3-n}Y_n$ ($n = 1,2$; X = Cl, Br)[82, 83], substituted alkyl halides CH_3AsX_2 and $(CH_3)_2$ AsX (X = Cl, Br)[80b] and the cyanide $As(CN)_3$ [84]. The crystal structure[85] of $CH_3As(CN)_2$ shows a strong intermolecular association through one strong As—N bond to form a chain polymer as does arsenic tricyanide[86], however,

$(CH_3)_2AsCN$ shows only a weak As—N interaction[85]. In contrast, $P(CN)_3$ forms a lattice polymer as a result of interactions of all three cyanide nitrogen atoms with adjacent phosphorus atoms[85].

Spin-lattice relaxation times in solid and liquid AsF_3 have been reported[87] as well as chemical shift, coupling constants and quadrupole relaxation of AsF_3 complexes[88]. N.Q.R. spectral data for ^{75}As and, where appropriate, ^{35}Cl and ^{37}Cl in a series of AsX_3 (X = Cl, CH_3, Et, C_6H_5, C_4H_9, C_3H_7) compounds has been given[89] and ^{35}Cl n.q.r. spectra have been used to demonstrate the existence of distinct complexes between $AsCl_3$ and hydrocarbons such as benzene, $C_{10}H_8$, $Me_2C_6H_4$ etc.[90]. Mass spectral studies of $AsCl_3$[91] and the perfluoroalkyl arsenicals $(CF_3)_2AsX$ (X = CF_3, H, $As(CF_3)_2$) and CF_3AsX_2 (X = H, Cl) have been reported[92]. All of the trifluoromethyl compounds gave As—F species under electron impact. A low resolution photoelectron spectrum of $AsCl_3$ has also been reported[93].

7.5.1.2 Physical properties

Other physical studies have provided electric dipole moments of $(CH_3)_n AsCl_{3-n}$ (n = 1, 2, 3) halides[94]; the surface tension of liquid $AsCl_3$[95] and the vapour pressure of AsI_3[96]. At high temperatures (600–1000 K) AsI_3 dissociates into its elements as described by the constant:

$$\log K_p = 7.54 - 10850/T \text{ giving } \Delta G° = 49600 - 34.5\, T(K)\ [96]$$

7.5.1.3 Synthesis

New chemical syntheses of the tribromide from $AsCl_3$ or As_2O_3, As_2O_5, As_2S_3 or As_2S_5[97] have been reported, and $AsCl_3$ has been prepared from As_2O_3 and HCl or NH_4Cl at 250 °C; the former in a flow reactor[98].

7.5.1.4 Chemical reactions

The relatively weak Lewis acidity of $AsCl_3$ has been investigated by n.m.r. spectroscopic studies of the α-H chemical shift in the $AsCl_3$·THF complex in tetrahydrofuran[99]. The acidity is also readily demonstrated by the formation of 1:1 aromatic hydrocarbon complexes[90,100], mentioned earlier, by the formation of 1:1 complexes with $(BuO)_3SiNH_2$[101] and by the formation of both 1:1 and 1:2 complexes of AsX_3 (X = Cl, Br) with bipyridyl, 1,10-phenanthroline and o-phenylene-bis-dimethylamine[102]. The acidity of $AsCl_3$ in $MeNO_2$ was evaluated by ΔH_{solv} and ΔH_{neut} measurements on the complexes formed by piperidine, α-picoline and dimethylaniline[113]. In organic solvents such as Et_2O, $AsCl_3$ reacts with HCl to form $AsCl_4^-$ complexes identified by i.r. spectroscopy[103]. No evidence for the formation of $AsCl_3Br^-$ or $AsBr_3Cl^-$ from $AsCl_3$ and HBr, or $AsBr_3$ and HCl, in diethyl ether was obtained in this system[103]. Salts of the mixed halogeno species AsX_3Y^- (X,Y = Cl, Br) have however been prepared and characterised by i.r. and Raman spectroscopy[104b]. Infrared and Raman spectral data for

$AsBr_4^-$ and $AsCl_4^-$ ions show that the former has C_{2v} symmetry and the latter C_{3v} [104], both systems are based on a trigonal-bipyramidal framework. The mixed halogeno species AsX_3Y^- also have structures based on a trigonal-bypyramid, and these ions possess C_s symmetry. Vibrational force constants and vibrational amplitudes have been calculated[105] for these species.

Arsenic trifluoride continues to find wide use as a fluorinating agent for saturated haloalkanes[106]. It converts VCl_4 into $VClF_3$ [107] forming $AsCl_3$ as the arsenic halide by-product. Vanadium pentafluoride reacts with AsF_3 to give VF_4 and $VF_3^+AsF_6^-$ [107]. Reaction with BCl_3 and H_2 gives borides AsB and AsB_6 [108]. AsF_3 is reported to react 'explosively' with P_4O_6 at room temperature, to give 'a wide variety of products'. The reaction could not be controlled by cooling or the use of solvents[109]. Unlike AsF_5 discussed below, AsF_3 did not form adducts with SeF_4, nor with TeF_4 or SF_4, although fluorine-exchange in SF_4 was catalysed by AsF_3 [127], and the presence of the AsF_2^+ or $As_2F_5^+$ ion has been suggested in AsF_3–SbF_5 solutions[144].

$AsCl_3$ reacts with the fluoro-olefin $CF_2{=}CFCl$ under the influence of u.v. irradiation to yield $Cl_2AsCF_2CFCl_2$ [110]. Reaction of $AsCl_3$ with VF_5 gave $VClF_3$ and Cl_2, or VF_4 if VF_5 was in excess[107]. Reaction of $AsCl_3$ with MoF_6 and UF_6 gave $Mo_2Cl_3F_6$ and UF_4 respectively, but no reaction occurred with WF_6 [111]. AsF_3 was formed as a by-product. Arsenic trichloride and SO_3 formed an inseparable mixture (by distillation), which was described as the adduct $AsCl_3{\cdot}SO_3$ in equilibrium with Cl_2AsOSO_2Cl [112]. Heating $AsCl_3$ with SO_3 at 70°C gave $(ClAsSO_4)_n$, a polymeric material which decomposed upon further heating to $AsCl_3$, SO_3 and $As_2(SO_4)_3$ [112]. Conductometric studies of the neutralisation of $AsCl_3$ by HSO_3F or H_2SO_4 in $MeNO_2$ were interpreted in terms of the auto-ionisation of $MeNO_2$ [114]; however, $AsCl_3$ reacts with $MeNO_2$ to yield $ClAs(CH_2NO_2)_2{\cdot}2MeNO_2$, which upon treatment with H_2S afforded $As_2S_3{\cdot}3MeNO_2$ [114b]. $AsCl_3$ was found to be a non-electrolyte in CH_3CN, and the solutions were also studied by i.r. spectroscopy[115]. $AsCl_3$ in dimethylformamide treated with silver perchlorate yielded the perchlorate salt $[As(DMF)_5]^{+3}[ClO_4^-]_3$ [116]. All of the AsX_3 halides (X = Cl, Br, I) gave adducts with $(BuO)_3PO$ [76], CH_3CONMe_2 and $(Me_2N)_3PO$ in $C_2H_4Cl_2$ solution[117]. No HX elimination was observed when AsX_3 (X = Cl, Br) reacted with 2-thioamidopyridine (HTP), but rather the salt HTP^+ was formed[118]. An extensive series of pentafluorophenylarsenic halides is formed from the fluorophenyl Grignard reagents and the arsenic halides[119], and this field has been recently reviewed[5, 119c]. $(CF_3)_3As$ undergoes an interesting step-wise replacement of CF_3 by $(CF_3)_2NO$ to produce $(CF_3)_2AsON(CF_3)_2$, $CF_3As[ON(CF_3)_2]_2$ and $As[ON(CF_3)_2]_3$ derivatives[120].

Chloromethyl arsenicals can be readily prepared from $Hg(CClH_2)_2$ and $AsCl_3$ or from $(C_6H_5)_2AsCH_2Cl$,

$$AsCl_3 + (C_6H_5)_2AsCH_2Cl \rightarrow Cl_2AsCH_2Cl$$

and halide replacement with LiBr or NaI gives the dibromo- or di-iodo-arsine derivatives[121]. The chloromethylarsine reacts with thiourea to give an adduct which hydrolyses in alkali to give $(CH_2S)_2As_2S$ and also the monoxide $(CH_2S)_2As_2O^{121}$ (see Section 7.7.2), both compounds having the bridged structure (I).

$$
\begin{array}{c}
\text{As} \\
CH_2 \diagdown \, | \diagdown \text{S} \\
| \quad \text{E} \quad | \qquad (E = O, S) \\
\text{S} \diagdown \diagup CH_2 \\
\text{As}
\end{array}
$$

(I)

7.5.2 Pentavalent arsenic fluorides

7.5.2.1 Structures and spectra

The structure of arsenic pentafluoride has been clearly established as trigonal-bypyramid by electron diffraction, and the axial bond length $((As{-}F)_{ax} = 1.711 \pm 0.0047 \text{ Å})$ shown to be greater than the equatorial bond length $((As{-}F)_{eq} = 1.656 \pm 0.0032 \text{ Å})$ [60]. The electronic structure of AsF_5 has been discussed [122]. Infrared vibrational spectra [123] (including force field analyses [122, 124]) confirm the D_{3h} symmetry. Vibrational studies in solid noble gas matrices at low temperature indicate monomeric species with the same symmetry as in the gas phase [125]. Thermodynamic functions [18, 124b] have been calculated. Much of the interest in the structural properties of this molecule stems from the recent interest in 5-coordinate molecules. The ^{19}F n.m.r. spectrum of neat AsF_5 or AsF_5 in hydrocarbon or chlorocarbon solvents shows only a single (but relatively broad) peak, which is consistent with a rapid substituent-averaging process such as that which has been suggested ('pseudo-rotation') for PF_5 [126]. The greater line-width in AsF_5 spectra could, in part, be due to the presence of the arsenic quadrupole rather than exchange. AsF_5 has been found to catalyse ^{19}F exchange in SF_4, as does AsF_3, possibly through the formation of intermediate associated species linked by fluorine-bridge bonds [127]. In the absence of basic solvents, or any solvent at all, exchange via associated $(AsF_5)_n$ intermediates cannot be ruled out, although the data presented to date do not resolve the problem.

7.5.2.2 Adduct and complex ion formation

Arsenic pentafluoride is a relatively good Lewis acid, as is demonstrated by its ability to interact with other molecules possessing basic character. For example, AsF_5 forms adducts with CH_3CN and $(CH_3)_3N$. While the latter adduct is non-conducting in solution, the former was reported to form a conducting solution in CH_3CN, and the species were formulated as $AsF_4L_2^+$ AsF_6^- [128, 129]. Muetterties and Tebbe [130] did not find any evidence in ^{19}F n.m.r. spectroscopic studies for auto-ionisation and formation of AsF_6^- in the $CH_3CN{-}AsF_5$ system, although the AsF_6^- ion is readily detected by its characteristic n.m.r. spectrum. They suggested that the observed conductivity was due to the susceptibility of the $F_5As{\cdot}CH_3CN$ adduct to hydrolysis which yields AsF_6^-, and demonstrated that this was occurring by a deliberate addition of water to the system [130]. Although the difference in behaviour may arise from the different concentrations[129] employed in the two studies, it is the dilute conductivity study[129, 130] which will be the most susceptible to trace contamination by water.

Readily dissociated complexes, detected by ^{19}F n.m.r. spectroscopy, are formed by SO_2FCl, CH_3SO_2F, SOF_2 and SO_2 which act as weak bases with AsF_5, but no complex was detected with SO_2F_2 [131]. Stronger fluoride donor bases tend to form AsF_6^-, and a considerable amount of AsF_5 chemistry is dominated by the use of the pentafluoride as a fluoride acceptor, resulting in the formation of salts in which the AsF_6^- ion stabilises the cationic species. For example, the formation of hexafluoroarsenates of Cl_3^+ [132] and Cl_2F^+ [132, 133a], ClO_2^+ [133c], Br_3^+ [134], O_2^+ [135], SeF_3^+ [66], F_2NO^+ [136, 137], $N_2F_3^+$ [137], NO_2^+ [133, 137], NO^+ [137] and xenon ions XeF^+ and $Xe_2F_3^+$ [138], all of which are produced by the general reaction

$$MF_x + AsF_5 \rightarrow MF_{x-1}^+ AsF_6^-$$

although in some cases the AsF_5 was produced in situ. In many cases the identity of the cation was deduced following the demonstration of the presence of AsF_6^- in the solid, typically by means of i.r.[134, 135, 137, 139], Raman[66, 134, 135, 137, 139], and n.m.r.[130, 140] spectroscopy.

Relatively complete vibrational analyses, including force constants and vibrational amplitudes[301], have been carried out on the AsF_6^- ion[66, 133, 134, 139]. The Raman spectra of the NO^+, NO_2^+, F_2NO^+ and $N_2F_3^+$ salts generally showed a splitting of the $\nu_5(F_{2g})$ band into a doublet, and the occasional observation of the ν_2 ((E_g), Raman-active) band in the infrared was interpreted as a relaxation of the mutual exclusion rule due to a deviation of the AsF_6^- ion from regular octahedral symmetry[137]. The crystal structure[145] of $pyH^+AsF_6^-$ showed that the AsF_6^- ion had O_h symmetry within experimental error (As—F = 1.77_7 Å; $<$ FAsF = 89.4 degrees), suggesting that the observed vibrational splitting may be due to a dynamic effect or to the effect of reduced site symmetry on the vibrational selection rules, which is an important consideration in the spectra of solids.

N.M.R. spectral[140, 143, 144] studies on fluoride solutions show that extensive exchange reactions occur, probably through the formation of fluorine-bridged polymers and ionic species such as $AsSbF_{11}^+$ in SO_2 or HF solvent[141] at $-78°C$ to give $As_2F_{11}^-$, which is less stable and more difficult to form than $Sb_2F_{11}^-$ from the more acidic SbF_5. Conductometric measurements in HSO_3F of AsF_5 and $AsF_5 \cdot SO_3$ [i.e. $F_4As \cdot OSO_2F$] with other fluorides, demonstrate acid strengths which decrease in the order

$$F_3As(OSO_2F)_2 > SbF_5 > F_4AsOSO_2F > AsF_5 \sim TiF_4 > NbF_5 \sim PF_5$$

Infrared and Raman spectra of the $F_3PO \cdot AsF_5$ adduct suggest that it is a donor–acceptor complex containing a P—O—As bridge, and not a POF_2^+ salt[142].

Trifluoromethylarsenic fluoro anions have been prepared[159] from the reactions:

$$(CF_3)_nAsF_{5-n} + CsF \xrightarrow{CH_3CN} Cs[(CF_3)_nAsF_{6-n}]$$
$$Ag[(CF_3)_2AsO_2] + 2SF_4 \longrightarrow Ag[(CF_3)_2AsF_4] + 2SOF_2$$

and their n.m.r. spectra analysed[159].

7.5.2.3 Fluorinations

AsF_5 reacts with VCl_4 to form $VClF_3$ and $AsCl_3$ [107]. Rather interesting reactions, involving both oxidative fluorination and the stabilisation of the product by the AsF_6^- ion formed in the system, are provided by the reactions of AsF_5 with (a) sulphur in HF solution[147] to give $S_8^{2+}(AsF_6^-)_2$ and $S_8^+AsF_6^-$; (b) with selenium[148], likewise to give $Se_8^{2+}(AsF_6^-)_2$ and (c) with antimony in SO_2 to give $(Sb_n^{n+})(AsF_6^-)_n$ [149]. AsF_3 was the reduction product. Chlorine dioxide is now known[150] to yield $ClO_2^+AsF_6^-$ upon reaction with AsF_5, presumably through oxidation of ClO_2 to ClO_2F followed by complex formation. The nitrogen dioxide 'adduct' $NO_2 \cdot AsF_5$, first reported many years ago, has been re-investigated[151, 152] and shown to be a mixture of NO^+ and NO_2^+ salts of AsF_6^-, with some inconclusive evidence for an $AsOF_4^-$ ion[152], probably again via oxidation and complexing reactions. The authentic nitrogen oxide salts can be readily prepared from AsF_5 and NO_xF ($x = 1,2$) in HF solution[153]. AsF_5 reacted with ClO_4^- salts in HF, AsF_3, BrF_5 or IF_5 solvents to yield ClO_3F [146], with the yield dependent on the Lewis acid strength of the medium.

7.5.2.4 Hydrolysis of AsF_6^-

AsF_6^- is not hydrolysed in alkaline solution, but acid converts it into $HAsF_6$ which reacts reversibly with water[154]

$$HAsF_6 + H_2O \rightleftarrows HAsF_5OH + HF$$

The acid $HAsF_5OH$ yields the AsF_5OH^- ion in solution, which hydrolyses in acid with first-order kinetics at constant pH[154].

7.5.2.5 Hydroxy anions

The free acids have been prepared, and the salts $KAsF_5OH$ and $KAsF_4(OH)_2$ can be isolated, and their crystal structures have been reported[155]. AsF_5OH^- reacts with $POCl_3$ or $POBr_3$ to yield AsF_5Cl^- or AsF_5Br^- respectively, which compounds have been isolated as ammonium salts[156]. The fluoro-hydroxy anions AsF_5OH^- and $AsF_4(OH)_2^-$ undergo ready condensation[157] to oxygen-bridged dimeric anions $F_5AsOAsF_5^{2-}$ and $F_4AsO_2F_4^{2-}$ based on octahedral coordination of arsenic. The infrared spectrum of the former has been reported[157b, 158].

7.5.3 Pentavalent arsenic chlorides

The quest for $AsCl_5$ continues, but it is becoming virtually certain that the compound does not exist. Several workers have shown[160, 161] that the solid formed by combination of $AsCl_3$ and chlorine has a spectrum which is a sum of that of the components, and not that of arsenic(V) chloride. Salts of the As^V chloride can be readily formed however. $SbCl_5$ gives $AsCl_4^+$

$SbCl_6^-$ [161, 162], and the $AsCl_4^+$ ion has been prepared with an $AlCl_4^-$ counterion. Both the $AsCl_4^+$ and $AsCl_6^-$ ions have been characterised by ^{35}Cl and ^{37}Cl n.q.r. spectroscopy[163], the latter at 77 K and at 197 K. PCl_5 and '$AsCl_5$' [162, 164] form $PCl_4^+AsCl_6^-$ which is only stable under high Cl_2 pressure[164]. Both ^{31}P n.m.r.[164] and vibrational spectroscopy[162, 165] were used to identify the PCl_4^+ ion in the salt. Photochemical reaction is thought[164] to be responsible for the formation of the PCl_6^- ion which is also found in the salt mixture. The infrared and Raman spectra of $AsCl_6^-$ in these solids have been recorded. At low temperatures the A_{1g} vibration is split, suggesting that the ion is deformed[165], or more likely, perhaps, that the site symmetry has been reduced. As discussed above, spectra of AsF_6^- show similar effects[137]. In the presence of phenanthroline, the complex $AsCl_5 \cdot SbCl_5 \cdot phen$ is formed which was identified[166] as $phen \cdot AsCl_4^+ SbCl_6^-$ by i.r. spectroscopic and conductometric evidence, although the results did not rule out the alternative formulation of the adduct as $phen \cdot SbCl_4^+ AsCl_6^-$.

7.5.4 Pentavalent chloro and fluoro organoarsenic compounds

The i.r. and n.m.r. spectral behaviour of pentavalent R_3AsX_2 compounds $(R = CH_3, C_6H_5; X = F, Cl, Br)$ is compatible with a bipyramidal structure[167]; the most complete evidence has been obtained for the fluoride. The present spectra disagree with some of the previous studies because the compounds were shown to react readily with moisture to yield oxy- and hydroxy-arsenic compounds[167]. The re-distribution reactions

$$R_3AsX_2 + R_3AsY_2 \rightleftharpoons R_3AsXY$$

have been studied by n.m.r. spectroscopy for the Y, X = F, Cl, Br, I series[168]. The F/Cl ratio deviated from the expected statistical scrambling, and varied with the nature of R $(R = CH_3, C_6H_5, C_6H_5CH_2)$. New spectral data have been provided for $(C_6H_5)_4AsX$ and $(C_6H_5)_3AsX_2$ compounds[169].

Triphenylarsine reacts with SO_2Cl_2 in toluene at $0°C$ to give a good yield of $(C_6H_5)_3AsCl_2$ [170], and the pentafluorophenylarsine $(C_6F_5)_3As$ reacts with Cl_2 or ICl in CH_3CN to give $(C_6F_5)_3AsCl_2$ (which is non-conducting in acetonitrile)[171]. Further reaction of dichloride with water gave the oxide $(C_6F_5)_3AsO$ [171]. Carbonyl chloride reacted with R_3As compounds[172], or with $(C_6H_5)_3AsO$ or $(C_6H_5)_3AsS$ [173], to give the dichloride $(C_6H_5)_3AsCl_2$. Acyl chloride and $(C_6H_5)_3AsO$ gave $(C_6H_5)_3AsCl_2$ and acetic anhydride. The product was contaminated with $(C_6H_5)_3As(OH)Cl$ because of the ready hydrolysis of the dichloride by traces of water, and it was for this reason that the reaction of ICl with $(C_6H_5)_3AsCl_2$ in acetonitrile gave[174] $[(C_6H_5)_3 AsOH]^+(Cl^-, ICl^-)$. These hydroxide-halides are weak electrolytes in methyl cyanide giving $(C_6H_5)_3AsOH^+$ ions in solution. Adducts and salts have been prepared and i.r. spectra recorded[175]. In a similar fashion, the pentavalent organoarsenic chlorides dontate a halide to form salts[176], e.g.

$$Me_3AsCl_2 + SbCl_5 \rightarrow Me_3AsCl^+ SbCl_6^-$$

and a recent re-investigation of the $MoCl_5 \cdot (C_6H_5)_3AsO$ adduct has shown that it should be formulated as $(C_6H_5)_3AsCl^+ MoOCl_4^-$, indicating that the molybdenum pentachloride has chlorinated triphenylarsine oxide and formed a salt of $(C_6H_5)_3AsCl_2$ [177].

7.6   ARSENIC OXIDES

7.6.1   Pentavalent compounds with one arsenic atom

7.6.1.1   *Arsine oxides*

A full vibrational Raman and i.r. spectroscopic study of $(CH_3)_3AsO$, including force constant calculations, has been reported[178]. The compound, and, in general, all arsine oxides (prepared in general by the direct oxidation of R_3As with HgO or H_2O_2), shows an As=O stretching absorption in the region 855–910 cm^{-1}, which correlates with 'shift constants' used in the correlation of $v(P\!=\!O)$ in the analogous phosphoryl compounds indicating a lower As=O bond polarity as compared to P=O [190, 181]. N.M.R. spectroscopic studies indicate less shielding of alkyl protons in arsine oxides and esters compared to the phosphorus analogues[180].

A crystal structure of the $(C_6H_5)_3AsO \cdot HX$ (X = Cl, Br) adducts[182] shows hydrogen bonding in the crystalline state with short hydrogen bonds (the O—H—X distance is 2.84 Å in the chloride and 3.02 Å in the bromide salt)[182]. N.M.R. spectroscopy shows that the hydrogen is very strongly associated[182] with oxygen, in a similar fashion to the adduct $(C_6H_5)_3AsO \cdot HO_2CCF_3$ (which is formulated as $(C_6H_5)_3AsOH^+O_2CCF_3^-$) and the halide salts in the preceding Section. The compounds $OAs(OR)_3$ (R = CH_3, C_2H_5) react with alkali metal alkoxides in ether to give a $M[O_2As(OR)_2]$ precipitate and $As(OR)_6^-$ ions in solution[183]. An extensive series of studies on alkoxy arsenic compounds continues to appear[183, 184].

7.6.1.2   *Penta-coordinate compounds*

The penta-coordinate ester $As(OCH_3)_5$, which has been known for some time, reacts with HX (X = Cl, F) to yield $XAs(OCH_3)_4$ and $F_2As(OCH_3)_3$, of which only $FAs(OCH_3)_4$ is reasonably stable at ordinary temperatures[184].

Recent n.m.r.[185, 186] spectroscopic studies of 5-coordinated esters XAs $[(OCR_2)_2]_2$ (X = OH [185]; OCH_3 [186]; C_6H_5 [185, 186]; CH_3, $C_6H_5CH_2$, p-$CH_3OC_6H_4$, p-$O_2NC_6H_4$ [185]; R = H, CH_3 [185, 186]; R_2 = O [185], propose a pseudo-rotation process to explain the observed ligand exchange. The barrier is estimated to be about 20 kcal mol^{-1} [185].

7.6.2   Pentavalent arsenic compounds with more than one arsenic atom

Triphenylarsine reacted with $(CH_3)_3SiOX$ (X = Cl, Br) to give $[(C_6H_5)_3As(X)]_2O$ compounds, and with $(C_6H_5)_3SiOX$ to give $(C_6H_5)_3As(X)OSi(C_6H_5)_3$ [187]. The latter hydrolysed to $(C_6H_5)_3SiOH$ and $(C_6H_5)_3As(X)OH$ [187].

Infrared spectral studies of pyroarsenates suggest that $As_2O_7^{2-}$ may have a linear As—O—As bridge[188]. Meta-arsenates ($As_3O_9^{3-}$ etc.) appear to have a cyclic structure with C_{3v} symmetry[189]. Spectra of $(AsF_5)_2O^{2-}$ and $(AsF_4O)_2^{2-}$ have also been analysed[158].

7.6.3   Trivalent arsenic compounds

Most of the studies involve sulphur compounds which are discussed below. Reaction of As_2O_3 with AsF_3, $AsCl_3$, $As(OCH_3)_3$ or $As(NMe_2)_3$ gives a mixture of scrambled products which were identified by n.m.r. spectroscopy[190]. The compound $((CH_3)_2As)_2O$ has, according to n.m.r. spectroscopy, an As—O—As bridge[190]. Single crystal and gas-phase laser Raman spectra of As_2O_3 indicate that the compound has the As_4O_6 cage structure[191]. Barriers to inversion of $XAs(ER)_2$ (E = O, S) esters and thioesters (including cyclic esters) have been calculated to be $\geqslant 25$ kcal mol^{-1}, and thus pyramidal structures should be stable at ordinary temperatures[192], in contrast to an earlier report that such compounds would be subject to rapid inversion. Infrared spectral studies of $As(OR)_3$ esters have also been reported[180b].

7.6.4   Arsenic acids, arsenites and arsenates

Because of space limitations, only a brief discussion of the extensive chemistry of these systems will be given. Reviews of condensed arsenates[193a] and polymeric arsenic(V) oxides[193b] have been published.

The crystal structure of the hemihydrate of the acid H_3AsO_4 shows that it contains tetrahedral AsO_4 units linked by extensive hydrogen bonding[194]. The ^{75}As n.q.r. spectral studies of $H_2AsO_2^-$ salts and their deuterated isomers also demonstrate the presence of hydrogen bonding, and suggest that the tetrahedral units are distorted from T_d symmetry[195]. Hydrogen bonding is also clearly observed in the infrared spectrum of KH_2AsO_4 [196].

The crystal structure[197] of the dialkylarsonic acid $(C_4H_9)_2As(O)OH$ also shows hydrogen bonding; in this case the result is an association to hydrogen-bonded dimers with a chair conformation.

The metal arsenate $Cu_3(AsO_4)_2$ possesses tetrahedral AsO_4^{3-} units which provide a 5-coordinate environment for the copper atom as established by crystal structure studies[198].

Symons and co-workers have generated the radical anion AsO_4^{2-} and measured the hyperfine coupling to arsenic[199].

The kinetics of the reduction of H_3AsO_4 by I^- [200] and the oxidation of AsO_3^{3-} by Cr^{VI} [201], I_3^- [202] and $Fe(CN)_6^{3-}$ [203] have been reported. All are complicated, and space limitations prevent a discussion here.

Aqueous solutions of As^{III} contain $As(OH)_3$ and AsO_3^{3-} ions according to Raman studies. No evidence for OAsOH in molecular or polymeric form was obtained[204, 205]. $AsCl_3$ gave orthoarsenites, and no chlorinated arsenic or oxyarsenic species[205].

Calcium forms a mixed chloride–arsenite[206]. New studies on the polymeric acid $[H_5As_3O_{10}]_x$ have been reported, and it has been suggested on the basis of infrared spectroscopic evidence that the above polymeric acid is really As_4O_{10} [207]. Thallous salts of $As_4O_{12}^{-4}$ have been prepared[208] and shown to contain the cyclic anion isostructural with $P_4O_{12}^{-4}$. New work on arsenates of Nb, Ta [209], W [210, 211], Mo [212, 213] and Cr [214] has been reported. H_3AsO_4 reacts with peroxide to give peroxyarsenates[215].

7.7 ARSENIC COMPOUNDS CONTAINING SULPHUR, SELENIUM AND TELLURIUM

7.7.1 Arsenic sulphides and selenides

The crystal structure of As_4S_3 shows the molecule to have C_{3v} symmetry. Three arsenic atoms form a basal triangular plane of a tetrahedron with As—As (2.45 Å) [216], and the apical arsenic is bridged by three sulphur atoms (As—S = 2.21 Å) [216]. A review of the crystallographic data on As_2S_3, As_2Se_3, As_2Te_3, and related glasses is also available[217]. The preparation, properties and x-ray structural studies on AsSI, AsSeI, and As_4Te_5I have been reported[218]. The ^{75}As n.q.r. spectrum shows that there are four non-equivalent sites for As in As_2S_3 [219], and two in $TlAsSe_2$ [220]. A number of similar compounds have been studied by ^{75}As n.q.r. methods[220]. Raman spectra of realgar (As_4S_4) and orpiment (As_2S_3) have been recorded[221] showing difference in intermolecular and intramolecular As—S bonds.

The $As_4S_6^{2-}$ ion [222] has been prepared as ammonium, Pb^{II} or Hg^{II} salts. Hydrolysis gave As_2S_5 and H_2S [222]. Thioarsenites AsS_4^{3-} can be prepared[223] from As_2S_5 or AsO_4^{3-} and Na_2S or H_2S. AsS_4^{3-} has a single electronic transition at 36 500 cm^{-1} which has been assigned to the $^1A_1 \rightarrow {}^1T_2$ (T_d symmetry) transition, and interpreted as indicative of the absence of π-bonding[224]. The sodium salt Na_3AsS_4 gives As_2S_5 and S in both basic[225] and acidic[226] solutions. The oxythioarsenates can be prepared from As_2S_5 and LiOH [227] or As_2O_5 and Na_2S [223]. The new $AsSe_3^-$ species has been reported as the product of the reaction of alkali metal carbonates and As_2Se_5 [228].

7.7.2 Substituted arsenic sulphides

In alcoholic KOH solution, As_2S_3 inserts CS_2 to form $As(S_2COR)_3$ [229]. Trithiocarbonates, R_2AsS_2CSR, or the corresponding xanthates [230], have been prepared by the reaction of R_2AsCl with NaS_2CSR or NaS_2COR. The analogous dithiocarbamates $As(S_2CNR_2)_3$ [229, 231, 232] or $R_2AsS_2CNR_2$ [223] have been prepared by the following reactions: (a) dithiocarbamic acid with As_2S_3 [230b], (b) As_2O_3, CS_2 and HNR_2, (c) AsH_3 with AgS_2CNR_2 [231], (d) $AsCl_3$, CS_2 and HNR_2 [232], or (e) $C_6H_5As(NR_2)_2$ with CS_2 [233]. The infrared spectrum of $As(S_2CNH_2)_3$ is similar to that of various trivalent transition metal analogues or the Ga^{III} analogue [234]. $As[S_2CN(C_2H_5)_2]_3$ crystallised from acetone with an approximate trigonal-prismatic molecular structure, with three short (2.35 Å) and three long (2.8–2.9 Å) As—S bonds at 90 degrees to each other[235]. Similarly, $BrAs(S_2CNEt_2)_2$ has two short (2.267, 2.321 Å) and two long As—S bonds, as well as a short (2.711 Å) As—Br bond[236]. $AsCl_3$

reacts with HS_2PFR ($R = CH_3$, C_2H_5) to eliminate HCl and form As $(S_2PFR)_3$ [237].

Various substituted thio-esters and sulphur-bridged compounds have been prepared from the arsenic(III) halides. $AsCl_3$ reacted with $Pb(SC_6Cl_5)_2$ to give $As(SC_6Cl_5)_3$ [238]. Elimination of HCl by the reaction of R'SH and R_nAsCl_{3-n} ($n = 0, 1, 2$) [239] gave a series of $R_nAs(SR')_{3-n}$ compounds. It is noteworthy that the corresponding oxides $Ph_nAs(O)(OR'')_{3-n}$ are reduced by BuSH to yield $Ph_nAs(SBu)_{3-n}$, BuOH, and Bu_2S [240]. Amine elimination has been employed to give $PhAs(SEt)_2$ and the heterocycles

$$C_6H_5As\begin{array}{c} E-CH_2 \\ | \\ E-CH_2 \end{array} \qquad (E = O, S)^{[241]}$$

Similar compounds have been discussed earlier[192] (Section 7.5.1.4), as was the reaction of thiourea with Cl_2AsCH_2Cl and aqueous base to give the bicyclic compounds[121].

Bridged compounds (II) were formed from $[Cl_2AsCH_2]_2$ and H_2E ($E = O, S$) followed by reaction with $AsCl_3$ [242].

$$\begin{array}{c} Cl \\ CH_2-\overset{|}{As} \\ | \qquad \diagdown \\ | \qquad \quad E \\ CH_2-\underset{|}{As} \diagup \\ Cl \end{array} \qquad (II)$$

The Cl atoms can be replaced by CH_3 groups[242].

In alcoholic solution, H_2S converts Me_2AsO_2Na to Me_2AsS_2Na, from which a number of transition metal complexes of the chelating ligand $Me_2AsS_2^-$ have been prepared[243].

7.8 ARSENIC NITROGEN COMPOUNDS

7.8.1 Aminoarsines and related compounds

Salt elimination [244-246] (e.g. equation (7.4)) has been employed to prepare silicon-[244, 245] or phosphorus-[246] substituted aminoarsines, including the interesting heterocycle illustrated:

$$O(SiMe_2NMeLi)_2 + AsCl_3 \longrightarrow O\begin{array}{c} SiMe_2NMe \\ \diagup \qquad \qquad \diagdown \\ \diagdown \qquad \qquad \diagup \\ SiMe_2NMe \end{array} AsCl + 2LiCl \qquad (7.4)$$

The reactions of arsenic halides with amines afford a wide variety of aminoarsines[233, 244, 247-249], iminoarsines[248], and hydrazinoarsines[249].

$$Bu_2AsCl + NH_3 \rightarrow Bu_2AsNH_2 \overset{\Delta}{\longrightarrow} (Bu_2As)_2NH$$

Aminotin compounds react with AsF_3 to yield $As(NMe_2)_3$ [250]. AsF_3 and propylamine gave unstable $[FAsNPr]_n$ ($n = 3, 4$) which rearranged in ether to AsF_3 and $As_4(NPr)_6$ [248].

Alcohol elimination occurs on aminolysis of $As(OMe)_nF_{5-n}$ ($n = 3, 4$) [251] to give $As(OMe)_3(NHPr)F$ which condenses to $[(MeO)_2AsF{=}NPr]_2$, probably containing the four-membered As_2N_2 ring[251]. Transamination reactions of $PhAs(NMe_2)_2$ with R_2NH gave a variety of products $PhAs(NR_2)_2$ [233]. Ammonolysis of Ph_3As by KNH_2 in liquid ammonia gives $K[As(NH_2)_2]$, which on heating loses NH_3 to leave $K_3As_3N_3$ [252].

Successive replacement of two CF_3 groups in $(CF_3)_3As$ by $(CF_3)_2N$ on reaction with $(CF_3)_2NCl$ occurs (equation (7.5)),

$$(CF_3)_3As \xrightarrow{(CF_3)_2NCl} (CF_3)_2As\overset{\text{'}}{N}(CF_3)_2 \xrightarrow{(CF_3)_2NCl} CF_3As[N(CF_3)_2]_2 \quad (7.5)$$

but elimination of $CF_3N{=}CF_2$ rather than displacement of a third CF_3 group takes place[253]. Chlorination of $(CF_3)_2AsN(CF_3)_2$ at $-50°C$ gave $Cl_2AsN(CF_3)_2$ with formation of CF_3Cl. At room temperatures the products were $AsCl_3$, $(CF_3)_2NCl$ and CF_3Cl [254].

In contrast to the above, the chloramination of $(C_6H_5)_3As$ or $[(C_6H_5)_2As]_2CH_2$ with R_2NCl ($R = H$, CH_3) yields[255] the amino-substituted arsonium salts $(C_6H_5)_3AsNR_2^+Cl^-$ or $[(C_6H_5)_2As(NR_2)]_2CH_2^{2+}2Cl^-$. The reaction of CF_3I with $As[N(CH_3)_2]_3$ gave $CF_3As[N(CH_3)_2]_2$ and the arsonium salt[256] $As[N(CH_3)_2]_4^+I^-$.

Addition of $RNCO$ ($R = Me$, $C_6H_5CH_2$) to Ph_3AsO, Ph_3AsBr_2 or Ph_3AsNCH_2Ph afforded the arsine imides $RCON{=}AsPh_3$ [257] respectively. Insertion of $PhNCO$ into $PhAs(NR_2)_2$ yielded the corresponding ureido derivatives[233].

Azides of As^{III}, As^V, and the azonium salt $[Me_3AsN_3][SbCl_6]$ have been prepared by the reactions of NaN_3 with $RR'As$-halides[258], Me_3AsCl_2 [259a], and $[Me_3AsCl][SbCl_6]$ [259b], respectively. Thiocyanogen oxidised R_3As to $R_3As(NCS)_2$ [260].

Aminoarsines may be lithiated by the reaction of alkyl-lithiums with primary aminoarsines[247]. These salts have been utilised in the preparation of diarsinylamines and silylaminoarsines[247]. With $MeCl$, $Me_3SiN(Li)AsBu_2$ gave the iminoarsine $Me_3SiN{=}AsMeBu_2$ [247], which could be methylated to the aminoarsonium ion[247].

The character of arsenic–nitrogen bonds in $(CF_3)_2AsNH_2$ and $[(CF_3)_2As]_2NH$ has been discussed in relation to $^{15}N{-}^1H$ coupling constants[261], while temperature-dependent n.m.r. spectroscopic studies enabled measurement of the barrier to rotation about the As—N bond in $PhAsClNMe_2$ [262]. Hindrance to rotation about the As—N bond in $Bu_2^tAsN(MMe_3)_2$ ($M = Si$, Ge) is steric in origin[263].

Whereas $OS(NHR)_2$ ($R = H$, Me) reacts with $As(NMe_2)_3$ to give bicyclic $As_2[NRSONR]_3$, $OC(NHR)_2$ yields $As[OC(NHR){=}NR]_3$ [264]. Displacement of Me_2N occurs on reaction of $ON(CF_3)_2$ with $(Me_2N)_3As$ to give $(Me_2N)_2AsON(CF_3)_2$, which was also prepared from $Hg[ON(CF_3)_2]_2$ and $(Me_2N)_2AsCl$ [265].

Amination with oxygen abstraction by $As(NMe_2)_3$ occurs with a series of aldehydes and ketones (e.g. equation (7.6))[266, 267]

$$RCH_2COR' + As(NMe_2)_3 \rightarrow RCH{=}CR'NMe_2 \quad (7.6)$$

In contrast, cyclic ethers or esters undergo ring opening to insert into the arsenic–nitrogen bond[267].

Nitrosyl chloride eliminated nitrogen on reaction with $(C_6H_5)_3As=NH$ to give $(C_6H_5)As(Cl)OH$ [268]. De-oxygenation of $PhNO_2$ by $As(NMe_2)_3$ yielded C_6H_5NO, Me_2NH, As_2O_3 and $(Me_2N)_2CH_2$ [269].

7.9 DIARSENIC COMPOUNDS, ARSINO–METALS AND METALLOIDS

7.9.1 Diarsenic compounds and arsino phosphines

The coordination chemistry of cyclic polyarsines has recently been reviewed[270]. The constitution of the cyclic polyarsines $(AsR)_n$ ($n = 5$, R = alkyl; $n = 6$, R = aryl) and the mixed Me/Ph polyarsines has been investigated by n.m.r. and mass spectrometric techniques[271]. For $(AsMe)_5$, a very rapid pseudo-rotation process was postulated[271]. In the equilibrium mixture of the reaction of As_2Me_4 with $(AsMe)_5$ the original reagents predominated[272]; however, some scrambling occurs involving As—As (but not As—C) bond cleavage, to give small amounts of species such as $(Me_2As)_2AsMe$. Diarsines were also formed in the reaction of R_2AsH with chloramines or Cl_2 and a HCl-acceptor ($RAsH_2$ gave tetra-substituted polyarsines)[273], or by the reaction of arsenic with alkyl halides and lithium in liquid ammonia[274].

Diarsines were also yielded in the reactions of $Me_3SiAsMe_2$ with Me_2AsCl, $AsCl_3$ (via $As(AsMe_2)_3$), or PCl_3 (via $P(AsMe_2)_3$) [275] or Me_3PbOMe with Me_2AsH [276]. The fluorinated arsinophosphines X_2PAsX_2 ($X = CH_3$, CF_3) and the diarsine $(CF_3)_2AsAsMe_2$ were obtained by the attack of Me_2MH (M = As or P) on the diarsine or diphosphine $[(CF_3)_2M]_2$ (M = As or P) [277]. $(CF_3)_2AsH$ attacked the fluorinated diphosphine to yield $(CF_3)_2PAs(CF_3)_2$ (which was also formed by the exchange of $[(CF_3)_2P]_2$ with $[(CF_3)_2As]_2$ at $25°C$), but the weaker base $(CF_3)_2PH$ did not attack $[(CF_3)_2As]_2$ [277]. In octahydrophenanthroline, As_2I_4 was formed from the elements[302]. This compound was thermally stable up to $150\,°C$, and then decomposed to yield AsI_3 and As [302].

The polyarsines and arsinophosphines $(Ph_2M)_3M'$, $(Ph_2M)_2M'Ph$, or Ph_2AsPPh_2 (M = As, M' = As, P; M = P, M' = As) were prepared by the reaction of the appropriate trimethylstannylarsine and phosphorus chloride[278]. Insertion of CF_3P to give $(Me_2As)_2PCF_3$ occurs in the reaction of $(CF_3P)_4$ with Me_4As_2 through a bimolecular exchange (^1H n.m.r. spectroscopy) [279].

7.9.2 Arsenic–boron and–aluminium compounds

Under mild conditions, B_2H_6 and $(CF_3)_2AsH$ give unstable $[(CF_3)_2AsBH_2]_2$, which is labile to attack by chlorine or $MeOH/HCl$ [281]. In acetone solution the conductivity of the adduct $C[CH_2AsPh_2·BX_3]_4$ (X = Cl, Br) indicates the formation of (III)[282].

$$\left[X_2B \overset{AsPh_2-CH_2}{\underset{AsPh_2-CH_2}{<}} C \overset{CH_2-AsPh_2}{\underset{CH_2-AsPh_2}{>}} BX_2 \right]^{2+} 2[BX_4]^- \qquad (III)$$

In tetrahydrofuran, $AsCl_3$ and $Na_3B_{10}H_{10}CH$ give $1,2-B_{10}H_{10}CHAs$ (IV) which rearranges at 505°C to $1,7-B_{10}H_{10}CHAs$, and at 575°C to $1,12-B_{10}H_{10}CHAs$ [283]. Piperidine reacts with (IV) to give 7,8- and $7,9-B_9H_{10}CHAs^-$ which, with C_5H_6 and $CoCl_2$ in Et_3N yields $CpCo(B_9H_{10}CHAs)$ [283a]. The salt $[Me_4N]^+[B_9H_{10}CHAs]^-$ reacts with $M(CO)_6$ (M = Cr, Mo, W) photolytically to afford $[Me_4N][B_9H_{10}CHAsM(CO)_5]$, in which M is σ-bonded to As [283a].

Arsine reacts with alkali metal aluminium hydrides to afford the arsinating agents $M[Al(AsH_2)_4]$ [284, 285].

7.9.3 Group IV metals

The chemistry of the compounds $R_3MER'_2$, etc., (M = Si, Ge, Sn, Pb; E = P, As, Sb, Bi) has been recently reviewed[286], and for this reason only the most recent advances will, therefore, be reported here. Arsination of halides of silicon[285, 286] or germanium by $Li[Al(AsH_2)_4]$ is the most efficient technique[285] for the preparation of H_3MAsH_2 compounds. Unexpectedly, Me_3SiBr and $Li[Al(AsH_2)_4]$ give $(Me_3Si)_2AsH$ in ether[284]. Transmetallation has been utilised to prepare germylarsines (equation (7.7)) [285a].

$$H_3SiAsH_2 + R_3GeCl \rightarrow R_3GeAsH_2 + H_3SiCl \qquad (7.7)$$

The reactions of arsines with tin chlorides[287] and germanium halides[288, 289] give the corresponding stannyl- and germyl-arsines.

The pyramidal structure of $(H_3Si)_3As$ has now been conclusively demonstrated by infrared spectroscopic[290] and electron diffraction[291] studies. The vibrational spectra of H_2AsSiH_3 [292], and the deuterated analogue[292], have been assigned and compared with the germyl analogues H_3GeAsH_2 [293] and $(H_3Ge)_2AsH$ [294] on a pyramidal model.

The infrared and Raman spectra of $(Me_3Si)_nAsH_{3-n}$, and calculated force constants for the system, imply an angle of 90–100 degrees for <SiAsSi [294]. Infrared studies of the methyl silicon analogues, $(Me_3Si)_nAs(H,D)_{3-n}$, also confirm the pyramidal structures. Extensive n.m.r. spectroscopic studies of the metallo-arsenic hydrides have been reported[285, 288, 289].

Whereas water cleaves Me_3SiAsH_2 to yield $(Me_3Si)_2O$ [285a], Me_3GeAsH_2 rearranges to give $(Me_3Ge)_3As$ and AsH_3 [285]. Likewise, H_3GeAsH_2 rearranges during hydrolysis to $(H_3Ge)_2AsH$ and $(H_3Ge)_3As$ [296]. Hydrogen sulphide gives H_3GeSH and $(H_3Ge)_2S$ [296]. In general, arsenic–silicon bonds are cleaved by alcohols or water, but not by amines[285].

Air reacts with $R_2AsSnR'_3$ to yield $R_2As(O)OSnR'_3$ [297].

7.9.4 Transition metals

The formation in solution of Ti^{III} species $Cp_2TiAsPh_2$ and $[Cp_2Ti(AsPh_2)_2]^-$ from Cp_2TiCl_2 and $NaAsPh_2$ is postulated on the basis of e.p.r. spectral patterns[298]. The reaction of $(Me_2N)_3TiBr$ with $LiAsEt_2$ to yield $(Me_2N)_3TiAsEt_2$ is more straightforward[299]. Reduction of $Ph_3MCo(B)(D)$ (M = Si, Ge, Sn; B = Base; D = dimethylglyoxime) with $NaBH_4$, and subsequent reaction with Me_2AsCl, gave $Me_2AsCo(D)$ [300].

References

1a. Doak, G. O., Freedman, L. D. and Long, G. G. (1963). *Kirk-Othmer Encyclopedia of Chemical Technology 2nd Ed.*, **2**, 718

1b. Sabine, D. (1969). *Chemistry*, **42**, 20

2a. Kolditz, L. (1965). *Adv. Inorg. Chem. Radiochem.*, **7**, 1;

2b. Kolditz, L. (1967). *Halogen Chemistry*, **2**, 115

3. Doak, G. O. and Freedman, L. D. (1969). *Organometal. Chem. Revs.*, **5**, 128

4. Booth, G. (1964). *Adv. Inorg. Chem. Radiochem.*, **6**, 1

5. Fild, M. and Glemser, O. (1969). *Fluorine Chem. Revs.*, **3**, 129

6. Schiferl, D. and Barrett, C. S. (1969). *J. Appl. Crystallogr.*, **2**, 30

7. Taylor, W. A., McCollum, D. C., Passenheim, B. C. and White, H. W. (1967). *Phys. Rev.*, **161**, 652

8. Verkin, B. I., Svechkarev, I. V. and Kuz'micheva, L. B. (1966). *Zh. Eksperim. i. Teor. Fiz.*, **50**, 1438 (*Chem. Abs.* (1966), **65**, 8163b)

9. Morino, Y., Ukaji, T. and Ito, T. (1966). *Bull. Chem. Soc. Jap.*, **39**, 64

10. Beattie, I. R., Ozin, G. A. and Perry, R. O. (1970). *J. Chem. Soc., A*, 2071

11. Perdigon, P., Martin, F. and d'Incan, J. (1970). *J. Mol. Spectrosc.*, **36**, 341

12. Perdigon, P., d'Incan, J. and Sfeila, J. (1969). *Compt. Rend Acad. Sci. (Paris)*, **268B**, 1432

13. Perdigon, P., d'Incan, J. and Janin, J. (1967). ibid., **265B**, 141

14. Mrozowski, S. and Santaram, C. (1967). *J. Opt. Soc. Amer.*, **57**, 522

15. d'Incan, J., Perdigon, P. and Janin, J. (1966). *Compt. Rend. Acad. Sci. (Paris)*, **262B**, 951

16. Topouzkhanian, A., Goure, J. P., Fiquet, J. and d'Incan, J. (1970). ibid., **270B**, 1676

17. Zijlstra, W. G., Henrichs, J. M., Mooy, J. H. M. and Van Voorst, J. D. W. (1970). *Chem. Phys. Lett.*, **7**, 553

18. O'Hare, P. A. G. (1969). *Nucl. Sci. Abstracts*, **23**, 34951 (*Chem. Abstr.* (1969) **71**, 33988x)

19. Rao, D. V. K. and Rao, P. T. (1970). *Proc. Phys. Soc. (London) (At. Mol. Phys.)*, **3**, 430

20. d'Incan, J. and Goure, J. P. (1969). *Compt. Rend. Acad. Sci. (Paris)*, **268B**, 1647

21. Shankar, R. and Singh, I. S. (1969). *Can. J. Phys.*, **47**, 1601

22. Shimanchi, M. (1969). *Sci. Light (Tokyo)*, **18**, 90

23a. Basco, N. and Yee, K. K. (1967). *Nature (London)*, **216**, 998

23b. idem. (1967). *Chem. Commun.*, 1255

24. Callear, A. B. and Oldman, R. J. (1968). *Trans. Faraday Soc.*, **64**, 840

25a. Dixon, R. N., Duxburg, G. and Lamberton, H. M. (1968). *Proc. Roy. Soc. (Ser. A)*, **305**, 271

25b. idem. (1966). *Chem. Commun.*, 460

25c. Dixon, R. N. and Lamberton, H. M. (1968). *J. Mol. Spectrosc.*, **25**, 12

26. Basco, N. and Yee, K. K. (1968). *Spectrosc. Lett.*, **1**, 17

27a. Jackel, G. S. and Gordy, W. (1968). *Phys. Rev.*, **176**, 443

27b. Morehouse, R. L., Christiansen, J. J. and Gordy, W. (1966). *J. Chem. Phys.*, **45**, 1747

28. Basco, N. and Yee, K. K. (1968). *Chem. Commun.*, 153

29. Yee, K. K., Liu, D. and Jones, W. E. (1970). *J. Mol. Spect.*, **35**, 153

30. Bellama, J. M. and MacDiarmid, A. G. (1968). *Inorg. Chem.*, **7**, 2070

31. Royen, P. and Rocktaeschel, C. (1966). *Z. Anorg. Allg. Chem.*, **346**, 290

32. Dobbie, R. C. and Cavell, R. G. (1967). *J. Chem. Soc. A*, 1308

33. Harvey, A. B. and Wilson, M. K. (1966). *J. Chem. Phys.*, **44**, 3535

34. Sarka, K., Papousek, D. and Rao, K. N. (1971). *J. Mol. Spectrosc.*, **37**, 1

35. Shih-Tung, K. and Overend, J. (1969). *J. Phys. Chem.*, **73**, 406

36. Rao, B. P. and Murthy, V. R. (1968). *Indian J. Pure Appl. Phys.*, **6**, 339

37. Belyavskaya, L. V., Shabur, V. N. and Morozov, V. P. (1968). *Opt. Spektrosk.*, **25**, 62

38. Ramaswamy, K. and Swaminathan, S. (1969). *Spectrosc. Mol.*, **18**, 7

39. Birchall, T. and Jolly, W. L. (1966). *Inorg. Chem.*, **5**, 2177

40. Ebsworth, E. A. V. and Sheldrick, G. M. (1967). *Trans. Faraday Soc.*, **63**, 1071

41. Sheldrick, G. M. (1967). *Trans. Faraday Soc.*, **63**, 1065

42. Helminger, P., Beeson, E. L., Jr. and Gordy, W. (1971). *Phys. Rev. A*, **3**, 122

43. Harvey, A. B. (1966). *J. Phys. Chem.*, **70**, 3370

44. Stevenson, P. E. and Lipscomb, W. N. (1970). *J. Chem. Phys.*, **52**, 5343

45. Cullen, W. R. and Leeder, W. R. (1970). *Can. J. Chem.*, **48**, 3757

46. Tzschach, A. and Schwarzer, R. (1967). *Ann. Chem.*, **709**, 248

47. Bruker, A. B., Grinshtein, E. I. and Soborovskii, L. Z. (1966). *Zh. Obshch. Khim.*, **36**, 1133

48. Kostyanovskii, R. G., Yakshin, V. V., Zimont, S. L. and Chervin, I. I. (1968). *Izv. Akad. Nauk. SSSR, Ser. Khim.,* 677
49. Nesmayanov, A. N., Borisov, A. E. and Kudryavtseva, L. V.(1969), *Izv. Akad. Nauk. SSSR, Ser. Khim.,* 1973
50. Cullen, W. R., Dhaliwal, P. S. and Styan, G. E. (1966). *J. Organometal. Chem.,* **6,** 364
51. Cullen, W. R., Dawson, D. S. and Styan, G. E. (1965). *Can. J. Chem.,* **43,** 3392
52. Cullen, W. R. and Leeder, W. R. (1969). *Can. J. Chem.,* **47,** 2137
53. Cullen, W. R. and Dhaliwal, P. S. (1967). *Can. J. Chem.,* **45,** 719
54. Cullen, W. R. and Waldman, M. C. (1970). *Can. J. Chem.,* **48,** 1885
55. Cullen, W. R. and Styan, G. E. (1966). *Can. J. Chem.,* **44,** 1225
56. Rheingold, A. L. and Bellama, J. M. (1969). *Chem. Commun.,* 1058
57. Tzschach, A. and Drohne, D. (1970). *J. Organometal. Chem.,* **21,** 131
58. Tzschach, A. and Schwartzer, R. (1968). *J. Organometal. Chem.,* **13,** 363
59. Konaka, S. (1970). *Bull. Chem. Soc. Jap.,* **43,** (a) 3107, (b) 1693
60. Clippard, F. B. and Bartell, L. S. (1970). *Inorg. Chem.,* **9,** 805
61. Trotter, J. (1965). *Z. Krist.,* **122,** 230
62. Morino, Y., Ukagi, T. and Ito, T. (1966). *Bull. Chem. Soc. Jap.,* **39,** 71
63. Brieux de Mandirola, O. (1969). *J. Mol. Struct.,* **3,** 465 (b) (1968) ibid., **1,** 203
64. Reichman, S. and Overend, J. (1970). *Spect. Acta.,* **26A,** 379
65. Levin, I. W. and Abramowitz, S. (1966). *J. Chem. Phys.,* **44,** 2562
66. Evans, J. A. and Long, D. A. (1968). *J. Chem. Soc. A,* 1688
67. Hoskins, L. C. (1966). *J. Chem. Phys.,* **45,** 4594
68. Reichman, S., Smith, D. F. Jr. and Overend, J. (1970). *Spect. Acta,* **26A,** 927
69. Timoshinim, V. S. and Godnev, I. N. (1970). *Optik. Spektrosk.,* **28,** 832
70a. Mirri, A. M. (1967). *J. Chem. Phys.,* **47,** 2823
70b. Mirri, A. M. (1968). *Corsi Semin. Chim.,* **14,** 75. *(Chem. Abs.* (1969). **71,** 75960n)
71. Loehr, T. M. and Plane, R. A. (1969). *Inorg. Chem.,* **8,** 73
72. Stufkens, D. J. (1970). *Rec. Trav. Chim. Pays-Bas,* **89,** 755
73. Hadni, A., Decamps, E. and Herbeuval, J. P. (1968). *J. Chem. Phys.,* **65,** 959
74. Hooper, M. A. and James, D. W. (1968). *Aust. J. Chem.,* **21,** 2379
75. Kiefer, W. (1970). *Z. Naturforsch.,* **25a,** 1101
76. Davies, J. E. D. and Long, D. A. (1968). *J. Chem. Soc. A,* 1757
77. Rai, S. N. and Thakur, S. N. (1970). *Aust. J. Chem.,* **23,** 881
78. Nagarajan, G. (1966). *Indian J. Pure Appl. Phys.,* **4,** 456
79. Pillai, M. G. K. and Pillai, P. P. (1968). ibid., **8.** 404
80. Claeys, E. G. and van der Kelen, G. P. (1966). *Spect. Acta,* **22A,** (a) 2095, (b) 2103
81. Cyvin, S. J. and Cyvin, B. N. (1969). *J. Mol. Struct.,* **4,** 341
82. Müller, A., Niecke, E., Krebs, B. and Glemser, O. (1968). *Z. Naturforsch.,* **23b,** 588
83a. Elvebredd, I., Vizi, B., Cyvin, S. J., Müller, A. and Krebs, B. (1968). *J. Mol. Struct.,* **2,** 158
83b. idem. ibid. p. 149
84. Nagarajan, G. (1966). *Acta Phys. Acad. Sci. Hung.,* **20,** 323. *(Chem. Abs.* (1966), **65,** 6318h)
85. Schlemper, E. O. and Britton, D. (1966). *Acta. Crystallogr.,* **20,** 777
86. Emerson, K. and Britton, D. (1963). *Acta Crystallogr.,* **16,** 113
87. Rhodes, M. and Aksnes, D. W. (1969). *Mol. Physics,* **17,** 261
88. Lunazzi, L. and Brownstein, S. (1969). *J. Mag. Resonance,* **1,** 119
89. Sverynn, V. I., Babrishkina, T. A., Schvedova, G. N., Kudryavtseva, L. V. and Seinin, G. K. (1970). *Izv. Akad. Nauk. SSSR. Ser. Khim.,* 482
90a. Biedenkapp, D. and Weiss, A. (1968). *Z. Naturforsch.,* **23b,** 172
90b. Grechishkin, V. S. and Yusupov, M. Z. (1970). *Zh. Fiz. Khim.,* **44,** 2933
91. Devyatykh, G. G., Rachkov, V. G. and Agafonov, I. L. (1968). *Zh. Neorg. Khim.,* **13,** 2907
92. Dobbie, R. C. and Cavell, R. G. (1967). *Inorg. Chem.,* **6,** 1450
93a. Lloyd, D. R. (1970). *Int. J. Mass Spectrom. Ion Phys.,* **4,** 500
93b. Potts, A. W., Lempka, H. J., Streets, D. G. and Price, W. C. (1970). *Phil. Trans. Roy. Soc. (London) Ser. A.,* **268,** 59
94. Claeys, E. G. (1966). *J. Organometal. Chem.,* **5,** 446
95. Pugachevich, P. P., Mogucheva, V. V. and Nisel'son, L. A. (1967). *Zh. Fiz. Khim.,* **41,** 595
96. Uchimura, K., Fujisawa, H., Murakami, T. and Yuizumi, M. (1967). *Denki Kagaku,* **35,** 95 *(Chem. Abstr.* (1968), **68,** 905243)

97a. Druce, P. M., Lappert, M. F. and Riley, P. N. K. (1967). *Chem. Commun.*, 486
97b. Lappert, M. F. and Prokai, B. (1967). *J. Chem. Soc. A*, 129
 98. Ivashentsev, Ya. I. and Kutakova, L. I. (1968). *Izv. Vyssh. Ucheb. Zaved., Khim, Khim Tekhnol.*, **11**, 845. (*Chem. Abs.*, (1968) **69**, 113081v)
 99. Deters, J. F., McCusker, P. A. and Pilger, R. C. Jr. (1968). *J. Amer. Chem. Soc.*, **90**, 4583
100. Olszer, R. and Siekierski, S. (1966). *J. Inorg. Nucl. Chem.*, **28**, 1991
101. Paul, R. C., Aggarval, V. K., Ahlunalia, S. C. and Narula, S. P. (1970). *Inorg. Nucl. Chem. Lett.*, **6**, 487
102. Deveney, M. J. and Webster, M. (1970). *J. Chem. Soc. A*, 1643
103. Davies, J. E. D. and Long, D. A. (1968). *J. Chem. Soc. A*, 1761
104. Ahlijah, G. E. B. Y. and Goldstein, M. (1970). *J. Chem. Soc. A*, (a) 326. (b) 2590
105. Nagarajan, G. (1966). *Czech. J. Phys.*, **16**, 157 (*Chem. Abs.* (1966), **64**, 10575d)
106. Pacini, H. A., Teach, E. G., Walker, F. H. and Pavlath, A. E. (1966). *Tetrahedron*, **22**, 1747
107. Canterford, J. H. and O'Donnell, T. A. (a) (1968). *Aust. J. Chem.*, **21**, 1421. (b) idem. (1967) *Inorg. Chem.*, **6**, 541
108. Krenev, V. A. and Evdokimov, V. I. (1968). *Zh. Neorg. Khim.*, **13**, 1279
109. Reiss, J. G. and Van Wazer, J. R. (1966). *Inorg. Chem.*, **5**, 178
110. Goldwhite, H. (1966). *Inorg. Nucl. Chem. Lett.*, **2**, 5
111a. O'Donnell, T. A. and Stewart, D. F. (1966). *Inorg. Chem.*, **5**, 1434
111b. O'Donnell, T. A., Stewart, D. F. and Wilson, P. (1966). *Inorg. Chem.*, **5**, 1438
112. Riesel, L. and Lehmann, H. A. (1969). *Z. Anorg. Allg. Chem.*, **371**, 289
113. Paul, R. C., Kaushal, R., Dhindsa, K. S., Pahil, S. S. and Ahluwalia, S. C. (1967). *J. Indian Chem. Soc.*, **44**, 964
114a. Paul, R. C., Kanshal, R. and Pahil, S. S. (1969). *J. Indian Chem. Soc.*, **46**, 26
114b. idem. (1967). ibid., **44**, 995
115. Schmulbach, C. D. and Ahmed, I. Y. (1968). *J. Chem. Soc. A*, 3008
116. Kellogg, F. D. and Matwiyoff, N. A. (1968). *Inorg. Nucl. Chem. Lett.*, **4**, 47
117. Gutmann, V. and Czuba, H. (1969). *Monatsh.*, **100**, 708
118. Sutton, G. J. (1966). *Aust. J. Chem.*, **19**, 733
119a. Green, M. and Kirkpatrick (1967). *Chem. Commun.*, 57
119b. idem. (1968). *J. Chem. Soc. A*, 483.
119c. Fild, M. and Glemser, O. (1969). *Fluorine Chem. Revs.*, **3**, 129
120. Ang, H. G. and Ho, K. F. (1969). *J. Organometal. Chem.*, **19**, P18
121. Sommer, K. (1970). *Z. Anorg. Allg. Chem.*, **377**, 128
122. Berry, R. S., Tamres, M., Ballhausen, C. J. and Johansen, H. (1968). *Acta Chem. Scand.*, **22**, 231
123a. Selig, H., Holloway, J. H., Tyson, J. and Claasen, H. H. (1970). *J. Chem. Phys.*, **53**, 2559
123b. Blanchard, S. (1967). *Comm. Energy At (Fr) Rapp.*, CEA R3195 (*Chem. Abstr.* (1968). **69**, 6621z)
124a. Cyvin, S. J. and Brunvoll, J. (1969). *J. Mol. Struct.*, **3**, 151
124b. Nagarajan, G. and Durig, J. R. (1967). *Bull. Soc. Roy. Sci. (Liege)*, **36**, 334 (*Chem. Abstr.* (1967), **67**, 111599e)
125. Aljiburg, A. K. and Redington, R. L. (1970). *J. Chem. Phys.*, **52**, 453
126. Muetterties, E. L., Mahler, W., Packer, K. J. and Schmutzler, R. (1964). *Inorg. Chem.*, **3**, 1298
127. Muetterties, E. L. and Phillips, W. D. (1967). *J. Chem. Phys.*, **46**, 2861
128. Kolditz, L. and Rehak, W. (1966). *Z. Anorg. Allg. Chem.*, **342**, 32
129. Kolditz, L. and Beierlein, I. (1967). *Z. Chem.*, **7**, 468
130. Tebbe, F. N. and Muetterties, E. J. (1967). *Inorg. Chem.*, **6**, 129
131. Brownstein, M. and Gillespie, R. J. (1970). *J. Amer. Chem. Soc.*, **92**, 2718
132. Gillespie, R. J. and Morton, M. J. (1970). *Inorg. Chem.*, **9**, 811
133a. Christe, K. O. and Sawodny, W. (1969). *Inorg. Chem.*, **8**, 212
133b. Christe, K. O., Schack, C. J., Pilipovich, D. and Sawodny, W. (1969). *Inorg. Chem.*, **8**, 2489
133c. Christe, K. O. and Pilipovich, D. (1969). *Inorg. Chem.*, **8**, 391
134. Glemser, O. and Smale, A. (1969). *Angew. Chem. Int. Edn.*, **8**, 517
135a. Shamir, J. and Binenboyun, J. (1968). *Inorg. Chim. Acta*, **2**, 37
135b. Beal, J. B. Jr., Pupp, C. and White, W. E. (1969). *Inorg. Chem.*, **8**, 828
136. Wamser, C. A., Fox, W. B., Sukornick, B., Holmes, J. R., Stewart, B. B., Juiwik, R., Vanderkooi, N. Jr. and Gould, D. (1969). *Inorg. Chem.*, **8**, 1249

137. Aureshi, A. M. and Aubke, F. (1970). *Can. J. Chem.*, **48**, 3117
138. See: Sladky, F. O., Bulliner, P. A. and Bartlett, N. (1969). *J. Chem. Soc. A*, 2179 and references contained therein
139. Begun, G. M. and Rutenberg, A. C. (1967). *Inorg. Chem.*, **6**, 2212
140a. Brownstein, S. (1969). *Can. J. Chem.*, **47**, 605
140b. Azeem, M., Brownstein, M. and Gillespie, R. J. (1969). *Can. J. Chem.*, **47**, 4159
141. Dean, P. A. W., Gillespie, R. J. and Hulme, R. (1969). *Chem. Commun.*, 990
142. Selig, H. and Aminadav, N. (1970). *Inorg. Nucl. Chem. Lett.*, **6**, 595
143. Gillespie, R. J., Ouchi, K. and Pez, G. P. (1969). *Inorg. Chem.*, **8**, 63
144. Davies, T. K. and Moss, K. C. (1970). *J. Chem. Soc. A*, 1054
145. Copeland, R. F., Conner, S. H. and Meyers, E. F. (1966). *J. Phys. Chem.*, **70**, 1288
146. Wamser, C. A., Fox, W. B., Gould, D. and Sukornick, B. (1968). *Inorg. Chem.*, **7**, 1933
147. Gillespie, R. J. and Passmore, J. (1969). *Chem. Commun.*, 1334
148. Gillespie, R. J. and Unmat, P. K. (1970). *Can. J. Chem.*, **48**, 1239
149. Dean, P. A. W. and Gillespie, R. J. (1970). *Chem. Commun.*, 853
150. Schack, C. J. and Pilipovich, D. (1970). *Inorg. Chem.*, **9**, 387
151. Morrow, S. I. and Young II, A. R. (1966). *Inorg. Nucl. Chem. Lett.*, **2**, 349
152. Peacock, R. D. and Wilson, I. L. (1969). *J. Chem. Soc. A*, 2030
153. Kuhn, S. J. (1967). *Can. J. Chem.*, **45**, 3207
154a. Lockhardt, W. L., Jones, M. M. and Johnston, D. O. (1969). *J. Inorg. Nucl. Chem.*, **31**, 407
154b. Gebala, A. E., Johnson, W. L. and Jones, M. M. (1969). *J. Inorg. Nucl. Chem.*, **31**, 3495
155. Dunken, H., Haase, W. and Schoenherr, F. (1967). *Z. Chem.*, **7**, (a) 399. (b) 422
156. Kolditz, L. and Krause, H. P. (1967). *Z. Chem.*, **7**, 157
157. Kolditz, L. and Gitter, M. (1967). (a) *Z. Chem.*, **7**, 202; (b) 240; (c) *Z. Anorg. Allg. Chem.*, **354**, 15
158. Griffith, W. P. (1969). *J. Chem. Soc. A*, 211
159. Chan, S. S. and Willis, C. J. (1968). *Can. J. Chem.*, **46**, 1237
160. Beattie, I. R. and Livingstone, K. M. (1969). *J. Chem. Soc. A*, 859
161. Brinkman, F. J. J., Gerding, H. and Olie, K. (1969). *Rec. Trav. Chim. Pays Bas*, **88**, 1358
162. Beattie, I. R., Gilson, T., Livingstone, K. M., Fawcett, V. and Ozin, G. A. (1967). *J. Chem. Soc. A*, 712
163. DiLorenzo, J. V. and Schneider, R. F. (a) (1967) *Inorg. Chem.*, **6**, 766. (b) (1968) *J. Phys. Chem.*, **72**, 761
164. Wieker, W. and Grimmer, A. R. (1967). *Z. Naturforsch.*, **22b**, 983
165. Brinkmann, F. J. J. and Gerding, H. (1970). *Rev. Chim. Minerale*, **7**, 729
166. Webster, M. and Deveney, M. J. (1968). *J. Chem. Soc. A*, 2166
167. O'Brien, M. H., Doak, G. O. and Long, G. G. (1967). *Inorg. Chim. Acta*, **1**, 34
168. Moreland, C. C., O'Brien, M. H., Douthit, C. E. and Long, G. G. (1968). *Inorg. Chem.*, **7**, 834
169. Mackay, K. M., Sowerby, D. B. and Young, W. C. (1968). *Spect. Acta*, **24A**, 611
170. Banister, A. J. and Moore, L. F. (1968). *J. Chem. Soc. A*, 1137
171. Harris, G. S. and Ali, M. F. (1968). *Inorg. Nucl. Chem. Lett.*, **4**, 5
172. Appel, R. and Rebhan, D. (1969). *Chem. Ber.*, **102**, 3955
173. Usacheva, G. M. and Kamai, G. (1970). *Izv. Akad. Nauk. SSSR. Ser. Khim.*, 1432
174. Ali, M. F. and Harris, G. S. (1969). *Inorg. Nucl. Chem. Lett.*, **5**, 701
175. Harris, G. S. and Ingles, F. (1967). *J. Chem. Soc. A*, 497
176. Schmidt, A. (1969). *Chem. Ber.*, **102**, 380
177. Kepert, D. L. and Mandyczewsky, R. (1968). *J. Chem. Soc. A*, 530
178. Choplin, F. and Kaufmann, G. (1970). *Spectrochim. Acta*, **26A**. 2113
179. Merijanian, A. and Zingaro, R. A. (1966). *Inorg. Chem.*, **5**, 187
180a. Durand, M. and Laurent, J. P. (1969). *Bull. Soc. Chim. Fr.*, 48.
180b. Laurent, J. P., Durand, M. and Gallais, F. (1967). *Compt. Rend. Acad. Sci. (Paris)*, **264C**, 1005
181. Shagidullin, R. R., Lamanova, I. A. and Urazgildeeva, A. K. (1967). *Dokl. Akad. Nauk. SSSR*, **174**, 1359
182. Ferguson, G. and Macauley, E. W. (1968). *Chem. Commun.*, 1288
183. Hass, D. (1968). *Z. Chem.*, **8**, 150
184. Hass, D. and Cech, I. (1969). *Z. Chem.*, **9**, 384
185. Mislow, K. and Casey, J. P. (1970). *Chem. Commun.*, 1410

186. Goldwhite, H. (1970). *Chem. Commun.*, 651
187. Dahlmann, J. and Austenat, L. (1970). *J. Prakt. Chem.*, **312**, 10
188. Hubin, R. and Tarte, P. (1967). *Spectrochim. Acta*, **23A**, 1815
189. Griffith, W. P. (1967). *J. Chem. Soc. A*, 905
190. Marsmann, H. C. and Van Wazer, J. R. (1970). *J. Amer. Chem. Soc.*, **92**, 3969
191. Beattie, I. R., Livingstone, K. M., Ozin, G. A. and Reynolds, D. J. (1970). *J. Chem. Soc. A*, 449
192. Casey, J. P. and Mislow, K. (1970). *Chem. Commun.*, 999
193a. Thilo, E. (1968). *Rev. Chim. Minerale*, **5**, 179
193b. Ladwig, G. and Grunze, H. (1968). *Monatsber. Deut. Akad. Wiss. Berlin*, **10**, 120
194. Worzala, H. (1968). *Acta. Crystallogr.*, **24B**, 987
195. Zhukov, A. P., Rez, I. S., Pakhomov, V. I. and Semin, G. K. (1968). *Phys. Status Solidi*, **27**, K129
196. Hill, R. M. and Ichiki, S. K. (1968). *J. Chem. Phys.*, **48**, 838
197. Smith, M. R., Zingaro, R. A. and Meyers, E. A. (1969). *J. Organometal. Chem.*, **20**, 105
198. Poulsen, J. J. and Calvo, C. (1968). *Can. J. Chem.*, **46**, 917
199. Subramanian, S. and Symons, M. C. R. (1970). *J. Chem. Soc. A*, 2367
200. Secco, F., Indelli, A. and Bonora, P. C. (1970). *Inorg. Chem.*, **9**, 337
201. Mason, J. G., Kowalak, A. D. and Tuggle, R. M. (1970). *Inorg. Chem.*, **9**, 847
202. Curtis, J. D. and Bruckenstein, S. (1968). *J. Amer. Chem. Soc.*, **90**, 6592
203. Krishna, B. and Singh, H. S. (1969). *J. Inorg. Nucl. Chem.*, **31**, 2964
204. Loehr, T. M. and Plane, R. A. (1968). *Inorg. Chem.*, **7**, 1708
205. Szymanski, H. A., Marabella, L., Hoke, J. and Harter, J. (1968). *Appl. Spectrosc.*, **227**, 29
206. Banks, E., Greenblatt, M. and Schwartz, R. W. (1968). *Inorg. Chem.*, **7**, 1230
207. Ladwig, G. (1968). *Monatsber. Deut. Akad. Wiss. Berlin*, **10**, 127
208. Dostal, K. and Kocman, V. (1969). *Z. Anorg. Allg. Chem.*, **367**, 92
209. Titora, Z. M. and Sharova, A. K. (1970). *Tr. Inst. Khim. Akad. Nauk. SSSR. Ural. Filial*, 120. (*Chem. Abs.* (1970), **73**, 124167m)
210. Towne, C., Tourne, G., Malik, S. A. and Weakley, T. J. R. (1970). *J. Inorg. Nucl. Chem.*, **32**, 3875
211. Preyssler, C. (1970). *Bull. Soc. Chim. Fr.*, 37
212. Polotebnova, N. A. and Radal, K. K. (1970). *Zh. Neorg. Khim*, **15**, 3149
213. Inoue, H. and Ito, H. (1969). *Nipon Kagaku Zasshi*, **90**, 193. (*Chem. Abstr.* (1969), **70**, 102556y)
214. Ronis, M. (1970). *Compt. Rend. Acad. Sci. (Paris)*, **271C**, 64
215. Krause, A., Domka, F. and Marciniec, B. (1969). *Rocz. Chem.*, **43**, 437
216. Whitfield, H. J. (1970). *J. Chem. Soc. A*, 1800
217. Vaipolin, A. A. (1970). *Zh. Strukt. Khim.*, **11**, 484
218. Dembrovskii, S. A. and Chernov, A. P. (1968). *Izv. Akad. Nauk. SSSR. Neorg Mater.*, **4**, 1229
219. Pen'kov, I. N. and Safin, I. A. (1968). *Kristallografia*, **13**, 330
220. Kravchenko, E. A., Dembovskii, S. A., Chernov, A. P. and Semin, G. K. (1969). *Phys. Status. Solidi*, **31**, K19
221. Schenermann, W. and Ritter, G. J. (1969). *Z. Naturforsch.*, **24a**, 409
222. Lauer, W., Becke-Goehring, M. and Sommer, K. (1969). *Z. Anorg. Allg. Chem.*, **371**, 193
223. Guerin, H. and Palazzi, M. (1967). *Bull. Soc. Chim. Fr.*, 1102
224. Müller, A., Diemann, E. and Leroy, M. J. F. (1970). *Z. Anorg. Allg. Chem.*, **372**, 113
225. Fridman, Ya. D. and Mikhailyuk, L. Ya. (1970). *Zh. Neorg. Khim.*, **15**, 3050
226. Thilo, E., Herzog, K. and Winkler, A. (1970). *Z. Anorg. Allg. Chem.*, **373**, 111
227. Remy, F. and Guerin, H. (1968). *Bull. Soc. Chem. Fr.*, 2327
228. Golovei, M. I., Semrad, E. E. and Luzhnaya, N. P. (1969). *Zh. Neorg. Khim.*, **14**, 2932
229. Kado, M., Maeda, T. (1966). *Jap. Patent* (a) 4969; (b) 4968
230a. Levskaya, G. S., Matyukhina, E. N., Kalutskii, L. A., Kolomiets, A. F., Bliznyuk, N. K. (1967). USSR Patent 196,830;
230b. Levskaya, G. S., Kalutskii, L. A., Matyukhina, E. N., Kolomiets, A. F., Bliznyuk, N. K. (1967). USSR Patent 196,816
230c. Levskaya, G. S., Matyukhina, E. N., Kolomiets, A. F., Bliznyuk, N. K., Vaoshavskii, J. L. (1967). USSR Patent 196,823..
231. Bode, H. and Hachmann, K. (1968). *Fresenius Z. Anal. Chem.*, **241**, 18
232. Manoussakis, G. E. and Karayannidis, P. (1969). *J. Inorg. Nucl. Chem.*, **31**, 2978

233. Fluck, E. and Jakobsen, G. (1969). *Z. Anorg. Allg. Chem.,* **369,** 178
234. Pilipenko, A. T. and Mel'nikova, N. V. (1970). *Ukr. Khim. Zh.,* **36,** 671
235. Colapietro, M., Domenicano, A., Scaramuzza, L. and Vaciago, A. (1968). *Chem. Commun.,* 302
236. Bally, R. (1970). *Compt. Rend. Acad. Sci. Paris,* **271C,** 1436
237. Roesky, H. W. and Dietl, M. (1970). *Z. Anorg. Allg. Chem.,* **376,** 230
238. Lucas, C. R. and Peach, M. E. (1970). *Can. J. Chem.,* **48,** 1869
239. Kamai, G. Kh., Chadaeva, N. A. and Mamakov, K. A. (1970). *Izv. Akad. Nauk. SSSR. Ser. Khim.,* 1092
240. Chadaeva, N. A., Kamai, G. and Mamakov, K. A. (1970). *Izv. Akad. Nauk. SSSR. Ser. Khim.,* 1640
241. Anderson, R. H. and Cragg, R. H. (1970). *Chem. Commun.,* 425
242. Sommer, K. (1970). *Z. Anorg. Allg. Chem.,* **376,** 150
243. Förster, M., Hertel, H. and Kuchen, W. (1970). *Angew. Chem. Int. Edn.,* **9,** 811
244. Wannagat, U., Bogusch, E. and Braun, R. (1969). *J. Organometal. Chem.,* **19,** 367
245. Wannagat, U. and Rabet, F. (1970). *Inorg. Nucl. Chem. Lett.,* **6,** 155
246. Scherer, O. J. and Janssen, W. M. (1969). *J. Organometal. Chem.,* **20,** 111
247. Scherer, O. J. and Janssen, W. M. (1969). *J. Organometal. Chem.,* **16,** P69
248. Hass, D. and Cech, I. (1969). *Z. Chem.,* **9,** 456
249. Peterson, L. K. and The, K. I. (1969). *Can. J. Chem.,* **47,** 339
250. George, T. A. and Lappert, M. F. (1969). *J. Chem. Soc. A,* 992
251. Hass, D. and Cech, I. (1969). *Z. Chem.,* **9,** 432
252. Schmitz-Dumont, O. and Ross, B. (1967). *Z. Anorg. Allg. Chem.,* **349,** 328
253. Ang, H. G. and Emeléus, H. J. (1968). *J. Chem. Soc. A,* 1334
254. Ang, H. G. (1969). *J. Inorg. Nucl. Chem.,* **31,** 3311
255a. Sisler, H. H. and Jain, S. R. (1968). *Inorg. Chem.,* **7,** 104
255b. Jain, S. R. and Sisler, H. H. (1969). *Inorg. Chem.,* **8,** 1243
256. Ang, H. G., Manoussakis, G. and El-Nigumi, Y. O. (1968). *J. Inorg. Nucl. Chem.,* **30,** 1715
257. Froeyen, P. (1969). *Acta. Chem. Scand.,* **23,** 2935
258. Revitt, D. M., Sowerby, D. B. (1969). *Inorg. Nucl. Chem., Lett.,* **5,** 459
259. Schmidt, A. (1968). *Chem. Ber.,* **101,** (a) 3976; (b) 4015
260. Wizeman, T., Müller, H., Seybold, D. and Dehnicke, K. (1969). *J. Organometal. Chem.,* **20,** 211
261. Cowley, A. H. and Schweiger, J. R. (1970). *Chem. Commun.,* 1492
262. Cowley, A. H., Dewar, M. J. S., Jackson, W. R. and Jennings, W. B. (1970). *J. Amer. Chem. Soc.,* **92,** 5206
263. Scherer, O. J. and Janssen, W. (1970). *Chem. Ber.,* **103,** 2784
264. Sommer, K. and Lauer, W. (1970). *Z. Anorg. Allg. Chem.,* **378,** 310
265. El-Nigumi, Y. O. and Emeléus, H. J. (1970). *J. Inorg. Nucl. Chem.,* **32,** 3213
266. Manoussakis, G. E. (1968). *J. Inorg. Nucl. Chem.,* **30,** 3100
267. Koketsu, J. and Ishii, Y. (1971). *J. Chem. Soc. C,* 2
268. MacCordick, J. and Appel, R. (1969). *Z. Naturforsch.,* **24b,** 938
269. Weingarten, H. and Miles, M. G. (1968). *J. Inorg. Nucl. Chem.,* **30,** 668
270. West, B. O. (1969). *Rec. Chem. Progr.,* **94,** 467
271. Elmes, P. S., Middleton, S. and West, B. O. (1970). *Aust. J. Chem.,* **23,** 1559
272. Knoll, F., Marsmann, H. C. and Van Wazer, J. R. (1969). *J. Amer. Chem. Soc.,* **91,** 4986
273. Krannich, L. K. and Sisler, H. H. (1969). *Inorg. Chem.,* **8,** 1032
274. Bogolyubov, G. M. and Petrov, A. A. (1970). *Zh. Obshch. Khim.,* **40,** 1759
275. Abel, E. W. and Illingworth, S. M. (1969). *J. Chem. Soc. A,* 1094
276. Abel, E. W., Hönigschmid-Grossich, R. and Illingworth, S. M. (1968). *J. Chem. Soc. A,* 2623
277. Cavell, R. G. and Dobbie, R. C. (1968). *J. Chem. Soc. A,* 1406
278. Schumann, H., Roth, A. and Stelzer, O. (1968). *Angew. Chem. Int. Edn.,* **7,** 218
279. Cowley, A. H. and Dierdorf, D. S. (1969). *J. Amer. Chem. Soc.,* **91,** 6609
280. Abel, E. W. and Crow, J. P. (1969). *J. Organometal. Chem.,* **17,** 337
281. Lane, A. P. and Burg, A. B. (1967). *J. Amer. Chem. Soc.,* **89,** 1040
282. Ellerman, J., Uller, W. (1970). *Z. Naturforsch.,* **25b,** 1180
283a. Todd, L. J., Burke, A. R., Silverstein, H. T., Little, J. L. and Wikholm, G. S. (1969). *J. Amer. Chem. Soc.,* **91,** 3376
283b. Todd, L. J., Burke, A. R., Garber, A. R., Silverstein, H. T. and Stornhoff, B. N. (1970). *Inorg. Chem.,* **9,** 2175

284. Glidewell, C. and Sheldrick, G. M. (1969). *J. Chem. Soc. A*, 350
285a. Drake, J. E. and Anderson, J. W. (1970). *J. Chem. Soc. A*, 3131
285b. Anderson, J. W. and Drake, J. E. (1969). *Inorg. Nucl. Chem. Lett.*, **5,** 887
286a. Abel, E. W. and Illingworth, S. M. (1970). *Organometal. Chem. Rev. A.* **5,** 143
286b. Drake, J. E. and Riddle, C. (1970). *Quart. Rev.*, **24,** 263
287. Schumann, H. and Roth, A. (1969). *Chem. Ber.*, **102,** 3713
288. Drake, J. E. and Riddle, C. (1968). *J. Chem. Soc. A*, 2452
289. Ebsworth, E. A. V., Rankin, D. W. H. and Sheldrick, G. M. (1968). *J. Chem. Soc. A*, 2828
290. McKean, D. C. (1968). *Spectrochim. Acta.*, **24A,** 1253
291. Beagley, B., Robiette, A. G. and Sheldrick, G. M. (1968). *J. Chem. Soc. A*, 3006
292a. Drake, J. E. and Simpson, J. (1968). *Spectrochim. Acta*, **24A,** 981
292b. Drake, J. E. and Riddle, C. (1970). *Spectrochim. Acta*, **26A,** 1697
293. Drake, J. E. and Riddle, C. (1969). *Inorg. Chim. Acta*, **3,** 136
294. Bürger, H., Götze, U. and Sawodny, W. (1970). *Spectrochim. Acta A*, **26,** 671
295. Bürger, H. and Götze, U. (1968). *J. Organometal. Chem.*, **12,** 451
296. Drake, J. E. and Riddle, C. (1968). *J. Chem. Soc. A*, 2709
297. Schumann, H., Roth, A. (1969). *Chem. Ber.*, **102,** 3725
298. Kenworthy, J. G., Myatt, J. and Todd, P. F. (1970). *J. Chem. Soc. B*, 791
299. Bürger, H. and Neese, H. J. (1970). *Inorg. Nucl. Chem. Lett.*, **6,** 299
300. Schrauzer, G. N. and Kratel, G. (1969). *Chem. Ber.*, **102,** 2392
301a. Singh, B. T., Pandey, A. N. and Singh, H. S. (1970). *Indian J. Pure and Appl. Phys.*, **8,** 193
301b. Rao, D. V. R., Thankur, S. N. and Rai, D. K. (1970). *Proc. Indian Acad. Sci. (Sect. A),* **71,** 42
302. Baudler, M. and Stassen, H. J. (1966). *Z. Anorg. Allg. Chem.*, **343,** 244

8
Inorganic Selenium Chemistry

J. W. GEORGE
University of Massachusetts

8.1 INTRODUCTION

Demarcation of the subject of this chapter from the closely related fields of (a) research on elementary selenium and metal selenides having a solid-state philosophy and orientation, (b) organoselenium chemistry, and (c) systematic x-ray crystallographic study of a wide variety of selenium-containing compounds is not readily accomplished. Nevertheless, in attempting to provide a modestly broad and representative description of recent developments in the inorganic chemistry of selenium it became necessary, for the most part, to exclude discussion of the many significant advances which have occurred in these allied areas. A good general account of selenium chemistry has been provided in a monograph[1], and another recent volume[2] gives detailed summaries of characteristics of elementary selenium and the selenides important to the solid-state physicist. A review of recent activity in organoselenium chemistry[3] is also available. The present article, organised in a classical way around the several types of selenium compounds, attempts to reflect the nature of some of the recent investigations in inorganic selenium chemistry and to summarise the results thereof.

8.2 CHEMISTRY OF HYDROGEN SELENIDE

As an important source for the introduction of selenium into chemical species, hydrogen selenide and its reaction chemistry have continued to be of interest. An 84% yield of digermyl selenide is obtained from a room-temperature, sealed-tube reaction of H_2Se and digermylcarbodi-imide[4]. The colorless liquid product has been fully characterised by mass, i.r., Raman, and n.m.r. spectra, and assignments of the vibrational spectroscopic data have been made. Both HI and HCl react completely with $(GeH_3)_2Se$ at room temperature within a few minutes to give H_2Se and the corresponding germyl halide. The reaction of disilyl selenide with germyl bromide also gives $(GeH_3)_2Se$ in high yield.

The cleavage of Ge—P and Ge—As bonds in germyl phosphine and germyl arsine by H_2Se has been studied[5]. The reactants were condensed in an n.m.r. tube at $-196°C$, allowed to warm to room temperature, and the ^1H n.m.r. spectrum was recorded. These spectra indicate the following reaction sequence:

$$2GeH_3MH_2 + 2H_2Se \rightarrow 2GeH_3SeH + 2MH_3 \qquad (M = P\ or\ As)$$
$$2GeH_3SeH \rightleftharpoons (GeH_3)_2Se + H_2Se$$
and, $2GeH_3MH_2 + H_2Se \rightarrow (GeH_3)_2Se + 2MH_3$

The reaction is rapid with germyl phosphine, being completed in about one hour, but is much slower with the arsine. After several hours standing there is indication of condensation of digermyl selenide yielding $(GeH_3Se)_n$ GeH_{4-n}, where $n = 2, 3$ or 4.

Rather small yields of boron selenohalides(BSeX) are obtained from the reaction at $50°C$ in CS_2, or at $100°C$ in n-heptane, of H_2Se and BCl_3, BBr_3 or BI_3 [6]. The BSeX products are white, microcrystalline solids and are insoluble in many common organic solvents but reactive toward alcohols, acetone, trimethylamine, dimethylformamide and dimethylsulphoxide. Cryoscopic measurements were not feasible, and thus comparison of the degree of association of BSeX with the trimeric sulphur analogues could not be made. Thermal degradation of BSeX, beginning slowly at $c.$ 170 °C but becoming very rapid at 350 °C, yields B_2Se_3 and BX_3. Infrared data and x-ray powder patterns are also provided.

Difluoroiodophosphine reacts smoothly with H_2Se in the presence of mercury at room temperature[7]:

$$PF_2I + H_2Se \xrightarrow{Hg} PF_2HSe + HI$$

Minor products included PF_3, SiF_4 and H_2. The ^{31}P, ^{19}F, and ^{1}H n.m.r. spectra, and the i.r. and mass spectra, are in accord with a trigonal-bipyramidal configuration in which Se and H atoms, and the lone electron pair, occupy equatorial positions around the phosphorus atom. In contrast to PF_2HO and PF_2HS which have high Trouton constants implying association, the Trouton constant for PF_2HSe is nearly normal. Nor are there any unusual features in the i.r. spectroscopic data which suggest association. The relatively low degree of association, if any, of the selenide may explain the greater stability to thermal decomposition compared to PF_2HO. The latter gives PF_3 and $HFPO_2H$ as principal decomposition products, presumably as a result of its —P—O—P—O— chain-like structure. The selenide, after standing at 25 °C for 3 months in a glass container, was 84% decomposed to SiF_4 and an unidentified insoluble red-grey solid.

The reaction of trisilylamine and excess H_2Se under pressure at room temperature proceeds according to[8]:

$$(SiH_3)_3N + H_2Se \rightarrow (SiH_3)_2Se + NH_4^+(SeSiH_3)^-$$

The white crystalline reaction product gave an i.r. spectrum appropriate to the presence of NH_4^+ and $SeSiH_3^-$ ions. The formation of this salt which is stable at room temperature under these conditions suggests that $HSeSiH_3$ is a stronger acid than H_2Se. Reaction of the white solid with methyl iodide yields CH_3SeSiH_3, which was identified by i.r. and n.m.r. spectra.

8.3 OXIDES AND OXYACIDS

8.3.1 Structure and physical properties

Analysis of isotopic shifts appearing in the i.r. spectrum of matrix-isolated SeO_2 results in a proposed bond angle of 110 ± 2 degrees[9]. Since this value differs only slightly from a microwave-determined value (113.5 degrees) it appears that matrix perturbations are relatively small. The presence of a dimeric SeO_2 species is indicated in the vapour phase over molten SeO_2.

A recent study of the 220–600 nm spectral region of SeO_2 at 20 °K focused on the detailed optical properties of the molecule, and has provided correlations and assignments for certain weak and diffuse gas-phase transitions[10].

Thermodynamic investigation of the sublimation of SeO_2, both pure and with excess selenium or oxygen, has given values for the standard enthalpy and entropy of formation for the compound, and has suggested that gaseous SeO_3 and SeO are unstable in the 200–320 °C range[11].

The electrical-conductance characteristics of selenic acid have continued to receive attention. The conductance of pure H_2SeO_4 is higher than that of any other liquid except some fused salts and perhaps HNO_3 [12]. Basic solutes, B, such as H_2O, HNO_3, N_2O_3, and N_2O_4 increase the specific conductance of H_2SeO_4 in proportion to the concentration of $HSeO_4^-$ ion[13, 14]:

$$B + H_2SeO_4 \rightarrow BH^+ + HSeO_4^-$$

Transport number measurements suggest $HSeO_4^-$ to be the chief current-carrying species in H_2SeO_4 solution. At higher concentrations of HNO_3 or nitrogen oxides the specific conductance changes in such a way as to indicate an upper limit for normal conductance behaviour in H_2SeO_4 at an $HSeO_4^-$ concentration of c. 1.5 M. In aqueous H_2SeO_4 the observed decrease in specific conductance with increasing electrolyte concentration is interpreted as indicative of a disruption in the normal chain structure within which conduction occurs via a proton-jump mechanism.

Density, viscosity and electrical conductance measurements on H_2SeO_4–H_2SO_4 solutions provided no evidence of interaction of these substances[15], but transport-number determinations suggest a weak acid–base reaction to give $H_3SeO_4^+$ and HSO_4^- ions. With phosphoric acid a 1 : 1 addition product of H_2SeO_4 is clearly indicated by the existence of maxima for relative temperature coefficients of viscosity and electrical conductance, and by the appearance of i.r. spectral absorption bands at 750 and 880 cm^{-1} associated with $HSeO_4^-$, and having their maximum intensity in the 1 : 1 composition region. With H_2SeO_4–HNO_3 solutions observations of a similar nature suggest the existence of 1 : 1 and 1 : 2 addition compounds.

8.3.2   Chemistry of the di- and tri-oxide

In a stream of gaseous PF_3 at 250°C SeO_2 is reduced to elementary Se [16]. Mass spectra of the gaseous products indicated m/e values in the range expected for various isotopic species of $PSeF_3$. It was also shown that this compound results from the direct reaction of Se and PF_3.

Direct reaction of SeO_3 with alkali-metal fluorides at elevated temperatures yields fluoroselenate salts[17, 18]. For example:

$$SeO_3 + CsF \xrightarrow{80\,°C} CsSeO_3F$$

Gaseous SeO_2F_2 and solid K_2SeO_4 were minor products. Fluoroselenates are also easily prepared in liquid SO_2 or liquid HF solution[18]. These alkali metal fluoroselenates decompose just above 200 °C and are of lower thermal stability than the corresponding fluorosulphates.

Nitromethane solutions of N_2O_4 and SeO_3 yield $(NO)_2Se_2O_7$ and, at low temperatures, a substance of empirical composition $N_2O_4 \cdot 3SeO_3$ [19]. X-ray powder patterns and i.r. spectra suggest this substance to be an equimolar mixture of $(NO)_2Se_3O_{10}$ and $(NO_2)_2Se_3O_{10}$. Since spectroscopic and cryoscopic measurements on nitromethane solutions of SeO_3 are in accord with the tetrameric nature of SeO_3 it is proposed that the above products result from a nucleophilic attack of N_2O_4 on the $(SeO_3)_4$ ring with the formation of $Se_4O_{13}^{2-}$, NO^+, and NO_2^+ ions. Subsequent decomposition of $Se_4O_{13}^{2-}$ yields $Se_2O_7^{2-}$ and/or $Se_3O_{10}^{2-}$ depending on the amount of N_2O_4 present.

The nitrosylselenate, $(NO)_2Se_3O_{10}$, also results from the reaction of SeO_3 and $(NO)_2Se_2O_7$ or N_2O_3 [20]. This product, as well as $NOHSe_2O_7$ from the reaction of SeO_3 and $NOHSeO_4$, was characterised by x-ray powder and i.r. spectroscopic data.

Similar experimental techniques applied to the 1 : 1 adduct of trimethylamine and SeO_3, prepared by reaction of the pyridine adduct of SeO_3 with $(CH_3)_3N$ in acetonitrile solution, result in data suggestive of the retention of the tetrameric SeO_3 ring structure in the complex[21]. No molecular-weight determinations were made, however.

The products of SeO_3 reactions with elementary sulphur and selenium have also been reported[22].

8.4 INTERCHALCOGENS

8.4.1 Selenium–sulphur compounds

The existence and stability of various interchalcogen compounds of selenium has been verified through recent research on selenium–sulphur and selenium–tellurium systems. Differential spectrophotometry of vapour-phase Se—S and Se—Te mixtures in the 400–1000 °C range gives absorptions at 310 and 370 nm respectively[23]. Since neither absorption is present in the spectra of the pure components the presence of SeS and SeTe compounds is proposed. Because the absorption maximum for SeS is essentially unchanged in the 700–1000 °C range considerable stability for the mixed species is indicated, but an observed absorption intensity decrease for the SeTe system around 900 °C supports some degree of dissociation.

Mass spectrometry, applied to the crystalline products obtained from the CCl_4 extract of Se and S mixtures (atomic ratios 1 : 1 to 1 : 3) which had been heated to 1000 °C and rapidly cooled, gave peaks associated with all members of the octatomic Se_nS_{8-n} series[24]. It is to be noted that the identified species are not necessarily unique since ring isomers are possible where more than one Se atom is involved. Thus, five isomers are possible for the Se_3S_5 ring, but those rings with adjacent Se atoms may be inherently less stable due to weakness of the Se—Se bond relative to the Se—S linkage.

Another study of Se—S mixtures applied a fractional-crystallisation procedure to a benzene extract of a chalcogen mixture which had been heated to 250–300 °C [25]. The solid products were examined by mass and absorption spectrophotometry, and by x-ray powder photography. Again, all species in the Se_nS_{8-n} series were found, but the possibility of using the fractional crystallisation technique for separation of unique molecular species was viewed pessimistically. The diffraction data show two distinct ranges of homogeneity, the first up to 14% Se, and the other in the range 14–50% Se. The compound SeS_7, isomorphous with orthorhombic sulphur, lies in the first region, and the latter range includes Se_2S_6, Se_3S_5 and Se_4S_4. The mass spectra, however, suggest that either range will contain small amounts of species from the other. Fragmentation and/or recombination processes could be responsible for this latter observation, however.

Chemical procedures for preparing certain members of the Se_nS_{8-n} series have been reported. Excess S_2Cl_2 reacts with H_2S_3 to give S_7Cl_2 [26]. The subsequent mixing of CS_2 solutions of H_2Se and S_7Cl_2 in diethyl ether leads to yellow crystals which, after recrystallisation from a benzene–carbon disulphide solution, analyse as $SeS_{7.4}$. The mass spectrum of the solid shows that SeS_7 is the most strongly represented mixed octatomic species, but the observation that the volatility of S_8 molecules from this material is greater than that from orthorhombic sulphur suggests the formation of S_8 from fragmentation and recombination processes involving SeS_7. Dimerisation of SeS_7 and elimination of S_2 is proposed as a source of $Se_2S_6^+$ peaks in the mass spectrum; molecules containing more than two Se atoms could be formed, in turn, from Se_2S_6.

With sulphur monochloride H_2Se reacts to give the eight-membered ring-compounds Se_2S_6 and Se_3S_5 in $1:2$ ratio[27]. Separation of these products was achieved through solubility differences. The i.r. spectral data gave no indication of Se—Se bonds in these substances. No six-membered ring species were obtained.

Laser Raman spectra of Se—S mixtures of $1:2$ and $1:19$ atom ratio at several temperatures show frequencies not associated with either pure Se or S, thus supporting further the existence of discrete Se_nS_{8-n} species[28].

8.4.2 Selenium–tellurium compounds

An equimolar mixture of selenium and tellurium was heated to 1000 °C, annealed for several days at 35 °C, and then extracted with CS_2 [29]. The crystalline product, in contrast to the Se—S system, gave no direct mass-spectral evidence for mixed octatomic species. However, a low-intensity peak corresponding to Se_6Te^+ may be due to cracking of a parent Se_7Te molecule, although fragmentation of some highly polymeric species is not ruled out. The high intensity of Se_5Te^+ and Se_6^+ peaks suggests that hexatomic species are stable in the vapour phase above the reaction product. No peaks corresponding to multi-tellurium-substituted molecules were observed.

8.5 HALIDES

8.5.1 Mono- and di-halides

8.5.1.1 Structure

A recent vibrational spectroscopic study, with particular emphasis on Laser Raman techniques in the examination of the darkly coloured, moisture-sensitive, and somewhat unstable Se_2Cl_2 and Se_2Br_2, has confirmed the symmetry of these species[30-32]. The appearance of several fundamentals in both the i.r. and Raman spectra rules out the trans-planar (C_{2h}) model to which the rule of mutual exclusion applies, and the i.r. activity and the polarisation of the torsion mode eliminates the *cis*-planar (C_{2v}) arrangement. General agreement has been reached on assignments for the *gauche* (C_2) symmetry, and a supporting normal coordinate analysis has been carried out[32].

8.5.1.2 Equilibrium studies

Some details concerning the equilibrium behaviour of the mono- and di-halides of selenium have been revealed from solubility, vapour pressure, and electrical-conductance studies of selenium–chlorine[33-35] and selenium–bromine[36] systems. Thus, for the system represented by the equation

$$Se_2Cl_2(l) \rightleftharpoons SeCl_2(g) + Se(s)$$

$\Delta H = 16.7$ kcal in the range 20–75 °C, and $\log K_{eq} = -1.5$ at 75 °C. Other equilibria for which $\log K_{eq}$ values at 75 °C were deduced include:

$$SeCl_4(s) \rightleftharpoons SeCl_2(g) + Cl_2(g); \log K_{eq} = -6.15$$
$$Se_2Cl_2(l) + Cl_2(g) \rightleftharpoons 2SeCl_2(g); \log K_{eq} = +3.57$$
$$Se(s) + Cl_2(g) \rightleftharpoons SeCl_2(g); \log K_{eq} = +5.0$$

The conductance behaviour of an equilibrium system of liquid selenium and gaseous SeX_2 at 280 °C suggests the presence of four ions per formula unit of SeX_2. Thus:

$$Se(l) + SeX_2(g) \rightleftharpoons 2Se^+ + 2X^-$$

It should be noted that these ion formulations do not reflect solvation effects; Se_n^+ and Se_nX^- are probably more representative of the ionic species. For the Se–Cl system, $\log K$ for the above reaction is approximately -5. For both chlorine and bromine systems the dependance of the total halogen concentration in the liquid phase on the partial pressure of $SeX_2(g)$ implies the dominance of a second reaction:

$$Se(l) + SeX_2(g) \rightleftharpoons 2SeX(l)$$

For this reaction $\log K_{eq}$ for the Se–Cl system is -0.09, and for the Se–Br system $+0.89$

8.5.1.3 Chemical behaviour

Neither the dichloride nor dibromide is known to exist in the solid state but, as indicated above, these substances are stable in the gaseous state. In an attempt to prepare tetramethylthiourea(tmtu) complexes of methylselenium trichloride and tribromide in methylene chloride solution 1 : 1 adducts of $SeCl_2$ and $SeBr_2$ with the ligand resulted[37], presumably because of the reaction:

$$CH_3SeX_3 + tmtu \rightarrow SeX_2 \, (tmtu) + CH_3X \qquad (X = Cl \text{ or } Br)$$

[1]H n.m.r. and cryoscopic data are in accord with the presence of undissociated monomeric $SeX_2(tmtu)$ in solution, and i.r. spectral data suggest that

no major structural changes occur between solution and solid state. The i.r. spectral data also indicate the presence of an S—Se bond and support, in general, a T-shaped heavy-atom skeletal model for the molecule. Some intermolecular interaction through bromine bridging in the solid bromide complex is inferred from the i.r. spectrum and from the observation that the complex exhibits surprisingly high stability in the presence of moist air.

8.5.2 Tetrahalides

8.5.2.1 Structural aspects

Of the tetrahalides of selenium there is least ambiguity concerning the nature of the fluoride. I.R. spectral data for solid and vapour phases[38], and Raman spectra for the liquid phase[39], show the C_{2v} symmetry for the molecular species. Association in the condensed phases is indicated by the low frequency displacement and pronounced broadening of a fundamental relative to the vapour phase, and this is confirmed by i.r. spectral data obtained from the matrix-isolated species. It should be noted that x-ray crystallographic results have demonstrated the associated character of the corresponding fluoride of tellurium[40].

Structural parameters and dipole-moment data for the SeF_4 molecule have been obtained[41] from analysis of the microwave spectra of $^{78}SeF_4$ and $^{80}SeF_4$.

In contrast to the relatively well understood tetrafluoride the tetra-chloride and -bromide are currently the subject of some controversy regarding their nature in the solid state. In addition, the character of selenium-containing species present in non-aqueous solutions of these halides is not entirely clear. Vibrational spectra of solid $SeCl_4$, $SeBr_4$ and two mixed halides, $SeClBr_3$ and $SeCl_3Br$, support a molecular MX_4 species of approximate C_{2v} symmetry for the binary tetrahalides, but with one axial halogen at a significantly greater distance from the Se atom than the other[42]. This molecular configuration with one elongated Se—X bond is not totally distinct from an MX_3^+, X^- formulation found compatible with the i.r. spectrum of $SeCl_4$ [43, 44] and $SeBr_4$ [43] in other studies.

A recent criticism of the MX_3^+, X^- proposal for $SeCl_4$ has stressed the substantial differences between Se—Cl stretching frequencies identified for species which almost certainly contain $SeCl_3^+$ ions (e.g., solid $SeCl_3^+AsF_5Cl^-$, $SeCl_3^+SO_3Cl^-$, and HSO_3F solutions of $SeCl_4$), and those of $SeCl_4$ in the solid state[45]. Thus, v_1 and v_3 (C_{3v} symmetry) lie at c. 415–430 cm^{-1} and 395 cm^{-1} for the former, but several studies agree that in solid $SeCl_4$ the corresponding vibrations appear c. 50 cm^{-1} lower. This displacement, coupled with some degree of complexity of the solid $SeCl_4$ spectrum, raises serious doubts about the validity of the $SeCl_3^+$, Cl^- formulation.

Resolution of these uncertainties awaits further x-ray examination of these substances; earlier x-ray studies of $SeCl_4$ did not determine the detailed molecular structure[46, 47]. Similar ambiguities with regard to $TeCl_4$, with which $SeCl_4$ is isotypic, have been significantly reduced by the elucidation of its molecular structure[48]. The positional parameters indicate a cubane-like, tetrameric structure in which $TeCl_3^+$ units are combined via bridging

Cl^- ions. There are three short (2.32 Å) Te—Cl bonds ($TeCl_3^+$), and three much longer (2.93 Å) Te—Cl interactions ($Cl_3Te^+ \cdots\cdots Cl^-$) for each Te atom.

A gas-phase Raman study of the dissociation products of solid $SeCl_4$ has confirmed the presence of the $SeCl_2$ species, the spectrum of which is assigned on the basis of C_{2v} symmetry[49].

8.5.2.2 Solution characteristics

In a variety of non-aqueous solvents $SeCl_4$ shows molar conductances substantially less than those expected for a 1 : 1 electrolyte[50]. Of the solvents examined only in dimethylformamide(DMF) does $SeCl_4$ display a conductance value which suggests significant ion concentrations. It is not known whether this is the result of coordination effects of the solvent, although no DMF complexes of $SeCl_4$ have been isolated[51], or due to the presence of solvent impurities. I.R. spectroscopy is of little help due to strong solvent absorption above 280 cm^{-1}. Cryoscopic measurements with acetonitrile and tetramethylene sulphone solutions gave results in accord with either a monomolecular or an ion-pair formulation for the $SeCl_4$ solute[50]. The i.r. spectra of $SeCl_4$ in these latter solvents, and also in benzene, differ from that of crystalline $SeCl_4$ primarily in the broad, intense band at c. 280 cm^{-1} found in the solution spectra. If solid $SeCl_4$ is similar to $TeCl_4$ in containing MX_3^+ and X^- ions, the spectroscopic differences between solid and solution could be interpreted in favour of the presence of monomolecular $SeCl_4$ units in solution. A study of the concentration dependence of the solution spectrum would be of interest.

In liquid hydrogen chloride $SeCl_4$ forms a poorly conducting solution with which BCl_3 gives no reaction [52]. In chlorosulphuric acid, however, $SeCl_4$ behaves as a strong base[45]:

$$SeCl_4 + HSO_3Cl \rightarrow SeCl_3^+ + SO_3Cl^- + HCl$$

The solution behaviour of $SeBr_4$ is substantially different from that of its chloride counterpart. In parallel with an earlier observation[53] of 100% dissociation of $SeBr_4$ in CCl_4 solution to an equilibrium mixture of Se_2Br_2, $SeBr_2$ and Br_2, i.r. and electronic spectroscopic examination of $SeBr_4$ in various solvents gives results which indicate that these solutions are also equilibrium systems of the same solute components[54, 55]. In addition, electrical conductance data for acetonitrile, tetramethylene sulphone, and nitrobenzene solutions of $SeBr_4$ are quite similar to those of 1 : 3 mole ratio solutions of Se_2Br_2 and Br_2. These findings suggest that conductance behaviour of these solutions cannot be explained by the presence of Se^{IV} ions. The conductance of acetonitrile solutions of S_2Cl_2 and SCl_2, although weak, has been reported[56]. Addition of pyridine to these solutions greatly enhanced their specific conductances.

8.5.2.3 Addition compounds

The characteristics of several adducts of SeF_4 have been described recently. With NbF_5 at 106 °C, or with TaF_5 at room-temperature, 1 : 1 adducts with identical space-groups and unit-cell dimensions are formed[57]. Single-crystal x-ray study of the NbF_5 adduct indicates that it is $SeF_3^+NbF_6^-$ with substantial fluorine bridging between ions resulting in tetrameric units. At room temperature the SeF_4—NbF_5 reaction system yields a $1:2$ adduct, x-ray examination of which results in the formulation $SeF_3^+Nb_2F_{11}^-$, but again the positional parameters suggest important fluorine-bridging interactions[58]. Vibrational and n.m.r. spectroscopic, conductometric, and cryoscopic investigations support the fluorine-bridging proposal for the above $1:1$ adducts, and distinguish comparable bridged ionic formulations for solid $1:1$ adducts of SeF_4 with BF_3, SbF_5 and AsF_5 [59]. Further, the bridging appears to persist in the molten state and, to some degree, in nitrobenzene solutions. Similar measurements on the $SeF_4·SO_3$ adduct in the condensed aggregation states and in nitrobenzene solution yield data interpreted in terms of an oxygen-bridged polymeric fluorosulphate structure for solid and melt, but partially disrupted in dilute solution to yield primarily dimeric species[60]. No evidence for adduct formation between SeF_4 and AsF_3 has been found[61].

Many of the corresponding $SeCl_4$ addition compounds have also been studied. Further details, in support of an earlier report[62], concerning the vibrational spectrum of $SeCl_4·SO_3$ [63] permit an unambiguous identification of the $SeCl_3^+$ and SO_3Cl^- ions as the essential components of the adduct. Similar ionic formulations have been proposed for the $1:1$ adducts of $SeCl_4$ with $GaCl_3$ and $AlCl_3$ [64]. A surprising feature in the vibrational spectrum of the $SeCl_4·SbCl_5$ adduct is the replacement of a characteristic $SeCl_3^+$ vibration (v_3) at 390 cm^{-1}, the antisymmetric Se—Cl stretching mode, by an absorption at $c.$ 350 cm^{-1}. (Note, in connection with the previous comments on $SeCl_3^+$ on page 236, that v_1, the symmetric stretching mode of $SeCl_3^+$ appears at 412 cm^{-1} in the $SeCl_4·SbCl_5$ adduct spectrum, and it thus does *not* appear to be displaced to lower frequencies). The shift to lower frequency of v_3 may reflect significant chlorine bridging in this adduct.

Phase study of the $SeCl_4$—$NbCl_5$ and $SeCl_4$—$TaCl_5$ systems has resulted in the identification of $1:1$ addition compounds[65]. Differential thermal analysis also supports the existence of a $1:1$ $SeCl_4$—$MoCl_5$ adduct[66].

Lewis-acid characteristics are also displayed by $SeCl_4$. The most recent examination of tetrachlorobis(pyridine)selenium(IV) raised some doubts about the earlier formulation of this complex as $SeCl_3(py)_2^+Cl^-$. The Raman spectrum and x-ray powder data of the adduct suggest it to be isomorphous with *trans*-octahedral $SnCl_4·2py$ [67]. I.R. spectral, conductance and cryoscopic data are also in accord with a molecular complex, but the complexity of the i.r. spectrum is greater than would be expected for a *trans*-species[68].

Mono- and bis-bipyridyl complexes of $SeBr_4$ have been prepared by precipitation from ether solution[51]. Each shows a molar conductance in acetonitrile appropriate to a strong electrolyte, and the mono-complex, on conductometric titration with pyridine in acetonitrile solution, gives a titration curve showing a substantially enhanced molar-conductance at a $2:1$ mole ratio of pyridine to the Se compound. This corresponds to the formation of a dissolved complex, $SeBr_2(bipyr)(py)_2^{2+},2Br^-$.

The most recent vibrational spectroscopic study of hexahalogenoselenates, by a laser Raman study of solid salts and hydrochloric and hydrobromic acid solutions of SeO_2, and i.r. spectroscopy of the salts, is in accord with the regular octahedral structure of the SeX_6^{2-} species[69]. However, another i.r. spectral examination of $SeCl_6^{2-}$ and $SeBr_6^{2-}$ offers evidence for the occurrence of the ubiquitous Jahn–Teller effect via vibronic coupling between ground and first excited states[70]. Two vibrational bands of these ions are much broader and of lower frequency than the corresponding absorptions found for similar complexes unaffected by Jahn–Teller effects. In addition, the Jahn–Teller-active modes display unusual intensities in the Raman spectra of these ions.

Spectra of both K_2SeBr_6 and SeO_2 in aqueous HBr give a number of absorption lines in addition to those usually associated with $SeBr_6^{2-}$ [69]. Although hydrolysis might be suspected there are no absorption frequencies appropriate to an Se—O vibration in the spectra. Accordingly, it was proposed that the species $SeBr_5^-$ might be an equilibrium component of such solutions. Absorption spectra of Se^{IV} in hydrochloric acid solution have been interpreted[71] as indicating the presence of $SeCl_5(H_2O)^-$, or perhaps $Se_2Cl_{10}^{2-}$, in addition to $SeCl_6^{2-}$.

8.6 OXYHALIDES

8.6.1 Oxyfluorides

Following the development of a new synthetic method for fluoroxypenta-fluoroselenium, SeF_5OF, some aspects of the chemistry of this species have been investigated. As compared to the original preparative method[72], both yield and purity of SeF_5OF are significantly enhanced by the potassium fluoride catalysed fluorination of $SeOF_2$ at room temperature[73]. In addition, this reaction also gives small amounts of trans-bis(fluoroxy)tetrafluoro-selenium, $SeF_4(OF)_2$. Better yields of this substance are obtained, however, from the reaction of $KSeOF_2$ and F_2 at $-78\,°C$ [74]. Several reactions, involving insertion of a reactant in the O—F bond, are indicated below; evidence for the nature of the products was obtained by chemical analysis, vapour density measurements, and mass, i.r., and n.m.r. spectroscopy.

$$SeF_5OF + c\text{-}C_5F_8 \xrightarrow[48\text{ h}]{\text{Room temp.}} SeF_5OC_5F_9$$

$$SeF_5OF + SF_4 \xrightarrow[10\text{ days}]{\text{Room temp.}} SeF_5OSF_5$$

$$SeF_5OF + CO \xrightarrow[12\text{ h}]{65\,°C} SeF_5O\overset{\displaystyle O}{\overset{\displaystyle \|}{C}}F$$

A by-product of the last reaction is believed to be SeF_5OSeF_5, although the limited amount of the material separated by fractional distillation was inadequate for chemical analysis. No products containing the SeF_5O group were identified when SeF_5OF was mixed with N_2F_4, SO_2, C_2F_4, C_3F_6, or C_2H_4; in those instances involving fluorine-containing reactants most of the SeF_5OF is converted to SeF_6.

While several different interaction modes of $SeOF_2$ with Lewis acids, e.g., coordination via lone pairs on Se, O or F, or fluoride donation, may be envisaged, the only report of a crystallographic study of such an adduct, $SeOF_2 \cdot NbF_5$, indicates that the oxygen atom occupies the sixth position in an octahedral coordination arrangement about niobium[75]. It should be noted that oxygen cannot be distinguished from fluorine by x-ray methods but, by analogy with results for $SeOCl_2 \cdot SbCl_5$ (see below), a satisfactory refinement was achieved with oxygen in the donor position. The selenium coordination sphere may be treated as either a distorted tetrahedron of two fluorines (1.68 Å), an oxygen (1.60 Å), and the Se lone pair, or, with the inclusion of three fluorines at distances of 2.69, 2.69 and 2.88 Å, an octahedron which is sufficiently distorted to correlate with stereochemical activity of the lone pair.

8.6.2 Oxychlorides

Selenium oxychloride is still interesting with respect to the varied types of complexes which $SeOCl_2$ may form with appropriate reactant species; several x-ray crystallographic investigations have illuminated the diverse adduct-forming behaviour of this substance[76].

Although 1,4-selenoxane, C_4H_8OSe, has been found[77], by use of i.r. and n.m.r. spectral data, to coordinate to various metal halides through Se rather than oxygen, there are no known examples in which $SeOCl_2$ utilises a selenium lone-pair in forming a coordinated covalent bond to another species. Thus, the classical Lewis-base behaviour of $SeOCl_2$ is limited to those examples in which oxygen is the coordinating atom. Adducts in this group include $SeOCl_2 \cdot SbCl_5$ [78] and $SnCl_4 \cdot 2SeOCl_2$ [79]. In the $SbCl_5$ complex both crystallographically different Se atoms achieve a tetragonal pyramidal coordination via a weak chlorine-bridging interaction involving two chlorine atoms, one from each of the two nearest octahedrally coordinated antimony groups. It is interesting that in this pyramidal grouping around Se a chlorine atom occupies the apical position. In $SnCl_4 \cdot 2SeOCl_2$ the two coordinating oxygen atoms occupy *cis*-positions in the octahedral coordination sphere of tin; a distorted octahedral arrangement about Se results from interaction with three chlorine atoms, two from one nearest-neighbour tin group, and a third from the other nearest neighbour. In both the $SnCl_4$ and $SbCl_5$ complexes the metal–oxygen distances (2.16 and 2.08 Å, respectively) are comparable to those observed in similar complexes involving oxygen as the donor atom[80]. In addition, the weak chlorine-bridging behaviour observed for both adducts emphasises that a degree of Lewis acidity is also displayed by $SeOCl_2$ in these systems.

When a strong Lewis-base is the other reactant in adduct formation, $SeOCl_2$ shows its Lewis-acid character more definitively; both $SeOCl_2 \cdot 2C_5H_5N$ [81] and the dimeric $[SeOCl_2 \cdot (C_6H_5)_3PO]_2$ [82] have this nature. In the latter adduct one of the phosphine oxide molecules is bonded to one $SeOCl_2$ molecule, the other is linked to both oxychloride molecules. The Se atoms again show a primarily tetragonal coordination involving two chlorines in the basal plane and three oxygens. For one Se atom the two oxygen atoms (at 2.47 and 2.59 Å) are donated by the two phosphine oxide molecules of the dimer; in the other Se atom coordination sphere the donated oxygen atom at 2.71 Å is from a phosphine oxide, and that at 3.05 Å is from the other $SeOCl_2$ unit of the dimer.

Coordination number five is also displayed by selenium in the $SeOCl_2$ adduct with 8-hydroxyquinolinium chloride[83]. This 1 : 1 complex consists of chains of $SeOCl_3^-$ with each tetragonally coordinated selenium bonded to an apical oxygen, two chlorines in *cis*-position at about 2.25 Å, and two chlorines at about 2.96 Å which serve as bridging chlorines with the next selenium-containing groups in the chain. A fifth chlorine at a distance (3.38 Å) somewhat shorter than the van der Waals distance may be considered to complete an octahedral coordination sphere for selenium. A somewhat similar structure, insofar as the Se coordination sphere is concerned, is reported for 2-amino-pyridinium(I) oxotrichloroselenate[84]. Dimeric $SeOCl_3^-$ units formed by two bridging chlorine atoms are present; the coordination sphere about each Se atom again involves an apical oxygen, two short and two long Se—Cl bonds. Again, a fifth chlorine at a distance of 3.45 Å links adjacent $(SeOCl_3^-)_2$ units and provides a distorted octahedral environment about the Se atoms.

In dipyridinium(II) oxotetrachloroselenate(IV), $C_{10}H_8N_2H_2^{2+}$ $SeOCl_4^{2-}$, the tetragonal-based pyramidal coordination about selenium is maintained, but no direct bridging interactions between $SeOCl_4^{2-}$ units is involved. Instead the chlorine at a distance of 2.99 Å from Se serves, through strong hydrogen bonding with nitrogen atoms of two adjacent cations, to link the anions with cations in a chain arrangement[85]. If this long Se—Cl bond is interpreted as an ion–dipole interaction the compound would be viewed as an assemblage of $C_{10}H_8N_2H_2^{2+}$, $SeOCl_3^-$, and Cl^- ions.

8.7 POLYATOMIC CATIONS

For almost 150 years it has been known that the dissolution of elementary chalcogens in strongly acidic solvents produces highly-coloured solutions. Thus, selenium dissolves in H_2SO_4, oleum, or SO_3 to give green and/or yellow solutions. From the green solutions of Se in SO_3 a solid substance of empirical composition $SeSO_3$ was isolated[86].

Very recently the nature of these solutions and their components have been subjected to close scrutiny. From observation of the colours, and changes thereof, produced by the dissolution of Se in acidic solvents, with and without added oxidising agents such as SO_3, SeO_2, or $K_2S_2O_8$, it was concluded that the green and yellow colours are due to species in which selenium displays a positive oxidation state less than 4, and that the yellow species involves Se in a higher oxidation state than that of the green form[87].

Progressive addition of the oxidising agent peroxydisulphuryl difluoride to a solution of selenium in HSO_3F results in maximum intensity for the yellow solution's characteristic absorption at 410 nm when the dissolved $Se : S_2O_6F_2$ ratio is 4 : 1. Further addition of the oxidising agent results, when this ratio is 1 : 2, in the complete disappearance of the 410 nm absorption; continued addition of $S_2O_6F_2$ converts all Se to colourless SeO_2. Since the oxidising ability of $S_2O_6F_2$ is due to reaction $S_2O_6F_2 + 2e^- \rightarrow 2SO_3F^-$, it is deduced that the yellow species must be associated with selenium of oxidation state $+\frac{1}{2}$. While Se_2^+, Se_4^{2+}, Se_6^{3+}, etc. are stoichiometrically possible, the diamagnetic nature of the yellow (and green) solutions eliminates odd-electron species. Further, oxygenated species such as Se_2O^-, $Se_4O_2^{2-}$ or $Se_8O_3^{2-}$ would probably be unstable in strongly acidic media. Electrical-conductance measurements of the HSO_3F–Se–$S_2O_6F_2$ system showed the number of SO_3F^- ions produced per Se atom to be $\frac{1}{2}$. Thus, the reaction may be represented:

$$nSe + n/4\ S_2O_6F_2 \rightarrow Se_n^{(\frac{1}{2}n+)} + n/2\,SO_3F^-$$

Cryoscopic data for the solution indicated that 0.75 particles were produced per Se atom, and this result is in better agreement with $n = 4$ than with $n = 2$, 6 or 8 in the above equation. Thus, the selenium species in yellow solutions is formulated as Se_4^{2+}.

The compounds $Se_4(SO_3F)_2$, $Se_4(HS_2O_7)_2$, $Se_4S_4O_{13}$ and $Se_4(Sb_2F_{11})_2$, which are stable in dry air, have been prepared from elementary Se and the appropriate oxidising agents in acidic solvents[88]. All dissolve readily in H_2SO_4, etc., and give absorption spectra characteristic of Se_4^{2+}. Vibrational spectroscopic data[89] for these Se_4^{2+} compounds are quite similar in the low frequency region showing characteristic Raman lines at 328 cm^{-1} and 188 cm^{-1}, and an i.r. absorption at 306 cm^{-1}. These are assigned within the D_{4h} symmetry of the planar ion deduced from a single-crystal x-ray examination of $Se_4(HS_2O_7)_2$ [90]. The somewhat shortened Se—Se distance of 2.28 Å (single-bond distance c. 2.34 Å) suggests some aromaticity in the four-membered Se ring.

The green Se_8^{2+} cation has been isolated in the compounds $Se_8(Sb_2F_{11})_2$ and $Se_8(AsF_6)_2$ [91]. The latter was prepared in quantitative yield in liquid HF by the reaction:

$$8Se + 3AsF_5 \rightarrow Se_8(AsF_6)_2 + AsF_3$$

Absorption, vibrational and ^{19}F n.m.r. spectra were in full accord with the indicated formulae. It is observed that when the green $Se_8(Sb_2F_{11})_2$ is warmed slightly above its melting range of 180–190 °C it produced a yellow melt. Presumably $Sb_2F_{11}^-$ oxidises Se_8^{2+} to yellow Se_4^{2+} when heated.

A phase study[92] of 80–97 mole % elementary selenium in $SeCl_4 + 4AlCl_3$ has demonstrated the formation of the compounds $Se_4(AlCl_4)_2$ and $Se_8(AlCl_4)_2$. The former substance showed i.r. spectral absorptions at 329 cm^{-1} and 310 cm^{-1}, plus the appropriate bands of $AlCl_4^-$, in support of the conclusion

that Se_4^{2+} is present. X-ray study of the latter compound has established the discrete character of the Se_8^{2+} ion[93]. As in elemental Se_8 the average Se—Se distance is 2.32 ± 0.03 Å. The Se_4—Se_5 distance of 2.83 Å is significantly shorter than the 3.3 Å observed for Se_6—Se_7 and Se_2—Se_3. Thus, an approximate bicyclic structure similar to bicyclo[3,3,0]octane is observed. The monoclinic elementary Se_8 ring is a 'crown' conformation; by loss of two electrons, a transannular closing of Se_4 and Se_5, and a ring flip of Se_1 to the observed position, the Se_8^{2+} conformation is attained. The i.r. spectra data for the $Se_8(AlCl_4)_2$ species shows a variety of weak absorptions in the 320–250 cm^{-1} region appropriate to Se—Se stretching motions.

Polyatomic cations of sulphur[94] and tellurium[95-97] are also of current interest.

8.8   SELENIUM-CONTAINING LIGANDS IN METAL ION COMPLEXES

8.8.1   Selenocyanate complexes

The mode of bonding of the selenocyanato-group in complex species continues to be a focus of research attention, with the interpretation of vibrational spectra the diagnostic approach most generally used. Relative to the integrated intensity of the C—N stretching frequency of the free SeCN$^-$ ion, an N-bonded group displays an increased intensity, an Se-bonded group a decreased value. Some caution in regard to solvent effects is necessary in applying this criterion since hydrogen-bonding solvents may also increase the intensity of this absorption. The use of the C—Se stretching frequency is somewhat more straightforward; the frequency increases for N-bonded and decreases for Se-bonded groups relative to the free-ion value of 558 cm^{-1} [98]. In addition, small increases in the frequency of the Se—C—N bending mode, again relative to the free ion, are detected in substances containing N-bonded groups. Some recent examples of compounds for which one or more of these criteria have been applied are listed in Table 8.1. In some instances the vibrational data are incomplete and the indicated interpretation should be regarded as tentative. Thus, for mixed-ligand rare-earth complexes involving SeCN groups and 1,10-phenanthroline, dipyridyl, etc., the C—Se stretching frequencies are obscured by other ligand frequencies, and only v(C—N) data, in themselves inconclusive, are available. In this instance, additional evidence for the proposed N-bonded monodentate SeCN groups was sought from x-ray powder data[126].

A test for the presence of bridging SeCN groups in associated metal-selenocyanato complexes has been suggested[99] based upon a high value of v(C—N), e.g., > 2100 cm^{-1}. There appears to be substantial overlap in the v(C—N) frequency ranges associated with various bonding modes of the SeCN group, and this criterion, in itself, is perhaps of limited usefulness.

Although the bonding mode of the thiocyanate group in its complexes with transition-metal ions has been shown to be dependent upon factors such as the metal-ion characteristics, solvent effects, and steric and electronic characteristics of other ligands attached to the metal ion, it has only recently been demonstrated that the bonding preference of the selenocyanate group is affected by other factors than the hard or soft character of the metal ion. The preparation and characterisation of the first example of selenocyanate linkage isomers[100], [Pd(Et$_4$dien)SeCN][B(C$_6$H$_5$)$_4$], has highlighted the importance of steric factors associated with non-coordinated counter ions in

Table 8.1 Vibrational frequency data, in cm^{-1}, for SeCN complexes with metal ions

Compound	ν(C—N) Frequency	ν(C—N) Intensity*	ν(C—Se)	δ(N—C—Se)	Remarks	Reference
K$^+$NCSe$^-$	2069[a]	3.1	558[b]	424, 416[b]	free ion values	[a]125 [b]98
[(C$_4$H$_9$)$_4$N]$_3$Pr(NCSe)$_6$	2058	10	613	425		127
[(C$_4$H$_9$)$_4$N]$_3$Sm(NCSe)$_6$	2063	11	614	425	N-bonded	127
[(C$_4$H$_9$)$_4$N]$_3$Er(NCSe)$_6$	2050	17	646	435		127
(C$_5$H$_5$)Fe(CO)[P(C$_6$H$_5$)$_3$]SeCN	2112	1.7	532		Se-bonded	101
(C$_5$H$_5$)Fe(CO)[P(C$_6$H$_5$)$_3$]NCSe	2107	5.3	663		N-bonded	101
Cd(SeCN)$_2$·2H$_2$O	2148, 2110, 2099		590	443	SeCN bridging group?	128
Hg(SeCN)$_2$	2140, 2130		538		N-bonded	128
K$_2$Bi(SeCN)$_5$	2080		543	400	Se-bonded	129
Ln(1.10-phenan)$_3$(NCSe)$_3$ (Ln = La, Ce, Pr or Nd)	2033–2098				N-bonded	126
Ln(Dipyridyl)$_3$NCSe)$_3$						
[Mn$_2$(CO)$_6$Cl$_2$(NCSe)$_2$]$^{2-}$	2119, 2091		634, 616		N-bonded and	130
[Mn$_2$(CO)$_6$(NCSe)$_4$]$^{2-}$	2105, 2083		634, 616		SeCN bridging groups	130
K$_3$[Cr(NCSe)$_6$]·3dioxane	2090		668	400	N-bonded	131
Ag$_3$[Cr(NCSe)$_6$]·5H$_2$O	2140		550, 655	430	SeCN bridge from Ag to Cr	131
Mn[(CH$_3$)$_2$SO]$_4$(NCSe)$_2$	2075		610		N-bonded	132
Cd[(CH$_3$)$_2$SO]$_2$(NCSe)$_2$	2131, 2115		580		SeCN bridge	132

* Integrated absorption intensity, calculated per mole of SeCN, 10^{-4}A, M^{-1} cm^{-2}.

determining whether N- or Se-bonding will occur in a given case. This species, in Se-bonded form, was isolated from solution at low temperature. When dissolved in dimethylformamide–water solution a slow isomerisation to the N-bonded form occurs. The solid N-bonded form recovered from solution then undergoes re-isomerisation to the Se-bonded form. Observation of changes in v(C—N) and v(C—Se), as well as an interpretation of the electronic spectra, are consistent with the indicated isomerisations. Thus, while the N-bonded isomer is favoured in solution, it is argued that the steric interaction of the Se atom with the large $B(C_6H_5)_4^-$ ion in the solid state is decisive in the $N \rightarrow Se$ isomerisation which occurs in the solid phase.

That electronic factors associated with other ligands present in the complex may have an effect in determining the bonding mode of SeCN is suggested by experiments which led to the other example of linkage isomerism for the selenocyanate group which is known at present[101]. Both Se- and N-bonded $(C_5H_5)Fe(CO)[P(C_6H_5)_3]SeCN$ may be isolated by alumina-column chromatography from the system which results when $(C_5H_5)Fe(CO)[P(C_6H_5)_3]$ $CH_2C_6H_5$ is treated with $Se(SeCN)_2$. In general the solubility of the Se-bonded isomer is somewhat greater than that of the N-bonded form in a variety of solvents, and each possesses considerable air and thermal stability. There is no evidence for isomerisation. When refluxed in benzene each form undergoes deselenation to the cyanide complex; this reaction is slower for the Se-bonded isomer.

Since $(C_5H_5)Fe(CO)_2CH_2C_6H_5$ and $Se(SeCN)_2$ give only an Se-bonded form of $(C_5H_5)Fe(CO)_2SeCN$ [102] the formation of the above linkage isomers indicates increased stability of the N-bonded isomer relative to the Se-bonded form as CO is replaced with ligands of reduced π-acceptor capacity.

8.8.2 Other selenium containing ligands in metal ion complexes

A wide variety of other selenium-containing ligands have been involved in studies of metal-ion complexes. The reaction of μ-dichlorobis[tetracarbonylrhenium(I)], $[Re(CO)_4Cl]_2$, with diethyl selenide gives both $Re(CO)_4$ $[Se(C_2H_5)_2]Cl$ and $Re(CO)_3[Se(C_2H_5)_2]_2Cl_2$ [103]. I.R. spectral data for reactants and products show that certain high-frequency carbonyl absorptions of $[Re(CO)_4Cl]_2$ disappeared during the cleavage reaction of the chlorine-bridged reactant. On the assumption that v(C—O) is an adequate measure of metal–carbon bond strength in octahedral carbonyl species, the spectra are interpreted as favouring, for the tricarbonyl complex, a configuration in which each CO lies $trans$ to Se or Cl. This ligand distribution would presumably maximise metal–carbon π-bonding interaction. Similarly, two carbonyl groups are $trans$ to Cl and Se in the tetracarbonyl complex. Halogen bridges are also ruptured in $[Rh(CO)_2Cl]_2$ when reaction with $(C_2H_5)_2Se$ occurs to yield mono-nuclear $trans$-$[Rh(CO)[(C_2H_5)_2Se]_2Cl]$ [104].

Both diphenyl selenide and triphenylphosphine selenide, as well as diethyl selenide, react with polymeric $[Ru(CO)_2I_2]_n$ in benzene to give mononuclear complexes of $Ru(CO)_2$ (selenide)$_2I_2$ [105]. Detailed examination of dipole moment and i.r. spectral data suggests C_{2v} symmetry for the complex with

carbonyl groups *cis* to each other, and Se groups in *trans* positions relative
to each other [106].

Displacement of carbonyl groups from $Mn(CO)_5Br$ by $(C_6H_5)_2Se$ has
been found[107].

$$Mn(CO)_5Br + 2(C_6H_5)_2Se \rightarrow Mn(CO)_3[Se(C_6H_5)_2]_2Br + 2CO$$

The resulting non-ionic selenium-containing complex shows three strong
$v(C—O)$ absorptions in its i.r. spectrum. A configuration in which each of the
three carbonyl groups is *trans* to either Br or Se is indicated.

Somewhat similar carbonyl complexes of Mn and Re have been prepared
by use of the bidentate 2,5-diselenohexane ligand:[108]

$$M(CO)_5X + C_2H_4(SeCH_3)_2 \rightarrow M(CO)_3[C_2H_4(SeCH_3)_2]X + 2CO$$
$$(M = Mn, Re; X = Cl, Br)$$

A related chromium complex was attained in good yield by the reaction of
this selenium ligand with bicyclo[2,2,1]heptadiene chromium tetracarbonyl:

$$C_7H_8Cr(CO)_4 + C_2H_4(SeCH_3)_2 \rightarrow Cr(CO)_4[C_2H_4(SeCH_3)_2]$$

I.R. and ^1H n.m.r. spectral data are consistent with structural formulations
which minimise the number of carbonyl groups *trans* to each other.

Several studies have provided further clarification of the role of *Se*-
containing ligands in affecting the magnetic and spectral characteristics of
transition-metal complexes. The diethyldiselenocarbamate, $(C_2H_5)NCSe_2^-$,
dsc^-, complexes of Fe^{III}, both octahedral $Fe(dsc)_3$ and square-pyramidal
$Fe(dsc)_2Cl$, are monomeric and low spin ($\mu = 2.4$ and 4.0, respectively, at
room temperature)[109]. The corresponding dithiocarbamates of Fe^{III} display
a broad range of magnetic behaviour which suggest a delicate balance
between ligand-field strength and pairing energy, and a corresponding close
approach to the $^6A_{1g}$–$^2T_{2g}$ crossover point for the d^5 system. It has been
proposed[110] that the distinct low-spin character of the diseleno-complexes, as
compared to the intermediate behaviour of dithio-complexes, is due primarily
to a dominance by the larger nephelauxetic effect for selenium; this out-
weighs the somewhat lower position of selenium ligands relative to sulphur
ligands in the spectrochemical series.

In agreement with this conclusion a comparison of the electronic spectra
of $Pt(SeCN)_6^{2-}$ and the corresponding thiocyanate complex suggests that
$SeCN^-$ occurs at a lower position in the spectrochemical series[111]. The high
intensities of ligand-field bands for $Pt(SeCN)_6^{2-}$ imply extensive ligand–metal
orbital mixing.

Spectroscopic and magnetic data for $M[CH_3)_3PSe]_4(ClO_4)_4$, $(M = Co,Ni,
Zn)$, have been compared with those for the corresponding trimethylphos-
phine sulphide complex; in each case coordination is through the chalcogen
atom[112]. The Δ-values of these complexes are similar, but the β-value of the
selenide is less than that of the sulphide.

A number of *N,N*-dimethyl-, *N,N*-diethyl- and *N*-methyl-*N*-phenyl-
diselenocarbamate complexes of tin(IV) have been studied[113]. Interpretation
of the i.r. spectral data suggests the bidentate behaviour of the diselenocarbamate group. Only a single sharp *N*-methyl ^1H n.m.r. signal is observed
in accord with two equivalent methyl groups. Other newly synthesised

diselenocarbamate chelates[114] include tris-complexes of Ga^{III}, Au^{III}, Fe^{III}, and Cr^{III}, bis-complexes of Ni^{II}, Cu^{II}, and Pt^{II}, and a tetrakis-complex of U^{IV}.

Olefinic ligands containing selenium have also been employed to form metal complexes[115]. Both Pd^{II} and Pt^{II} form complexes of general formula MCl_2L with but-3-enyl butyl selenide and but-3-enyl phenyl selenide. Platinum complexes of the same formula type are also formed by n-butyl pent-4-enyl selenide and pent-4-enyl phenyl selenide. The reduction in the frequency of the C—C stretching mode from 1640 cm^{-1} in the free ligand to the $1500–1540 \text{ cm}^{-1}$ region for the complexes is indicative of coordination of the olefinic double bond to the metal. Further, two absorptions in the $300–330 \text{ cm}^{-1}$ region of the spectra of these complexes are associated with metal–chlorine stretching frequencies and imply cis-coordination for chlorine. A broad band at c. 220 cm^{-1} in the spectra of some of these complexes is believed to be the metal–selenium stretch. These complexes are thus formulated as involving bidentate behaviour (double-bond and selenium-atom coordination) of these ligands.

Dimethylselenoxide is found to form complexes with a wide variety of transition and non-transition metals, e.g., Mn, Co, Ni, Cu, Zn, Cd, Mg, Cr, Fe, Ag, and UO_2. In all cases the shifts in $v(Se—O)$ from 820 cm^{-1} in the free ligand to $760–800 \text{ cm}^{-1}$ in the complex indicates coordination by oxygen[116].

Complexes of $CH_3SeC_6H_4P(C_6H_5)_2$, CH_3Se—P, with Ni^{II} and Pd^{II} undergo rapid Se-demethylation when refluxed in butanol[117]. The resulting $Ni(Se—P)_2$ complex is a square monomeric species, but the insolubility of the $Pd(Se—P)_2$ compound in common solvents suggests a polymeric substance. The i.r. spectrum of the palladium complex is interpreted in terms of Se-bridging between Pd atoms in a dimeric species. It was pointed out in this study that this synthetic route should enable the preparation of organo-selenide complexes while avoiding the use of toxic and malodorous selenol ligands. Single-crystal x-ray study of the nickel complex confirms the monomeric square-planar configuration of the demethylation product[118].

Other selenium-containing ligands which have received recent attention for their metal complex forming ability include O,O'-diethyl diselenophosphate[119], 1,2-bis(isopropylseleno)ethane[120], 2,2-dicyanoethylene-1,1-diselenol[121], selenourea[122, 123], and bis(trifluoromethyl)-1,2-diselenetene[124]. For the paramagnetic Ni^{II} complexes of the last ligand the e.s.r. spectrum displays a distinct g-value anisotropy. This indicates a planar configuration for the complex with a b_{3g} ground-state. Ligand hyperfine-structure due to ^{77}Se ($I = \frac{1}{2}$) suggests about 80% ligand character for the unpaired electron. The absence of any ligand hyperfine-interaction in the e.s.r. spectrum of the $VSe_6C_6(CF_3)_6^{2-}$ anion requires that the single unpaired electron is localised in a molecular orbital, largely metal d_{z^2} in character, in a probable trigonal-prismatic complex ion.

8.9 SELENIUM-CONTAINING DONOR MOLECULES

In Section 8.8 the donor behaviour of various selenium-containing ligands towards metal ions was discussed. In this section attention is directed to several studies of the interaction of Se-containing donor molecules with classical Lewis-acid molecules.

8.9.1   Diorgano-selenides

Interpretation of the i.r. spectrum of $(CH_3)_2Se \cdot BF_3$ is reported to be relatively straightforward due to the heavy Se atom and the lack of significant coupling of vibrations primarily associated with the donor or acceptor species[133]. Little effect on the vibrational frequencies of the $(CH_3)_2Se$ group is observed on complexation but, for the fundamentals of the BF_3 portion of the adduct, pronounced decreases are observed. These changes are, of course, appropriate to a shift from the planar geometry of free BF_3 to the pyramidal configuration of this four-atom group in the adduct. While the decreases in frequency for $(CH_3)_2Se \cdot BF_3$ and for the corresponding sulphide complex are approximately equal for the antisymmetric B—F stretching-frequency, the decreases in the other three fundamentals are slightly greater for the selenide adduct, a reflection, perhaps, of the greater steric requirement of the $(CH_3)_2Se$ molecule.

    A comparative study of the ^1H n.m.r. shifts relative to the observed resonances of free ligand for boron trihalide complexes of $(CH_3)_2Se$, $(C_2H_5)_2Se$, $(i-C_3H_7)_2Se$ and $(C_6H_5)_2Se$ has been reported[134]. For $(CH_3)_2Se \cdot BX_3$ the shift in ^1H frequency in the complexes is greater for BBr_3 and BI_3 than for BCl_3 in accord with the greater Lewis-acidity of the heavier BX_3 molecules. Comparison with similar data for $(CH_3)_2S$ and $(CH_3)_2Te$ establishes the order of donor strength toward BX_3 as $(CH_3)_2S > (CH_3)_2Se > (CH_3)_2Te$.

    The adduct-forming ability of $(CH_3)_2Se$ has been employed in boron isotope exchange studies[135]. For the reaction

$$^{10}BF_3(g) + (CH_3)_2X \cdot ^{11}BF_3(l) \rightleftharpoons {}^{11}BF_3(g) + (CH_3)_2X \cdot ^{10}BF_3(l)$$
$$(X = Se, S \text{ or } O)$$

the equilibrium constants are such that the lighter boron isotope concentrates in the liquid phase. Comparison of log K_{eq} values at 30 °C for $(CH_3)_2O$ (0.0110), $(CH_3)_2S$ (0.0153) and $(CH_3)_2Se$ (0.0137) suggests an anomalous position for the selenide K_{eq}, e.g., less than that for the sulphide, but greater than that of the oxide. This was rationalised on the basis of bond-strength differences arising from an interplay of electron-donating capacity and polarisability differences for O, S and Se.

    With tungsten hexafluoride $(C_2H_5)_2Se$ forms an orange liquid of low volatility, $[(C_2H_5)_2Se]_2WF_6$, which dissociates slightly at 20 °C [136]. In parallel observations on other donor–WF_6 complexes the ^{19}F n.m.r. spectrum is a single peak, presumably due to rapid intramolecular exchange. However, unlike the single proton-resonance found for $[(C_2H_5)_2O]_2WF_6$, the selenide adduct gives a ^1H n.m.r. spectrum composed of two complex multiplets in a 3 : 2 intensity ratio reflecting the non-equivalence of the ethyl group protons. The exchange rate of the selenide groups between two possible coordination sites is reported to be slow even at the decomposition temperature of 100 °C.

8.9.2   Triarylphosphine selenides

The donor properties of triphenyphosphine selenides toward BCl_3 and BBr_3 have also been examined and compared with those of the corresponding sulphide[134]. With either Lewis acid the selenide is a weaker base than the sulphide; both phosphine chalcogenides are weaker donors than any dimethyl chalcogenide.

By means of absorption spectrophotometry the equilibrium constants for the formation of several triaryphosphine selenide–iodine charge-transfer complexes in chloroform solution have been evaluated[137]. Comparison of these data with those for analogous phosphine oxides and sulphides established a stability order toward I_2 of $P = Se > P = S > P = O$. The bonding description involves σ-donation from Se to I, supplemented by back donation from a 5p-orbital of I to an empty 4d level of Se. Energetically this π-interaction may be a significantly better match than the corresponding donation to the 3d-orbital of sulphur. Alternatively, it is pointed out that the stability order towards I_2 is in accord with expectations based on mutual polarisability of the large Se and I atoms. Consistent with this view it is found that complexes of $P = Se$ and $As = S$ are formed with polarisable metal ions such as Ag^I, Hg^{II}, Cd^{II}, Pd^{II}, and Pt^{II} [138], but these same ligands give no complexes with ions of limited polarisability, e.g., Co^{II}, Ni^{II} or Cu^{II}.

8.9.3 Methyl-substituted selenium tetrahalides

As noted in Section 8.5.2.3 the selenium tetrahalides form addition compounds without utilising the lone-pair electrons associated with the selenium atom. Further examination of the possible involvement of the lone pair of a 4-coordinated selenium atom in coordinate bond formation has focused on the molecular organo-substituted selenium halide species, $(CH_3)_2SeX_2$, (X = Cl, Br), and CH_3SeCl_3 [139, 140]. The former react with BX_3 to give 1:1 adducts, but i.r. spectra clearly demonstrate an ionic formula, $(CH_3)_2$ $SeX^+BX_4^-$. Treatment of a methylene chloride solution of CH_3SeCl_3 with BCl_3, however, gave no detectable adduct, and a similar negative result was obtained from an attempted sealed-tube reaction of these materials. In CH_2Cl_2 solution CH_3SeCl_3 exists as an associated species, probably involving chlorine-bridging between selenium atoms. Such a formulation, which may be regarded as an intermolecular donor–acceptor interaction, may be correlated with the failure of another acceptor species to remove a chloride ion.

It thus appears that tetracoordinated selenium with one or more halogens as ligand atoms exhibits basic properties only in the sense of halide-ion transfer. Although no simple tetra-alkyl or -aryl derivatives of selenium are known, a species with four Se—C bonds, bis-2,2'-biphenylyleneselenium, has been reported as a somewhat unstable substance[141]. It would be interesting to determine its behaviour toward strong Lewis-acids.

References

1. Bagnall, K. W. (1966). *The Chemistry of Selenium, Tellurium, and Polonium.* (Amsterdam: Elsevier Publishing Co.)
2. Chizhikov, D. M. and Shchastlivyi, V. P. (1968). *Selenium and Selenides.* (London: Collet's Publishers Ltd.)
3. Jensen, K. A. (1970). *Quart. Rep. Sulfur Chem.*, **5**, 45

4. Cradock, S., Ebsworth, E. A. V. and Rankin, D. W. H. (1969). *J. Chem. Soc. A*, 1628
5. Drake, J. E. and Riddle, C. (1969). *J. Chem. Soc. A*, 1573
6. Cueilleron, J. and Hillel, R. (1968). *Bull. Soc. Chim. Fr.*, 3635
7. Centofanti, L. F. and Parry, R. W. (1970). *Inorg. Chem.*, **9**, 744
8. Angus, H. F., Cradock, S. and Ebsworth, E. A. V. (1969). *Inorg. Nucl. Chem. Lett.*, **5**, 717
9. Hastie, J. W., Hauge, R. and Margrave, J. L. (1969). *J. Inorg. Nucl. Chem.*, **31**, 281
10. Voigt, E. M., Meyer, B., Morelle, A. and Smith, J. J. (1970). *J. Mol. Spectrosc.*, **34**, 179
11. Sonin, V. I., Novikov, G. I. and Polyachenok, O. G. (1969). *Russ. J. Phys. Chem.*, **43**, 1680
12. Nour, M. M. and Wasif, S. (1968). *J. Chem. Soc. A*, 3024
13. Nour, M. M. and Wasif, S. (1969). *J. Chem. Soc. A*, 474
14. Wasif, S. (1970). *J. Chem. Soc. A*, 118
15. Fialkov, Y. Y. and Yakovleva, A. V. (1969). *Russ. J. Inorg. Chem.*, **14**, 1141
16. Chaigneau, M. and Santarromana, M. (1969). *C. R. Acad. Sci., Ser. C*, **269**, 1643
17. Dostal, K. and Cernik, M. (1966). *Z. Chem.*, **6**, 424
18. Edwards, A. J., Mouty, M. A. and Peacock, R. D. (1967). *J. Chem. Soc. A*, 557
19. Kempe, G. (1968). *Z. Chem.*, **8**, 152
20. Kempe, G. (1968). *Z. Anorg. Allg. Chem.*, **363**, 273
21. Touzin, J. and Kratochvila, J. (1969). *Coll. Czech. Chem. Commun.*, **34**, 1080
22. Ruzicka, A. and Dostal, K. (1967). *Z. Chem.*, **7**, 394
23. Chernozubov, Y. S. and Selivanov, G. K. (1970). *Russ. J. Phys. Chem.*, **44**, 465
24. Cooper, R. and Culka, J. V. (1967). *J. Inorg. Nucl. Chem.*, **29**, 1217
25. Ailwood, C. R. and Fielding, P. E. (1969). *Aust. J. Chem.*, **22**, 2301
26. Cooper, R. and Culka, J. V. (1970). *J. Inorg. Nucl. Chem.*, **32**, 1857
27. Schmidt, M. and Wilhelm, E. (1970). *Z. Naturforsch. B*, **25**, 1348
28. Ward, A. T. (1968). *J. Phys. Chem.*, **72**, 4133
29. Cooper, R. and Culka, J. V. (1969). *J. Inorg. Nucl. Chem.*, **31**, 685
30. Hendra, P. J. and Park, P. J. D. (1968). *J. Chem. Soc. A*, 908
31. Frankiss, S. G. (1968). *J. Mol. Struct.*, **2**, 271
32. Forneris, R. and Hennies, C. E. (1970). *J. Mol. Struct.*, **5**, 449
33. Lundqvist, M. and Sillen, L. G. (1966). *Acta. Chem. Scand.*, **20**, 1723
34. Lundqvist, M. (1968). *Acta Chem. Scand.*, **22**, 281
35. Lundqvist, M. and Lellep, M. (1968). *Acta Chem. Scand.*, **22**, 291
36. Hogberg, K. and Lundqvist, M. (1970). *Acta Chem. Scand.*, **24**, 255
37. Wynne, K. J. and Pearson, P. S. (1971). *161st Amer. Chem. Soc. Meeting, Los Angeles*, Inorg. Div. paper 172
38. Aynsley, E. E., Dodd, R. E. and Little, R. (1962). *Spectrochim. Acta*, **18**, 1005
39. Rolfe, J. A., Woodward, L. A. and Long, D. A. (1953). *Trans. Faraday Soc.*, **49**, 1388
40. Edwards, A. J. and Hewaidy, F. I. (1968). *J. Chem. Soc. A*, 2977
41. Bowater, I. C., Brown, R. D. and Burden, F. R. (1968). *J. Mol. Spectrosc.*, **28**, 454
42. Hayward, G. C. and Hendra, P. J. (1967). *J. Chem. Soc. A*, 643
43. George, J. W., Katsaros, N. and Wynne, K. (1967). *Inorg. Chem.*, **6**, 903
44. Beattie, I. R. and Chudzynska, H. (1967). *J. Chem. Soc. A*, 984
45. Robinson, E. A. and Ciruna, J. A. (1968). *Can. J. Chem.*, **46**, 3197
46. Cordes, A. W., Kruh, R. F., Gordon, E. K. and Kemp, M. P. (1964). *Acta Crystallogr.*, **17**, 756
47. Shoemaker, C. B. and Abrahams, S. C. (1965). *Acta Crystallogr.*, **18**, 296
48. Buss, B. and Krebs, B. (1970). *Angew. Chem.*, **82**, 446
49. Ozin, G. A. and Vander Voet, A. (1970). *Chem. Commun.*, 896
50. Katsaros, N. and George, J. W. (1969). *Inorg. Chim. Acta.*, **3**, 165
51. Couch, D. A., Elmes, P. S., Fergusson, J. E., Greenfield, M. L. and Wilkins, C. J. (1967). *J. Chem. Soc. A*, 1813
52. Peach, M. E. (1969). *Can. J. Chem.*, **47**, 1675
53. Tideswell, N. W. and McCullough, J. D. (1956). *J. Amer. Chem. Soc.*, **78**, 3026
54. Katsaros, N. and George, J. W. (1968). *Chem. Commun.*, 662
55. Katsaros, N. and George, J. W. (1969). *Inorg. Chem.*, **8**, 759
56. Heal, H. G. and Kane, J. (1967). *J. Inorg. Nucl. Chem.*, **29**, 1539
57. Edwards, A. J. and Jones, G. R. (1970). *J. Chem. Soc. A*, 1891
58. Edwards, A. J. and Jones, G. R. (1970). *J. Chem. Soc. A*, 1491
59. Gillespie, R. J. and Whitla, W. A. (1970). *Can. J. Chem.*, **48**, 657

60. Gillespie, R. J. and Whitla, W. A. (1969). *Can. J. Chem.*, **47**, 4153
61. Evans, J. A. and Long, D. A. (1968). *J. Chem. Soc. A*, 1688
62. Gerding, H. (1956). *Recl. Trav. Chim. Pays-Bas*, **75**, 589
63. Gerding, H., Stufkens, D. J. and Gijben, H. (1970). *Recl. Trav. Chim. Pays-Bas*, **89**, 619
64. Gerding, H. and Stufkens, D. J. (1969). *Rev. Chim. Minerale*, **6**, 795
65. Chikanov, N. D. (1968). *Russ. J. Inorg. Chem.*, **13**, 1483
66. Chikanov, N. D. (1970). *Izv. Vyssh. Ucheb. Zaved., Khim. Khim. Tekhnol.*, **13**, 1378; *Chem. Abstr.*, **74**, 35199b
67. Beattie, I. R., Milne, M., Webster, M., Blayden, H. E., Jones, P. J., Killean, R. C. G. and Lawrence, J. L. (1969). *J. Chem. Soc. A*, 482
68. Katsaros, N. and George, J. W. (1969). *J. Inorg. Nucl. Chem.*, **31**, 3503
69. Hendra, P. J. and Jovic, Z. (1968). *J. Chem. Soc. A*, 600
70. Stufkens, D. J. (1970). *Recl. Trav. Chim. Pays-Bas*, **89**, 1185
71. Kolesnikova, N. M. and Iofa, B. Z. (1966). *Vestn. Mosk. Univ., Ser. II, Khim.*, **21**, 67; *Chem. Abstr.*, **66**, 6807
72. Mitra, G. and Cady, G. H. (1959). *J. Amer. Chem. Soc.*, **81**, 2646
73. Smith, J. E. and Cady, G. H. (1970). *Inorg. Chem.*, **9**, 1442
74. Smith, J. W. and Cady, G. H. (1970). *Inorg. Chem.*, **9**, 1293
75. Edwards, A. J. and Jones, G. R. (1969). *J. Chem. Soc. A*, 2858
76. Hermodsson, Y. (1969). *Ark. Kemi*, **31**, 199
77. Baker, K. L. and Fowles, G. W. A. (1968). *J. Chem. Soc. A*, 801
78. Hermodsson, Y. (1967). *Acta Chem. Scand.*, **21**, 1313
79. Hermodsson, Y. (1969). *Ark. Kemi*, **31**, 73
80. Lindqvist, I. (1963). *Inorganic Adduct Molecules of Oxo-Compounds.* (New York: Academic Press Inc.)
81. Lindqvist, I. and Nahringbauer, G. (1959). *Acta Crystallogr.*, **12**, 638
82. Hermodsson, Y. (1968). *Ark. Kemi*, **30**, 15
83. Cordes, A. W. (1967). *Inorg. Chem.*, **6**, 1204
84. Dill, E. D. (1969). *Diss. Abst. B*, **30**, 1023
85. Wang, B.-C. and Cordes, A. W. (1970). *Inorg. Chem.*, **9**, 1643
86. Divers, E. and Shimoze, M. (1884). *J. Chem. Soc.*, 194
87. Barr, J., Gillespie, R. J., Kapoor, R. and Malhotra, K. C. (1968). *Can. J. Chem.*, **46**, 149
88. Barr, J., Crump, D. B., Gillespie, R. J., Kapoor, R. and Ummat, P. K. (1968). *Can. J. Chem.*, **46**, 3607
89. Gillespie, R. J. and Pez, G. P. (1969). *Inorg. Chem.*, **8**, 1229
90. Brown, I. D., Crump, D. B., Gillespie, R. J. and Santry, D. P. (1968). *Chem. Commun.*, 853
91. Gillespie, R. J. and Ummat, P. K. (1970). *Can. J. Chem.*, **48**, 1239
92. Prince, D. J., Corbett, J. D. and Garbisch, B. (1970). *Inorg. Chem.*, **9**, 2731
93. McMullan, R. K., Prince, D. J. and Corbett, J. D. (1969). *Chem. Commun.*, 1438
94. Barr, J., Gillespie, R. J. and Ummat, P. K. (1970). *Chem. Commun.* 264
95. Barr, J., Gillespie, R. J., Pez, G. P., Ummat, P. K. and Vaidya, O. C. (1971). *Inorg. Chem.*, **10**, 362
96. Lokken, D. A., Couch, T. W. and Corbett, J. D. (1970). *160th Amer. Chem. Soc. Meeting, Chicago*, Inorg. Div. paper 18
97. Bjerrum, N. J. (1970). *Inorg. Chem.*, **9**, 1965
98. Morgan, H. W. (1961). *J. Inorg. Nucl. Chem.*, **16**, 367
99. Kharitonov, Y. Y. and Skopenko, V. V. (1965). *Russ. J. Inorg. Chem.*, **10**, 984
100. Burmeister, J. L., Gysling, H. J. and Lim, J. C. (1969). *J. Amer. Chem. Soc.*, **91**, 44
101. Jennings, M. A. and Wojcicki, A. (1969). *Inorg. Chim. Acta.*, **3**, 335
102. Jennings, M. A. and Wojcicki, A. (1968). *J. Organometal. Chem.*, **14**, 231
103. Faraone, F., Sergi, S. and Pietropaolo, P. (1970). *J. Organometal. Chem.*, **24**, 453
104. Faraone, F., Pietropaolo, P. and Sergi, S. (1970). *J. Organometal. Chem.*, **24**, 797
105. Hieber, W. and John, P. (1970). *Chem. Ber.*, **103**, 2161
106. John, P. (1970). *Chem. Ber.*, **103**, 2178
107. Hieber, W. and Stanner, F. (1969). *Chem. Ber.*, **102**, 2930
108. Abel, E. W. and Hutson, G. V. (1969). *J. Inorg. Nucl. Chem.*, **31**, 3333
109. Cervone, E., Diomedi Camassei, F., Luciani, M. L. and Furlani, C. (1969). *J. Inorg. Nucl. Chem.*, **31**, 1101
110. Furlani, C., Cervone, E. and Diomedi Camassei, F. (1968). *Inorg. Chem.*, **7**, 265

252 GROUPS V AND VI

111. Swihart, D. L. and Mason, W. R. (1970). *Inorg. Chem.*, **9**, 1749
112. Brodie, A. M., Rodley, G. A. and Wilkins, C. J. (1969). *J. Chem. Soc. A*, 2927
113. Kamitani, T., Yamamoto, H. and Tanaka, T. (1970). *J. Inorg. Nucl. Chem.*, **32**, 2621
114. Lorenz, B., Kirmse, R. and Hoyer, E. (1970). *Z. Anorg. Allg. Chem.*, **378**, 144
115. Goodall, D. C. (1969). *J. Chem. Soc. A*, 890
116. Paetzold, R. and Bochmann, G. (1969). *Z. Anorg. Allg. Chem.*, **368**, 202
117. Meek, D. W. (1969). *Inorg. Nucl. Chem. Lett.*, **5**, 235
118. Curran, R., Cunningham, J. A. and Eisenberg, R. (1970). *Inorg. Chem.*, **9**, 2749
119. Krishnan, V. and Zingaro, R. A. (1969). *Inorg. Chem.*, **8**, 2337
120. Greenwood, N. N. and Hunter, G. (1969). *J. Chem. Soc. A*, 929
121. Jensen, K. A. and Krishnan, V. (1970). *Acta Chem. Scand.*, **24**, 1090, 1092
122. Domiano, P., Manredotti, A. G., Crossoni, G., Nardelli, M. and Tani, M. E. V. (1969). *Acta Crystallogr., B*, **25**, 591
123. Shul'man, V. M. and Tyuleneva, L. I. (1970). *Izv. Akad. Nauk SSSR, Ser. Khim.*, 1189; *Chem. Abstr.*, **73**, 62164x
124. Davidson, A. and Shawl, E. T. (1970). *Inorg. Chem.*, **9**, 1820
125. Pecile, C. (1966). *Inorg. Chem.*, **5**, 210
126. Golub, A. M., Kopa, M. V., Skopenko, V. W. and Zinzadse, G. (1970). *Z. Anorg. Allg. Chem.*, **375**, 302
127. Burmeister, J. L. and Deardoff, E. A. (1970). *Inorg. Chim. Acta*, **4**, 97
128. Kharitonov, Yu. Ya., Tsintsadze, G. V. and Tsivadze, A. Yu. (1970). *Russ. J. Inorg. Chem.*, **15**, 204
129. Golub, A. M., Skopenko, V. V. and Zhumabaev, A. (1969). *Russ. J. Inorg. Chem.*, **14**, 1579
130. Farona, M. F., Frazee, L. M. and Bremer, N. J. (1969). *J. Organometal. Chem.*, **19**, 225
131. Brusilovets, A. I., Skopenko, V. V. and Tsintsadze, G. V. (1969). *Russ. J. Inorg. Chem.*, **14**, 239
132. Skopenko, V. V., Tsintsadze, G. V. and Brusilovets, A. I. (1970). *Ukr. Khim. Zh.*, **36**, 474; *Chem. Abstr.*, **73**, 62129q
133. LeCalve, J. and Lascombe, J. (1968). *Spectrochim. Acta A*, **24**, 737
134. Schmidt, M. and Block, H. D. (1970). *Chem. Ber.*, **103**, 3705
135. Palko, A. A. and Drury, J. S. (1967). *J. Chem. Phys.*, **46**, 2297
136. Noble, A. M. and Winfield, J. M. (1968). *Inorg. Nucl. Chem. Lett.*, **4**, 339
137. Condray, B. R., Zingaro, R. A. and Kudchadker, M. V. (1968). *Inorg. Chim. Acta*, **2**, 309
138. Nicpon, P. and Meek, D. W. (1966). *Chem. Commun.*, 398
139. Wynne, K. J. and George, J. W. (1965). *J. Amer. Chem. Soc.*, **87**, 4750
140. Wynne, K. J. and George, J. W. (1969). *J. Amer. Chem. Soc.*, **91**, 1649
141. Hellwinkel, D. and Fahrbach, G. (1968). *Liebig's Ann. Chem.*, **715**, 68

9
Complexes of Metals with Nitrogen Containing Pseudohalide Ligands

W. BECK and W. P. FEHLHAMMER

Institut für Anorganische Chemie der Universität, München

9.1 INTRODUCTION

In this review we intend to survey aspects of the chemistry of the nitrogen containing pseudohalide transition metal complexes, i.e. complexes with N_3^-, CNO^-, NCO^-, NCS^-, $NCSe^-$, $NCNR^-$, CN^-, $N(CN)_2^-$ and $C(CN)_3^-$ ligands. The most prominent feature of these anionic species is their tendency to form stable coordination compounds, also one of the criteria of Birckenbach[1] who introduced the term 'pseudohalide'. Today numerous research groups are involved in studies of these compounds. Here the work of J. L. Burmeister, F. Basolo, K. Dehnicke, A. M. Golub, A. Wojcicki, A. H. Norbury, H. H. Schmidtke and G. V. Tsintsadze on NCO-, NCS- and NCSe-complexes, and of H. Köhler on $N(CN)_2$- and $C(CN)_3$-complexes should be mentioned. Research in our Munich laboratory has been concentrated on fulminate and azide compounds. Present interests seem to be mainly directed towards the ambivalent nature of these ligands and their ability to form bridges. These topics will be particularly emphasised in this article. A great deal of work has also been devoted to gaining an understanding of the bonding in, and the structure of these compounds by studying the vibrational and electronic spectra, and performing x-ray structural determinations, but a discussion of these aspects must be deferred to a later date. Similarly, a consideration of the problems of the reactivity of coordinated pseudohalides must also be deferred.

In the past, reviews have appeared on cyanides[2-6], azides[7-9], thiocyanates and selenocyanates[10, 11], percyanocarbon and cyano-amide complexes[12, 13] and fulminates[14]. Organometallic pseudohalides[15, 16], pseudohalide complexes in non-aqueous solutions[17] and the behaviour of pseudohalides as ambidentate ligands [18-20] have also been reviewed.

It appears to be beyond the scope of this article to give an exhaustive review on recent developments in the chemistry of cyano complexes. Similarly, no kinetic data are presented, nor was it attempted to survey all existing pseudohalide complexes containing other ligands, such as amines, etc. Furthermore, no comment is made on the vast field of coordination chemistry in solution.

9.2 HOW PSEUDOHALIDES BOND TO A TRANSITION METAL

All of the linear triatomic anions (NCO^-, NCS^-, $NCSe^-$, CNO^-, N_3^-, $NCNR^-$) and CN^- with 16 and 10 valence electrons, respectively, can coordinate through both ends, thus giving rise to linkage isomers. Linearity of the triatomic species is maintained on coordination in all cases. Many compounds are known containing pseudohalide bridges. In particular, the pseudohalides N_3^-, SCN^- and CN^- are extremely good bridging ligands, as is to be presumed by simple inspection of the various canonical structures of these anionic ligands, with lone pairs being available on each end atom.

Table 9.1 reveals the great variety in the modes of coordination found in pseudohalide compounds; examples are given for each listed mode of bonding.

Table 9.1 Possible modes of coordination of pseudohalides

Ligand	Mode of bonding a	Examples
Cyanide	M—C—N*	all stable cyano complexes[3, 21], e.g. [Fe (CN)_6]^{4-}
	M—N—C	assumed as intermediates[22, 24], [Co(trien) (NC)_2]ClO_4 [25]
	M—C—N—M*	AgCN·2AgNO_3 [26], Ni(CN)_2·NH_3·X (X = benzene, thiophene, aniline, etc)[27], [R_2Au CN]_4 [28, 29]

Fulminate   M—C—N—O $[Pt(CNO)_4]^{2-}$, $[Au(CNO)_2]^-$ [30, 31]

$$
\begin{array}{c}
O \\
N \\
C \\
\diagup \quad \diagdown \\
M \quad\quad M^\star
\end{array}
$$

AgCNO [32]

Cyanate   M—N—C—O* π-$C_5H_5Cr(NO)_2NCO$ [33], π-$C_5H_5Mo(CO)$ $(PPh_3)_2NCO$ [34]

$$
\begin{array}{c}
\quad O—C—N \\
\diagup \\
M
\end{array}
$$

$[Re(OCN)_6]^{2-}$, $[Re(OCN)_6]^-$, $[Mo(OCN)_6]^{3-}$ [35] $(C_5H_5)_2M(OCN)_2$ (M = Ti^{VI}, Zr^{IV}, Hf^{IV} [36]

$$
\begin{array}{c}
O \\
C \\
N \\
\diagup \quad \diagdown \\
M \quad\quad M^\star
\end{array}
$$

AgNCO [37], $[(Ph_3P)_2Pt(NCO)_2Pt(PPh_3)_2]^{2+}$ [38] $[(CH_3)_2AuNCO]_2$ (cf. ref. 29)

Thiocyanate   M—N—C—S* $[(C_2H_5)_4N]_4[U(NCS)_8]$ [39], $K_3[Mo(NCS)_6]\cdot H_2O\cdot CH_3COOH$ [40] Ni(NNNAs) $(NCS)_2^+$ [41].

$$
\begin{array}{c}
\quad S—C—N^\star \\
\diagup \\
M
\end{array}
$$

$Cu(en)_2(SCN)_2$ [42], $[Cu(trien)SCN]$ NCS [43].

$$
\begin{array}{c}
\quad S—C—N \\
\diagup \quad\quad\quad \diagdown \\
M \quad\quad\quad\quad M^\star
\end{array}
$$

AgSCN [44], $(n$-$C_3H_7)_3PAgSCN$ [45], $Cd(etu)_2$ $(NCS)_2^+$ [46], $Cr_2(CO)_{10}SCN$ [47, 48].

$$
\begin{array}{c}
\quad\quad\quad\quad M \\
\quad\quad\quad\quad \diagup \\
M—N—C—S \\
\quad\quad\quad\quad \diagdown \\
\quad\quad\quad\quad M
\end{array}
$$

$[(CH_3)_3PtNCS]_4$ [49]

$$
\begin{array}{c}
N \\
C \\
S \\
\diagup \diagdown \quad N \\
M \quad M \quad C
\end{array}
$$

assumed in $[Et_2MSCN]_3$ (M = Al, Ga, In) [50], assumed as intermediates [51, 52].

$$
\begin{array}{c}
\quad\quad\quad\quad S \\
M\text{-----}\quad\quad \diagdown \\
\quad S \quad\quad\quad M^\star
\end{array}
$$

$K_2[Pd(SCN)_4]$ [53]

$$
\begin{array}{c}
C \\
N \\
\diagup \quad \diagdown \\
M \quad\quad M
\end{array}
$$

discussed as intermediates [54, 55] (however, cf. Ref. 56)

Azide
$$
\begin{array}{c}
\quad N—N—N^\star \\
M \quad N \\
\quad\quad N
\end{array}
$$
Cu(Et$_4$dien)N$_3$Br$^+$ [57], $[AsPh_4]_2[Fe(N_3)_5]$ [58], $[Ru(N_3)$ (N_2) $(NH_2CH_2CH_2NH_2)_2]$ $[PF_6]$ [59]

$$
\begin{array}{c}
N \\
\diagup \diagdown \\
M \quad M^\star
\end{array}
$$

$[AsPh_4]_2[(N_3)_2Pd(N_3)_2Pd(N_3)_2]$ [60], $[(CO)_3Mn(N_3)_3Mn(CO)_3]^-$ [61], $[BCl_2$ $N_3]_3$ [62], $Cu(N_3)_2$ [63].

$$
\begin{array}{c}
\quad N—N—N \\
\diagup \quad\quad\quad \diagdown \\
M \quad\quad\quad\quad M^\star
\end{array}
$$

$[(Ph_3P)_2CuN_3]_2$ [64], assumed as intermediates [65].

Cyanamide   M—N—C—NR $[Pd(NCNH)_4]^{2-}$, $[Pd(NCNPh)_4]^{2-}$ [66].

$$
\begin{array}{c}
\quad NR—C—N \\
\diagup \\
M
\end{array}
$$

Table 9.1 Continued

Tricyanomethanide   M—N—C—C (structure with C—N groups) $Ph_3SnNCC(CN)_2$ [67], $(py)_4Ni[NCC(CN)_2]_2$ [68].

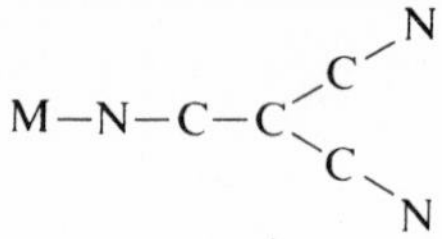

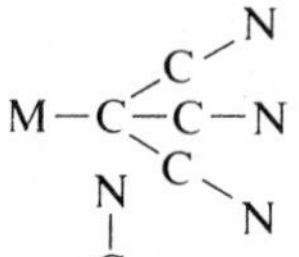

M—C—C—N (structure) $[(OC)_5WC(CN)_3]^-$ [69, 70], $trans\text{-}[(Ph_3P)_2Pt(H)C(CN)_3]$, $cis\text{-}[(Ph_3P)_2Pt(Cl)C(CN)_3]$ [70]

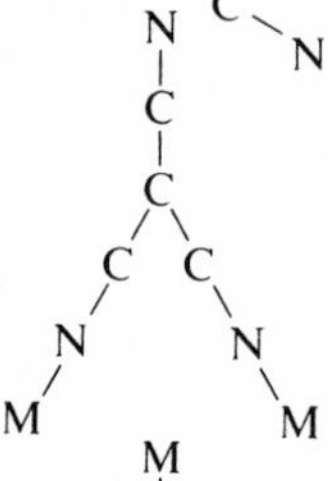

(branched structure with C, N, M) $[Co(NO)_2C(CN)_3]_n(?)$ [71]

(branched structure with M, N, C, M*) $Cu[C(CN)_3]_2$ [72], $Ni[C(CN)_3]_2$ [73]

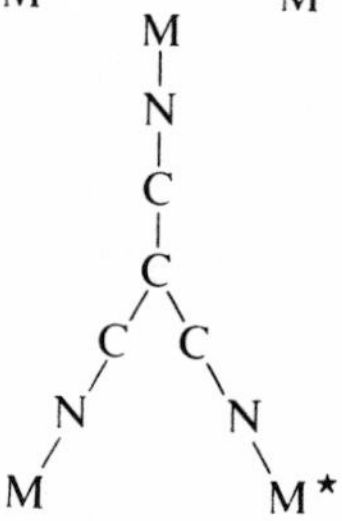

Dicyanoamide   M—N—C—N (structure with C—N groups) $Co\{N(CN)_2\}_4]^2$ [74], $Ph_3SnN(CN)_2$ [75]

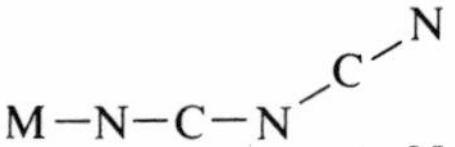

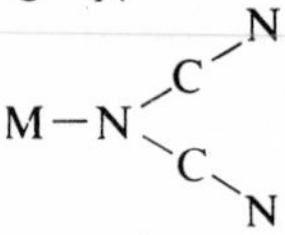

M—N (structure with C—N groups)   (Ref. 12)

(branched structure with M, N, C) $\{(CH_3)_2Sn[N(CN)_2]_2\}_n$
$[(CH_3)_3SnN(CN)_2]_n$ [430]

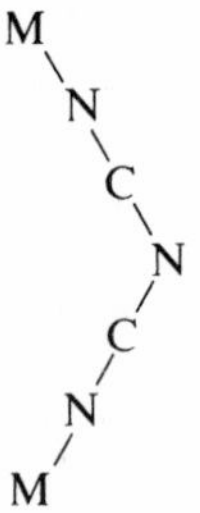

Nitrosodicyanomethanide   M (structure with N, C, N, O, M) $(py)_4Ni[ONC(CN)_2]_2$ [76]

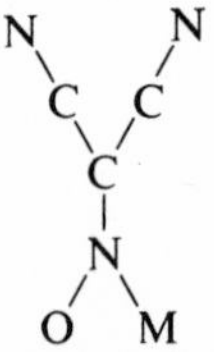

---

a The modes of coordination which have been established by x-ray structural determination are marked by *

† NNNAs is $(Et_2NCH_2CH_2)_2N(CH_2CH_2AsPh_2)$; etu is $SC(NHCH_2)_2$; Et_4dien is $Et_2NC_2H_4NHC_2H_4NEt_2$

9.3   METHODS OF PREPARATION

Reports of several representatives of the class of pseudohalide complexes were made early in the nineteenth century; in fact, the first known co-ordination complex, Prussian Blue[77], was discovered as early as 1704. Since then cyano complexes of most transition metals have been prepared, the cyanide ion being the most thoroughly studied pseudohalogen ligand. On the other hand, a number of the less familiar fulminato and azido metal compounds can also be found in the older literature, e.g. $Na_4[Fe(CNO)_6]\cdot$ $18H_2O$ [78], $Na[M(CNO)_2]$ (M = Ag, Au) [79] and $Na_2[M(CNO)_4]\cdot5H_2O$, (M = Ni, Pd, Pt) [79], as well as $M(N_3)_2$ (M = Ni, Co, Zn, Mn), and $Fe(N_3)_3$, all containing varying amounts of crystal water[80]. In 1824, J. Liebig and Gay-Lussac discovered that silver fulminate AgCNO had the same com-position as silver isocyanate AgNCO, providing the first example of isomerism[81]. Isolation of $Co(N_3)_4^{2-}$ was achieved by Senise[82] in 1959, and Wiberg and Michaud[83] reported the preparation of $Sn(N_3)_6^{2-}$. Later in 1964 hexa-azidoferrate(III) was synthesised by Kröhnke and Sander[84]. It was only recently that the interest in the chemistry of both pseudohalides has risen again, and that extensive studies have been carried out particularly with regard to their coordination behaviour[30, 31, 85, 86]. Although the selenocyanato complexes of platinum[87], gold[87] and mercury[88] have long been known, very little has since been reported on selenocyanato complexes until the past few years. The only transition metal cyanato complexes which had been described prior to 1964 were $K_2[Co(NCO)_4]$ [89], $K[M(NCO)_3]$ (M = Cu, Cd), and $[M\{(CH_2)_6N_4\}_2(H_2O)_4]$ $[M(NCO)_4]$ (M = Co, Ni, Cu) [90].

In contrast, much work has been directed also by earlier authors[91] to complex thiocyanates. The metals were mainly from the first transition series and the preparation and analytical characterisation of complexes, such as $Na_2[Co(NCS)_4]\cdot8H_2O$, $K_4[Ni(NCS)_4]\cdot4H_2O$, $Na_3[Cr(NCS)_6]\cdot$ $12H_2O$, $K_3[Fe(NCS)_6]\cdot4H_2O$ [92] and $(enH)_2Fe(NCS)_4\cdot4H_2O$ [93], were des-cribed.

To date, as far as the preparation particularly of homogeneous pseudo-halide complexes is concerned, the 'new' idea is to isolate products free from 'water of crystallisation'. Thus the possibility of the water molecules partially replacing the pseudohalide ligands in the coordination sphere is ruled out. Numerous new homogeneous pseudohalide complexes, meeting these requirements, have since been obtained by carrying out the preparations in non-aqueous solvents, together with the use of large quaternary ammonium, phosphonium, and arsonium cations, such as $N(C_2H_5)_4^+$ and $M(C_6H_5)_4^+$ (where M is P,As). Particularly the bis(triphenylphosphine)iminium cation, $[(C_6H_5)_3P]_2N^+$, first synthesised by Appel and Hauss[94] and recently prepared by Ruff and Schlientz[95] by a simpler method, proved to be extremely useful for stabilising unstable complex anions[96]. The complex salts of these large cations are largely insoluble in water and isolation of the complex anions from aqueous solution is thus considerably simplified. They also show good crystallisation properties from organic solvents. Further, the application of large counter ions, which also include complex cations of the type $[ML_3]^{n+}$ (M = Cr^{III}, Ni^{II}, Co^{II}, L = o-phenanthroline,2,2'-bipyridyl,

Table 9.2 Homogeneous azido complexes

$V(N_3)_6^{3-}$ [86]		
$VO(N_3)_4^{2-}$ [85]		
$Cr(N_3)_6^{3-}$ [85, 103]		
$Mn(N_3)_4^{2-}$ [85]		
$Fe(N_3)_6^{3-}$ [84]	$Ru(N_3)_6^{3-}$ [86]	$Os(N_3)_6^{3-}$ (?) [86]
$Fe(N_3)_5^{2-}$ [85]		
$Co(N_3)_4^{2-}$ [82, 100]	$Rh(N_3)_6^{3-}$ [85, 86]	$Ir(N_3)_6^{3-}$ [86]
$Ni(N_3)_6^{4-}$ [104]	$Pd(N_3)_4^{2-}$ [85, 86]	$Pt(N_3)_6^{2-}$ [85, 86]
	$Pd_2(N_3)_6^{2-}$ [85]	
$Cu(N_3)_6^{4-}$ [105]		$Au(N_3)_4^{-}$ [85, 86]
$Cu(N_3)_4^{2-}$ [105, 106]		$Au(N_3)_2^{-}$ [85]
$Cu(N_3)_3^{-}$ [85, 105, 106]		
$Cu_2(N_3)_5^{-}$ [105, 106]		
$Cu(N_3)_2^{-}$ [85]		
$Zn(N_3)_4^{2-}$ [85, 100, 107]	$Cd_2(N_3)_5^{-}$ [85]	$Hg(N_3)_4^{2-}$ [85]
		$Hg(N_3)_3^{-}$ [85]

$UO_2(N_3)_4^{2-}$ [85]

$B(N_3)_4^{-}$ [85]	$Al(N_3)_5^{2-}$ [83]		
	$Al(N_3)_4^{-}$ [83]		
$C(N_3)_3^{+}$ [99]		$Sn(N_3)_6^{2-}$ [83, 85, 100]	$Pb(N_3)_6^{2-}$ [85]
	$P(N_3)_6^{-}$ [101]		
	$P(N_3)_4^{+}$ [102]		

Table 9.3 Homogeneous fulminato complexes

$Fe(CNO)_6^{4-}$ [30, 78]	$Ru(CNO)_6^{4-}$ [31]	
$Co(CNO)_6^{3-}$ [30, 97]	$Rh(CNO)_6^{3-}$ [31]	$Ir(CNO)_6^{3-}$ [31]
$Ni(CNO)_4^{2-}$ [30, 79, 97]	$Pd(CNO)_4^{2-}$ [31, 79, 108]	$Pt(CNO)_4^{2-}$ [31, 79, 108]
$Cu(CNO)_2^{-}$ [79]	$Ag(CNO)_2^{-}$ [79, 109, 110]	$Au(CNO)_2^{-}$ [31, 79]
$Cu(CNO)_3^{2-}$ [79]		
$Zn(CNO)_4^{2-}$ [31]	$Cd(CNO)_4^{2-}$ [31]	$Hg(CNO)_4^{2-}$ [111]

ethylenediamine)[31], proved to be particularly advantageous for the safe synthesis of both azido[82, 85, 86] and fulminato[30, 31, 97] complexes. The presence of these bulky groups 'dilute' and/or stabilise[98] (see below) the more or less explosive complex anions, leading to thermally stable compounds that are easy to handle*. The homogeneous anionic azido and fulminato complexes so far known are listed in Tables 9.2 and 9.3.

In the most commonly used preparative method, metal pseudohalides, e.g. $Ni(NCS)_2$ [92], CuN_3 [85], or other simple metal salts such as $Pd(NO_3)_2$ [85], $MnCl_2 \cdot 4H_2O$ [112] (or even covalent compounds, e.g. $SnCl_4$ [83] Tables 9.2 and 9.3) serving as a starting material, are treated with alkali or ammonium pseudohalides. Complete reaction can be ensured by addition of excess

---

*Alkali salts of azido complexes such as $Na_2Pt(N_3)_4$ and $Na_2[P+(N_3)_6]$ (T. Theophanides and Y. Lafortune (1968). *Canad. Spectroscopy*, **13**, 3) are extremely explosive and their isolation should be avoided.

pseudohalide ion. Besides water, many solvents have been used as reaction media — liquid ammonia, alcohols, acetone, nitromethane, acetonitrile, and tetrahydrofuran, being perhaps the most common. The resulting complex anions are subsequently precipitated on addition of the large cations. In this connection, it should be mentioned, that often the stereochemistry, and sometimes even the mode of linkage (see below) of the resulting complexes depends strongly on the nature of the chosen cation. An example[113] is the precipitation of triazidomercurate(II), $[Hg(N_3)_3]^-$ (with a $P(C_6H_5)_4^+$ cation), or tetra-azidomercurate(II), $[Hg(N_3)_4]^{2-}$ (using $N[P(C_6H_5)_3]_2^+$), from aqueous solutions of the same molar ratios $Hg^{2+}:N_3^-$. Similar observations have been made by Funk and Böhland[114] when various ammonium cations, such as quinolinium, piperidinium, and p-toluidinium, precipitate complexes of tungsten with a varying content of NCS^-. The stabilising effect of large counter ions, preferably those of the same but opposite charge, is best demonstrated by the recent isolation[98, 115] of $Ni(CN)_5^{3-}$ (the alkali salts of which are not stable at room temperature) as the salts of $[Cr(NH_3)_6]^{3+}$ and $[Cr(en)_3]^{3+}$.

Mixing concentrated solutions of the metal pseudohalide and the large cation pseudohalide in the correct molar ratio represents a more direct route to homogeneous pseudohalide complexes, e.g.

$Ni(NCS)_2 + 4Cat^+NCS^- \rightarrow (Cat^+)_4[Ni(NCS)_6]$
$(Cat^+ = \text{e.g. } (CH_3)_4N^+, (C_2H_5)_4N^+$ [116]
$C_9H_8N^+ \text{ (quinolinium) [117])}$

$M(NCS)_2 + 2Ga(NCS)_3 \rightarrow M[Ga(NCS)_4]_2$ [118]
$(M = Ca,Sr,Ba)$

$2AgCNO + KBr \rightarrow K[Ag(CNO)_2] + AgBr$ [14, 79]

A widely used method for the preparation of mixed ligand complexes is the addition of neutral ligands (amines or phosphines) to the anionic complex species in solution, followed by immediate precipitation of the 'mixed ligand' pseudohalides[85, 108, 119]:

$$[M(Ps)_4]^{2-} + 2L \xrightarrow[-2Ps^-]{H_2O/ethanol} \text{cis- or trans-}[L_2MPs_2]$$

$(M = Ni,Pd,Pt; Ps(pseudohalide) = N_3,CNO,NCS;$
$L = P(C_6H_5)_3, As(C_6H_5)_3; 2L = bipyridyl)$

Again, the bulky group L provides additional stabilisation for complex azides and fulminates.

Extensive studies have been directed by Behrens and co-workers[120] towards the substitution reactions of metal carbonyls and nitrosyls with cyanide and thiocyanate, e.g.

$$Co(NO)(CO)_3 + 3CN^- \rightarrow [Co(NO)(CN)_3]^{3-} + 3CO$$

This reaction, which also yielded the two intermediates $[CoNO(CO)(CN)_2]^{2-}$ and $[CoNO(CO)_2CN]^-$, was carried out in liquid ammonia. Substitution of both CO and NO occurs in the reaction of $Mn(NO)_3CO$ with alkali cyanide[121] (also in liquid ammonia) yielding the dimeric species $[(NC)_2(ON)_2 Mn-Mn(NO)_2(CN)_2]^{4-}$. By reacting the dinuclear carbonyl metalate $[Cr_2(CO)_{10}]^{2-}$ with $FeCl_3$ in the presence of potassium thiocyanate according to

$$[Cr_2^{-1}(CO)_{10}]^{2-} + 2Fe^{3+} + 2SCN^- \rightarrow 2[Cr^0(CO)_5NCS]^- + 2Fe^{2+}$$

thiocyanato pentacarbonyl chromate(0) is produced[47]. This compound was also obtained by Wojcicki and Farona by a completely different approach[122].

$$M(CO)_6 + [(CH_3)_4N]SCN \xrightarrow[-CO]{\text{diglyme}} [(CH_3)_4N][M(CO)_5NCS]$$
$$(M = Cr, Mo, W)$$

Also substitution products $[M(CO)_3ophenX]^-$ $(X = CN, NCS, N_3 S_4)$ have been reported[429].

In the dimeric species $Cr_2(CO)_{10}SCN$ [47], $[Cr_2(CO)_{10}SCN]^-$ [123] and $[Cr_2(CO)_{10}CN]^-$ [123], the presence of thiocyanato and cyano bridges has been ascertained by i.r. spectroscopy. SCN bridging linkages have also been established by i.r. methods[124] in the polymeric thiocyanatonitrosyl compounds of iron, cobalt and nickel $[Fe(NO)_2SCN]_n$, $[Co(NO)_2SCN]_n$, and $[Ni(NO)SCN]_n$ prepared by Hieber et al.[125]. Recently, the investigation of the preparation and the stereochemistry of pseudohalo carbonyl metalates has been further extended by Ruff[126] who by photolysis of mixtures of bis (triphenylphosphine)iminium pseudohalides and Group VI metal carbonyls or iron-pentacarbonyl produced stable salts of mononuclear and dinuclear carbonyl anions, e.g.

$$Fe(CO)_5 + CN^- \xrightarrow{h\nu} [Fe(CO)_4CN]^- + CO$$
$$[Fe(CO)_4CN]^- + Fe(CO)_5 \rightarrow [Fe_2(CO)_8CN]^- + CO$$

Metathetical reactions starting from chloro complexes, or mixed ligand chloro complexes and alkali pseudohalides in water, or in suitable non-aqueous solvents (heterogeneous reaction) provide another method frequently used in the synthesis of pseudohalide complexes,

e.g. $[(n-C_4H_9)_4N]_2[Re_2Cl_8] + 8SCN^- \xrightarrow[\text{methanol}]{\text{in acidified}}$

$$[(n-C_4H_9)_4N]_2[Re_2(NCS)_8] + 8Cl^- \text{ [127]}$$

$$LMCl_2 + 2NaN_3 \rightarrow LM(N_3)_2 + 2NaCl \text{ [128]}$$

(where M is Ni Pd, Pt, and L is a bidentate ligand, such as bipy, o-phen, or diphos) [129]

$$trans-[Ir(en)_2Cl_2]ClO_4 \xrightarrow[H_2O]{N_3^-, PF_6^-} trans-[Ir(en)_2(N_3)_2]PF_6 \text{ [129]}$$

Similarly, a number of both monomeric and dimeric pseudohalo carbonyl manganates(I) and rhenates(I) have been obtained from pentacarbonyl metal halides with SCN^-, $SeCN^-$ and N_3^-, e.g.

$[Re(CO)_3(NCS)_3]^{2-}$ [130, 131], $[Mn(CO)_4(NCS)_2]^-$ [131] $[Mn_2(CO)_6(NCSe)_4]^{2-}$ [132], $[Mn_2(CO)_6(N_3)_3]^-$ and $[Re_2(CO)_6(N_3)_2(NCO)_2]^{2-}$ [67]

A mixed ligand complex containing both cyano and azido groups attached to the same metal[133], $K[Au(CN)_2(N_3)_2]$, was obtained by mixing methanolic solutions of $K[Au(CN)_2Cl_2]$ and KN_3. Oven-dried potassium thiocyanate in acetone was used for the conversion by metathetical reaction of organotin halides to the corresponding isothiocyanates, $R_2Sn(NCS)_2$ [134].

A series of cyanoammine and cyanodiamine complexes of cobalt(III)[135], cis- and trans-$[Co(CN)_2(NH_3)_4]^+$, trans-$[Co(CN)_2(en)_2]^+$ and mer-[Co(CN)_3dien] were synthesised by the reaction of potassium cyanide with $[Co(NH_3)_6]^{3+}$, $[Co(en)_3]^{3+}$, or $[Co(dien)_2]^{3+}$ in the presence of activated charcoal at low temperatures (0–5 °C). In a few cases metathesis is accompanied by reduction of the central metal ion

$$[Ru^{IV}Cl_6]^{2-} \xrightarrow[H_2O]{SCN^-} [Ru^{III}(NCS)_6]^{3-} \text{ [136]}$$

$$W^{VI}Cl_6 \xrightarrow{NH_4SCN} [W^V(OH)_3(NCS)_3]^- \text{ [114]}$$

Air oxidation is involved in the transformation of $[C_5H_5Fe(CO)_2]_2$ into the two isomers $C_5H_5Fe(CO)_2SCN$ and $C_5H_5Fe(CO)_2NCS$ in the presence of potassium thiocyanate and a strong acid containing a non-coordinating anion, namely HPF_6 [137].

For the preparation of moisture sensitive complexes (e.g. $[Sn(N_3)_6]^{2-}$ [85]), the use of highly concentrated aqueous solutions of the pseudohalide anion has proved useful. $[Pb(N_3)_6]^{2-}$, but on the other hand, could only be obtained by extraction of a dry mixture of $(NH_4)_2[PbCl_6]$, NaN_3, and $[As(C_6H_5)_4]Cl$ with CH_2Cl_2 containing a trace of water in order to start the reaction[85]. (See also Ref. 136.) The preparation of the $[Re(SCN)_6]^{2-}$ complex (see also Ref. 127) was achieved in a 'fused salt' reaction by dissolving K_2ReCl_6 in molten $KSCN$ [138]. Reactions of several transition metals in molten potassium cyanide with mercury(II) cyanide yielding some unusual cyano complexes like $KMn(CN)_3$, $K_2Cu(CN)_3$, or $K_3Co(CN)_5$ have been described by Kleinberg et al.[139]. The same authors also were able to obtain certain cyano complexes of metals in lower oxidation states ($K_4[Ni(CN)_4]$, $K_4[Ni_2(CN)_6]$) by dissolving the standard cyano complexes ($K_2[Ni(CN)_4]$) in a KCN-melt at >500 °C [140].

Reduction of the central metal ion was also observed in the reaction of ruthenate(VI) with fulminate ions, yielding hexafulminato-ruthenate(II) [31]. The fluoride ligand in $(OC)FM(PPh_3)_2$ ($M = Ir^I$, Rh^I) is an excellent leaving group, i.e. it is easily replaced by univalent anionic ligands such as pseudohalides, giving rise to a series of new d^8 complexes[141]. From i.r.-studies (vCO) an electronegativity scale for anionic ligands was derived[141].

A variation of the above method involves the use of heavy metal pseudohalides as the reactive material, in particular silver salts.

$$[(CH_3)_3PtI]_4 + 4AgSCN \rightarrow [(CH_3)_3PtSCN]_4 + 4AgI \quad [142]$$

$$[CuBr_4]^{2-} + 4AgN(CN)_2 \rightarrow [Cu(N(CN)_2)_4]^{2-} + 4AgBr \quad [12]$$

The synthesis by this method of a variety of trimethyl- and triphenyl-antimony pseudohalides of the type R_3SbX_2 and $(R_3SbX)_2O$ ($X = N_3$, NCO, NCS) from the corresponding organoantimony dihalides and the appropriate silver salt was recently reported[143]. The isocyanato compounds of palladium and platinum,

$$L_2MCl_2 + 2AgNCO \rightarrow L_2M(NCO)_2 + 2AgCl$$

($M = Pd, Pt; L = py, PPh_3, AsPh_3, SbPh_3; L_2 = bipy$)

must be prepared in the absence of water[144]. Generally, reactions employing the more covalent silver pseudohalides occur faster, sometimes being connected with a redox process, e.g.

$$[Fe^{II}Br_4]^{2-} + 5AgN_3 \rightarrow [Fe^{III}(N_3)_5]^{2-} + Ag^0 + 4AgBr \quad [85]$$

If direct substitution of a halide is not possible, reaction may be achieved via a more reactive intermediate (often an aquo complex), e.g.[145]

$$trans\text{-}[RuCl_2(NH_2CH_2CH_2NH_2)_2]Cl \xrightarrow[\text{2.N}_3^-, \text{PF}_6^-]{\text{1.silver p-toluenesulphonate}}$$

$$\rightarrow trans\text{-}[Ru(N_3)_2(NH_2CH_2CH_2NH_2)_2]PF_6 \text{ or}$$
$$trans\text{-}[Ru(N_3)(N_2)(NH_2CH_2CH_2NH_2)_2]PF_6.$$

In an analogous manner, the special synthetic problems encountered in the syntheses of (iso)selenocyanato (formation of phosphine selenides[146]) and isocyanato complexes (formation of a carbamato complex species) are circumvented by the following reaction sequences[147],

$$L_2PdCl_2 + 2AgNO_3 \xrightarrow[-2AgCl]{H_2O} [PdL_2(H_2O)_2]^{2+} + 2NO_3^-$$

$$\xrightarrow{NaSeCN} L_2Pd(SeCN)_2 \text{ (}L = \text{e.g. } 4\text{-}NH_2\text{-pyridine, } (n\text{-}C_4H_9)_3P$$
$$2L = 5\text{-}NO_2\text{-phenanthroline, ethylendiamine)}$$

In other cases, success was achieved by using N,N-dimethylformamide as the solvent in the reaction, which then proceeded via a $[PdL_2(HCON(CH_3)_2)_2]^{2+}$ intermediate[148].

$$[Co(NH_3)_5N_3]^{2+} \xrightarrow[OP(OC_2H_5)_3]{NOClO_4} [Co(NH_3)_5OP(OC_2H_5)_3]^{3+} + N_2 + N_2O$$

$$\xrightarrow[DMSO]{[(C_6H_5)_4As]NCO} [Co(NH_3)_5NCO]^{2+} + OP(OC_2H_5)_3 \xrightarrow[H_2O]{Br^-}$$
$$[Co(NH_3)_5NCO]Br_2$$

This exchange of an azide ligand for another ligand L requiring a reagent that contains the NO^+ species, proved to be a method of general applicability[149]. For example, by use of the same method, the thermodynamically labile linkage isomer, isothiocyanato-pentacyano cobaltate(III) was made accessible[150] (see Section 9.4, however)

$$[Co(NH_3)_5N_3]^{2+} \xrightarrow{CN^-} [Co(CN)_5N_3]^{3-} \xrightarrow{HNO_2} [Co(CN)_5H_2O]^{2-}$$
$$\xrightarrow{SCN^-} [Co(CN)_5NCS]^{3-}$$

Other methods reported – mostly limited in scope – involve the reaction of pseudohalogens and inter(halogen)pseudohalogens with:

(a) carbonylmetalate anions[151]
$$Na[Mn(CO)_5] + ClSCN \rightarrow NaCl + Mn(CO)_5SCN$$

(b) carbonyl halides[152]
$$Cr(CO)_5I + PsI \rightarrow Cr(CO)_5Ps + I_2$$
$$(Ps = CN, SCN)$$

(c) hydrido complexes[137]
$$C_5H_5Mo(CO)_3H + (SCN)_2 \rightarrow C_5H_5Mo(CO)_3SCN + HSCN$$

Another interpseudohalogen reagent, which was recently introduced by Dehnicke[8] for a variety of azidisation reactions, is ClN_3, e.g.

$$R_2Hg + ClN_3 \rightarrow RHgN_3 + RCl \ ^{153}$$
$$(R = cyclopropyl, - pentyl, and - hexyl)$$
$$TiCl_4 + ClN_3 \rightarrow TiCl_3N_3 + Cl_2 \ ^{154}$$

$$3BCl_3 + 3ClN_3 \xrightarrow{0\,°C} [BCl_2N_3]_3 + 3Cl_2 \ ^{155}$$

Many examples in the older literature use cyanogen and its derivatives as pseudohalogenating agents of organometallic compounds[15]. $(CN)_2$ and ICN in ether or tetrahydrofuran were also used to introduce the cyano group into $Na_2[Cr(CO)_5]$ to give $Cr(CO)_5CN$ [156]. The previously reported oxidative addition of cyanogen to some d^{10} transition metal phosphine complexes of nickel(0), palladium(0) and platinum(0), to give the (cis) dicyano compounds, is of interest[157, 158]. A different product, $RhCl(C_2N_2)[P(C_6H_5)_3]_2$ was obtained by the reaction of $RhCl[P(C_6H_5)_3]_3$ with $(CN)_2$ in benzene[158]. Bridging dicyano groups of the type $(OC)_5MNC\text{-}CNM(CO)_5$ (M = Cr, Mo, W) were described by Guttenberger[159]. Oxidative addition of HN_3, or of acyl- or aryl-azides to tetrakis(triphenylphosphine)platinum(0) or to $Pt(PPh_3)_2O_2$ results in the formation of cis-$Pt(PPh_3)_2(N_3)_2$ [160]

$$(Ph_3P)_4Pt + 2HN_3 \xrightarrow[-2PPh_3]{} (Ph_3P)_2Pt(N_3)_2 + H_2$$

$Pt(PPh_3)_4$ and dicyanoketenimine afford the hydrido-tricyanomethanido platinum(II) complex:

$$Pt(PPh_3)_4 + HN{=}C{=}C(CN)_2 \rightarrow (PPh_3)_2Pt(H)C(CN)_3$$

which, from its ^{14}N-n.m.r. spectrum, presumably contains $C(CN)_3$ bonded via the carbon atom to the metal[161].

The oxidative addition of nitromethane to $(Ph_3P)_4Pt$ followed by dehydration provides a convenient and safe method for the preparation of difulminato-bis(triphenylphosphine)platinum(II), $(Ph_3P)_2Pt(CNO)_2$ [162]. Since nitromethane is believed to react as an acid, the presence of polar solvents is essential. Similarly, gold(III) complexes were prepared from gold(I) complexes using the oxidative addition method.

$$LAuNCO + Br_2 \rightarrow LAuBr_2(NCO) \text{ [163]}$$
$$(L = PPh_3, AsPh_3)$$
$$[Au(CNO)_2]^- + Br_2 \rightarrow [Au(CNO)_2Br_2]^- \text{ [31]}$$

Thiocyanogen was added to the linear chloro(triphenylphosphine)gold(I) [163],

$$Ph_3PAuCl + (SCN)_2 \rightarrow Ph_3PAu(SCN)_2Cl$$

and to various square planar-carbonyl complexes of Rh^I,

$$[RhX(CO)(Ph_3P)_2] + (SCN)_2 \rightarrow [RhX(SCN)_2(CO)(Ph_3P)_2] \text{ [164]}$$
$$(X = F, Cl, Br, I, \text{ or } SCN)$$

to produce the respective thiocyanato complexes of the trivalent metals.

In several cases, where other methods have failed, hydrogen pseudo-halides have been effectively used to introduce the pseudohalide ligand via acid-base reaction:

$$NiCO_3 + 2HN_3 + 4N_3^- \rightarrow Ni(N_3)_6^{4-} + H_2O + CO_2 \text{ [104]}$$

$$ZnCO_3 + 4HN_3 \xrightarrow{\text{KOH}} K_2[Zn(N_3)_4] \text{ [107]}$$

$$[(C_6H_5)_3P]_2PtCO_3 + 2HN_3 \rightarrow \textit{cis-}[(C_6H_5)_3P]_2Pt(N_3)_2 \text{ [165]}$$

$$LiBH_4 + 4HN_3 \rightarrow Li[B(N_3)_4] + 4H_2 \text{ [83]}$$

$$Ph_3PAuOAc + HNCO \xrightarrow[20\,°C]{\text{ether}} Ph_3PAuNCO + AcOH \text{ [166]}$$

$$Ph_3PbOH + HN{=}C{=}C(CN)_2 \rightarrow Ph_3PbN{=}C{=}C(CN)_2 + H_2O \text{ [67]}$$

Organotin cyanides of the type R_3SnCN or $R_2Sn(CN)_2$ have been obtained by the reaction in pentane of $R_3SnN(C_2H_5)_2$ or $R_2Sn[N(C_2H_5)_2]_2$ with liquid hydrogen cyanide[167]. In place of free fulminic acid, nitromethane has been successfully used with phenylmercury hydroxide or bis(methylmercury) oxide to give organo mercury fulminates under mild conditions[168], e.g.

$$C_6H_5HgOH + CH_3NO_2 \xrightarrow[20\,°C]{-H_2O} \left[C_6H_5{-}Hg\underset{O}{\overset{CH_2}{\diagup}}N{=}O \right] \xrightarrow{-H_2O} C_6H_5HgCNO$$

Also selenium selenocyanate and selenium dicyanide have been reacted with organometallic and carbonyl compounds, e.g.

$$Pb(C_6H_5)_4 + Se(SeCN)_2 \rightarrow (C_6H_5)_3PbSeCN + C_6H_5SeCN + Se \text{ [169]}$$
$$C_5H_5Fe(CO)[P(C_6H_5)_3]CH_2C_6H_5 + Se(SeCN)_2 \rightarrow$$
$$C_5H_5Fe(CO)[P(C_6H_5)_3]SeCN + C_5H_5Fe(CO)[P(C_6H_5)_3]NCSe + Se \text{ [170]}$$
$$(C_6H_5)_4Pb + Se(CN)_2 \rightarrow (C_6H_5)_3PbCN + C_6H_5SeCN \text{ [169]}$$

By contrast, action of $Se(SeCN)_2$ on hexaphenyldistannane results in the cleavage of the Sn—Sn bond to give the isoselenocyanate, $(C_6H_5)_3SnNCSe$[169].

In a few cases isocyanates or (iso)thiocyanates could be prepared by fusion of organometallic hydroxides with respectively urea[171] or thiourea[172]. The preparation of N-bonded cyanatopentammine cobalt(III) by the reaction of $(NH_3)_5CoOH_2^{3+}$ with urea, several N-substituted ureas, and urethane has been reported[173]. Evidence is presented that the reaction proceeds through an N-bonded urea complex which decomposes to $(NH_3)_5CoNCO^{2+}$ and NH_4^+. Crystals of silver isocyanate could be grown from an aqueous solution containing silver nitrate and urea[174].

Introduction of oxygen, sulphur (at elevated temperatures), and selenium (at room temperature) into trialkylcyano-silanes and germanes (= chalcogenation) leads respectively to the formation of isocyanates, (iso)thiocyanates, and (iso)selenocyanates[175]. Similar reactions do not occur with tellurium. Conversely, triphenylphosphine will remove oxygen, sulphur or selenium respectively from fulminates, isothiocyanates or isoselenocyanates:

$$(Ph_3P)_2Pt(CNO)_2 + 2PPh_3 \rightarrow (Ph_3P)_2Pt(CN)_2 + 2Ph_3PO \text{ [162, 176]}$$
$$R_3MNCSe + PPh_3 \rightarrow R_3MCN + Ph_3PSe \text{ [146, 175]}$$

Oxidising reagents, such as H_2O_2 [177], Ce^{IV} in acidic solution[178], or $S_2O_8^{2-}$ at low acidity[178] behave similarly, e.g.

$$[Co(NH_3)_5NCS]^{2+} \xrightarrow{\text{Ox}} [Co(NH_3)_5CN]^{2+}$$

It has been pointed out that such 'oxidative substitution' reactions might be of considerable utility in preparing specific cyano complexes which are presently unknown[178]. Furthermore, the formation of isocyanato, thiocyanato and azido complexes by reaction of coordinated carbonyl and nitrosyl ligands with some N-containing nucleophiles is of interest. Isocyanato complexes of transition metals have been obtained from neutral or cationic metal carbonyls, such as $M(CO)_6$ (M = Cr,Mo,W), $C_5H_5Fe(CO)_3^+$, $C_5H_5Ru(CO)_3^+$, $Re(CO)_6^+$, by reaction with azide ions[179], hydrazine[180], chloramine[181], or hydroxylamine[181], e.g.

$$W(CO)_6 + N_3^- \rightarrow [W(CO)_5NCO]^- + N_2 \text{ [179]}$$
$$[Re(CO)_6]^+ + N_3^- \rightarrow Re(CO)_5NCO + N_2 \text{ [182]}$$
$$[(Ph_3P)_2PtClCO]^+ + N_3^- \rightarrow (Ph_3P)_2PtCl(NCO) + N_2 \text{ [38]}$$
$$[C_5H_5Fe(CO)_3]^+ + N_2H_4 \rightarrow C_5H_5Fe(CO)_2NCO + NH_4^+ \text{ [180]}$$

Kinetic studies[179, 183] suggest that the reactions proceed by attack of the nucleophilic agent on the carbon atom of a CO group to produce an intermediate, which rearranges with loss of N_2, amine HCl, or H_2O respectively, yielding the stable isocyanato complexes.

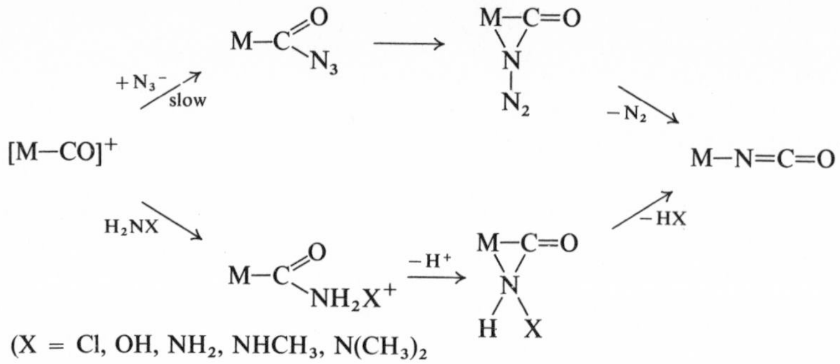

$$[\text{M-CO}]^+ \xrightarrow[\text{slow}]{+\text{N}_3^-} \quad \text{M-C}\overset{O}{\underset{N_3}{\diagup}} \longrightarrow \begin{array}{c} \text{M-C=O} \\ \diagdown \diagup \\ \text{N} \\ | \\ \text{N}_2 \end{array} \xrightarrow{-\text{N}_2} \text{M-N=C=O}$$

$$[\text{M-CO}]^+ \xrightarrow{\text{H}_2\text{NX}} \quad \text{M-C}\overset{O}{\underset{NH_2X^+}{\diagup}} \xrightarrow{-\text{H}^+} \begin{array}{c} \text{M-C=O} \\ \diagdown \diagup \\ \text{N} \\ \diagup \diagdown \\ \text{H} \quad \text{X} \end{array} \xrightarrow{-\text{HX}} \text{M-N=C=O}$$

(X = Cl, OH, NH$_2$, NHCH$_3$, N(CH$_3$)$_2$

In the reactions of the cationic metal carbonyls, $[\text{C}_5\text{H}_5\text{Fe(CO)}_3]^+$, $[\text{C}_5\text{H}_5\text{Ru(CO)}_3]^+$, $[\text{Re(CO)}_5\text{NH}_2\text{CH}_3]^+$ with hydrazine and methyl substituted hydrazines, the unstable carbazoyl derivatives could be isolated[180]. It may be noted that the decomposition of the suggested intermediates in the reaction of metal carbonyls with azide, hydroxylamine, and chloramine corresponds to the Curtius, Lossen, or Hofmann degradation of organic acylazides, hydroxamic acids, or carboxylic amides, respectively.

Similar reactions of the azide ion and hydrazine have been recently carried out with the thiocarbonyl cation $[\text{C}_5\text{H}_5\text{Fe(CO)}_2\text{CS}]^+$ yielding the isothiocyanate complex $\text{C}_5\text{H}_5\text{Fe(CO)}_2\text{NCS}$ [184]. The formation of isocyanato complexes seems to be generally favoured; thus u.v. irradiation of a benzene solution of triphenylphosphine and dicarbonyl-π-cyclopentadienylnitrosyl-molybdenum yields $(\pi\text{-C}_5\text{H}_5)\text{Mo(CO)(PPh}_3)_2\text{NCO}$ [34]. An organometallic nitrene which then captures carbon monoxide to give isocyanate is suggested as an intermediate:

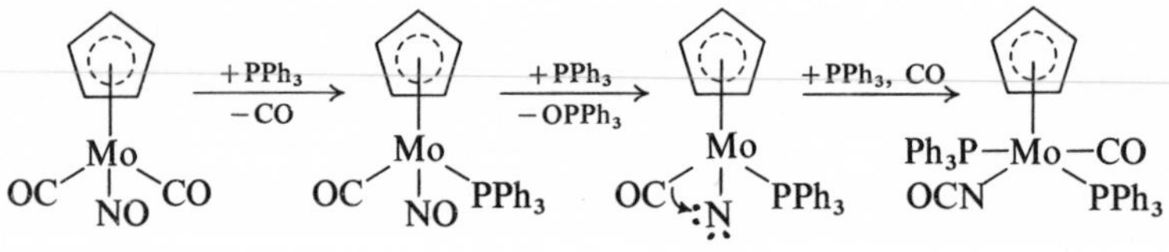

An azido complex of ruthenium(II) is formed by reaction of $[\text{Ru(das)}_2\text{ClNO}]^{2+}$ (das is o-phenylenebis(dimethylarsine)) with hydrazine, involving nucleophilic attack at the nitrosyl nitrogen atom[185].

$$3\text{N}_2\text{H}_4 + [\text{Ru(das)}_2\text{ClNO}]\text{Cl}_2 \rightarrow [\text{Ru(das)}_2\text{ClN}_3] + 2\text{N}_2\text{H}_5\text{Cl} + \text{H}_2\text{O}$$

Similarly, besides the nitrogen and dinitrogen oxide complexes $[\text{Ru(NH}_3)_5\text{N}_2]^{2+}$ and $[\text{Ru(NH}_3)_5\text{N}_2\text{O}]^{2+}$, azidopentammineruthenium(III) $[\text{Ru}^{3+}(\text{NH}_3)_5\text{N}_3]^{2+}$ has been obtained from $[\text{Ru(NH}_3)_5\text{NO}]^{3+}$ and hydrazine hydrate[186].

A number of azide complexes, preferably those of d^8 metals, e.g. $(\text{PPh}_3)_2$ $\text{M(N}_3)_2$ (M = Pd, Pt), $(\text{PPh}_3)_2\text{M(CO)N}_3$ (M = Rh, Ir) can be converted under mild conditions with carbon monoxide to the corresponding iso-cyanate complexes[238]*

9.4   LINKAGE ISOMERISM

9.4.1   General remarks

From inspection of Table 9.4 it becomes obvious, that in the thiocyanato compounds the SCN groups may be bonded to the metal either through the

Table 9.4   Homogeneous thiocyanato complexes

(a) d-elements

$Sc(NCS)_6^{3-}$ [187]	$Y(NCS)_6^{3-}$ [194]	*$La(NCS)_6^{3-}$ [206]
$Ti(NCS)_6^{3-}$ [188]	$Zr(NCS)_6^{2-}$ [195]	$Hf(NCS)_6^{2-}$ [195]
$Ti(NCS)_6^{2-}$ [189]	*$Zr(NCS)_8^{4-}$ [196]	*$Hf(NCS)_8^{4-}$ [196]
$V(NCS)_6^{3-}$ [190]	$Nb(NCS)_6^{2-}$ [197]	$Ta(NCS)_6^{-}$ [197, 198]
	$Nb(NCS)_6^{-}$ [197–199]	
$Cr(NCS)_6^{3-}$ [92, 136]	$Mo(NCS)_6^{3-}$ [136, 200]	*$W(NCS)_6$ [207]
$Mn(NCS)_4^{2-}$ [112, 117]	$Tc(NCS)_6^{-}$ [201]	$Re(NCS)_6^{2-}$ [127]
$Mn(NCS)_6^{4-}$ [112, 117]		$Re_2(NCS)_8^{2-}$ [127]
		$Re(SCN)_6^{2-}$ [138]
		$Re(SCN)_6^{-}$ [208]
$Fe(NCS)_4^{2-}$ [112]		$Os(NCS)_6^{3-}$ [136]
$Fe(NCS)_6^{4-}$ [92]		
$Fe(NCS)_6^{3-}$ [92, 112, 136]	$Ru(NCS)_6^{3-}$ [136]	
$Co(NCS)_4^{2-}$ [92, 116, 191]		
$Ni(NCS)_4^{2-}$ [92, 116]	$Rh(SCN)_6^{3-}$ [136]	$Ir(SCN)_6^{3-}$ [136]
$Ni(NCS)_6^{2-}$ [116]	$Pd(SCN)_4^{2-}$ [192, 202]	$Pt(SCN)_4^{2-}$ [192, 202]
$Cu(NCS)_4^{2-}$ [116]		$Pt(SCN)_6^{2-}$ [136, 202]
$Zn(NCS)_4^{2-}$ [192, 193]	$(AgSCN)(SCN)^{-}$ [203]	$Au(SCN)_4^{-}$ [136, 192, 202]
$Zn(NCS)_3^{-}$ [193]	$Cd(SCN)_4^{2-}$ [192, 204]	$Hg(SCN)_4^{2-}$ [92, 205, 209]
$Zn(NCS)_6^{4-}$ [193]	$Cd(SCN)_6^{4-}$ [204]	$Hg(SCN)_3^{-}$ [92]
	$Cd(SCN)_{6-n}(NCS)_n^{4-}$ [204]	
	$Cd(SCN)_2(NCS)_2^{2-}$ [205]	

(b) p-elements

$Be(NCS)_4^{2-}$ [210]	$Si(NCS)_4$ [213]	$P(NCS)_3$ [213]
$Al(NCS)_6^{3-}$ [189]	$Sn(NCS)_3^{-}$ [214]	$As(NCS)_3$ [213]
$Ga(NCS)_4^{-}$ [118]	$Sn(SCN)_4^{2-}$	$Se(SCN)_2$ [217]
*$In(NCS)_6^{3-}$ [211]	*$Pb(NCS)_4^{2-}$ [212]	$I(SCN)_2^{-}$ [218]
*$Tl(NCS)_2^{-}$ [212]	*$Pb(NCS)_6^{4-}$ [215]	
	*$Pb(NCS)_8^{6-}$ [216]	

(c) f-elements

$Pr(NCS)_6^{3-}$ [194, 219]	$Tb(NCS)_6^{3-}$ [219]	$Th(NCS)_8^{4-}$ [220]
$Nd(NCS)_6^{3-}$ [194, 206, 219]	$Dy(NCS)_6^{3-}$ [219]	$U(NCS)_8^{4-}$ [220, 221]
$Sm(NCS)_6^{3-}$ [194, 219]	$Ho(NCS)_6^{3-}$ [194, 219]	
$Eu(NCS)_6^{3-}$ [194, 219]	$Er(NCS)_6^{3-}$ [194, 219]	
$Gd(NCS)_6^{3-}$ [219]	$Tm(NCS)_6^{3-}$ [194]	
	$Yb(NCS)_6^{3-}$ [194, 219]	

*No information is available on ligand–metal bonding. In these cases it is common practice to employ the notation MCNS, thus formulating the unknown thiofulminate. We feel this atom inversion, against chemical knowledge, should be strictly avoided since it might give rise to serious confusion.

*The reactions of the coordinated azide will be reviewed elsewhere.

nitrogen atoms (to give isothiocyanates) or through the sulphur atoms (to give thiocyanates). In the case of a cadmium complex, it is suggested that both types of bonding are present. A similar variation is observed for the homogeneous selenocyanato/isoselenocyanato complexes (Table 9.5), and, though very limited in number, even O-bonded cyanato metal complexes have been detected besides a wide number of well established metal compounds of the iso-form (Table 9.6). By way of contrast, for the isoelectronic fulminate ion, CNO^-, coordination appears to occur solely through the carbon atom (Table 9.3). No isofulminato metal complexes have as yet been reported, and it seems unlikely that these isomers can exist. The same marked preponderance of metal carbon bonds holds for the diatomic cyanide species almost exclusively forming cyano compounds, but there is some evidence that the unstable isocyano form exists as reaction intermediate (see below).

In this connection it should be noted, that all of the linkage isomers in transition metal chemistry cited above are well known among the more

Table 9.5 Homogeneous selenocyanato complexes

	$Y(NCSe)_6^{3-}$ [225]	
$V(NCSe)_6^{3-}$ [222]		$In(NCSe)_6^{3-}$ [230]
$Cr(NCSe)_6^{3-}$ [223]	$Mo(NCSe)_6^{3-}$ [136]	$Pb(SeCN)_3^-$ [228]
$Mn(NCSe)_4^{2-}$ [224]		$Pb(SeCN)_6^{4-}$ [228]
$Mn(NCSe)_6^{4-}$ [224, 225]		$Bi(SeCN)_5^{2-}$ [231]
$Fe(NCSe)_4^{2-}$ [224, 225]		$Bi(SeCN)_6^{3-}$ [231]
$Fe(NCSe)_6^{4-}$ [224]		
$Fe(NCSe)_6^{3-}$ [136, 225]		
$Co(NCSe)_4^{2-}$ [224, 226, 227]	$Rh(SeCN)_6^{3-}$ [136, 225, 229]	
$Co(NCSe)_6^{4-}$ [224]		
$Ni(NCSe)_4^{2-}$ [224]	$Pd(SeCN)_4^{2-}$ [224, 225, 229]	$Pt(SeCN)_4^{2-}$ [224, 229]
$Ni(NCSe)_6^{4-}$ [224, 225]		$Pt(SeCN)_6^{2-}$ [202, 226]
$Cu(SeCN)_3^{2-}$ [228]	$Ag(SeCN)_2^-$ [228]	$Au(SeCN)_2^-$ [228]
$Cu(SeCN)_4^{3-}$ [228]	$Ag(SeCN)_3^{2-}$ [228]	$Au(SeCN)_2^-$ [136]
	$Ag(SeCN)_4^{3-}$ [228]	
$Zn(NCSe)_4^{2-}$ [224, 225]	$Cd_2(SeCN)_6^{2-}$ [225]	$Hg(SeCN)_3^-$ [88, 228]
	$Cd(SeCN)_4^{2-}$ [228]	$Hg(SeCN)_4^{2-}$ [136, 226]
	$Cd(SeCN)_6^{4-}$ [228]	
$Pr(NCSe)_6^{3-}$ [232]	$Nd(NCSe)_6^{3-}$ [232]	$Sm(NCSe)_6^{3-}$ [232]
$Dy(NCSe)_6^{3-}$ [232]	$Ho(NCSe)_6^{3-}$ [232]	$Er(NCSe)_6^{3-}$ [232]

Table 9.6 Homogeneous cyanato complexes

	$Mo(OCN)_6^{3-}$ [235]	
$Mn(NCO)_4^{2-}$ [233]		$Re(OCN)_6^{2-}$ [235]
		$Re(OCN)_6^-$ [235]
$Fe(NCO)_4^{2-}$ [233]		
$Fe(NCO)_4^-$ [233]		
$Co(NCO)_4^{2-}$ [233]		
$Ni(NCO)_4^{2-}$ [233]	$Pd(NCO)_4^{2-}$ [236]	$Pt(NCO)_4^{2-}$ [144]
$Cu(NCO)_4^{2-}$ [233]	$Ag(NCO)_2^-$ [428]	$Au(NCO)_2^-$ [238]
$Cu(NCO)_3^-$ [234]		
$Zn(NCO)_4^{2-}$ [233]	$Cd(NCO)_4^{2-}$ [233]	
	$Cd(NCO)_3^-$ [234, 237]	$Sn(NCO)_6^{2-}$ [236]

covalent organometallic compounds as well as among pure organic reagents. On the other hand, in accordance with the complete lack of any metal isofulminate species, no esters of isofulminic acid RONC have been described so far, but, unlike isocyanometal compounds, organic isocyanides (isonitriles) form a well established class of compounds.

9.4.2 Thiocyanato complexes

It is essentially the ability of the linear triatomic species of the general form NCX^- (where X = O, S, Se) to function as an ambidentate ligand, i.e., to coordinate to a metal ion through both the N or the X end, which caused the recent interest in the bonding pattern, in particular, of the thiocyanato metal complexes. Consequently numerous investigations have been carried out primarily directed at the elucidation of the various factors determining the modes of metal thiocyanate attachment. In general, the homogeneous thiocyanato complexes of the first row transition metals are known from x-ray data to be N-bonded. Isothiocyanates also predominate amongst the main group elements. As was originally depicted by Mitchell and Williams[239] the change from M—NCS to M—SCN bonding closely parallels the Ahrland–Chatt–Davies[240] and Schwarzenbach[241] classification of metals. Class a type metals, or 'hard acids' according to the Pearson[242] concept, form N-bonded isothiocyanates, and class b metals (or 'soft acids') form S-bonded thiocyanato metal complexes. This coordination behaviour requires – to fit Pearson's theory of hard bases preferably coordinating to hard acids and vice versa – that the ambidentate nature of the thiocyanate ligand be interpreted in terms of sulphur being a soft base and the nitrogen end being a hard base.

The situation is not as clear cut for borderline class a/class b metals. Cadmium is such a metal and in aqueous solution both Cd—NCS and Cd—SCN linkages are present in the same molecule. This unique behaviour of Cd^{II} is in contrast to its congeners Zn^{II} (which is unequivocally N-bonded) and Hg^{II} (which exclusively binds to the S end), and was first reported by Tramer[243] on the basis of infrared and Raman spectral studies, and confirmed by recent ^{14}N n.m.r. measurements[244].

This pattern is further amplified by consideration of mixed ligand thiocyanato complexes, some representative examples of which are given in Table 9.7. Turco and Pecile[245] were the first of several investigators to point out that the presence of other ligands in a thiocyanate complex strikingly affects the preferred mode of thiocyanate coordination. Thus, using infrared evidence, they observed a change from an M—S to an M—N bond, when two SCN groups of $[M(SCN)_4]^{2-}$ (where M is Pd^{II}, Pt^{II}) were replaced by two phosphine molecules, but found M—S bonding to be maintained, when the same SCN groups were substituted by two ammonia molecules, giving $M(NH_3)_2(SCN)_2$. In order to explain these experimental findings, it has been suggested by the same authors that the M SCN linkage is stabilised with regard to the M—NCS linkage by π-bonding which is formed

Table 9.7 Effect of other ligands on the thiocyanate bonding mode

$K_2[Pd(SCN)_4]$ [245]	$[Pd(PEt_3)_2(NCS)_2]$ [245]
$K_2[Pt(SCN)_4]$ [245]	$[Pd(PPh_3)_2(NCS)_2]$ [247]
	$[Pd(AsBu_3)_2(NCS)_2]$ [247]
trans-$[Pd(NH_3)_2(SCN)_2]$ [245]	*$[Pd(AsPh_3)_2(NCS)_2]$ [119]
cis-$[Pt(NH_3)_2(SCN)_2]$ [245]	$[Pd(SbPh_3)_2(SCN)_2]$ [247]
$[Pd(py)_2(SCN)_2]$ [246]	$[Pt(PEt_3)_2(NCS)_2]$ [245]
$[Pd(\gamma\text{-pic})_2(SCN)_2]$† [247]	$[Pt(PPh_3)_2(NCS)_2]$ [247]
$[Pd(4\text{-cyanopy})_2(SCN)_2]$ [248]	$[Pt(AsPh_3)_2(NCS)_2]$ [247]
$[Pd(4\text{-nitropy})_2(SCN)_2]$ [248]	$[Pt(SbPh_3)_2(NCS)_2]$ [247]
cis-$[Pt(py)_2(SCN)_2]$ [246]	$[Pd(dien)SCN]^+$† [249]
trans-$[Pt(py)_2(SCN)_2]$ [246]	*$Pd(Et_4dien)NCS]^+$ [249]
*$[Pd(bipy)(NCS)_2]$ [119]	$[Pd(PAs)(NCS)(SCN)]$† [250]
$[Pd(4,4'\text{-dimethyl-bipy})$	$[Pd(PC_2N)(NCS)(SCN)]$† [250]
$(NCS)(SCN)]$ [248]	$[Pd(PC_3N)(NCS)(SCN)]$† [250]
	$[Pd(PP)(NCS)(SCN)]$† [250]
$[Pd(ophen)(SCN)_2]$ [247]	
$[Pd(5\text{-nitrophen})(NCS)_2]$ [248]	$[Pd(AsPS)(SCN)_2]$† [250]
$[Pd(5\text{-methyl-6-nitrophen})(SCN)_2]$ [248]	$[Pd(SP)(SCN)_2]$† [250]
*$[Pd(4,7\text{-diphenylphen})(SCN)_2]$ [248]	$[Cr(CO)_5NCS]^-$ [122]
	$[Mo(CO)_5NCS]^-$ [122]
$[Mn(CO)_5SCN]$ [151]	$[W(CO)_5NCS]^-$ [122]
$[Mn(CO)_4(PPh_3)SCN]$ [251]	$[Cr(CO)_5NCS]$ [152]
$[Mn(CO)_4(AsPh_3)SCN]$ [251]	
$[Mn(CO)_4(SbPh_3)SCN]$ [251]	$[\pi\text{-}C_5H_5Cr(NO)_2NCS]$ [137, 252]
	*$[\pi\text{-}C_5H_5Mo(CO)_3NCS]$ [137]
cis-$[Mn(CO)_3(AsPh_3)_2SCN]$ [251]	*$[\pi\text{-}C_5H_5Fe(CO)_2NCS]$ [137]
trans-$[Mn(CO)_3(AsPh_3)_2NCS]$ [251]	$[\pi\text{-}C_5H_5W(CO)_3SCN]$ [137]
cis-$[Mn(CO)_3(SbPh_3)_2SCN]$ [251]	
trans-$[Mn(CO)_3(SbPh_3)_2NCS]$ [251]	$[Mo(CO)_2(PPh_3)_2(NCS)_2]$ [253]
cis-$[Mn(CO)_3(py)_2NCS]$ [251]	$[W(CO)_2(PPh_3)_2(NCS)_2]$ [253]
$[Co(NH_3)_5NCS]^{2+}$ [256]	$(\pi\text{-}C_5H_5)_2Ti(NCS)_2$ [36]
$[Co(en)_2(NCS)_2]^+$ [256]	$(\pi\text{-}C_5H_5)_2V(NCS)_2$ [254]
$[Co(dmg)_2(py)NCS]^+$† [257]	$(\pi\text{-}C_5H_5)_2Mo(NCS)_2$ [255]
	$(\pi\text{-}C_5H_5)_2W(NCS)_2$ [255]
$[Co(CN)_5SCN]^{3-}$ [258]	
$[Co(bipy)_2(SCN)_2]^+$ [259]	

*Stable linkage isomer.
†γ-pic is γ-picoline; dien is diethylenetriamine; PAs is diphenyl(o-diphenylarsinophenyl)phosphine; PC₂N is diphenyl(2-dimethyl-aminoethyl)phosphine; PC₃N is diphenyl(3-dimethylaminopropyl)phosphine; PP is 1,2-bis(diphenylphosphino)ethane; AsPS is diphenyl(o-diphenylarsinophenyl)phosphine sulphide; SP is diphenyl(o-methylthiophenyl)phosphine; dmg is dimethyl-glyoximato.

between the electrons of the metals and the empty antibonding π-orbital, located on the S atom of the thiocyanate groups. The availability of the metal d electrons for the M—SCN π-bond will, however, be reduced by their participation in π-back bonding with any strong π-acceptor organic ligands (e.g. phosphines) also present in the complex to such an extent that the M—NCS bond may be preferred. Another way of describing the situation is that π-bonding ligands in these systems tend to reduce the electron density on the metal, and thereby change class b, or soft metals, to class a, or hard metals. This is accompanied by a change in M—SCN bonding to M—NCS, respectively.

A large amount of data (Table 9.7) supports the π-bonding hypothesis, but Basolo and Burmeister[247] drew attention to a striking sensitivity of the mode of coordination of SCN towards even slight changes in the ligand L. Thus, the change from N-bonding in $Pd(py)_2(NCS)_2$ to S-bonding in $Pd(\gamma\text{-pic})_2(SCN)_2$ by the mere introduction of a methyl group in the *para* position of the coordinated pyridine ring was cited by Burmeister and Basolo as an unequivocal example, free of complicating steric effects (see below), of the directive force of electronic factors present in the other ligands, such as the methyl group donating electron density into the aromatic ring system through an inductive effect or hyperconjugation. From infrared spectra, however, Sabatini and Bertini[246] came to the conclusion that the thiocyanate bonding assignment in the pyridine complex was incorrect.

Nevertheless, similar arguments – in terms of the above theory – may be quoted to explain the difference in thiocyanate bonding in the [Pd(phen)(SCN)$_2$] and [Pd(5-nitrophen)(NCS)$_2$] complexes, where the ability of the second ligand to form π-bonds is increased by the presence of a nitro group due to its well known inductive and mesomeric electron withdrawing effects[248]. A temperature dependence of the thiocyanate bonding mode was established for the *cis*-palladium thiocyanate complex of 4,7-diphenyl-*o*-phenanthroline which, in accordance with the above, gives rise to an N-bonded compound at lower temperatures (0 °C); at higher temperatures, however, thermal motion of the phenyl rings reduces the effectiveness of their conjugation with the rings of the *o*-phenanthroline, and, as a consequence, Pd—S bonding is stabilised. On the other hand, it is not as easily understood why the complex with bipyridyl in place of *o*-phenanthroline is N-bonded, and the explanation referring to the acid dissociation constants pK (of which bipy has the lower value) is no longer valid for [Pd(5-CH$_3$-6-nitrophen)(SCN)$_2$] with this ligand having the same pK as 5-nitrophen-anthroline[248]. Nor is it plausible, why even in the presence of strong acceptor groups Pd—S bonding is maintained, as in the $Pd(4\text{-cyanopy})_2(SCN)_2$ and $Pd(4\text{-nitropy})_2(SCN)_2$ complexes, though in view of the probable *trans*-configuration of these complexes a weakening of the directive influence of the other ligand on the *cis*-positioned thiocyanato group might be expected[248].

A further complication arises from the fact that steric effects in bulky ligands also may alter the nature of the thiocyanate coordination. The point is that M—SCN bonding, because of the angular structure of M—S—C has a larger steric requirement than does the (more) linear structure of M—N—C. Selected bond angles from reported x-ray structure determinations of both thiocyanate and isothiocyanate metal compounds are given in Table 9.8. Steric factors have been kept essentially constant in the examples discussed so far, and the electronic effects could be evaluated to a certain extent, that way. On the other hand, in most instances, it turned out to be impossible to assess the importance of each individual factor, electronic or steric, to the type of SCN-bonding, in particular if both factors operate in the same direction, which is true, for example, with bulky phosphine ligands PR$_3$ (R = phenyl, alkyl). In this connection it is of interest to state that a larger ligand

Table 9.8   M—N—C, M—S—C, and N—C—S angles in some isothiocyanato and thiocyanato complexes

Compound	Stereochemistry of central atom	M—N—C M—S—C (degrees)	N—C—S (degrees)
[Zn(N₂H₄)₂(NCS)₂] [260]	octahedral, hydrazine-bridged	161.2(9)	170.7(1.6)
[Ni{SC(NH₂)₂}₂(NCS)₂] [261]	octahedral, linked by thiourea	162.2(1.0)	178.0(1.0)
[Cu(tren)(NCS)](SCN)* [262]	trigonal-bipyramidal	163.3	177.4
[Zn(tren)(NCS)](SCN)* [263]	trigonal-bipyramidal	166.6	178.5
[Co(bdme)(NCS)₂]* [264]		172.9(1.1)	176.7(1.3)
[Co(en)₂(SO₃)NCS]·2H₂O [265]	distorted trigonal-bipyramidal	161.1(1.1)	176.9(1.2)
[Ni(bddae)(NCS)₂]* [41]	octahedral	170.7(1.0)	177.5(1.2)
	almost square-pyramidal	169.9(1.0) 178.1(1.0)	178.0(1.1) 177.1(1.2)
[(C₄H₉)₄N]₃[Er(NCS)₆] [266]	octahedral	174(2)	176(3)
K₃[Mo(NCS)₆]·H₂O·CH₃ COOH [40]	octahedral	162.9 176.4	173.5 179.5
[(π-C₃H₅)Mo(CO)₂(bipy) NCS] [267]	octahedral	175.8(5)	176.4(7)
[(C₄H₇)Mo(CO)₂(ophen)NCS] [268]	octahedral	161.1(8)	179.2(1.0)
CoHg(SCN)₄ [269]	Hg and Co both tetrahedrally coordinated	97.4(5)	177.9(1.7)
[Cu(en)₂(SCN)₂] [42]	distorted octahedral	79.9	176.9
[Cu(trien)SCN]NCS [43]	square-pyramidal	89.5(1)	178.0(6)
[PdPh₂PCH₂CH₂CH₂N (CH₃)₂(SCN)(NCS)] [270]	square-planar	—NCS 177.6(5) —SCN 107.3(3)	178.6(8) 173.0(8)

*Tren is N(CH₂CH₂NH₂)₃; bdme is (Et₂NCH₂CH₂)₂NCH₂CH₂SMe; bddae is (Et₂NCH₂CH₂)₂NCH₂CH₂AsPh₂; trien is triethylenetetramine.

Table 9.9   Thiocyanate linkage isomers

Stable linkage isomer	Remarks
[Pd(AsPh₃)₂(NCS)₂] [247] ⎫ [Pd(bipy)(NCS)₂] [247] ⎬	S-bonded kinetic products isolated at low temperatures
[Pd(Et₄dien)NCS]⁺ [249, 273]	Isomerisation was found to be anion-dependent
[Pd(AsBuⁿ₃)₂(SCN)₂] [246]	Partial isomerisation to N-bonded form in the melt
[Pd(diphenylphen)(SCN)₂]* [248]	N-bonded isomer obtained at 0 °C
[Pd{P(OCH₃)₃}₂(NCS)₂] [147]	Mixture of isomers isolated at room temperature
[Cu(tripyam)(NCS)₂]* [274]	N-form obtained by heating the S-form or the mixed N- and S-bonded complex at 220 °C and 140 °C resp.
[Au(CN)₂(SCN)₂]⁻ [275]	N-bonded isomer initially obtained from aqueous solutions
[Rh(NH₃)₅NCS]²⁺ [276] ⎫ [Ir(NH₃)₅NCS]²⁺ [276, 277] ⎬	Both S-isomers are remarkably stable
Co(dmg)₂py(NCS)* [257]	
[Co(CN)₅(SCN)]³⁻ [150, 258, 278]	Cation-induced linkage isomerism established
[Mn(CO)₅SCN] [151]	S-bonded in the solid state, almost completely N-bonded in CH₃CN
[(C₅H₅)Fe(CO)₂NCS] [137]	Isomerisation S → N preferably occurs in the solid
[(C₅H₅)Mo(CO)₃NCS] [137]	Relatively high kinetic stability of S-isomer
[Cr(H₂O)₅NCS]²⁺ [55]	Unstable isomer detected in solution, but not isolated

*Et₄dien is N,N,N′,N′-tetraethyldiethylenetriamine; diphenylphen is 4,7-diphenyl-1,10-phenanthroline; tripyam is tri(2-pyridyl)amine; dmg is dimethylglyoximato.

does not necessarily mean, that this ligand exerts the higher steric repulsion, when introduced into the complex. Thus, undoubtedly, the overcrowding about the central atom will be somewhat lowered if the smaller phosphorus in $[(C_6H_5)_3P]_2Pd(NCS)_2$, which places the phenyl groups nearer to the metal and offers a greater steric hindrance at the metal, is replaced by the larger arsenic or antimony[247]. Besides, PPh_3 also is the better π-acceptor[271] (or the weaker σ-donor, resp.)*. As a consequence, Pd—S bonding was established for the $[(C_6H_5)_3Sb]_2Pd(SCN)_2$ complex containing triphenyl-stibine[247]. In contrast, an overcompensation of the electronic trend by the oppositely directed steric hindrance of four ethyl groups is demonstrated for the compounds $[Pd(dien)SCN]^+$ and $[Pd(Et_4dien)NCS]^+$ which have been recently described[249]. Both kinetic and thermodynamic evidence strongly support the steric effect in the Et_4dien complexes[272].

In view of the strong dependence of the coordination mode of the SCN group on the nature of the other ligands present, it is not surprising that there exist some 'borderline ligands' where the energy difference between the M—SCN and M—NCS isomer is relatively small permitting the isolation of isomeric pairs. Several isomeric pairs have been synthesised up to now (see Table 9.9). For the first two examples, $(Ph_3As)_2Pd(NCS)_2/(Ph_3As)_2$ $Pd(SCN)_2$ and $(bipy)Pd(NCS)_2/(bipy)Pd(SCN)_2$, the S-bonded isomers were isolated at low temperatures and found to isomerise to the N-bonded species, when the solid was heated[247]. Dissolving these compounds in various solvents also had an effect on the bonding; solvents having higher dielectric constants, such as pyridine, acetone and some nitriles, were shown to promote Pd—SCN bonding, whereas benzene, chloroform, etc., mostly solvents of low polarity, gave rise to mixtures containing the Pd—SCN, Pd—NCS, and Pd—SCN—Pd species[279]. Precisely this observed solvent control of the SCN bonding mode is to be anticipated following Klopman's treatment concerning chemical reactivity and the concept of charge and frontier controlled reactions[280] (see below). The sensitivity of the SCN linkage to subtle changes in the physical environment of the complex is further demonstrated by the complex $trans$-$(Bu_3^nAs)_2Pd(SCN)_2$, which in the solid, forms only the S-bonded species, but undergoes some isomerisation to an N-bonded thiocyanate in the melt. Similarly, reversible isomerisation $S \rightarrow N$ in a CH_2Cl_2 solution of the chelate complex $[Pd(das)_2(SCN)_2]$, containing the bidentate bis(diphenylarsino)ethane (das), has been reported[250].

To the list of directive influences on the bonding mode of SCN^- two more have been added recently. These are: (a) the influence of the non-coordinated anion[281], which was observed for the complex species $[Pd(Et_4 dien)NCS](BPh_4)$ as compared to $[Pd(Et_4dien)NCS](NCS)$, and (b) a cation-induced linkage isomerism which was established for the complexes $K_3[Co(CN)_5SCN]$, $K_3[Co(CN)_5NCS]$ and $(Bu_4^nN)_3[Co(CN)_5NCS]$ [278]. Both studied examples exhibit completely opposite isomerisation tendencies depending on their physical state: with $[Pd(Et_4dien)SCN]^+$ a $S \rightarrow N$ 'normal' isomerisation takes place in solution, but a $N \rightarrow S$-bonded solid state isomerisation is induced by the tetraphenylborate anion, possibly a

*The π-acceptor ability of phosphines is still a matter of discussion.

crystal packing effect. Similarly, $[Co(CN)_5SCN]^{3-}$ is believed to be the stable species in solution and in solid state (as potassium salt) as well, but with n-butylammonium it is apparently not possible to prepare pure samples of $(Bu_4^nN)_3 [Co(CN)_5SCN]$, due to its ready isomerisation to the N-bonded complex. Electronic effects like the stabilisation of either the free nitrogen end of the S-bonded thiocyanate by K^+, or the soft sulphur end by the hydro-carbon environment of $Bu_4^nN^+$ were regarded as responsible for the observed coordination pattern[278].

More recently, a number of examples of complex species containing both

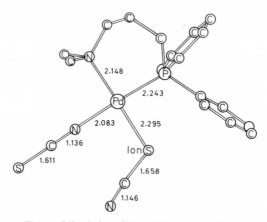

Figure 9.1 A view of the isothiocyanato-thiocyanato (1-diphenylphosphino-3-dimethylaminopropane)palladium(II) molecule normal to the plane of the four donor atoms. The estimated standard deviations for the bond lengths are:
Pd—S or P, 0.002 Å; P—N, 0.007 Å; C—S, 0.008 Å; and C—N, 0.011 Å
(Reprinted from J. Amer. Chem. Soc. **92**, 1077 (1970). Copyright 1970 by the American Chemical Society. Reprinted by permission of the copyright holder.)

N- and S-bonded thiocyanate in the same molecule have been reported, among these Pd(PN)(SCN)(NCS) (where PN is Ph_2P—CH_2—CH_2—CH_2—$N(CH_3)_2$), the structure of which is depicted in Figure 9.1 [270].

The authors point out that the mixed thiocyanate coordination might be rationalised on the basis of a competition for π-back bonding between the amine and phosphine donors, and the N and S ends of the thiocyanate ions, with the 'soft' atoms finally being *trans* to the 'hard' ones. Steric effects are also taken into consideration; possibly steric hindrance by the *cis*-positioned diphenylphosphine group could cause the SCN group to bend away, thus minimising the interaction of the π-orbitals on the phenyl- and SCN-group. The importance of steric over electronic effects becomes evident from the crystal structure of isothiocyanato-thiocyanato-[1,2-bis(diphenylphosphine)-ethane] palladium(II)[282]. Here mixed thiocyanate bonding also occurs, although the ligand has only soft equivalent phosphorus atoms. The same arguments hold for the previously reported [Pd(4,4'-dimethylbipy) (NCS)

(SCN)] with two hard donors[248]. These findings again explicitly illustrate the delicate balance of electronic and/or steric forces that govern whether a given thiocyanate group is *N*- or *S*-bonded. X-ray studies of Cu(tren)(thiocyanate)$_2$ (tren = (H$_2$NCH$_2$CH$_2$)$_3$N), originally interpreted as another example of a mixed thiocyanato-isothiocyanato complex[283], revealed the presence of two thiocyanate groups involved in different types of bonding. However, only one was coordinated to the metal (through N), while the other acted as a mere counter-anion[262]. Analogous trigonal-bipyramidal cations with SCN$^-$ counter-anions were found to exist in the crystal structure of [Zn(tren)NCS]$^+$SCN$^-$ [263].

(1)   (2)   (3)

Meek *et. al.*[250, 284] prepared and characterised several new palladium(II) complexes, emphasising the use of ligands that contain two different donor atoms. These ligands, such as diphenyl(*o*-diphenylarsinophenyl)phosphine (1), diphenyl(*o*-methylthiophenyl)phosphine (2), or diphenyl(*o*-methylthiotetrafluorophenyl)phosphine (3) were expected to exert different electronic influences on the thiocyanate groups that were *trans* to different types of ligand atoms*. Indeed, several molecular complexes that contain both N and S bonded thiocyanate groups resulted from these studies. Interpretation of the results required a consideration of steric interactions in addition to π-back bonding effects. The only known example of an isolation of all three possible thiocyanate linkage isomers of a complex containing two thiocyanated groups namely [Cu(tripyam)(NCS)$_2$], [Cu(tripyam)(SCN)$_2$] and [Cu(tripyam)(NCS)(SCN)] (tripyam = tris(2-pyridyl)amine), has been reported by Kulasingam and McWinnie[274]. Very recently, other squareplanar d^8 pseudohalide complexes of both higher and lower oxidation states than Pd^{II} and Pt^{II} have been introduced in the discussion of linkage isomerism by DeStefano and Burmeister[163], who from the strict N coordination of the thiocyanates and even of the selenocyanates (see below) came to the conclusion that cooperative and electronic ligand effects due to π-acceptor ligands are more pronounced in complexes of low valent metals. S-coordination was established for the thiocyanate complexes of class *b* metal ion gold(III), diphosAu$_2$Br$_4$(SCN)$_2$ and Ph$_3$PAuCl(SCN)$_2$, and also for some phosphine and arsine compounds of gold(I), which as a d^{10} system is an even softer Lewis acid. In accordance with these findings are the assignments (made on the basis of i.r. data) of Au—N and Au—S bonding, respectively, in the phosphine bridged dinuclear species[285]:

$$NCS-Au-P(Ph)_2-C\equiv C-P(Ph)_2-Au-SCN$$

$$\text{and} \quad SCN-Au \overset{\diagup P(Ph)_2-C\equiv C-P(Ph)_2}{\underset{\diagdown P(Ph)_2-C\equiv C-P(Ph)_2}{\diagdown}} Au-NCS$$

*The π-acceptor ability of a ligand should be transmitted to the coordinated pseudohalide group much more effectively when the ligand is *trans* to the pseudohalide than when it is *cis* to it.

Obviously, the π-acceptor ability of only one phosphorus atom (DPPA or diphos) coordinated to gold(I) is insufficient to switch gold(I) from a soft acid to a hard acid; therefore, S-bonding is observed. On the other side, with three π-acceptor ligands incorporated into the coordination sphere, Au—N bonding is accomplished. As in these cases, π-bonding was also considered the major factor controlling the type of metal–thiocyanate linkage in some isothiocyanate complexes of rhodium(I), e.g. $[Rh(CO)_2(NCS)_2]^-$, $Rh[P(C_2H_5)_3]_2(CO)(NCS)$, $Rh[P(OC_6H_5)_3]_3(NCS)$, reported previously by Jennings and Wojcicki[271].

So far, the discussion of effects of other ligands on the bonding mode of SCN was restricted solely to a few metals belonging to class b or showing characteristics intermediate between class a and class b. The picture, however, changes completely if typical class a metals are included. Here any explanation involving π-bonding obviously is contradicted, for example, by cobalt(III) as well as rhodium(III) and iridium(III) compounds. N-coordinated iso-thiocyanates are formed in the presence of ammonia, $[M(NH_3)_5NCS]^{2+}$ [276] or ethylenediamine, $[Coen_2(NCS)_2]^+$ [256] while S-coordination occurs if ligands are present which are believed to take part easily in π-bonding, e.g. $[Co(CN)_5SCN]^{3-}$ [258]. Maintenance of N-coordination has also been established for numerous mixed ligand thiocyanato complexes of other class a metal ions containing amines, e.g. $[Ni(1,3\text{-}pn)_2(NCS)_2]$ (1,3-pn = 1,3-diaminopropane), $Ni(trien)(NCS)_2$ (trien = triethylenetetramine)[286], $[Ni(NCS)_2(\alpha\text{-naphthylamine})_2]$ [287]; phosphines, e.g. $[V(NCS)_3(Et_2PhP)(MeOH)_2]$ [288], $[NiL_2NCS]ClO_4$ (L = diphenyl(o-diphenylarsinophenyl)phosphine)[289] (p. 275); amino acid anions, e.g. $K_2[Cr(NCS)_4(gly)]$, $K_2[Cr(NCS)_4(ala)]$[290]; N-heterocycles, e.g. $[Yb(NCS)_2(ophen)_2]SCN$ [291], $[M(NCS)_2(ophen)_2]$ where M = Fe^{II}, Co^{II}, Ni^{II}, Cu^{II}, Zn^{II} [292], $[Ni(\gamma\text{-pic})_4(NCS)_2]$ [293], $[CoL_2(NCS)_2]$ (L = 2-pyridinalphenylimine)[294], $[ScL_2(NCS)_2]SCN$, (L = bipy, ophen)[295], $CH_3Sn(NCS)_3(ophen)$, $(C_2H_5)_2Sn(NCS)_2(bipy)$ [134]; urea/thiourea, e.g. $[M(thiourea)_2(NCS)_2]$ where M = Mn^{II}, Fe^{II}, Co^{II}, Ni^{II}, Cd^{II} [296]; or oxygen donor ligands such as pyridine N-oxide, e.g. $Sc(NCS)_3(pyO)_3$ [297], $[Co(2,6\text{-dimethylpyO})_2(SCN)_2]$ [298], $[Cd(4\text{-cyanopyO})_2(NCS)_2]$ [299], and phosphine(arsine)oxides, e.g. $[M(NCS)_3(OPPh_3)_3]$, $[M(NCS)_3(OPPh_3)_4]$, $[M(NCS)_3(OAsPh_3)_3]$ where M = La, Ce, Pr, Nd [300], in the coordination sphere.

On the other hand, according as the π-hypothesis fails, Jörgensen's concept of symbiosis[301], depicting the tendency of flocking together of like ligands in the same complexes, offers an alternative description, accounting for the behaviour of typical class a metals. A marked solvent control of the bonding mode of SCN^- coordinated to the hard acid Co^{3+} was observed such that solvents with high dielectric constants promote the formation of metal–NCS bonds, thus contrasting the behaviour in various solvents of the class b Pd^{2+}/SCN^- [302]. However, it has been shown very recently by Hassel and Burmeister[303], that the different bonding modes exhibited by the complex $[Co(dmg)_2py(SCN)]$ (dmg = dimethylglyoximato) on its dissolution in high or low dielectric constant solvents, are kinetically rather than thermodynamically controlled. The isolation of the other isomer, $[Co(dmg)_2py(NCS)]$, was reported previously[257]. Both isomeric forms of $[Rh(NH_3)_5SCN]^{2+}$ and $[Rh(NH_3)_5NCS]^{2+}$ were isolated and found to be remarkably stable, the N-form being the thermodynamically favoured isomer[276]. This is in accord with the prediction from the Ahrland–Chatt–Davies classification of these compounds rendering the $[M(NH_3)_5]^{3+}$ radical (M = Rh^{III}, Co^{III}) a hard acid, whereas for the Rh^{III} ion rather a soft character follows from its S-coordination in $[Rh(SCN)_6]^{3-}$ (Table 9.4). Actually, as it was often pointed out, coordination by the ambidentate SCN$^-$ ion can be used as a test case for the classification a or b, of a certain metal ion.

The coordination behaviour of metal carbonyl thiocyanates reported by Wojcicki and Farona[151, 251, 304] is in general agreement with that of cobalt(III), i.e. 'symbiosis' of ligands is established for most of these compounds with the metals in low oxidation states. Thus, of the two linkage isomers [Mn(CO)$_5$SCN] and [Mn(CO)$_5$NCS], the one containing S-bonding is the more stable, and substitution of two CO groups by weaker π-bonding ligands such as amines and phosphines generally yields N-bonded complexes, [Mn(CO)$_3$L$_2$(NCS)] [251]. In order to explain apparent exceptions, i.e. cis-[Mn(CO)$_3$(AsPh$_3$)$_2$SCN] and cis-[Mn(CO)$_3$(SbPh$_3$)$_2$SCN], it is suggested that steric factors play a major role in stabilising the angular Mn—SCN linkage in these compounds (see above). Both linkage isomers are reported to exist for

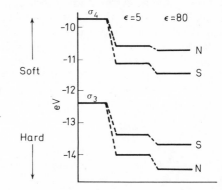

Figure 9.2 Softness (hardness) character of N and S in thiocyanate orbitals (see text) (From Norbury[305], by courtesy of the Chemical Society)

the complexes π-C$_5$H$_5$Fe(CO)$_2$(thiocyanate) and π-C$_5$H$_5$Mo(CO)$_3$(thiocyanate), the former representing the first example of linkage isomerisation taking place solely in the solid phase, presumably via an intermolecular reaction path involving an activated complex with either single or double SCN bridges[137]. On the other hand, π-C$_5$H$_5$Cr(NO)$_2$NCS, with chromium in the formal oxidation state zero, exhibits only Cr—N bonding, thus supporting the suggestion of Wojcicki[137], that an increase in negative charge on the metal results in a switch to N-bonding. The same is true for the chromium carbonyl derivative Cr(CO)$_5$NCS and the anionic species [Cr(CO)$_5$NCS]$^-$, both with strict N-coordination[151]. But, as Norbury and Sinha pointed out, the complete absence of any S-bonded isomer in these chromium compounds, could be as well understood in terms of chromium being closer to class a than are its congeners molybdenum and tungsten. In the prevailing situation, where various explanations – π-bonding, symbiosis, steric hindrance, and electron density on the metal – have been offered for the bonding behaviour described above, an approach was undertaken by Norbury[305] towards a consistent explanation for all types of behaviour, on the basis of Klopman's polyelectron perturbation theory[280]. The most interesting results of this straightforward treatment considering only the two most accessible thiocyanate orbitals σ_3 and σ_4, are presented in Figure 9.2. This reveals four softness parameters depending on which atom is about to coordinate, and on which orbital is considered. In accordance with the experimental findings, the hardest acids (class a), especially those in the presence of hard ligands, e.g. $[Co(NH_3)_5NCS]^{2+}$ are thus predicted to bond (in a 'truly symbiotic manner') with the hardest nucleophilic centre available (N of σ_4). On the other hand, surrounding the same class a metal by soft ligands, $[Co(CN)_5 SCN]^{3-}$, results in coordination to nucleophiles of intermediate character, i.e. to the sulphur atom (σ_4). Similarly for the low energy-increasing softness side of the diagram, metal-to-nitrogen (σ_3) coordination is predicted for a 'soft-soft' combination of a class b metal and its ligational environment, $Pd[P(C_6H_5)_3]_2(NCS)_2$, as is S-coordination for the intermediate case, $Pd(NH_3)_2(SCN)_2$.

Redox reactions involving pseudohalides which can act as bridging groups between two metal ions in an inner sphere mechanism (see Section 9.5) proved to be another source of linkage isomers. However, most of the unstable isomers, in which the 'wrong end' of the ambidentate ligand is attached to the metal, isomerise too rapidly to permit their isolation. Thus, the product of the reaction between $[Co(NH_3)_5NCS]^{2+}$ and Cr^{II} was found to be $[Cr(H_2O)_5NCS]^{2+}$ and not $[Cr(H_2O)_5SCN]^{2+}$, which might have been expected on the assumption that the bridged-activated complex was of the form $[Co\text{-}NCS\text{-}Cr]^{4+}$ [306]. On the other hand, the proper choice of the other ligands present in the coordination sphere of the reductant, should have a stabilising effect on the primary product. This was confirmed by the successful isolation of $[Co(CN)_5SCN]^{3-}$ from the reaction between $[Co(NH_3)_5 NCS]^{2+}$ and $[Co(CN)_5]^{3-}$ [285]. The system $Co(NH_3)_5NCS^{2+}/Cr^{2+}$ (with Cr^{II} in excess) was re-examined by Shea and Haim[52], who came to the conclusion that the reaction proceeds quantitatively by the remote-attack mechanism.

First direct evidence for adjacent attack in inner sphere reactions with polyatomic bridging ligands has been obtained by the same authors who, for the reaction of the isomeric thiocyanato complex $[CoC(NH_3)_5SCN]^{2+}$ with chromium(II) ion, give the following scheme[52]:

$$Co(NH_3)_5SCN^{2+} + Cr^{2+} \longrightarrow \begin{cases} [(NH_3)_5CoSCr]^{4+} \underset{N}{\overset{C}{|}} \longrightarrow CrSCN^{2+} \\ [(NH_3)_5CoSCNCr]^{4+} \longrightarrow CrNCS^{2+} \end{cases}$$

Green solutions containing the $[Cr(H_2O)_5SCN]^{2+}$ ion were prepared by Haim and Sutin[54, 55] by the reaction of chromium(II) with $FeNCS^{2+}$ or with $trans\text{-}[Co(en)_2(H_2O)(NCS)]^{2+}$. They were relatively stable in the absence of excess chromium(II), which catalysed the isomerisation. Possible mechanisms for the conversion of $CrSCN^{2+}$ into the purple $CrNCS^{2+}$ include spontaneous and mercury catalysed isomerisation and aquation[56] (cf. Ref. 307).

Both spontaneous aquation and isomerisation of $CrSCN^+$ are suggested to proceed via formation of the same 'intimate ion pair', $Cr(H_2O)_5^{3+} \cdot SCN^-$, in which rotation of the thiocyanate can readily occur[56]. The same type of dissociative mechanism has been proposed for the isomerisation of $[(Et_4 dien)PdSCN]^+$ mentioned above[273]. A further possible mechanism which involves bending of the thiocyanate group to form a 7-coordinated transition state was abandoned because of the large amount of energy required[56, 273]. However, as pointed out recently by Buckingham, Creaser and Sargeson[308], formation of an intermediate in which the metal is bonded to a thiocyanate π-orbital, may be regarded as only a minor alteration, if the MSCN angle in the starting material is close to 90 degrees (cf. Table 9.8).

9.4.3 Cyanato complexes

Of the homogeneous anionic transition metal complexes formed by the cyanate ion (Table 9.6), only three compounds, namely $[Re(OCN)_6]^{2-}$, $[Re(OCN)_6]^-$ and $[Mo(OCN)_6]^{3-}$ have been reported to contain O-bonded cyanate groups[235]. The others are regarded as isocyanato complexes with metal–N bonding. Alkaline earth isocyanato complexes are likely to exist in KBr pellets containing small amounts of barium(II) or calcium(II) together with NCO^-, as is indicated by significant frequency shifts of the v_1 and v_3 infrared absorptions compared to the free anion[309]. The bis-(cyclopentadienyl)-titanium(IV), -zirconium(IV), and -hafnium(IV) dicyanates, previously described as N-bonded species[310], were recently shown by i.r. and mass spectrometry to contain metal–oxygen bonds[36]. The Ti—O coordination in $(C_5H_5)_2Ti(OCN)_2$ has been confirmed by ^{14}N n.m.r. measurements in our laboratory[311]. $(C_5H_5)_2TiNCO^{312}$, $(C_5H_5)_2V(NCO)_2$ [254] and $(C_5H_5)_2W(NCO)_2$ [255], however, form N-bonded complexes[36], as do a variety of other transition metal complexes, e.g. trans-$M(PPh_3)_2(CO)NCO$ where M = Rh^I or Ir^I [163], LAuNCO (L = $P(C_6H_5)_3$, $As(C_6H_5)_3$) [163, 166], $Cu(NCO)_2L_2$ (L = methyl substituted pyridine)[313], $M(NCO)_2L_2$ where M = Mn^{II}, Co^{II}, Ni^{II}, Zn^{II}, Cd^{II} (L = bipy and ophen)[314], $M(NCO)_2(H_2O)_2$ (Urt) where M = Mn^{II}, Co^{II}, Ni^{II} (Urt = hexamethylenetetramine)[315]. Ablov, Popova and Samus[316] have reported the dimethylglyoxime (= dmg) cobalt(III) complexes $[(dmg)_2Co(NCO)(OH_2)]$, $[(dmg)_2Co(NCO)X]^-$ (X = I, NO_2), and $[(dmg)_2Co(NCO)_2]^-$. Previously, Nelson et al.[317] studied a series of cobalt and nickel isocyanate complexes containing heterocyclic aromatic amines. Complexes of hexamethylphosphoramide, $[(CH_3)_2N]_3 PO$, (HMPA), with cobalt(II) and zinc(II) isocyanates, $Co(NCO)_2(HMPA)_2$ and $Zn(NCO)_2(HMPA)_2$ have been reported[318]. Extensive experimental work was performed by Norbury and Sinha[144, 319] to elucidate the effects of ligands with varying σ and π-bonding properties on the coordination behaviour of the cyanate group towards palladium(II) and platinum(II). In complexes of the type $[ML_2(NCO)_2]$ (L = e.g. pyridine, α- and γ-picoline, PPh_3, $AsPh_3$, or $SbPh_3$), without exception bonding was through the nitrogen atom. Similarly, Burmeister et. al.[237] investigating the cyanato complexes of copper(II) and nickel(II) containing various aromatic π-acceptor ligands, concluded that the mode of coordination of the cyanate ion was rather insensitive to variations in the electronic environment of the metal. Introduction of particularly strong π-acceptor ligands, however, resulted in the formation of some NCO bridged species[237, 320] (see Section 9.5). For a time NCO bridging was thought to involve coordination at the oxygen[320], M—N—C—O—M, but subsequent research established that the bridging was of the type M—N(CO)—M, again avoiding metal–O bonding[321]. The same type of bridging has been ascertained by x-ray methods for crystalline AgNCO [174]. On the other hand, evidence was found for the presence of [Pt(PEt$_3$)$_2$(H)OCN] in solutions of [Pt(PEt$_3$)$_2$(H)NCO] from n.m.r. data[322] and chromium(II) is believed to be coordinated by cyanate groups through the oxygen in melts of CrCl$_3$ in KNCS[323]; yet, none of these cyanate species has been isolated. Similarly, although presumed as intermediates in inner sphere oxidation-reduction processes involving isocyanate complexes, no O-bonded metal cyanates could be isolated from these reactions[324, 325]. Instead, in a subsequent reaction, water is added to the initially formed M—OCN group and carbamate is found[325]. In this connection, reference should be made to the limited stability of organic cyanates[241, 326] with respect to their decomposition to isocyanates, and to the fact, that no O-bonded organometallic cyanate has been reported. A possible exception is H$_3$GeOCN, traces of which were detected in H$_3$GeNCO [327].

The strong tendency of NCO$^-$ to bond at nitrogen rather than oxygen is ascribed to the accumulation of negative charge on the N atom of the cyanate anion. Similar calculations reveal a more equitable distribution for NCS$^-$, accounting for its chemical versatility[328]. However, recent INDO calculations on NCO$^-$ and CNO$^-$, which also included σ-bonding, showed almost equal electron densities on both ends of these anions[329]. Thus, it is felt, that the marked preference for coordination via nitrogen or carbon with these ligands is rather due to the different expansion and orientation (i.e. to the different availability) of the bonding orbitals on both end atoms. With the fulminate ion, the latter factors are assumed to differ markedly for the C and O atoms, thus explaining the non-existence of metal–O bonding.

It is obvious that all of the metal ions [Re^{IV}, Re^V, Mo^{III}, Ti^{IV}, Hf^{IV}, Zr^{IV}] which have been found to coordinate to the oxygen atom of the cyanate exist in high oxidation states. They thus have either vacant or only partly filled dπ orbitals to interact favourably with the filled π-orbitals of NCO, and it is the O-end which most efficiently counterbalances the high positive charge on these class a metals. From the foregoing it becomes plausible, that at least in the square-planar d^8 complexes of palladium(II) and platinum(II) or of other class b metals, no $N \rightarrow O$ (soft-hard) bonding reversal is to be expected upon introduction even of strong π-acceptor ligands.

9.4.4 Selenocyanato complexes

In its ambivalent nature, the selenocyanate ion is to be placed somewhere in between the cyanate and the thiocyanate ions. Thus, the SeCN anion closely follows the 'normal' class a–isoselenocyanato, class b–selenocyanato bonding pattern in complexes which contain no other ligands (Table 9.5). Its bonding mode, in contrast to that of the thiocyanate ion, is largely unaffected by changes in the electronic character of the other ligand in the coordination sphere. For example, the type of metal thiocyanate attachment in palladium(II) and platinum(II) complexes is known to be changed by the presence of other ligands (see Table 9.7), but the changes in electron density on the palladium metal, brought about by introduction of the same ligands into the coordination sphere of the corresponding selenocyanate complexes, proved to be insufficient to effect a Pd—Se to Pd—N switch[147]. The marked preference of the selenocyanate ion for coordination through the selenium atom obviously parallels that of the cyanate to coordinate through its nitrogen atom (see above). This may be rationalised similarly in terms of charge distribution, which places most of the negative charge on the Se atom[144, 328]. However there is still sufficient charge left on the nitrogen atom (considerably more than on the oxygen atom of the cyanate ion) to allow coordination through that atom in certain circumstances. Favourable circumstances to initiate Pd—NCSe coordination were found to exist in the presence of very bulky ligands, such as N,N,N',N'-tetraethyl-diethylentriamine ($= Et_4dien$). Here the steric requirements promote the formation of the linear Pd—NCSe linkage at the expense of the angular Pd—SeCN linkage, thus giving rise to the synthesis of the first example of selenocyanate linkage isomers, [Pd$(Et_4dien)SeCN][B(C_6H_5)_4]/[Pd(Et_4dien)NCSe][B(C_6H_5)_4]$[330]. The former, obtained from $[Pd(SeCN)_4]^{2-}$, Et_4dien, and sodium tetraphenyl borate at $-70\,°C$ in methanol, was shown to isomerise completely to the N-bonded form over a period of several hours on dissolution in DMFA[330]. Qualitatively, the rate of isomerisation decreased with decrease in the dielectric constant of the solvent, suggesting the operation of a dissociative process[331]. The corresponding diethylentriamine complex, $[Pd(dien)SeCN]B(C_6H_5)_4$, on the other hand, has been reported to be stable with respect to isomerisation. Interestingly, as in the case of the thiocyanate analogues $[(Pd(Et_4dien)SCN]^+X^-$ ($X = NCS,B(C_6H_5)_4,PF_6$), a marked dependence of the bonding mode of the selenocyanate ligand on the nature of the non-coordinated group, $B(C_6H_5)_4^-$, was established in that the N-bonded isomer $[Pd(Et_4dien)NCSe]$ $[B(C_6H_5)_4]$, once isolated, was found to re-isomerise slowly to the Se-bonded form at room temperature in the solid state[331]. This coordination behaviour of the selenocyanate anion, however, is contrasted by that of the cyanate ion which was shown to be completely unaffected by the same steric influences giving only isocyanate complexes $[Pd(dien)NCO]^+$ and $[Pd(Et_4dien)NCO]^+$[332]. The relative sensitivity of the SeCN bonding mode toward steric and, to a lesser extent, electronic factors in the environment of the metal atom is further demonstrated by the preparation of another pair of linkage isomers π-$C_5H_5Fe(CO)(PPh_3)SeCN$ and π-$C_5H_5Fe(CO)(PPh_3)NCSe$ by Jennings and Wojcicki[170]. Reaction of π-$C_5H_5Fe(CO)_2CH_2C_6H_5$ with $Se(SeCN)_2$ gives only the Se-bonded isomer π-$C_5H_5Fe(CO)_2SeCN$ while with the triphenylphosphine derivative π-$C_5H_5Fe(CO)(PPh_3)CH_2$ C_6H_5 both isomers cited above are obtained. Both are stable at room temperature with respect to interconversion, and at higher temperatures, deselenation to give π-$C_5H_5Fe(CO)(PPh_3)CN$, takes place in preference to isomerisation.

Cooperative electronic ligand control of the bonding mode of the selenocyanate ion in particular has been further observed in the low valent d^8 complexes[163], $trans$-$[Rh(PPh_3)_2(CO)NCSe]$ and $trans$-$[Ir(PPh_3)_2(CO)$

NCSe]. Rh^{III} is a typical borderline ion with slightly prevailing b character and gives the 6-coordinate species, $[Rh(SCN)_6]^{3-}$ and $[Rh(SeCN)_6]^{3-}$. The switch to N-bonding in the Rh^I complex is then due most probably to the greater π-withdrawal by the *trans* CO-group rather than to steric factors; this follows from a comparison of the complexes in which essentially the same

$$\begin{array}{ccc} NCSe & & P(C_6H_5)_3 \\ & Pd & \\ (C_6H_5)_3P & & SeCN \end{array} \qquad \begin{array}{ccc} OC & & P(C_6H_5)_3 \\ & Rh & \\ (C_6H_5)_3P & & NCSe \end{array}$$

steric factors are operative. In this connection, attention should again be drawn to the opposing trends in complexes exhibiting a truly symbiotic (class a) behaviour, namely $Fe(NCSe)_4^{2-}$ [225] and π-$C_5H_5Fe(CO)_2SeCN$ [170], as against those showing class b—antisymbiotic behaviour, such as Pd $(SeCN)_4^{2-}$ [225] and $M(PPh_3)_2(CO)NCSe$ ($M = Rh^I, Ir^I$) [163].

Only a few complex selenocyanates both homogeneous and mixed ligand, were known before 1965. Since then extensive work has been directed toward the synthesis of new complexes, such as those of the noble metals[136, 229, 333] and the rare earth metals $[Ln(NCSe)_6]^{3-}$ [232]. Recently reported studies include selenocyanato complexes of main group elements such as bismuth(III) [205], indium(III) [230] as well as mixed thiocyanato–selenocyanato complexes of Cd^{II} [334]. The coordination behaviour of the cadmium(II) ion, forming only $[(n-C_4H_9)_4N]_2$ $[Cd_2(NCSe)_6]$ with bridging groups of the type that the metal is coordinated to both nitrogen and selenium atoms again lies clearly between that of class a $(Zn(NCSe)_4^{2-})$ and class b metals $(Hg(SeCN)_4^{2-})$ [225] (cf. the system Cd^{2+}/SCN^-). Mixed ligand selenocyanato complexes recently reported include those of rare earth metals, such as $Ln(ophen)_3(NCSe)_3$, or $Ln(bipy)_3(NCSe)_3$, 70 new compounds altogether[335], as well as some iso-selenocyanato-pentammine complexes of chromium(III) and cobalt(III) [336]. Diselenocyanatobis(2,2'-bipyridyl)iron(II) was shown from magnetic and x-ray measurements to represent a possible iron(II) analogue of the Lifschitz nickel(II) complexes; a crystal structure showed that two iron(II) ions were spin-paired, while the third was spin-free[337]. The synthesis of a new monobasic acid with Co—NCSe—bonds, $H[Co(NioxH)_2(NCSe)_2]$ (where $NioxH_2$ = 1,2-cyclohexanedione-dioxime), and of 34 new complex salts of this acid has been described[338].

Se-coordination is maintained in all reported gold(I) selenocyanates[163], namely LAuSeCN (where L = triphenylphosphine, triphenylarsine) and $L'Au_2(SeCN)_2$ (L' = 1,2-bis(diphenylphosphino)ethane), while Ti^{IV} in Cp_2Ti $(NCSe)_2$ tends to bond to the nitrogen atom of SeCN [339]. That the bonding mode of the selenocyanate group is relatively insensitive toward ligand effects is further demonstrated by the isolation of the complexes $[Co(NH_3)_5 NCSe]^{2+}$, $[Co(NH_3)_4(CN)NCSe]^+$ and $[Co(CN)_5NCSe]^{3-}$ [340]. No linkage isomers, e.g. $[Co(CN)_5SeCN]^{3-}$ have been obtained from oxidation-reduction reactions which according to

$$[Co(NH_3)_5NCSe]^{2+} + [Co(CN)_5]^{3-} \rightarrow [(H_3N)_5Co\text{-}NCSe\text{-}Co(CN)_5]^- \rightarrow$$
$$[Co(CN)_5OH_2]^{2-} + [Co(OH_2)_6]^{2+} + NH_3 + NCSe^-$$

only yield aquo complexes[340].

9.4.5   Cyanides and isocyanides

Several reviews exist on the chemistry of the cyano group, its structural features, and its ambivalent nature[2-6]. Generally, coordination through the carbon atom is assumed to occur in the anionic transition metal cyano complexes, $[M(CN)_n]^{m-}$, when CN^- is acting as a monodentate ligand. There are, however, a number of well-established examples in which the cyano group acts as a bridging ligand of the kind that the nitrogen is involved in coordination as well, i.e. M—C≡N—M'. Bridges of the same type are also formulated in redox reactions, which are known to proceed via an inner-sphere activated bridge mechanism involving cyano compounds M—CN and a reducing agent M'.

Subsequent splitting of the bridge previously formed by 'remote' attack of M' at the coordinate cyanide group could lead to the formation of an N-bonded cyanide (isocyano) complex. Halpern and Nakamura[341] have studied the kinetics of the reaction in which M—CN was $[Co(NH_3)_5CN]^{2+}$ and M' was $[Co(CN)_5]^{3-}$; the final product was $[Co(CN)_6]^{3-}$. It was found that the disappearance of $[Co(NH_3)_5CN]^{2+}$ coincided with the appearance of a transient absorbing more strongly than $[Co(CN)_6]^{3-}$ at a wavelength of 320 nm. This is possibly a pentocyano-isocyano cobaltate(III) complex which subsequently underwent first-order decay to $[Co(CN)_6]^{3-}$ with a half-life of 1.6 s. Another transition metal isocyanide intermediate, $[Cr(H_2O)_5 NC]^{2+}$ with a half-life of 1.78 min. at 15 °C, has been detected by Espenson and Birk, in the chromium(II) reduction of $[Co(NH_3)_5CN]^{2+}$, trans-$[Co(NH_3)_4(H_2O)CN]^{2+}$, and trans-$[Co(en)_2(H_2O)CN]^{2+}$ in acidic perchlorate solution[23]. Likewise both inner-sphere electron exchange reactions in aqueous perchlorate solution between the complexes cis-$[Cr(H_2O)_4 (CN)_2]^+$ and 1,2,3-$[Cr(H_2O)_3(CN)_3]$ on the one hand, and $[Cr(H_2O)_6]^{2+}$ on the other hand were shown to produce quantitative yields of $[Cr(H_2O)_5 CN]^{2+}$. However, spectral evidence was obtained for the presence of metastable reaction intermediates and the possible role of the isocyano complex $CrNC^{2+}$ as an unstable intermediate in the exchange reaction has been discussed[24]. An isocyano intermediate is also likely in the base deprotonation reaction of the hydrogen cyanide derivatives of hexacarbonyls[342]:

$$M(CO)_5NCH \ (M = Cr, W) \xrightarrow{-H^+} [M(CO)_5CN]^-$$

The question concerning the existence of isocyano compounds has been the subject of lasting controversy and discussion, particularly in organometallic chemistry. In organic chemistry isonitriles, but not nitriles, may be oxidised to the corresponding isochalcogenides, and in an effort to explain some of the typical reactions given by organometallic cyanides, for example

$$R_3SiCN + S \rightarrow R_3SiNCS$$

the presence of an 'iso' form R_3SiNC (in equilibrium with the 'normal' form) was postulated[343, 344]. Parenthetically it is of interest to note that the same type of isomer is also likely to be involved (as an intermediate) in reverse reactions, such as desulphuration[178] or deselenation[170].

$$[Co(NH_3)_5NCS]^{2+} \xrightarrow[\text{or } S_2O_8^{2-}]{Ce^{IV}/H^+} [Co(NH_3)_5CN]^{2+}$$

$$[\pi\text{-}CpFe(CO)(PPh_3)NCSe] \rightarrow [\pi\text{-}CpFe(CO)(PPh_3)CN] \ (Cp = C_5H_5)$$

As Thayer[175] pointed out, however, such a species is not necessarily required for an understanding of these reactions, if one considers the bonding between the CN group and the rest of the molecule. This is covalent and unreactive in organic nitriles (or isonitriles) but polar and quite labile in the organo-silicon and -germanium derivatives. This relative lability of the silicon–cyanide bond has also been put forward to explain the preferred formation of isonitrile derivatives, such as $(CH_3)_3SiN\equiv C$—$Fe(CO)_4$, upon introduction of $(CH_3)_3SiCN$ into metal carbonyls[345, 346]. Here, as in the case of hydrogen cyanide above, an isomerisation mechanism involving a Si—C bond cleavage in the coordinated nitrile, $R_3SiCNFe(CO)_4 \rightarrow R_3SiNCFe(CO)_4$, might be operative[342]. Nevertheless, convincing evidence meanwhile has been obtained for the presence of small amounts of $(CH_3)_3SiNC$ in the isomeric cyanide[347]. Furthermore, sodium isocyano-trihydroborate, Na^+ BH_3NC^-, has been isolated in mixtures (1:4) with the 'normal' salt, and characterised by ^{11}B n.m.r. and i.r. methods[348]. Isomerisation of this salt to $NaBH_3CN$ has been investigated and found to be H^+ and CN^- catalysed. The existence of two unexpectedly stable di-isocyano complexes of cobalt(III), cis-α-$[Co(trien)(NC)_2]$ and cis-β-$[Co(trien)(NC)_2]$ (where α and β refer to the different types of coordination of the tetradentate ligand trien) has been claimed recently by Kuroda and Gentile[25] using i.r. and vis-u.v. evidence. The available data on these complexes, however, appear not to be sufficient at present to confirm the proposed structures, and other authors have criticised the interpretation of these data as well as the absence of any rationale for the formation and stability of these unusual compounds[19, 135].

9.4.6   Complexes with tricyanomethanide, dicyanamide, cyanamide and related ligands

The planar tricyanomethanide ion according to Birkenbach[1] is also a member of the class of pseudohalides and can coordinate to a metal through its cyano nitrogen atom or its central sp^2 carbon atom (see Table 9.1). It appears that most complexes are of the keteniminato form M—N = C = $C(CN)_2$ such as $Ph_3MNCC(CN)_2$ (M = Si, Ge, Sn, Pb)[67, 349], $Co(NO)_2$ $PR_3NCC(CN)_2$[350], $Nipy_4[NCC(CN)_2]_2$[68, 351], $Pd(dien)NCC(CN)_2$[352], Ir $(CO)(PPh_3)_2NCC(CN)_2]_2$[353]. The reactions of MBr_4 (M = Sn, Ti, Zr) with the 'inter(halogen)pseudohalogen' $BrC(CN)_3$ proceed similar to that with chlorine azide.

$$MBr_4 + BrC(CN)_3 \rightarrow Br_2 + Br_3MNCC(CN)_2 \quad [354]$$

The reactions of pentacarbonylmanganese halides with tricyanomethanide ion, previously reported[355] to give $Mn(CO)_5C(CN)_3$, have been shown[356] to afford a polymeric tricarbonyl species, which reacts with phosphines to give $Mn(CO)_3(PR_3)_2C(CN)_3$.

Other related keteniminato complexes have been prepared by 1,4 'insertion' of tetracyanoethylene into metal–R groups (R = H, alkyl), e.g.

$$IrH(CO)_2(PPh_3)_2 + 2(CN)_2C = C(CN)_2 \rightarrow$$
$$TCNE(PPh_3)_2(CO)Ir—N = C = C—C(CN)_2H \quad [357]$$
$$\qquad\qquad\qquad\qquad\qquad\qquad\qquad CN$$

The keteniminato linkage in this iridium complex was ascertained by x-ray analysis[358].

Recent ^{14}N n.m.r. and i.r. studies indicate[161] that in the platinum complexes, trans-$HPt(PPh_3)_2C(CN)_3$ and cis-$Pt(PPh_3)_2ClC(CN)_3$, and the carbonyl metalate anion[69] $[W(CO)_5C(CN)_3]^-$, the $C(CN)_3$ ligand is bonded through the carbon to the metal. This linkage isomerism of $C(CN)_3^-$ should, however, be confirmed by other methods. Evidence of the existence of isomeric 1,2 and 1,4 insertion products obtained from $C_5H_5Fe(CO)_2CH_2C_6H_5$ and tetra-cyanoethylene has been presented[359]. The oxidative addition of 1,1,1-tricyanoethane to $Pt(PPh_3)_4$ involves cleavage of a carbon–carbon bond and yields the complex $(Ph_3P)_2Pt(CN)[C(CN)_2Me]$ [352].

For the dicyanamido complexes studied so far, i.e. $Co[N(CN)_2]_4^{2-}$ [74, 360], $Ph_3SnN(CN)_2$ [75], $(CH_3)_2Sn[N(CN)_2]_2$ [361] $Nipy_4[N(CN)_2]_2$ [68, 351], $(Ph_3P)_2$ $(CO)IrN(CN)_2$ [353], it is supposed that of the two possible donor sites of this ligand (see Table 9.1) only the cyanide nitrogen is involved in coordination[12].

The nitroso- and nitro-dicyanomethanide ions, $[ONC(CN)_2]^-$ and $[O_2NC(CN)_2]^-$, which may be regarded as nitrite and nitrate analogues[362], provide a number of possible donor sites. For complexes such as $\{M amine_4[ONC(CN)_2]_2\}$ and $\{M amine_4[O_2NC(CN)_2]_2\}(M = Ni, Co, Cu)$ coordination, respectively, through the nitrosyl nitrogen or the oxygen was suggested[76, 363].

Of the four possible isomers of general formula CH_2N_2, i.e.

$$H_2C=N=N \quad H_2N—C\equiv N \quad HN=C=NH \quad H_2N—N\equiv C$$

diazomethane, cyanamide and isodiazomethane (N-isocyanamine)[364] are well established; while only organic derivatives of carbodi-imide are known. Metal derivatives of the mono and dianionic forms of diazomethane such as Me_3SiCHN_2, $(Me_3Sn)_2CN_2$ [365, 366], $(MeHg)_2CN_2$, $HgCN_2$ [367] have been described. In our group cyanamido complexes of palladium(II) $[Pd(NRCN)_4]^{2-}$ (R = H, Ph) have been prepared[66], but the site of attachment to the palladium (carbodi-imide or cyanoimide form) is still a matter of discussion.

9.5 PSEUDOHALIDE BRIDGES

X-ray structural determinations have shown that fulminate and cyanate form bridges between two metal atoms via only one atom (the carbon and nitrogen atoms respectively), whereas in cyanide, thiocyanate, and seleno-cyanate complexes both terminal atoms of the pseudohalide are involved in the bridge function. In principle, a close relationship can be seen between the coordination behaviour of a given pseudohalide as a terminal ligand and its preferred mode of linking two metals, as a bridging ligand. Thus, whenever linkage isomers are likely to occur (SCN,SeCN) end-to-end bridging obviously is favoured, as is bridging via only one atom in cases where the pseudohalide is known to strongly prefer one terminal atom for coordination (NCO, CNO, CO). For the highly symmetrical azide ion, however, both end-to-end bridging and bridging through a terminal nitrogen atom have been established.

9.5.1   Azide, cyanate and fulminate bridged complexes

The occurrence of azide bridged intermediates in redox processes of metal complexes involving group transfer is widely accepted[368]. The large accelerating effect of the azide ion on the rate of such processes, e.g.

$$Cr(N_3)_2{}^+ + {}^*Cr^{2+} \longrightarrow \left[Cr \underset{N=N=N}{\overset{N=N=N}{\diagup\diagdown}} Cr\right]^{3+} \longrightarrow Cr^{2+} + {}^*Cr(N_3)_2{}^+$$

(Ref. 369)

classifies N_3^- as an extremely good bridging ligand. Also, the relative rates of reduction of azido and thiocyanato complexes has been used to distinguish between inner-sphere and outer-sphere mechanisms, based on the stability of the azide bridged transition state. Usually, a symmetrical end-to-end bridged intermediate has been formulated, though the alternative bridging mode via one nitrogen atom cannot be ruled out. Actually the latter mode of bridging seems to be favoured in discrete dinuclear complexes, as has been shown from recent x-ray structural studies of $[Pd_2(N_3)_6]^{2-}$ [60], $[Mn_2(CO)_6$ $(N_3)_3]^-$ [61], and $[BCl_2N_3]_3$ [62]. In contrast to the previously reported $[(Ph_3P)_2$ $CuN_3]_2$ [64], these complexes with azide bridges of the N-diazonium type are also completely stable in solution. Both terminal and bridging azide groups are linear and exhibit a pronounced asymmetry in the two N—N bond lengths, indicative of a contribution of the resonance form:

Terminal and bridging azide groups in $[Pd_2(N_3)_6]^{2-}$ reveal a close similarity in the corresponding N—N and Pd—N bond lengths, which is also demonstrated by the appearance of only one ^{14}N n.m.r. signal[370]. On the other hand, bridging and terminal azide ligands can easily be distinguished from the position of their antisymmetric N_3-stretching vibrations, azide bridges showing the higher $\nu_{as}N_3$ frequencies[61, 85, 371]. Obviously $\nu_{as}N_3$ frequencies are particularly sensitive towards small changes in bonding. Of interest are the comparatively small metal–nitrogen–metal angles, apparently due to the steric requirements of the bridgehead nitrogen atoms, whereas the monobridged $[Cl_2BN_3]_3$ and presumably $[(Ph_3P)_2MCO]_2N_3^+$ [61] (M = Rh, Ir) [61] with remarkably higher $\nu_{as}N_3$ frequencies show bond angles in the range expected for sp^2 hybridised nitrogen atoms (see Table 9.10).

Table 9.10   Bonding parameters of azide bridged compounds

	N_α—N_β (Å)	N_β—N_γ (Å)	$\sphericalangle$M—N—N (deg.)	$\sphericalangle$M—N—M (deg.)	$\nu_{as}N_3$ (cm^{-1})	Ref.
$[Pd_2(N_3)_6]^{2-}$	1.24	1.14	128.4	103.2	2060	60, 85
$[Mn_2(CO)_3(N_3)_3]^-$	1.22^a	1.16^a	135^a	88.6^a	2070	61
$[Cl_2BN_3]_3$	1.26^a	1.09^a	116^a	127.5^a	2219, 2210 2160	62, 155, 372
$Cu(N_3)_2$	1.21^a	1.10^a	127^a	102.5		373
AgNCO			128	98		174
AgCNO (orthorhomb.)			139	82.2		32

a = average value.

An *N*-diazonium type of azide bridge can also be assumed for other dinuclear complexes[371, 374], such as

$$Ph_3P(N_3)M \overset{N_3}{\underset{N_3}{\diagdown\diagup}} M(N_3)PPh_3 \quad \text{and} \quad [(Ph_3P)_2M \overset{N_3}{\underset{N_3}{\diagdown\diagup}} M(PPh_3)_2]^{2+} \quad (M = Pd, Pt)$$

Both azide and cyanate bridged complexes of palladium and platinum are formed by abstraction of one pseudohalide ligand per metal, using oxonium or nitrosonium salts from complexes such as $(Ph_3P)_2MX_2$, where X is NCO or N_3 [38, 375].

Furthermore, isocyanate bridged complexes, e.g.

$$Ph_3P(NCO)Pd \overset{\overset{\displaystyle O \atop \parallel \atop \displaystyle C \atop \parallel \atop \displaystyle N}{}}{\underset{\underset{\displaystyle N \atop \parallel \atop \displaystyle C \atop \parallel \atop \displaystyle O}{}}{\diagdown\diagup}} Pd(NCO)PPh_3 \quad \text{(Ref. 238)} \qquad LRh \overset{\overset{\displaystyle O \atop \parallel \atop \displaystyle C \atop \parallel \atop \displaystyle N}{}}{\underset{\underset{\displaystyle N \atop \parallel \atop \displaystyle C \atop \parallel \atop \displaystyle O}{}}{\diagdown\diagup}} RhL \quad (L = CO, 1/2 \text{ dien})$$

(Ref. 376)

can also be prepared by reaction of corresponding azide bridged complexes with carbon monoxide. The extremely explosive dimeric dimethylgoldazide is precipitated from an aqueous solution of $[(CH_3)_2Auen]N_3$ on addition of hydrochloric acid according to the following equation:

$$2[(CH_3)_2Au \; en]N_3 + 2HCl \longrightarrow \overset{\overset{\displaystyle N_2}{\underset{\displaystyle |}{}}}{\underset{\underset{\displaystyle N_2}{\underset{\displaystyle |}{}}}{H_3C \atop H_3C}} \overset{N}{\underset{N}{Au \diagup\diagdown Au}} \overset{CH_3}{\underset{CH_3}{}} + 2 \; enH^+Cl^-$$

Azide bridges presumably occur in the $Cd_2(N_3)_5^-$ and $Hg(N_3)_3^-$ complex anions[85]. By reaction of chlorine azide with metal carbonyls, Dehnicke and co-workers[377] prepared polynuclear azide bridged derivatives $M(CO)_2(N_3)Cl$ (M = Fe,Mo,W), for which the following structures have been suggested.

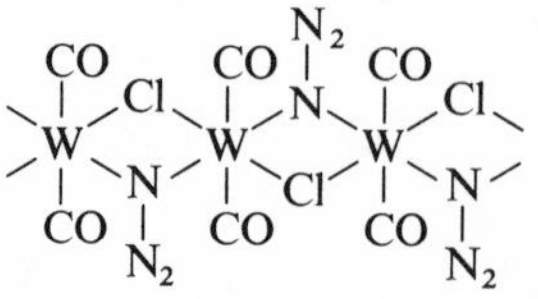

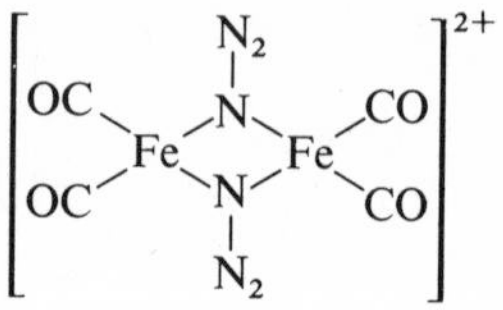

Interestingly, the iron compound represents the first example of a paramagnetic iron(II) carbonyl derivative. Evidence for association in the solid state via azide bridges has been obtained[154] for some mixed chloro-azido metal compounds, e.g. $VOCl_2N_3$, $TiCl_3N_3$.

Elaborate studies were performed by Dehnicke, Paetzold and Wiberg on azide bridged compounds of main group elements, $[BX_2N_3]_3$ (X = Cl, Br) [155, 372], $[SbCl_4N_3]_2$ [378, 379], $[SnCl_2(N_3)_2]_n$ [380], $[RZnN_3]_n$ [381], $[R_2 MN_3]_3$ [382, 383], $[(Me_3M)_2N_3]^-$ (M = Al,Ga) [384] and $[R_2InN_3]_2$ [383].

Copper diazide was shown to possess a polymeric structure with azide bridging via the N_α atom together with an additional weak $Cu-N_\gamma$ interaction[373]. A structural example of an azide bridge through the terminal nitrogens was recently reported by Dori and Eisenberg for the copper(I) complex $[(Ph_3P)_2CuN_3]_2$, which in the solid state forms an eight membered nonplanar ring, yet appears to be monomeric in solution[64]. In contrast to covalent azides the N—N distances are all equal and close to those values observed in the azide ion, indicating a more ionic copper–azide bonding. Structural results obtained so far seem to indicate that stable azide bridges in molecules are of the N-diazonium type, whereas end-to-end bridging is restricted to the solid state. This is also true for $Cdpy_2(N_3)_2$ where in the crystal the cadmium atom is octahedrally coordinated, with the octahedra linked together by end-to-end bonded symmetrical azide bridges[385].

The only established example of fulminate bridging is represented by silver fulminate, which occurs in two polymorphic forms, an infinite zig-zag C—Ag—C—Ag chain and a cyclic hexamer[32]. For an interpretation of the bonding, especially of the very small Ag—C—Ag angle (see Table 9.10) a description in terms of a three-centre Ag—C—Ag bond, involving an sp orbital on C and sp orbitals on the Ag atoms was suggested[32].

The closely related structure of the isomeric silver isocyanate with infinite zigzag Ag—N—Ag—N chains, however, shows a larger Ag—N—Ag angle which can be interpreted in terms of an sp^2 nitrogen atom. It appears that the same type of NCO bridges via the nitrogen is also verified in other isocyanato complexes, namely $ML_2(NCO)_2$ (M = Mn,Fe,Co,Ni; L = 3- or 4-cyanopyridine) [321], Cu(quinoline) $(NCO)_2$, Ni(4-carbomethoxypyridine)$_2$ $(NCO)_2$, $K[Cd(NCO)_3]$ [237], $[(CH_3)_2AuNCO]_2$ [386], $[Mn_2(CO)_6(NCO)_3]^-$ [61].

The preferred formation of azide bridges to isocyanate bridges is clearly indicated by the i.r. spectrum of $[(NCO)(CO)_3 Re(N_3)_2 Re(CO)_3(NCO)]^{2-}$ showing the presence of bridging azide and terminal isocyanate ligands[61].

In contrast to azide, ν_{as}NCO absorptions of NCO bridges appear at lower frequencies than those of terminal NCO groups[61, 321].

9.5.2 Cyanide bridges

Investigations of cyano bridged binuclear complexes are of special importance in view of recent interest in structural, redox, and substitution studies with the ambidentate ligand. Numerous cyanide bridged complexes have been reported. In all these compounds both carbon and nitrogen are involved in bridge formation; bridging via the carbon atom, analogous to that of carbonyl bridged complexes, is unknown. A bridge of the second type has been discussed for $Ni_2(CN)_6^{2-}$, but a recent structural investigation has revealed that the two $Ni(CN)_3$ units are linked together by a Ni—Ni bond[387]. On the other hand, bridges of this type have been established in isonitrile bridged systems[388].

Cyano bridges, which are easily detected by their high νCN frequencies, are known to occur in simple salts such as AgCN, AuCN, AgCN·2Ag NO$_3$ [6, 26], CuCN·N$_2$H$_4$ [389], Hg(CN)$_2$·AgNO$_3$·2H$_2$O [390], where infinite linear chains form the skeleton of the structures. Two dimensional sheet-like structures have been observed in Ni(CN)$_2$ [391], Ni(CN)$_2$·NH$_3$·L (L = e.g. benzene, thiophene) [392] and are assumed in Pd(CN)$_2$ [393]. Three-dimensional polymeric structures are verified in Prussian Blue and its analogues which are still subjects of extensive investigations[394–397]. Of particular interest is the linkage isomerisation of the cyano bridge observed in some Prussian Blue analogues[398, 399], Fe$_3$[Mn(CN)$_6$]$_2$, Co$_3$[Cr(CN)$_6$]$_2$ and Fe$_3$[Cr(CN)$_6$]$_2$. A plausible mechanism of these solid state reactions implies the 'flip' of a CN bridging group M—C≡N—M′ → M—N≡C—M′.

The role of cyanide bridges in electron transfer reactions is well established[341, 400] and some of the bridged complexes assumed as intermediates have been isolated, e.g.

$$[(CN)_5Fe^{2+}-CN-Co^{3+}(CN)_5]^{6-}\ ^{401},$$
$$[(NH_3)_5Co^{3+}-NC-Co^{3+}(CN)_5]\ ^{402}.$$

In the latter the two cobalt atoms are bridged by a CN group with the carbon atom bonded to the Co atom with all cyanide ligands[403]. Interestingly, the Co—C—N—Co groups turned out to be non-linear. Deviation of M—N—C—M linkages from linearity is not uncommon with polymeric cyanides (cf. K[Cu$_2$(CN)$_3$] [404]), possibly due to steric requirements. Dialkyl gold cyanides are tetrameric and are usually formulated with the (AuCN)$_4$ skeleton possessing C_{4h} symmetry [405]. Recent ^{1}H n.m.r. spectra indicate that in some of the molecules disordering of the cyanides along the Au—Au axis occurs[386].

Cyanide bridged dinuclear carbonyl metalates of iron, chromium and tungsten, [(CO)$_4$FeCNFe(CO)$_4$]$^-$ and [(CO)$_5$M—CN—M(CO)$_5$]$^-$ (M = Cr, W) have been reported by Behrens[47, 48] and Ruff[126]. Furthermore, a number of complexes have been prepared in which cyanide groups act as bridges between transition metals and main group Lewis acids[406], e.g.

$K_4[Fe(CNBF_3)_6]$, $K_4[Mo(CNBF_3)_8]$, $Fephen_2(CN)_2 \cdot nGeCl_4$ ($n = 1, 2, 3$), $Fe(CNR)_4(CNBX_3)_2$ (X = F, Cl; R = H, Me). Addition of aluminium and gallium alkyls to the cyanide ion[384] yields the complexes $[Me_3M-CN-MMe_3]^-$.

9.5.3 Thiocyanate and selenocyanate bridges

All the x-ray structural determinations carried out so far show that the thiocyanate bridge is of the type

$$M^{\diagdown S-C-N}{\diagdown}_M$$

though for $(Et_2MSCN)_3$ [407], $[(Me_3M)_2SCN]^-$ (M = Al, Ga), and $[(Me_3Al)_2 SeCN]^-$ [384] bridges via only the chalcogen atom have also been suggested (see Table 9.1). An interesting type of isomerism occurs in the planar thiocyanate bridged platinum compounds $[Pr_3PPtClSCN]_2$ [408], one isomer having the phosphine groups *trans* to the nitrogen atoms, the other *trans* to the sulphur atom of the linear SCN group[409]. From the differences in the Pt—N and Pt—S bond lengths, the *trans* weakening effect of the phosphine groups is clearly indicated.

Other doubly bridged systems with approximately planar eight membered rings have been found in $(Ph_3P)_2Cu(SCN)_2Cu(PPh_3)_2$ [410], $[en_2Ni(SCN)_2 Nien_2]^{2+}$ [411], $C_3H_5Pd(SCN)_2PdC_3H_5$ [412], $(CH_3)_2Au(NCS)_2Au(CH_3)^2$ [413]. The M—S—C—N and M—N—C—S bond angles found in the bridged compounds are close to those of terminal M—NCS and M—SCN groups (see Table 9.8). Thiocyanate (selenocynate) bridges have been established by i.r. methods in $[Fe(NO)_2SCN]_n$, $[Co(NO)_2SCN]_n$, $[Ni(NO)SCN]_n$, $[Ni (NO)(P(C_6H_{11})_3)SCN]_2$ [124], $[Mn_2(CO)_6(SCN)_4]^{3-}$, $[Mn_2(CO)_6(Se CN)_4]^{3-}$ [132]. The high $\nu_{as}SCN$ stretching frequencies are indicative of SCN bridges. The rhodium complexes $(PR_3)_3RhNCS$ dissociate in solution with the resultant formation of dinuclear SCN bridged compounds of the type $(PR_3)_2Rh(SCN)_2Rh(PR_3)_2$ [271]. Single bridges are present in the anionic carbonyl complexes $[(CO)_5M-SCN-M(CO)_5]^-$ (M = Cr, W) [48, 123, 126, 414] and in the unusual paramagnetic species $(CO)_5CrSCNCr(CO)_5$ [123, 414].

Crystal structures of a number of other SCN bridged polymeric compounds have been determined, e.g. AgSCN [44], $Cu_2(NCS)_3(NH_3)_3$ [415], $(CH_3)_2Sn(NCS)_2$ [416], $(CH_3)_3SnNCS$ [417], (ethylenethiourea)$_2Cd(NCS)_2$ [46], (2-thioimidazolidine)$_2Ni(NCS)_2$ [418]. A 3-coordinated sulphur atom as in $[Pr_3PAgSCN]_n$ [45] and $K_2Pd(SCN)_4$ [53] is also assumed for the tetrameric $[(CH_3)_3PtSCN]_4$ [49, 419]. The structures of $Pr_3PAgSCN$ and $[(CH_3)_3PtSCN]_4$ can be regarded as basically endless zigzag chains respectively a dimer, which are cross linked by additional longer metal sulphur bonds.

From the pioneering work of Taube it is known that SCN acts as a bridge in inner sphere electron transfer reactions[368, 420]. Dinuclear complexes, $CrNCSHg^{4+}$ [307], $[L_5CoNCSHg]^{n+}$, and $[L_5CoNCS]_2Hg^{m+}$ [421], which play a major role in mercury(II) assisted aquation of complexes, have been detected in solution. HgSCNCo bridges also exist in polymeric $Hg[Co (NCS)_4]$, where mercury is surrounded exclusively by sulphur atoms and cobalt by nitrogen atoms[422, 423], in accordance with the respective class *a* and class *b* behaviour of these metals. A high spin–low spin equilibrium of $Co(PEt_3)_2(NCS)_2$ in solution has been reported [424]. The low spin species is dimeric in solution and it has been suggested that the cobalt atoms are linked by SCN groups to give 5-coordinated metal ions.

9.5.4 Tricyanomethanide and dicyanamide bridges

Bridges of the planar tricyanomethanide ion occur in transition metal salts $M[C(CN)_3]_2$ (M = Mn, Fe, Co, Ni, Cu) which possess a polymeric octahedral structure[68, 351, 425]. From an x-ray analysis of $Cu[C(CN)_3]_2$, every copper atom is surrounded by six N atoms belonging to six different $C(CN)_3$ groups[426]. Polymeric octahedral structures with either $C(CN)_3$ or $N(CN)_2$ bridges[12] have to be assumed also in the compounds $M[N(CN)_2]_2$ [68, 351], $ML_2[N(CN)_2]_2$, $ML_2[C(CN)_3]_2$ (M = e.g. Ni, Co, Cu; L = DMSO, pyridine)[427], $[Co(NO)_2C(CN)_3]_n$ and $[Ni(NO)C(CN)_3]_n$ [350]. The cobalt salt $Co[C(CN)_3]_2$ provides an interesting example of isomerism; it exists in two forms, a red polymeric octahedral, and a blue polymeric tetrahedral structure[68, 351]. Addition of trimethyl aluminium to the $C(CN)_3^-$ ion affords the complex $[C(CN)_3(Me_3Al)_3]^-$ [384]. The crystal structure of $(CH_3)_2Sn[N(CN)_2]_2$ consists of an infinite two-dimensional network of tin atoms and bridging dicyanamide groups, that of $(CH_3)_3Sn[N(CN)_2]$ is composed of infinite chains —Sn—NCNCN—Sn—. In both structures bridging occurs through the cyano nitrogen atoms[430]. An infinite helical network of planar trimethyltin groups linked by linear NCN units was established for crystalline $[(CH_3)_3Sn]_2N_2C$ [431].

Usually halide as well as pseudohalide bridges are easily split on addition of neutral ligands such as phosphines, amines, isonitriles or anions. These bridge splitting reactions proved in several cases to be useful for the preparation of monomeric species and have been extensively applied by Chatt, Hieber and other research groups.

Notes added in proof

Reviews have appeared on pseudohalogens NCX (X = O, S, Se, Te) in coordination chemistry (Kuroda, K. (1971). *Kagaku no Ryoiki*, **25**, 80 (Japan) and on i.r. spectra of thiocyanate and related (NCO, NCSe, CNO) complexes (Bailey, R. A., Kozak, S. L., Michelsen, T. W., and Mills, W. N. (1971). *Coord. Chem. Rev.*, **6**, 407).

Stable salts of the tellurocyanate ion have been reported (Austad, R., Songstad, J., and Ase, K. (1971). *Acta Chem. Scand.*, **25**, 331).

The reactions of a number of halogenophosphine carbonyl rhenium(I) and -manganese(I) complexes with anhydrous hydrazine yielding isocyanate compounds have been studied. (Moelwyn-Hughes, J. T., Garner, A. W. B., and Howard, A. S. (1971). *J. Chem. Soc. A.*; 2361, 2370). Eight new complexes of the type $PdL_2(SCN)_2$ (L = substituted pyridines have been synthesised and characterised (Craciunescu, D., Ben-Bassat, A. H. I. (1971). *J. Less-Common Metals*, **25**, 11).

Square-planar, N-bonded complexes of the type $[M(Et_4dien)NCX]$ $[B(C_6H_5)_4]$ (M = Ni^{II}, Cu^{II}; X = O, S, Se) and the complexes $M(Et_4$ dien) $(NCX)_2$ (M = Co^{II}, Ni^{II}), have been prepared. The effects of the nature of the counteranion on the bonding mode of the pseudohalogens have been discussed (Burmeister, J. L., O'Sullivan, T. P. and Johnson, K. A. (1971). *Inorg. Chem.*, **10**, 1803).

A number of $[MX_4]^{2-}$ (M = Co, Ni, Cu, Zn; X = RNCN, NCO, NCS, NCSe) have been synthesised; their electronic structure has been investigated (Hollebone, B. R. and Nyholm, R. S. (1971). *J. Chem. Soc. A.*, 332, 481). Chlorine isocyanate has been used to introduce the isocyanate ligand into organometallic compounds of silicon and mercury (Leimeister, H. and Dehnicke, K. (1971). *J. Organometal. Chem.*, 31, C3).

References

1. Birckenbach, L. and Kellermann, K. (1925). *Chem. Ber.*, **58**, 786, 2377
2. Griffith, W. P. (1962). *Quart. Rev. Chem. Soc.*, **16**, 188
3. Chadwick, B. M. and Sharpe, A. G. (1966). *Advan. Inorg. Chem. Radiochem.*, **8**, 83
4. Shriver, D. F. (1966), *Structure and Bonding*, **1**, 32
5. Halpern, J. (1970). *Accounts Chem. Res.*, **3**, 386
6. Britton, D. (1967). *Perspectives in Structural Chem.*, **1**, 109
7. Yoffe, A. D. (1966). *Developments in Inorganic Nitrogen Chemistry*, (Amsterdam: Elsevier)
8. Dehnicke, K. (1967). *Angew. Chem.*, **79**, 253; *Angew. Chem. Int. Edn.*, **6**, 240
9. Thayer, J. S. (1966). *Organometal. Chem. Rev.*, **1**, 157
10. Livingstone, S. E. (1965). *Quart. Rev. Chem. Soc.*, **19**, 386
11. Golub, A. M. and Skopenko, V. V. (1965). *Russ. Chem. Rev.*, **34**, 907
12. Köhler, H. (1969). *Wiss. Z. Univ. Halle*, **18**, 33, (1967); *Z. Naturforsch.*, **22b**, 238
13. Baddley, W. H. (1968). *Inorg. Chim. Acta Rev.*, **2**, 7
14. Beck, W. *Organometal. Rev.*, to be published
15. Thayer, J. S. and West, R. (1967). *Advan. Organometal. Chem.*, **5**, 169 and cited references therein
16. Lappert, M. F. and Pyszora, H. (1966). *Advan. Inorg. Chem. Radiochem.*, **9**, 133
17. Gutmann, V. (1968). *Coordination Chemistry in Nonaqueous solutions*, (Wien: Springer Verlag)
18. Burmeister, J. L. (1966). *Coord. Chem. Rev.* **1**, 205
19. Burmeister, J. L. (1968). *Coord. Chem. Rev.*, **3**, 225
20. Norbury, A. H. and Sinha, A. I. P. (1970). *Quart. Rev. Chem. Soc.*, **24**, 69
21. Tullberg, A. and Vannerberg, N-G. (1971). *Acta Chem. Scand.*, **25**, 343
22. Espenson, J. H. and Birk, J. P. (1965). *J. Amer. Chem. Soc.*, **87**, 3280
23. Birk, J. P. and Espenson, J. H. (1968). *J. Amer. Chem. Soc.*, **90**, 1153
24. Birk, J. P. and Espenson, J. H. (1968). *J. Amer. Chem. Soc.*, **90**, 2266
25. Kuroda, K. and Gentile, P. S. (1967). *Inorg. Nucl. Chem. Lett.* **3**, 151
26. Britton, D. and Dunitz, J. D. (1965). *Acta Crystallogr.*, **19**, 815
27. Rayner, J. H. and Powell, H. M. (1952). *J. Chem. Soc.*, 319
28. Phillips, R. F. and Powell, H. M. (1939). *Proc. Roy. Soc. A*, **173**, 147
29. Stocco, F., Stocco, G. C., Scovell, W. M. and Tobias, R. S. *J. Amer. Chem. Soc.* (in press)
30. Beck, W. (1962). *Z. Naturforsch.*, **17b**, 130
31. Beck, W., Swoboda, P., Feldl, K. and Schuierer, E. (1970). *Chem. Ber.*, **103**, 3591
32. Britton, D. and Dunitz, J. D. (1965). *Acta Crystallogr.*, **19**, 662
33. Bush, M. A. and Sim, G. A. (1970). *J. Chem. Soc. A*, 605
34. McPhail, A. T., Knox, G. R., Robertson, C. G. and Sim, G. A. (1971). *J. Chem. Soc. A*, 205

35. Bailey, R. A. and Kozak, S. L. (1969). *J. Inorg. Nucl. Chem.*, **31**, 689
36. Burmeister, J. L., Deardorff, E. A. and Van Dyke, C. E. (1969). *Inorg. Chem.*, **8**, 170; Burmeister, J. L., Deardorff, E. A., Jensen, A. and Christiansen, V. H. (1970). *Inorg. Chem.*, **9**, 58
37. Britton, D. and Dunitz, J. D. (1965). *Acta Crystallogr.*, **18**, 424
38. Beck, W. and von Werner, K. (1971). *Chem. Ber.*, **104**, 2901
39. Countryman, R. and McDonald, W. S. (1971). *J. Inorg. Nucl. Chem.*, **33**, 2213
40. Knox, J. R. and Eriks, K. (1968). *Inorg. Chem.*, **7**, 84
41. Di Vaira, M. and Sacconi, L. (1969). *Chem. Commun.*, 10; *J. Chem. Soc. A*, (1971), 148
42. Brown, B. W. and Lingafelter, E. C. (1964). *Acta Crystallogr.*, **17**, 254
43. Marongiu, G., Lingafelter, E. C. and Paoletti, P. (1969). *Inorg. Chem.*, **8**, 2763
44. Lindqvist, I. (1957). *Acta Crystallogr.*, **10**, 29
45. Turco, A., Panattoni, C. and Frasson, E. (1960). *Nature (London)*, **187**, 772
46. Cavalca, L., Nardelli, M. and Fava, G. (1960). *Acta. Crystallogr.*, **13**, 125
47. Behrens, H., Schwab, R., Herrmann, D. (1966). *Z. Naturforsch.*, **21b**, 590
48. Lindner, E. and Behrens, H. (1967). *Spectrochim. Acta*, **23A**, 3025
49. Homan, J. M., Kawamoto, J. M. and Morgan, G. L. (1970). *Inorg. Chem.*, **9**, 2533
50. Dehnicke, K. (1967). *Angew. Chem. Int. Edn.*, **6**, 947
51. Mason, W. R., Berger, E. R. and Johnson, R. C. (1967). *Inorg. Chem.*, **6**, 248
52. Shea, C. and Haim, A. (1971). *J. Amer. Chem. Soc.*, **93**, 3055
53. Mawby, A. and Pringle, G. E. (1970). *Chem. Commun.*, 385
54. Haim, A. and Sutin, N. (1965). *J. Amer. Chem. Soc.*, **87**, 4210
55. Haim, A. and Sutin, N. (1966). *J. Amer. Chem. Soc.*, **88**, 434
56. Orhanović, M. and Sutin, N. (1968). *J. Amer. Chem. Soc.*, **90**, 538, 4286
57. Dori, Z. (1968). *Chem. Commun.*, 714
58. Drummond, J. and Wood, J. S. (1969). *Chem. Commun.*, 1373
59. Davis, B. R. and Ibers, J. A. (1970). *Inorg. Chem.*, **9**, 2768
60. Fehlhammer, W. P. and Dahl, F. *J. Amer. Chem. Soc.*, in press
61. Mason, R., Rusholme, G. A., Beck, W., Engelmann, H., Joos, K., Lindenberg, B. and Smedal, H. S. (1971). *Chem. Commun.*, 496
62. Müller, U. (1971). *Z. Anorg. Allg. Chem.*, **382**, 110
63. Agrell, I. (1967). *Acta Chem. Scand.*, **21**, 2647
64. Ziolo, R. F., Gaughan, A. P., Dori, Z., Pierpont, C. G. and Eisenberg, R. *J. Amer. Chem. Soc.*, **92**, 738 (1970); *Inorg. Chem.*, **10**, 1289 (1971)
65. Ball, D. L. and King, E. L. *J. Amer. Chem. Soc.*, **80**, 1091 (1958); Snellgrove, R. and King, E. L. *J. Amer. Chem. Soc.*, **84**, 4609 (1962)
66. Bock, H. PhD. Thesis, München, 1970
67. Beck, W,, Smedal, H. S. and Köhler, H. (1967). *Z. Anorg. Allg. Chem.*, **354**, 69
68. Köhler, H. (1964). *Z. Anorg. Allg. Chem.*, **331**, 237
69. Beck, W., Nitzschmann, R. E. and Smedal, H. S. (1967). *J. Organometal. Chem.*, **8**, 547
70. Becker, W., Schorpp, K., Schlodder, R. and Beck, W., unpublished results
71. Beck, W., Hieber, W. and Neumair, G. (1966). *Z. Anorg. Allg. Chem.*, **344**, 285
72. Biondi, C., Bonamico, M., Torelli, L. and Vaciago, A. (1965). *Chem. Commun.*, 191
73. Enemark, J. H. and Holm, R. H. (1964). *Inorg. Chem.*, **3**, 1516
74. Köhler, H. and Seifert, B. (1966). *Z. Anorg. Allg. Chem.*, **344**, 63
75. Köhler, H. and Beck, W. (1968). *Z. Anorg. Allg. Chem.*, **359**, 241
76. Köhler, H. and Seifert, B. (1968). *Z. Anorg. Allg. Chem.*, **360**, 137; Köhler, H. and Seifert, B. (1970). *Z. Anorg. Allg. Chem.*, **379**, 1
77. Anon. (1710). *Miscellanea Berolinensia ad Incrementum Scientiarum*, Berlin, p. 277
78. Nef, J. U. (1894). Liebigs. *Ann. Chem.*, **280**, 334
79. Wöhler, L., Weber, A. and Berthmann, A. (1929). *Ber. Dtsch. Chem. Ges.*, **62**, 2742, 2748
80. Wöhler, L. and Martin, F. (1917). *Ber. Dtsch. Chem. Ges.*, **50**, 586
81. Gay-Lussac, J. L. and Liebig, J. (1824). *Annales de chimie et de physique*, **25**, 285
82. Senise, P. (1959). *J. Amer. Chem. Soc.*, **81**, 4196
83. Wiberg, E. and Michaud, H. (1954). *Z. Naturforsch.*, **9b**, 495
84. Kröhnke, F. and Sander, B. (1964). *Z. Anorg. Allg. Chem.*, **334**, 66
85. Beck, W., Fehlhammer, W. P., Pöllmann, P., Schuierer, E. and Feldl, K. (1967). *Chem. Ber.*, **100**, 2335
86. Schmidtke, H.-H. and Garthoff, D. (1967). *J. Amer. Chem. Soc.*, **89**, 1317; (1969). *Z. Naturforsch*, **249**, 126

87. Clarke, J. W. and Dudley, W. L. (1878). *Ber. Dtsch. Chem. Ges.*, **11**, 1325; *Gmelins Handbuch der Anorganischen Chemie*, (1942), 8th Edn., Syst. No. 68, "Platin" [C], 212, Verlag Chemie, Berlin
88. Rosenheim, A. and Pritze, M. (1909). *Z. Anorg. Chem.*, **63**, 275
89. Blomstrand, C. W. (1871). *J. Prakt. Chem.*, **3**, 221
90. Ripan, R. (1930). *Chem. Zentral*, **101**, I, 967
91. See, e.g. Williams, H. U. (1948). *Cyanogen Compounds*, 2nd Edn. (London:Arnold)
92. Rosenheim, A. and Cohn, R. (1901). *Z. Anorg. Chem.*, **27**, 280
93. Grossmann, H. and Schück, B. (1906). *Z. Anorg. Chem.*, **50**, 1
94. Appel, R. and Hauss, A. (1961). *Z. Anorg. Allg. Chem.*, **311**, 290
95. Ruff, J. K. and Schlientz, W. J., submitted for publication.
96. See, e.g. Ruff, J. K. (1967). *Inorg. Chem.*, **6**, 2080
97. Beck, W. and Schuierer, E. (1962). *Chem. Ber.*, **95**, 3048
98. Basolo, F. (1968). *Coord. Chem. Rev.*, **3**, 213
99. Müller, U. and Dehnicke, K. (1966). *Angew. Chem. Int. Ed. Engl.*, **5**, 841
100. Forster, D. and Horrocks, W. D., Jr. (1966). *Inorg. Chem.*, **5**, 1510
101. Roesky, H. W. (1967). *Angew. Chem. Int. Ed. Engl.*, **6**, 363
102. Schmidt, A. (1970). *Chem. Ber.*, **103**, 3923
103. Oliveri-Mandalá, E. and Comella, G. (1922). *Gazz. Chim. Ital.*, **52**, 112
104. Fritzer, H. P. and Torkar, K. (1968). *Monatsh Chem.*, **99**, 2333
105. Straumanis, M. and Cirulis, A. (1944). *Z. Anorg. Allg. Chem.*, **252**, 9, 121
106. Cirulis, A. and Straumanis, M. (1943). *Chem. Ber.*, **76**, 825
107. Krischner, H. and Fritzer, H. P. (1970). *Z. Anorg. Allg. Chem.*, **376**, 162
108. Beck, W. and Schuierer, E. (1965). *Chem. Ber.*, **98**, 298
109. Swoboda, P. (1970). PhD. Thesis, München
110. Wieland, H. (1909). *Ber. Dtsch. Chem. Ges.*, **42**, 820
111. Beck, W. and Schuierer, E. (1966). *Z. Anorg. Allg. Chem.*, **347**, 304
112. Forster, D. and Goodgame, D. M. L. (1965). *J. Chem. Soc.*, 268
113. Pöllmann, P. (1968). Ph.D. Thesis, TU München
114. Funk, H. and Böhland, H. (1962). *Z. Anorg. Allg. Chem.*, **318**, 169
115. Raymond, K. N. and Basolo, F. (1966). *Inorg. Chem.*, **5**, 949
116. Forster, D., Goodgame, D. M. L. (1965). *Inorg. Chem.*, **4**, 823
117. Grossmann, H. and Hünseler, F. (1905). *Z. Anorg. Allg. Chem.*, **46**, 361
118. Mikheeva, L. M., Komissarova, L. N. and Elfimova, G. I. (1970). *Zh. Neorg. Khim.*, **15**, 3215; (1971). *Zh. Neorg. Khim.* **16**, 2111
119. Basolo, F., Burmeister, J. L. and Poe, A. J. (1963). *J. Amer. Chem. Soc.*, **85**, 1700
120. Behrens, H., Lindner, E. and Schindler, H. (1966). *Chem. Ber.*, **99**, 2399; Behrens and co-workers, *Chem. Ber.* (1963), 2220, (1966). *Chem. Ber.* **99**, 2745; (1967). *Z. Anorg. Allg. Chem.*, **361**, 125; **354**, 184; **349**, 251
121. Behrens, H., Lindner, E. and Schindler, H. (1969). *Z. Anorg. Allg. Chem.*, **365**, 119
122. Wojcicki, A. and Farona, M. F. (1964). *J. Inorg. Nucl. Chem.*, **26**, 2289
123. Behrens, H. and Herrmann, D. (1966). *Z. Naturforsch.*, **21b**, 1234
124. Beck, W. and Lottes, K. (1965). *Z. Anorg. Allg. Chem.*, **335**, 258
125. Hieber, W., Bauer, I. and Neumair, G. (1965). *Z. Anorg. Allg. Chem.*, **335**, 250
126. Ruff, J. K. (1969). *Inorg. Chem.*, **8**, 86
127. Cotton, F. A., Robinson, W. R., Walton, R. A. and Whyman, R. (1967). *Inorg. Chem.*, **6**, 929
128. Bowman, K. and Dori, Z. (1970). *Inorg. Chem.*, **9**, 395
129. Bauer, R. A. and Basolo, F. (1969). *Inorg. Chem.*, **8**, 2231
130. Hawkes, M. J. and Ginsberg, A. P. (1969). *Inorg. Chem.*, **8**, 2189
131. Treichel, P. M. and Douglas, W. M. (1969). *J. Organometal. Chem.*, **19**, 221
132. Farona, M. F., Frazee, L. M. and Bremer, N. J. (1969). *J. Organometal. Chem.*, **19**, 225
133. Negroiû, D. and Baloiu, L. M. (1970). *Z. Anorg. Allg. Chem.*, **374**, 105
134. Holloway, J. H., McQuillan, G. P. and Ross, D. S. (1969). *J. Chem. Soc. A*, 2505
135. Konya, K., Nishikawa, H. and Shibata, M. (1968). *Inorg. Chem.*, **7**, 1165
136. Schmidtke, H.-H. and Garthoff, D. (1967). *Helv. Chim. Acta.*, **50**, 1631
137. Sloan, T. E. and Wojcicki, A. (1968). *Inorg. Chem.*, **7**, 1268
138. Bailey, R. A. and Kozak, S. L. (1967). *Inorg. Chem.*, **6**, 419
139. Reinstein, J. G., Griswold, E. and Kleinberg, J. (1969). *Inorg. Chem.*, **8**, 2499; Magnuson, W. L., Griswold, E. and Kleinberg, J. (1964). *Inorg. Chem.*, **3**, 88

140. v. Winbush, S., Griswold, E. and Kleinberg, J. (1961). *J. Amer. Chem. Soc.*, **83**, 3197
141. Vaska, L. and Peone, J. (1971). *Chem. Commun.*, 418
142. Tobias, R. S., to be published
143. Goel, R. G. and Ridley, D. R. (1971). *Inorg. Nucl. Chem. Lett.*, **7**, 21
144. Norbury, A. H. and Sinha, A. I. P. (1968). *J. Chem. Soc. A*, 1598
145. Kane-Maguire, L. A. P., Sheridan, P. S., Basolo, F. and Pearson, R. G. (1968). *J. Amer. Chem. Soc.*, **90**, 5295
146. Nicpon, P. and Meek, D. W. (1966). *Inorg. Chem.*, **5**, 1297
147. Burmeister, J. L. and Gysling, H. J. (1967). *Inorg. Chim. Acta*, **1**, 100
148. Burmeister, J. L. and DeStefano, N. J. (1970). *Inorg. Chem.*, **9**, 972
149. See, e.g. Jordan, R. B., Sargeson, A. M. and Taube, H. (1966). *Inorg. Chem.*, **5**, 1091
150. Stotz, I., Wilmarth, W. K. and Haim, A. (1968). *Inorg. Chem.*, **7**, 1250
151. Farona, M. F. and Wojcicki, A. (1965). *Inorg. Chem.*, **4**, 857
152. Behrens, H. and Herrmann, D. (1966). *Z. Naturforsch.*, **21b**, 1236
153. Shihada, A. F. and Dehnicke, K. (1971). *J. Organometal. Chem.*, **26**, 157
154. Dehnicke, K. (1965). *J. Inorg. Nuclear Chem.*, **27**, 809
155. Paetzold, P. I., Gayoso, M. and Dehnicke, K. (1965). *Chem. Ber.*, **98**, 1173
156. Behrens, H. and Köhler, J. (1959). *Z. Naturforsch.*, **14b**, 463
157. Argento, B. J., Fitton, P., McKeon, J. E. and Rick, E. A. (1969). *Chem. Commun.*, 1427
158. Bressan, M., Favero, G., Corain, B. and Turco, A. (1971). *Inorg. Nucl. Chem. Lett.*, **7**, 203
159. Guttenberger, J. F. *Angew. Chem.* (1967), **79**, 1071; *Angew. Chem. Int. Ed. Engl.* (1967), **6**, 1081
160. Beck, W., Bauder, M., La Monica, G., Cenini, S. and Ugo, R. (1971). *J. Chem. Soc. A*, 113
161. Schorpp, K., Becker, W. and Beck, W., unpublished results
162. Beck, W., Schorpp, K. and Kern, F. (1971). *Angew. Chem.*, **83**, 43
163. DeStefano, N. J. and Burmeister, J. L. (1971). *Inorg. Chem.*, **10**, 998
164. Singh, M. M. and Varshavskii, Yu. S. (1969). *Russ. J. Inorg. Chem.*, **14**, 1278
165. Beck, W. and Kreutzer, P., unpublished results
166. Nichols, D. I. and Charleston, A. S. (1969). *J. Chem. Soc. A*, 2581
167. Lorberth, J. (1965). *Chem. Ber.*, **98**, 1201
168. Kashutina, M. V. and Okhlobystin, O. Y. (1967). *J. Organometal. Chem.*, **9**, 5
169. Aynsley, E. E., Greenwood, N. N., Hunter, G. and Sprague, M. J. (1966). *J. Chem. Soc. A*, 1344
170. Jennings, M. A. and Wojcicki, A. (1969). *Inorg. Chim. Acta*, **3**, 335
171. Stamm, W. (1965). *J. Org. Chem.*, **30**, 693
172. Cummins, R. A. and Dunn, P. (1964). *Aust. J. Chem.*, **17**, 411
173. Balahura, R. J. and Jordan, R. B. (1970). *Inorg. Chem.*, **9**, 1567
174. Britton, D. and Dunitz, J. D. (1965). *Acta Crystallogr.*, **18**, 424
175. Thayer, J. S. (1968). *Inorg. Chem.*, **7**, 2599
176. Schorpp, K. and Beck, W., unpublished results
177. Caldwell, S. M. and Norris, A. R. (1968). *Inorg. Chem.*, **7**, 1667
178. Schug, K., Miniatas, B., Sadowski, A. J., Yano, T. and Ueno, K. (1968). *Inorg. Chem.*, **7**, 1669
179. Beck, W. and Smedal, H. S. (1966). *Angew. Chem.*, **78**, 267; Beck, W., Werner, H., Engelmann, H. and Smedal, H. S. (1968). *Chem. Ber.*, **101**, 2143; Werner, H., Beck, W. and Engelmann, H. *Inorg. Chim. Acta* (1969), **3**, 331
180. Angelici, R. J. and Busetto, L. (1969). *J. Amer. Chem. Soc.*, **91**, 3197; Kruse, A. E. and Angelici, R. J. (1970). *J. Organometal. Chem.*, **24**, 231; Angelici, R. J. and Kruse A. E. (1970). *J. Organometal. Chem.*, **22**, 461
181. Beck, W., Lindenberg, B. (1970). *Angew. Chem.*, **82**, 701
182. Angelici, R. J. and Faber, G. C. (1971). *Inorg. Chem.*, **10**, 514
183. Graziani, M., Busetto, L. and Palazzi, A. (1971). *J. Organometal. Chem.*, **26**, 261
184. Busetto, L., Graziani, M. and Belluco, U. (1971). *Inorg. Chem.*, **10**, 78
185. Douglas, P. G., Feltham, R. D. and Metzger, H. G. (1971). *J. Amer. Chem. Soc.*, **93**, 84
186. Bottomley, F. and Crawford, J. R. (1971). *Chem. Commun.*, 200
187. Sas, T. M., Komissarova, L. N., Gulia, V. G. and Grigor'ev, A. I. (1967). *Russ. J. Inorg. Chem.*, **12**, 1090; Michalski, E. and Walewski, L. (1961). *Chem. Analyt.*, **6**, 273
188. Lenz, W., Schläfer, H. L. and Ludi, A. (1969). *Z. Anorg. Allg. Chem.*, **365**, 55

189. Schmitz-DuMont, O. and Ross, B. (1964). *Angew. Chem. Int. Ed. Engl.*, **3**, 315, 586
190. Böhland, H. and Malitzke, P. (1967). *Z. Anorg. Allg. Chem.*, **350**, 70
191. Cotton, F. A., Goodgame, D. M. L., Goodgame, M. and Sacco, A. (1961). *J. Amer. Chem. Soc.*, **83**, 4157
192. Forster, D. and Goodgame, D. M. L. (1965). *Inorg. Chem.*, **4**, 715
193. Kharitonov, Yu. Ya., Tsintsadze, G. V. and Tsivadze, A. Yu. (1970). *Russ. J. Inorg. Chem.*, **15**, 776, 614, 204, 1513
194. Martin, J. L., Thompson, L. C., Radonovich, L. J. and Glick, M. D. (1968). *J. Amer. Chem. Soc.*, **90**, 4493
195. Bailey, R. A., Michelsen, T. W. and Nobile, A. A. (1970). *J. Inorg. Nucl. Chem.*, **32**, 2427
196. Golub, A. M. and Sergunkin, V. N. (1966). *Russ. J. Inorg. Chem.*, **11**, 419
197. Knox, G. F. and Brown, T. M. (1969). *Inorg. Chem.*, **8**, 1401; (1971). *Inorg. Syn.* **13**, 226
198. Brown, T. M. and Knox, G. F. (1967). *J. Amer. Chem. Soc.*, **89**, 5296
199. Boehland, H., Tiede, E. and Zenker, E. (1968). *J. Less-Common Metals*, **15**, 89
200. Lewis, J., Nyholm, R. S. and Smith, P. W. (1961). *J. Chem. Soc.*, 4590
201. Schwochau, K. and Pieper, H. H. (1968). *Inorg. Nucl. Chem. Lett.*, **4**, 711
202. Sabatini, A. and Bertini, I. (1965). *Inorg. Chem.*, **4**, 959
203. Lindqvist, I. and Strandberg, B. (1957). *Acta Crystallogr.*, **10**, 173
204. Kharitonov, Yu. Ya., Tsintsadze, G. V. and Tsivadze, A. Yu. (1970). *Russ. J. Inorg. Chem.*, **15**, 484, 364, 931, 1811
205. Taylor, K. A., Long, T. V. and Plane, R. A. (1967). *J. Chem. Phys.*, **47**, 138
206. Golub, A. M., Olevinskii, M. I. and Zhigulina, N. S. (1966). *Russ. J. Inorg. Chem.*, **11**, 841
207. Funk, H. and Böhland, H. (1963). *Z. Anorg. Allg. Chem.*, **324**, 168; Ulko, N. V. and Savchenko, R. A. *Russ.* (1967). *J. Inorg. Chem.*, **12**, 169
208. Bailey, R. A. and Kozak, S. L. (1967). *Inorg. Chem.*, **6**, 2155
209. Ciavatta, L. and Grimaldi, M. (1970). *Inorg. Chim. Acta*, **4**, 312
210. Pochkarev, T. I., Mikheeva, L. M., Grigor'ev, A. I. and Ganem, A. (1970). *Russ. J. Inorg. Chem.*, **15**, 45
211. Tuck, D. G. (1966). *Coord. Chem. Rev.*, **1**, 286
212. Bertazzi, N., Silvestri, A. and Barbieri, R. (1971). *J. Inorg. Nucl. Chem.*, **33**, 799
213. Sowerby, D. B. (1961). *J. Inorg. Nucl. Chem.*, **22**, 205
214. Chamberlain, B. R. and Moser, W. (1969). *J. Chem. Soc. A*, 354
215. Golub, A. M., Romanenko, L. I. and Samoilenko, V. M. (1959). *Ukrain. Khim. Zh.*, **25**, 50
216. Golub, A. M. and Romanenko, L. J. (1961). *Ukrain. Khim. Zh.*, **27**, 11
217. Ohlberg, S. M. and van der Meulen, P. A. (1953). *J. Amer. Chem. Soc.*, **75**, 997
218. Long, C. and Shoog, D. A. (1966). *Inorg. Chem.*, **5**, 206
219. Burmeister, J. L., Patterson, S. D. and Deardorff, E. A. (1969). *Inorg. Chim. Acta*, **3**, 105
220. Grey, I. E. and Smith, P. W. (1969). *Aust. J. Chem.*, **22**, 311
221. Markow, V. P. and Traggeim, E. N. (1961). *Russ. J. Inorg. Chem.*, **6**, 1175
222. Skopenko, V. V. and Ivanova, E. I. (1969). *Russ. J. Inorg. Chem.*, **14**, 388
223. Michelson, K. (1963). *Acta Chem. Scand.*, **17**, 1841
224. Forster, D. and Goodgame, D. M. L. (1965). *Inorg. Chem.*, **4**, 1712
225. Burmeister, J. L. and Williams, L. E. (1966). *Inorg. Chem.*, **5**, 1113
226. Turco, A., Pecile, C. and Nicolini, M. *Proc. Chem. Soc.* (1961), 213; *J. Chem. Soc.* (1962), 3008
227. Cotton, F. A., Goodgame, D. M. L., Goodgame, M. and Haas, T. E. (1962). *Inorg. Chem.*, **1**, 565
228. Golub, A. M. and Skopenko, V. V. (1965). *Russ. Chem. Rev.*, **34**, 901
229. Schmidtke, H.-H. (1966). *J. Inorg. Nucl. Chem.*, **28**, 1735
230. Skopenko, V. V., Mikitchenko, V. F. and Tsintsadze, G. V. (1969). *Russ. J. Inorg. Chem.*, **14**, 939
231. Golub, A. M., Skopenko, V. V. and Zhumabaev, A. (1969). *Russ. J. Inorg. Chem.*, **14**, 1579; *Ukr. Khim. Zh.* (1971). **37**, 528
232. Burmeister, J. L. and Deardorff, E. A. (1970). *Inorg. Chim. Acta*, **4**, 97
233. Forster, D. and Goodgame, D. M. L. *J. Chem. Soc.* (1965), 262; *ibid.* (1964), 2790
234. Tsivadze, A. Yu., Tsintsadze, G. V., Kharitonov, Yu. Ya., Golub, A. M. and Mamulashvili, A. M. (1970). *Russ. J. Inorg. Chem.*, **15**, 934
235. Bailey, R. A. and Kozak, S. L. (1969). *J. Inorg. Nucl. Chem.*, **31**, 689

236. Forster, D. and Goodgame, D. M. L. (1965). *J. Chem. Soc.*, 1286
237. Burmeister, J. L. and O'Sullivan, T. P. (1969). *Inorg. Chim. Acta*, **3**, 479
238. Beck, W., Fehlhammer, W. P., Pöllmann, P. and Schächl, H. (1969). *Chem. Ber.*, **102**, 1976
239. Mitchell, P. C. H. and Williams, R. J. P. (1960). *J. Chem. Soc.*, 1912
240. Ahrland, S., Chatt, J. and Davies, N. R. (1958). *Quart. Rev. Chem. Soc.*, **12**, 265
241. Schwarzenbach, G. (1961). *Advan. Inorg. Chem. Radiochem.*, **3**, 257
242. Pearson, R. G. (1963). *J. Amer. Chem. Soc.*, **85**, 3533
243. Tramer, A. (1962). *J. Chim. Phys.*, **59**, 232; Tramer, A. (1964). *Theory and Structure of Complex Compounds*, 225, B. Jezowska-Trzebiatowska (New York: Pergamon Press)
244. Howarth, O. W., Richards, R. E. and Venanzi, L. M. (1964). *J. Chem. Soc.*, 3335
245. Turco, A. and Pecile, C. (1961). *Nature (London)*, **191**, 66
246. Sabatini, A. and Bertini, I. (1965). *Inorg. Chem.*, **4**, 1665
247. Burmeister, J. L. and Basolo, F. (1964). *Inorg. Chem.*, **3**, 1587
248. Bertini, I. and Sabatini, A. (1966). *Inorg. Chem.*, **5**, 1025
249. Basolo, F., Baddley, W. H. and Burmeister, J. L. (1964). *Inorg. Chem.*, **3**, 1202
250. Meek, D. W., Nicpon, P. E. and Imhof Meek, V. (1970). *J. Amer. Chem. Soc.*, **92**, 5351
251. Farona, M. F. and Wojcicki, A. (1965). *Inorg. Chem.*, **4**, 1402
252. Piper, T. S. and Wilkinson, G. (1956). *J. Inorg. Nucl. Chem.*, **2**, 38
253. Colton, R. and Scollary, G. R. (1968). *Aust. J. Chem.*, **21**, 1435
254. Doyle, G. and Tobias, R. S. (1968). *Inorg. Chem.*, **7**, 2479
255. Green, M. L. H. and Lindsell, W. E. (1969). *J. Chem. Soc. A*, 2150
256. Chamberlain, M. M. and Bailar, J. C. (1959). *J. Amer. Chem. Soc.*, **81**, 6412
257. Norbury, A. H. and Sinha, A. I. P. (1968). *Inorg. Nucl. Chem. Lett.*, **4**, 617
258. Burmeister, J. L. (1964). *Inorg. Chem.*, **3**, 919
259. Maki, N. and Sakuraba, S. (1969). *Bull. Chem. Soc. Japan*, **42**, 579
260. Ferrari, A., Braibanti, A., Bigliardi, G. and Lanfredi, A. M. (1965). *Acta Crystallogr.*, **18**, 367
261. Nardelli, M., Fava Gasparri, G., Giraldi Battistini, G. and Domiano P. (1966). *Acta Crystallogr.*, **20**, 349
262. Jain, C. and Lingafelter, E. C. *J. Amer. Chem. Soc.* (1967), **89**, 724; *ibid.* (1967), **89**, 6131 6131
263. Andreetti, G. D., Jain, P. C. and Lingafelter, E. C. (1969). *J. Amer. Chem. Soc.*, **91**, 4112
264. Dapporto, P., Di Vaira, M. and Sacconi, L. (1969). *Chem. Commun.*, 153; (1971). *J. Chem. Soc. A*, 1891
265. Baggio, S. and Becka, L. N. (1969). *Acta Crystallogr.*, **B25**, 946
266. Martin, J. L., Thompson, L. C., Radonovich, L. J. and Glick, M. D. (1968). *J. Amer. Chem. Soc.*, **90**, 4493
267. Graham, A. J. and Fenn, R. H. (1969). *J. Organometal. Chem.*, **17**, 405
268. Graham, A. J. and Fenn, R. H. (1970). *J. Organometal. Chem.*, **25**, 173
269. Jeffery, J. W. (1963). *Acta Crystallogr. Suppl.*, **16**, A66
270. Clark, G. R., Palenik, G. J. and Meek, D. W. (1970). *J. Amer. Chem. Soc.*, **92**, 1077; *Inorg. Chem.*, **9**, 2754
271. Jennings, M. A. and Wojcicki, A. (1967). *Inorg. Chem.*, **6**, 1854
272. Hewkin, D. J. and Poe, A. J. (1967). *J. Chem. Soc. A*, 1884
273. Basolo, F., Baddley, W. H. and Weidenbaum, K. J. (1966). *J. Amer. Chem. Soc.*, **88**, 1576
274. Kulasingam, G. C. and McWhinnie, W. R. *Chem. Ind. (London)*, (1966), 2200; *J. Chem. Soc. A*. (1968), 254
275. Negóiu, D. and Baloiu, L. M. (1971). *Z. Anorg. Allg. Chem.*, **382**, 92
276. Schmidtke, H.-H. *J. Amer. Chem. Soc.*, (1965), **87**, 2522; *Z. Phys. Chem. (Frankfurt)*, (1965), **45**, 305
277. Schmidtke, H.-H. (1966). *Inorg. Chem.*, **5**, 1682
278. Guttermann, D. F. and Gray, H. B. (1969). *J. Amer. Chem. Soc.*, **91**, 3105
279. Burmeister, J. L., Hassel, R. L. and Phelan, R. J. (1970). *Chem. Commun.*, 679; (1971). *Inorg. Chem.*, **10**, 2032
280. Klopman, G. (1968). *J. Amer. Chem. Soc.*, **90**, 223
281. Burmeister, J. L. and Lim, J. C. (1968). *Chem. Commun.*, 1346
282. Beran, G. and Palenik, G. J. (1970). *Chem. Commun.*, 1354
283. Raymond, K. N. and Basolo, F. (1966). *Inorg. Chem.*, **5**, 1632
284. Nicpon, P. and Meek, D. W. (1967). *Inorg. Chem.*, **6**, 145

285. Carty, A. J. and Efraty, A. (1969). *Canad. J. Chem.*, **47**, 2573
286. Curtis, N. F. and Curtis, Y. M. (1966). *Aust. J. Chem.*, **19**, 1423
287. Carbacho, H., Ungerer, B. and Contreras, G. (1970). *J. Inorg. Nucl. Chem.*, **32**, 579
288. Hackel-Wenzel, B. and Thomas, G. (1971). *J. Less-Common Metals*, **23**, 185
289. DuBois, T. D. and Meek, D. W. (1967). *Inorg. Chem.*, **6**, 1395
290. Contreras, G. and Schmidt, R. (1970). *J. Inorg. Nucl. Chem.*, **32**, 127, 1295
291. Hart, F. A. and Laming, F. P. (1964). *J. Inorg. Nucl. Chem.*, **26**, 579
292. Schilt, A. A. and Fritsch, K. (1966). *J. Inorg. Nucl. Chem.*, **28**, 2677
293. Nelson, S. M. and Shepherd, T. M. (1965). *J. Chem. Soc.*, 3276
294. Schmauss, G. and Barth, P. (1970). *Z. Naturforsch.*, **25b**, 799
295. Crawford, N. P. and Melson, G. A. (1969). *J. Chem. Soc. A*, 427
296. Flint, C. D. and Goodgame, M. (1969). *Inorg. Chem.*, **8**, 1833
297. Crawford, N. P. and Melson, G. A. (1969). *J. Chem. Soc. A*, 1049
298. Ramaswamy, H. N. and Jonassen, H. B. (1965). *Inorg. Chem.*, **4**, 1595
299. Ahuja, I. S. and Rastogi, P. (1970). *J. Inorg. Nucl. Chem.*, **32**, 1381
300. Cousins, D. R. and Hart, F. A. (1968). *J. Inorg. Nucl. Chem.*, **30**, 3009
301. Jörgensen, C. K. (1964). *Inorg. Chem.*, **3**, 1201
302. Norbury, A. H., Shaw, P. E. and Sinha, A. I. P. (1970). *Chem. Commun.*, 1080
303. Hassel, R. L. and Burmeister, J. L. (1971). *Chem. Commun.*, 568
304. Wojcicki, A. and Farona, M. F. (1964). *Inorg. Chem.*, **3**, 151
305. Norbury, A. H. (1971). *J. Chem. Soc. A*, 1089
306. Carlin, R. L. and Edwards, J. O. (1958). *J. Inorg. Nucl. Chem.*, **6**, 217
307. Armor, J. N. and Haim, A. (1971). *J. Amer. Chem. Soc.*, **93**, 867
308. Buckingham, D. A., Creaser, I. I. and Sargeson, A. M. (1970). *Inorg. Chem.*, **9**, 655
309. Conant, D. R. and Decius, J. C. (1967). *Spectrochim. Acta*, **23A**, 2931
310. Coutts, R. S. P. and Wailes, P. C. (1966). *Aust. J. Chem.*, **19**, 2069
311. Becker, W. and Beck, W., unpublished results
312. Coutts, R. and Wailes, P. C. (1967). *Inorg. Nucl. Chem. Lett.*, **3**, 1
313. Kohout, J., Quastlerová-Hvastijová, M., Kohútová, M. and Gazo, J. (1971). *Monatsh., Chem.*, **102**, 350
314. Golub, A. M., Tsintsadze, G. V. and Mamulashvili, A. M. (1969). *Russ. J. Inorg. Chem.*, **14**, 1589
315. Tsintsadze, G. V., Mamulashvili, A. M. and Demchenko, L. P. (1970). *Russ. J. Inorg. Chem.*, **15**, 145
316. Ablov, A. V., Popova, A. A. and Samus, N. M. *Zh. Neorg. Khim.*, (1969), **14**, 994; *ibid.*, (1971), **16**, 411
317. Nelson, S. M. *Proc. Chem. Soc.*, (1961), 372; Nelson, S. M. and Shepherd, T. M. *Inorg. Chem.*, (1965), **4**, 813
318. Schafer, M. and Curran, C. (1965). *Inorg. Chem.*, **4**, 623
319. Norbury, A. H. and Sinha, A. I. P. (1967). *Inorg. Nucl. Chem. Lett.*, **3**, 355
320. King, H. C. A., Körös, E. and Nelson, S. M. (1962). *Nature (London)*, **196**, 572
321. Nelson, J. and Nelson, S. M. (1969). *J. Chem. Soc. A*, 1597
322. Powell, J. and Shaw, B. L. (1965). *J. Chem. Soc.*, 3879
323. Kerridge, D. H. and Mosley, M. (1967). *J. Chem. Soc. A*, 1874
324. Burmeister, J. L. and DeStefano, N. J. (1970). *Inorg. Chem.*, **9**, 972
325. Balahura, R. J. and Jordan, R. B. (1971). *Inorg. Chem.*, **10**, 198
326. Hoyer, H. (1961). *Chem. Ber.*, **94**, 1042
327. Griffiths, J. E. (1968). *J. Chem. Phys.*, **48**, 278
328. Wagner, E. L. (1965). *J. Chem. Phys.*, **43**, 2728
329. Holsboer, F. J. and Beck, W. (1970). *Chem. Commun.*, 262
330. Burmeister, J. L. and Gysling, H. J. (1967). *Chem. Commun.*, 543
331. Burmeister, J. L., Gysling, H. J. and Lim, J. C. (1969). *J. Amer. Chem. Soc.*, **91**, 44
332. Burmeister, J. L. and DeStefano, N. J. (1969). *Inorg. Chem.*, **8**, 1546
333. Schmidtke, H.-H. (1967). *Ber. Bunsengesellschaft Phys. Chem.*, **71**, 1138
334. Tsintsadze, G. V., Kharitonov, Yu. Ya., Tsivadze, A. Yu., Golub, A. M. and Managadze, A. S. (1970). *Russ. J. Inorg. Chem.*, **15**, 1210
335. Golub, A. M., Kopa, M. V., Skopenko, V. W. and Tsintsadze, G. V. *Z. Anorg. Allg. Chem.*, (1970), **375**, 302; *Russ. J. Inorg. Chem.*, (1969), **14**, 1444
336. Duffy, N. V. and Kosel, F. G. (1969). *Inorg. Nucl. Chem. Lett.*, **5**, 519
337. König, E., Madéja, K. and Böhmer, W. H. (1969). *J. Amer. Chem. Soc.*, **91**, 4582

338. Várhelyi, Cs., Finta, Z. and Zsakó, J. (1970). Z. Anorg. Allg. Chem., 374, 326
339. Köpf, H., Block, B. and Schmidt, M. (1967). Z. Naturforsch., 22b, 1077
340. Burmeister, J. L. and Al-Janabi, M. Y. (1965). Inorg. Chem., 4, 962
341. Halpern, J. and Nakamura, S. (1965). J. Amer. Chem. Soc., 87, 3002
342. Guttenberger, J. F. (1968). Chem. Ber., 101, 403
343. McBride, J. J. Jr. and Beachell, H. C. (1952). J. Amer. Chem. Soc., 74, 5247
344. Bither, T. A., Knoth, W. H., Lindsey, R. V. Jr. and Sharkey, W. H. (1958). J. Amer. Chem. Soc., 80, 4151
345. Seyferth, D. and Kahlen, N. (1960). J. Amer. Chem. Soc., 82, 1080
346. Wannagat, U. and Seyffert, H. (1965). Angew. Chem. Int. Ed. Engl., 4, 438
347. Booth, M. R. and Frankiss, S. G. (1968). Chem. Commun., 1347
348. Wade, R. C., Sullivan, E. A., Berschied, J. R. Jr. and Purcell, K. F. (1970). Inorg. Chem., 9, 2146
349. Köhler, H. and Maushake, K. H. (1968). J. Organometal. Chem. 14, 103
350. Beck, W., Hieber, W. and Neumair, G. (1966). Z. Anorg. Allg. Chem., 344, 285
351. Köhler, H. Z. Anorg. Allg. Chem., (1964), 331, 237; Köhler, H., Hartung, H. and Seifert, B., ibid., (1966), 347, 30
352. Burmeister, J. L. and Edwards, L. M. (1971). J. Chem. Soc. A, 1663
353. Schlodder, R. and Beck, W., unpublished result
354. Köhler, H. and v. Mülmann, E. (1970). Z. Anorg. Allg. Chem., 373, 222
355. Beck, W., Nitzschmann, R. E. and Neumair, G. (1964). Angew. Chem., 76, 346
356. Beck, W., Smedal, H. S. and Weis, J. C., unpublished results
357. Ricci, J. S., Ibers, J. A., Fraser, M. S. and Baddley, W. H. (1970). J. Amer. Chem. Soc., 92, 3489
358. Ricci, J. S. and Ibers, J. A. (1971). J. Amer. Chem. Soc., 93, 2391
359. Su, S. R., Hanna, J. A. and Wojcicki, A. (1970). J. Organometal. Chem., 21, P 21
360. Köhler, H. and Seifert, B. (1967). Z. Naturforsch., 22b, 238
361. Köhler, H. and Seifert, B. (1968). J. Organometal. Chem., 12, 253
362. Köhler, H., Eichler, B. and Salewski, R. (1970). Z. Anorg. Allg. Chem., 379, 183
363. Kolbe, A. and Köhler, H. (1970). Z. Anorg. Allg. Chem., 373, 230
364. Müller, E., Kästner, P., Bentler, R., Rundel, W., Suhr, S. and Zech, B. (1968). Liebigs Ann. Chem., 713, 87
365. Lappert, M. F., Lorberth, J. and Poland, J. S. (1970). J. Chem. Soc. A, 2954
366. Seyferth, D., Dow, A. W., Menzel, H. and Flood, T. C. (1968). J. Amer. Chem. Soc., 90, 1080
367. Lorberth, J. (1971). J. Organometal. Chem.
368. Basolo, F. and Pearson, R. G. (1967). Mechanisms of Inorganic Reactions (New York: J. Wiley)
369. Snellgrove, R. and King, E. L. (1962). J. Amer. Chem. Soc., 84, 4609
370. Beck, W., Becker, W., Chew, K. F., Derbyshire, W., Logan, N., Revitt, D. M. and Sowerby, D. B. J. Chem. Soc. A, in press
371. Beck, W., Fehlhammer, W. P., Pöllmann, P. and Tobias, R. S. (1968). Inorg. Chim. Acta, 2, 467
372. Paetzold, P. I. (1963). Z. Anorg. Allg. Chem., 326, 47
373. Agrell, I. and Lamnerik, S. (1967). Acta Chem. Scand., 21, 2647; (1968), 22, 2038
374. Fehlhammer, W. P., Beck, W. and Pöllmann, P. (1969). Chem. Ber., 102, 3003
375. Beck, W., Kreutzer, P. and v. Werner, K. (1971). Chem. Ber., 104, 528
376. Busetto, L., Palazzi, A. and Ros, R. (1970). Inorg. Chem., 9, 2792
377. Lange, G. and Dehnicke, K. (1966). Z. Anorg. Allg. Chem., 344, 167
378. Wiberg, N. and Schmid, K. H. (1967). Chem. Ber., 100, 741
379. Müller, U. and Dehnicke, K. (1967). Z. Anorg. Allg. Chem., 350, 113
380. Wiberg, N. and Schmid, K. H. (1967). Chem. Ber., 100, 748
381. Müller, H. and Dehnicke, K. (1967). J. Organometal. Chem., 10, P1
382. Wiberg, N., Joo, W.-Ch. and Henke, H. (1967). Inorg. Nucl. Chem. Lett., 3, 267
383. Dehnicke, K. Strähle, J., Seybold, D. and Müller, J. J. Organometal. Chem. (1966), 6, 298; Z. Anorg. Allg. Chem., (1966), 348, 261; J. Organometal. Chem. (1967). 7, P 1; ibid., (1968), 12, 37
384. Weller, F., Wilson, I. L. and Dehnicke, K. (1971). J. Organometal. Chem., 30, C 1
385. Agrell, I. (1970). Acta Chem. Scand., 24, 3575
386. Stocco, F., Stocco, G. C., Scovell, W. M. and Tobias, R. S. Inorg. Chem. (in press)

387. Jarchow, O., Schulz, H. and Nast, R. *Angew. Chem.*, (1970), **82**, 43; Jarchow, O. *Z. Anorg. Allg. Chem.*, (1971), **383**, 40
388. Joshi, K. K., Mills, O. S., Pauson, P. L., Shaw, B. W. and Stubbs, W. H. (1965). *Chem. Commun.*, 181
389. Cromer, D. T., Larson, A. C. and Roof, R. B. gr. (1966). *Acta Crystallogr.*, **20**, 279
390. Mahon, C. and Britton, D. (1971). *Inorg. Chem.*, **10**, 586
391. Weiss, A. (1958). *Chem. Ber.*, **91**, 487
392. Rayner, J. H. and Powell, H. M. (1952). *J. Chem. Soc.*, 319
393. Wells, A. F. (1962). *Structural Inorganic Chemistry*, (Oxford. Clarendon Press)
394. Wilde, R. E., Ghosh, S. N. and Marshall, G. J. (1970). *Inorg. Chem.*, **9**, 2512
395. Ludi, A., Güdel, H.-U. and Rüegg, U. (1970). *Inorg. Chem.*, **9**, 2224
396. Izatt, R. M., Watt, G. D., Bartholomew, C. H. and Christensen, J. H. (1970). *Inorg. Chem.*, **9**, 2019
397. Ayers, J. B. and Waggoner, W. H. (1971). *J. Inorg. Nucl. Chem.*, **33**, 721
398. Brown, D. B., Shriver, D. F. and Schwartz, L. H. (1968). *Inorg. Chem.*, **7**, 77
399. House, J. E. Jr. and Bailar, J. C. Jr. (1969). *Inorg. Chem.*, **8**, 672
400. Burmeister, J. L. and Sutherland D. (1965). *Chem. Commun.*, 175
401. Haim, A. and Wilmarth, W. K. (1961). *J. Amer. Chem. Soc.*, **83**, 509
402. De Castelló, R. A., Mac-Coll, C. P., Egen, N. B. and Haim, A. (1969). *Inorg. Chem.*, **8**, 699
403. Wang, B.-C., Schaefer, W. P. and Marsh, R. E. (1971). *Inorg. Chem.*, **10**, 1492
404. Cromer, D. T. and Larson, A. C. (1962). *Acta Crystallogr.*, **15**, 397
405. Phillips, R. F. and Powell, H. M. (1939). *Proc. Roy. Soc. A*, **173**, 147
406. Shriver, D. F. (1966). *Structure and Bonding*, **1**, 32
407. Dehnicke, K. (1967). *Angew. Chem.*, **79**, 942
408. Chatt, J. and Hart, F. A. (1961). *J. Chem. Soc.*, 1416
409. Gregory, U. A., Jarvis, J. A. J., Kilbourn, B. T. and Owston, P. G. (1970). *J. Chem. Soç. A*, 2770
410. Gaughan, A. P., Ziolo, R. F. and Dori, Z. (1970). *Inorg. Chim. Acta*, **4**, 640
411. Shvelashvili, A. E., Porai-Koshits, M. A., Antsishkina, A. S., Tsintsadze, G. V. and Shchedrin, B. M. (1966). *Zh. Strukt. Khim.*, **7**, 810; *Chemical Abstracts*, (1967), **66**, 147189
412. Tibbetts, D. L. and Brown, T. L. (1969). *J. Amer. Chem. Soc.*, **91**, 1108
413. Scovell, W. M., Stocco, G. C. and Tobias, R. S. (1970). *Inorg. Chem.*, **9**, 2682
414. Behrens, H. and Schwab, R. (1964). *Z. Naturforsch*, **19b**, 768
415. Garaj, J. (1968). *Inorg. Chem.*, **8**, 304
416. Chow, Y. M. (1970). *Inorg. Chem.*, **9**, 794
417. Forder, R. A. and Sheldrick, G. M. (1970). *J. Organometal. Chem.*, **21**, 115
418. Nardelli, M., Gaspari, G. T., Musetti, A. and Manfredotti, A. (1966). *Acta Crystallogr.*, **21**, 910
419. Stocco, G. C. and Tobias, R. S. (in press)
420. Taube, H. (1955). *J. Amer. Chem. Soc.*, **77**, 4481
421. Falk, L. C. and Linck, R. G. (1971). *Inorg. Chem.*, **10**, 215
422. Jeffrey, J. W. (1963). *Acta Crystallogr.*, *Suppl.*, **16**, A 66
423. Porai-Koshits, M. A. (1963). *Acta Crystallogr.*, *Suppl.*, **16**, A 42
424. Nicolini, M., Boschi, T., Pecile, C. and Turco, A. (1966). *Coord. Chem. Rev.*, **1**, 133, 269
425. Enemark, J. H. and Holm, R. H. (1964). *Inorg. Chem.*, **3**, 1516
426. Biondi, C. (1965). *Chem. Commun.*, 191
427. Köhler, H. (1965). *Z. Anorg. Allg. Chem.*, **336**, 245
428. Norberry, A. B. and Sinha, A. I. P. (1971). *J. Inorg. Nucl. Chem.*, **33**, 2683
429. Behrens, H., Lindner, E. and Lehmert, G. (1970). *J. Organometal. Chem.*, **22**, 665
430. Show, Y. M. (1971). *Inorg. Chem.*, **10**, 1938
431. Forder, R. F. A., Sheldrick, G. M. (1970). *Chem. Commun.*, 1023

10
Metal Complexes Containing Group VIB Donor Atoms

J. A. McCLEVERTY

University of Sheffield

10.1 INTRODUCTION

During the past 10 years there has been a considerable growth in the study of transition metal complexes containing Group VIB donor atoms, particularly sulphur. The reasons for this are not entirely clear, although there are obvious parallels to be drawn between metal complexes of thioethers and of trialkyl and triarylphosphines. The latter have assumed great importance in the study of homogeneously catalysed reactions. Sulphur and its congeners are, however, remarkably versatile elements, and may commonly be found in di-, tetra- or hexapositive states, and may also readily adopt either trigonal or tetrahedral geometries in their compounds with transition metals.

Perhaps the most significant areas of development have been transition metal dithiolene complexes, i.e. those species derived from cis-1,2-disubstituted ethylene dithiolates and their benzene or substituted benzene analogues, and in the study of the previously quite well known complexes of ligands derived from CS_2, e.g. dithiocarbamates, xanthates, etc. Largely, developments in these two areas have provided the impetus for a continuing and expanding interest in the chemistry of transition metal sulphur complexes, and it is expected that this will lead to discoveries of utility in the area of biological inorganic chemistry. The development of analogous chemistry involving selenium or tellurium has been slower, perhaps due to difficulties in obtaining Se or Te analogues of the sulphur ligands.

This review tries to be critical and, for brevity, must be selective. Consequently, it has been necessary to omit discussion of compounds containing thioethers, thioureas and their derivatives, dithio-oxalates, thiophosphonates etc., and many other species whose ligands contain, in addition to S, Se or Te, other donor atoms. Binary and ternary sulphides are reviewed in Chapter 5.

A number of useful reviews in this area of chemistry have recently appeared. These cover dithiolene compounds[1,2], metal complexes of ligands derived from CS_2[3], and some structural aspects of the structures of metal 1,1- and 1,2-dithiolates[4]. Other useful surveys of various aspects of transition metal sulphur, selenium and tellurium complex chemistry have been presented by Livingstone[5], Harris and Livingstone[6], Jørgensen[7], Gray et al.[8], Martin et al.[9], and Abel and Crosse[10].

10.2 TRANSITION METAL DITHIOLENE COMPLEXES

10.2.1 Pure dithiolene complexes

In 1962, Schrauzer and Mayweg[11] reported the preparation of an unusual nickel complex $[Ni(S_2C_2Ph_2)_2]$, in the reaction between $Ni(CO)_4$, sulphur and diphenylacetylene. This nickel complex was an intense green colour, volatile, diamagnetic, and quite unlike any other nickel complex known at that time. Almost concurrently, Gray and his co-workers[12] reported the formation of metal complexes of $\{S_2C_2(CN)_2\}^{2-}$, of the type $[M\{S_2C_2(CN)_2\}_2]^{2-}$ (M = Cu, Ni, Pd and Co), which were apparently planar. Shortly thereafter, Davison et al. observed[13] that there was only a two-electron difference between the neutral $[Ni(S_2C_2R_2)_2]$ and the dianionic $[Ni(S_2C_2R_2)_2]^{2-}$ (ignoring R), and reasoned that it should be possible to effect an oxidation of the dianion to the neutral species, or reduction of the neutral compound to a dianion, possibly via an intermediate monoanionic species $[Ni(S_2C_2R_2)_2]^{-}$. They discovered that this was possible electrochemically, perhaps their most important contribution to dithiolene chemistry, and proceeded to synthesise and characterise a series of different monoanionic species of the type $[M(S-S)_2]^-$ (M = Ni, Pd or Pt; S—S represents a bidentate 1,2-dithiolato ligand). Not long after the publication of this work, Gray and his colleagues[14] announced that the most stable complexes formed between Ni, Co and Fe, and toluene-3,4-dithiol, a well-known analytical reagent, were monoanionic, i.e. $[M(S_2C_6H_3Me)_2]^-$, the nickel compound having a spin-doublet ground state, and the other two reputedly having spin-quartet ground states.

The structures of six bisdithiolene complexes have been determined crystallographically. These are $[Me_4N]_2[Ni\{S_2C_2(CN)_2\}_2]$[15], $[Ph_3MeP]$ $[Ni\{S_2C_2(CN)_2\}_2]$[16], $Ni(S_2C_2Ph_2)_2$[17], $[Bu_4^nN]_2^-[Co\{S_2C_2(CN)_2\}_2]$[18], $[Bu_4^nN]$ $[Co(S_2C_6H_3Me)_2]$[19] and $[Bu_4^nN]_2$ $[Cu\{S_2C_2(CN)_2\}_2]$[20]. The symmetry of these planar anions was essential D_{2h}, and in $[Ni\{S_2C_2(CN)_2\}_2]^{2-}$, $[Ni(S_2C_2Ph_2)_2]$, $[Co\{S_2C_2(CN)_2\}_2]^{2-}$ and $[Co(S_2C_6H_3Me)_2]^-$, the metal atoms were well separated from each other in the crystal lattices. However, in $[Cu\{S_2C_2(CN)_2\}_2]^-$ there are two stacks of anions per unit cell with the four Cu atoms in each stack almost exactly above each other. In $[Ni\{S_2C_2(CN)_2\}_2]^-$, the anions are associated in pairs, so that there are two Ni...S interactions between adjacent pairs of anions, this effectively rendering each Ni atom 5-coordinate with respect to S. It is highly probable that this Ni...S — Ni interaction is responsible for the unusually low magnetic moments of $[Ni\{S_2C_2(CN)_2\}_2]^-$, and its Pd and Pt analogues in the solid state[21]. Of particular interest was the structural data obtained from the three nickel complexes since a useful comparison could be made of the changes in bond angles and lengths with alterations in overall charge. Thus, as the charge decreases $(-2\rightarrow0)$ the 'ethylenic' bond increases in length while the Ni—S and S—C distances decrease. It was concluded that as the nickel-dithiolene complexes were progressively oxidised, the sulphur ligand gradually lost its dithiolate character and became more 'dithioketonic', that is, that oxidation of the ligand system rather than of the metal took place.

A systematisation of the half-wave potential data for the one electron transfer reactions which the bisdithiolenes, and most of their analogues, undergo, has been made[22]. Thus, generalisations could be made about the synthetic chemistry of the complexes, and about their stability towards redox reagents. It has been shown that the simple bisdithiolenes can, in general, exist as part of a four-membered electron transfer series[23], namely $[M(S-S)_2]^{2-} \leftrightarrow [M(S-S)_2]^- \leftrightarrow [M(S-S)_2]^0 \leftrightarrow [M(S-S)_2]^+$. Although monocationic species have not yet been isolated, they have been detected voltammetrically[24]. In these complexes, the order of decreasing oxidative stability of the dianions, $[M(S-S)_2]^{2-}$, that is the order of increasing negative potentials for the couple $[M(S_2C_2R_2)_2]^- + e^- \rightleftharpoons$ $[M(S_2C_2R_2)_2]^{2-}$, was shown[25] to be R = CN > CF$_3$ > Ph > H > Me > Et > Pr^n > Pr^i > Bu^n. In other words, the dianions were the most stable to oxidation when the ligand substituents, R, were electron-releasing groups. It was shown that there was a linear relationship; between the $E_{\frac{1}{2}}$-values for the electron transfer reaction connecting the mono- and di-anions, and for the process $[M(S_2C_2R_2)_2] + e^- \rightleftharpoons [M(S_2C_2R_2)_2]^-$, where M = Ni, Pd or Pt[25], and Taft's σ^* substituent constant[26]. These results were believed to be in good agreement with the proposed electronic structures of the complexes[27]. Similar substituent effects were observed in the polarographic data obtained from the arene–dithiolene complexes $[M(S_2C_6X_2Y_2)_2]^z$, such that the order of decreasing potentials for any given electrode reaction (M = Co, Ni or Cu) [28] was $S_2C_6Cl_4 > S_2C_6H_4 > S_2C_6H_3Me > S_2C_6H_2Me_2 > S_2C_6Me_4$. In the overall ordering, the arene-dithiolato ligands come between R = CF_3 and R = Ph, and these general trends are observed in all types of dithiolene complexes[23].

It has been shown that $[Ni\{S_2C_2(CF_3)_2\}_2]$ is a powerful one-electron oxidising agent[29] and that treatment of such dianions as $[Pd\{S_2C_2(CN)_2\}_2]^{2-}$ with it rapidly afforded $[Pd\{S_2C_2(CN)_2\}_2]^-$. The use of this type of reaction synthetically presupposed that little, or no, ligand exchange took place, and, indeed, early preparative work supported this assumption. However, later studies indicated that such reactions between neutral and dianionic species, or even pairs of monoanions, under certain conditions, did result in the formation of mixed ligand species, e.g. $[Ni(S_2C_2R_2)_2]^{0/-} + [Ni(S_2C_2R'_2)_2]^{2-/-}$ $\rightarrow [Ni(S_2C_2R_2)(S_2C_2R'_2)]^-$. Voltammetry was a particularly useful technique in detecting the occurrence and extent of ligand reorganisation. It was established[30] that the rate of reorganisation was both solvent and temperature dependent, and it proved possible to adjust the conditions so that ligand exchange proceeded to near completion, thereby permitting isolation of the monoanionic mixed ligand species.

There have been three major attempts to understand and explain the electronic structures of the planar bisdithiolene complexes. Two of these used simple modified Hückel[27] and Wolfsberg–Helmholtz[32] theories and applied the resultant MO schemes to an interpretation of the spectral, electrochemical and chemical properties of this class of molecules. The third approach[31], perhaps the most rigorous, was developed from a *prima facie* consideration of the e.s.r. spectra of $[M(S\text{---}S)_2]^{2-}$, and $[Ni(S\text{---}S)_2]^-$ (M = Cu, Co or Rh; S—S here is $S_2C_2(CN)_2$). The result of these three different approaches was to produce, for $[Ni(S\text{---}S)_2]^-$, three superficially different ground state assignments, but all of these indicated that the orbital of the unpaired electron contained significant $S(\pi\text{---})$ ligand character, i.e. that $[Ni(S\text{---}S)_2]^-$ could not be described correctly as containing Ni^{III}. It now appears that two of the calculations provided, essentially, the same 'answer', so that agreement was reached as to the symmetry of the ground state half-filled orbital in $[Ni(S\text{---}S)_2]^-$, and this was confirmed by an independent e.s.r. study[33] of ^{33}S hyperfine splittings in $[Ni\{S_2C_2(CN)_2\}_2]^-$. It is now conceded that the remaining calculation[32] contained some ambiguities. As a result of the Hückel calculation[27] it has been suggested that the monoanionic and neutral complexes, particularly of nickel, have highly delocalised ground states, visualised as in Figure 6.1. The dependence of the $E_{\frac{1}{2}}$ values on the S-ligand substituents is entirely consistent with the general MO descriptions of the complexes.

During the characterisation of the planar bisdithiolene complexes, it was shown that an oxidised cobalt compound, $[Co\{S_2C_2(CF_3)_2\}_2]_2$, was dimeric[34]. The crystal structure revealed that each metal atom was 5-co-ordinate with respect to S and that there was no direct Co—Co bond. Subsequent polarographic and conductivity measurements[24, 35] on various iron and cobalt anionic bisdithiolene species revealed that they too were dimeric, and this has been confirmed by x-ray studies of $[Bu_4^nN]_2$ $[Fe\{S_2C_2(CN)_2\}_2]_2$ [38] and $[Bu_4^nN]_2[Co(S_2C_6Cl_4)_2]_2$ [39] (the latter is only dimeric in the solid).

The polarographic and voltammetric studies of these compounds clearly demonstrated that the discrete dimeric dianions and the neutral dimeric species could undergo several one-electron transfer reactions, and one two-electron reduction. The general behaviour of the binuclear species can be summarised as $[M(S—S)_2]_2 \leftrightarrow [M(S—S)_2]_2^- \leftrightarrow [M(S—S)_2]_2^{2-} \leftrightarrow 2[M(S—S)_2]^{2-}$.

The realisation that some of the dimeric dianionic iron and cobalt species dissociated into monomers in polar solvents giving monoanionic species (with or without solvent coordination) provided an explanation of the apparently anomalous magnetic data obtained from many of the complexes in solution[38, 39]. Thus, it was established that the species $[Co(S_2C_2 R_2)_2]_2^{2-}$ (R = CN or CF$_3$) was diamagnetic in the solid, and in most solvents,

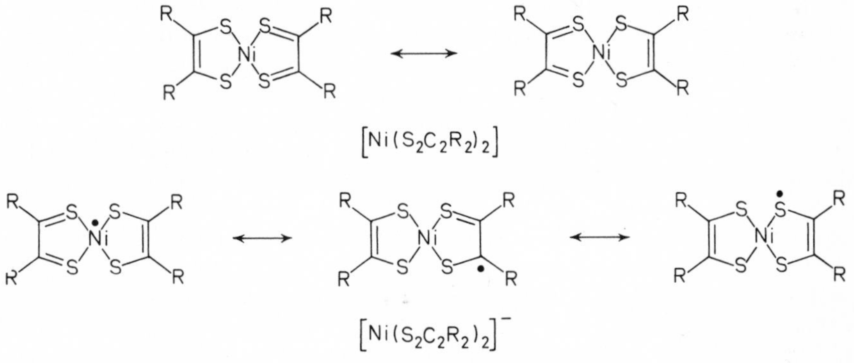

$$[Ni(S_2C_2R_2)_2]$$

$$[Ni(S_2C_2R_2)_2]^-$$

Figure 10.1

and $[Co(S_2C_6X_2Y_2)_2]^-$ (X = Y = H or Me; X = H, Y = Me) was always monomeric and high-spin (S = 1) both in the solid state and in solution. The magnetic moments of the iron complexes $[Fe(S—S)_2]_n^{n-}$, (n = 1 or 2), in strongly donating solvents (e.g. pyridine, DMF, DMSO, etc.) were consistent with quartet ground states whereas, at low temperatures, in the solid state, the parent species were probably diamagnetic.

Following shortly on the discovery of the planar bisdithiolene complexes, the synthesis of $[Mo\{S_2C_2(CF_3)_2\}_3]$ was reported[40]. This stimulated a successful search for other 6-coordinate dithiolene complexes, and to the development of a chemistry of these species which is broadly similar, in electrochemical terms, to that of their bisdithiolene analogues.

The trisdithiolene complexes belong to at least a six-membered electron transfer series, namely $[M(S—S)_3]^{4-} \leftrightarrow [M(S—S)_3]^{3-} \leftrightarrow [M(S—S)_3]^{2-} \leftrightarrow [M(S—S)_3]^- \leftrightarrow [M(S—S)_3]^0 \leftrightarrow [M(S—S)_3]^+$. In general, as in the bisdithiolene series, the potential data for the couple $[M(S—S)_3]^z + e^- \rightleftharpoons [M(S—S)_3]^{z-1}$ could be correlated linearly[25] with Taft's σ^* substituent constant. A comparative study of the electron transfer reactions of

$[M\{S_2C_2(CN)_2\}_3]^z$, $[M\{S_2C_2(CF_3)_2\}_3]^z$ $[M(S_2C_6Cl_4)_3]^z$ and $[M(S_2C_6H_3Me)_3]^z$ confirmed[43] the predicted positions of the arene-dithiolato ligands in the half-wave potential series (page 304).

Perhaps the most significant result obtained from this group of complexes was the discovery that many of the most highly oxidised species, e.g. $[M(S\!-\!S)_3]^0$, had trigonal prismatic geometries both in the solid state and in solution. Three such complexes have been studied by x-ray methods, $[Re(S_2C_2Ph_2)_3]$ [42], $[Mo(S_2C_2H_2)_3]$ [43] and $[V(S_2C_2Ph_2)_3]$ [44], and all of these have essential D_{3h} symmetry – there was a slight distortion in the vanadium complex resulting in three different interligand S—S distances. In the Mo complex, the Mo—S bond distances were comparable with those in MoS_2 [45]. A point of considerable interest was what happened to this trigonal prismatic geometry when the complexes were reduced. This problem has not been completely solved, but in the dianions $[V\{S_2C_2(CN)_2\}_3]^{2-}$ [46] and $[Mo\{S_2C_2(CN)_2\}_3]^{2-}$ [47] the overall structure of the complex anions is best described as the result of a compromise between trigonal prismatic coordination (preferred by the ligands) and octahedral geometry (preferred by the metal). It is noticeable that the inter- and intra-S—S distances in the trigonal prismatic complexes are close to 3.05 Å, whereas those reported in the vanadium dianion vary between 3.12 and 3.18 Å. Although no details are as yet available, it is known[48] that $[Ph_3MeP]_2[Fe\{S_2C_2(CN)_2\}_3]$ adopts a regular (D_{3d}) octahedral geometry. Spectral studies suggested[44] that $[V(S_2C_2Ph_2)_3]^-$, which is isoelectronic with $[Cr(S_2C_2Ph_2)_3]$ (itself isomorphous with $[V(S_2C_2Ph_2)_3]$), was also trigonal prismatic.

Molecular orbital calculations performed on the neutral species $[M(S\!-\!S)_3]$, showed[49, 50] that there was delocalisation of electron density involving both the π- and σ-systems, thereby accounting for the 'aromaticity' of some of the prismatic complexes. Indeed, Friedel-Crafts alkylation of $[Mo(S_2C_2H_2)_3]$ afforded $[MoS_6C_6H_nR_{6-n}]$. Attempts were made also to estimate the oxidation state of the metal ions in the neutral species. Valence bond representations could be drawn[49] which implied an oxidation number of IV for Cr, Mo or W complexes, and of III for V complexes.

Some thought has been given to the reasons for the stability of trigonal prisms in this system[50]. No theory has been proposed, but it was observed that there is a near constancy in the interligand S—S distances. This could be taken as evidence for some form of interligand bonding which would give added stability to the trigonal prisms.

10.2.2 Dithiolene complexes containing phosphines

The dianionic and neutral dimeric dithiolene complexes of Co and Fe, $[M(S\!-\!S)_2]_2^{0,-2}$, undergo dissociation in the presence of Lewis bases such as phosphines, phosphites, arsines and stibines, generally forming 5-coordinate adducts with these bases[39, 51–53].

It seems that the phosphine and related adducts undergo at least three one-electron transfer reactions, namely $[M(L)(S\!-\!S)_2]^{2-} \leftrightarrow [M(L)(S\!-\!S)_2]^- \leftrightarrow [M(L)(S\!-\!S)_2]^0 \leftrightarrow [M(L)(S\!-\!S)_2]^+$, and in at least one case a dicationic species has been identified electrochemically[53]. The half-wave potentials for the electron transfer reactions depend on three factors – the nature of the metal atom, the sulphur ligand substituents, and the nature of the substituents attached to the Lewis base donor atom (P, As or Sb); there was no obvious relationship between the donor atom and the $E_{\frac{1}{2}}$ values[39, 51]. In the bisdiaryldithiolene complexes $[M(L)(S_2C_2Ar_2)_2]^z$ (Ar = Ph, 4-MeC_6H_4 and 2-, 3- or 4-$MeOC_6H_4$), the redox potentials became increasingly negative, with respect to a given L, in the order Ar = 3-MeO < H < 4-Me < 4-MeO < 2-MeO, which correlated quite well with the known electronic directing effects of Me and MeO groups in o-, m-, and p-positions on a benzene ring[53]. Of the complexes containing di-substituted benzene rings the 3,4-$CH_2OC_6H_3$ derivatives had $E_{\frac{1}{2}}$ values reasonably close to the mean of the $E_{\frac{1}{2}}$ values of the 3- and 4-$MeOC_6H_4$ derivatives.

The effect of the phosphine or arsine substituents on the half-wave potentials for any one series of complexes having the same sulphur ligand substituents was considerable[39]. Generally, the half-wave potentials for any given couple tended to become more negative (or less positive), as the basicity, or electron-releasing ability, of the phosphine increased. The precise way in which the $E_{\frac{1}{2}}$ values were influenced by R and R' in $[M(PR_3)(S_2C_2R'_2)_2]^z$ was of great significance in the synthetic chemistry of these complexes. Thus, the combination of strongly electron-releasing substituents on the sulphur ligand with a strongly electron-releasing phosphine pushed the potentials for the couple $[M(L)(S_2C_2R_2)_2]^+ + e^- \rightleftharpoons [M(L)(S_2C_2R_2)_2]^0$ close to 0.0 V (v. SCE) thereby permitting the formation of some monocationic Lewis-base-metal dithiolene adducts, e.g. $[Co\{P(NMe_2)_3\}\{S_2C_2(4\text{-}MeOC_6H_4)_2\}_2][PF_6]$, to be accomplished very easily from the corresponding neutral species using mild oxidising agents[53]. Thus, by careful substituent control, monocationic, neutral and monoanionic species could easily be obtained.

From a consideration of the e.s.r. spectral data obtained from $[Fe(PR_3)(S-S)_2]^-$ and $[Co(PR_3)(S-S)_2]^0$, it was concluded[39, 54] that these 5-coordinate species could be considered as having highly-delocalised ground states, and, in $[Co(PEt_3)\{S_2C_2(CN)_2\}_2]^0$, it was estimated[39] that the orbital of the unpaired electron had only 20–25% metal character. The dependence of $E_{\frac{1}{2}}$ on the sulphur ligand substituents, but not on the Lewis base donor atom substituents, could be rationalised in terms of the proposed MO scheme[39].

Other phosphine dithiolene complexes which have been reported include $[Pt\{P(Bu^n)_3\}_2(S_2C_2Ph_2)_2]$, $[Pt(diphos)(S_2C_2Ph_2)_2]$ (diphos = $Ph_2PCH_2CH_2PPh_2$)[55] and $[M(diphos)(S_2C_2Ph_2)_2]$ (M = Mo or W)[56]. Only the latter were studied polarographically, and they underwent two, reversible, one-electron reduction steps. However, the platinum complexes readily eliminated one sulphur ligand giving the 4-coordinate phosphine dithiolene complexes; related nickel and palladium compounds have also been described[55, 57].

10.2.3 Dithiolene complexes containing cyanide, isonitriles, azide, etc

It was found that $[Fe\{S_2C_2(CN)_2\}_2]_2^{2-}$ was cleaved[39] by CN^-, CNPh, N_3^- and NCO^-. With the first two, low-spin 5-coordinate species were formed, e.g. $[Fe(CN)\{S_2C_2(CN)_2\}_2]^{2-}$, and with the last two high-spin $(S = 3/2)$ $[Fe(N_3)\{S_2C_2(CN)_2\}_2]^{2-}$ and $[Fe(NCO)\{S_2C_2(CN)_2\}_2]^{2-}$. The azide and cyanate adducts were oxidised by air giving $[Fe(NO)-\{S_2C_2(CN)_2\}_2]^{2-}$ which was identified by its e.s.r. spectrum. In a related reaction, PPh_3 could be oxidised[60] in the presence of $[Fe\{(S_2C_2R_2)_2\}_2]^{2-}_2$ $(R = CN$ or $CF_3)$ in solution giving the high-spin $[Fe(OPPh_3)(S_2C_2R_2)_2]^-$; the crystal structure of the complex with $R = CN$ $(Ph_3MeP^+$ salt) has been determined[60] and has confirmed the presence of the Fe—O—P bond and 5-coordinate geometry.

10.2.4   Nitrosyl dithiolene complexes

Nitric oxide also caused the dissociation of the dimeric iron and cobalt dithiolene complexes, and 5-coordinate species $[Fe(NO)(S—S)_2]^{0,-1}$ and $[Co(NO)(S—S)_2]^{-1,-2}$ could be obtained[61]. The crystal structure determination of $[Et_4N]_2[Fe(NO)\{S_2C_2(CN)_2\}_2]$ [62] has shown that the Fe—N—O bond angle could deviate from linearity by up to 12 degrees.

The iron complexes existed as part of a five-membered electron transfer series, namely $[Fe(NO)(S—S)_2]^{3-} \leftrightarrow [Fe(NO)(S—S)_2]^{2-} \leftrightarrow [Fe(NO)(S—S)_2]^- \leftrightarrow [Fe(NO)(S—S)_2]^0 \leftrightarrow [Fe(NO)(S—S)_2]^+$, and there was some evidence for the existence of a dication $[Fe(NO)\{S_2C_2(2\text{-MeOC}_6H_4)_2\}_2]^{2+}$ [63]. The order of stability of the reduced species in the general couple $[Fe(NO)(S—S)_2]^z + e^- \rightleftharpoons [Fe(NO)(S—S)_2]^{z-1}$ was very similar to that found in the bis-, tris- and phosphine-dithiolenes.

The dianionic and neutral iron complexes, and the monoanionic cobalt species, were paramagnetic $(S = 1/2)$ and exhibited characteristic e.s.r. spectra[61, 63]. It was shown that the NO stretching frequencies were very dependent on the charge on the complexes, and on the nature of the sulphur ligand substituents. The original MO scheme which was used to rationalise[61] these data was shown subsequently[39] to be incorrect, and a revised scheme was presented. This showed that the unpaired electron in $[Fe(NO)(S—S)_2]^{2-}$ was largely located in an orbital of metal and σ-NO character, whereas, in $[Co(NO)(S—S)_2]^-$ it was in an orbital predominantly of sulphur-ligand $(\pi\text{-})$ character. The unique electron in $[Fe(NO)(S—S)_2]^0$ was thought to be in a highly delocalised orbital of sulphur-ligand $(\pi\text{-})$ type. There was a clear electronic structural relationship between the 5-coordinate phosphine and nitrosyl complexes which was reflected in their respective MO schemes[39].

The manganese compounds, $[Mn(NO)(S—S)_2]^{2-}$, were prepared by treatment[64] of $[\pi\text{-C}_5H_5Mn(NO)(S—S)_2]^{0,-1}$ with sulphur ligand in the presence of strong bases, or by reaction of $[Mn(S—S)_2]^{2-}$ with NO. These complexes were much less stable than their iron or cobalt analogues, and could not be investigated electrochemically.

The 6-coordinate nitrosyls $[M(NO)_2\{S_2C_2(CN)_2\}_2]^{2-}$ $(M = Mo$ or W), had cis-octahedral structures[65]; the Cr analogue was obtained[66] by treatment of $[Cr(NO)_2(NCMe)_4]^{2+}$ with $Na_2S_2C_2(CN)_2$, and the Mo and W analogues by similar treatment of $[M(NO)_2Cl_2]_n$. The related $[M(NO)_2(S_2C_6Cl_4)_2]^{2-}$ and $[M(NO)_2\{S_2C=C(CN)_2\}_2]^{2-}$ were obtained similarly. The complexes underwent one one-electron oxidation step, giving the corresponding monoanions.

When equimolar amounts of $[Fe(NO)\{S_2C_2(CN)_2\}_2]^-$ and $[Fe(NO)-(S_2C_2Ph_2)_2]^-$ were mixed[67], a rapid electron exchange occurred, giving $[Fe(NO)\{S_2C_2(CN)_2\}_2]^{2-}$ and $[Fe(NO)(S_2C_2Ph_2)_2]$. This reaction was followed by a slow ligand exchange and electron re-exchange which afforded $[Fe(NO)(S_2C_2Ph_2)\{S_2C_2(CN)_2\}]^-$. These reactions were followed by a combination of i.r, e.s.r. spectral and voltammetric techniques.

10.2.5 Carbonyl dithiolene complexes

Reactions of $Mn(CO)_5Br$ with $Na_2S_2C_2(CN)_2$, $Na_2S_2C=C(CN)_2$, $Na_2S_2C_6$ H_3Me and $Na_2S_2C_6Cl_4$ afforded the 1,2-dithiolato complexes $[Mn(CO)_4-(S—S)]^-$ and $[Mn(CO)_4S_2C=C(CN)_2]^-$ [68]. The carbonyl stretching frequencies were dependent on the nature of the sulphur ligands, but the complexes did not appear to undergo simple redox reactions. Treatment of $Mn(CO)_5Br$ with $Na_2S_2C_2H_2$ gave[69] $Mn_2(CO)_6S_2C_2H_2$ which contains a bridging di (μ-mercaptide) ligand. Reaction of $Mn(CO)_5Br$ with toluene-3,-4-dithiol gave[70] the paramagnetic monomer $Mn(CO)_3\{(HS)_2C_6H_3Me\}$ and the diamagnetic dimer $[Mn(CO)_3S_2C_6H_3Me]_2$.

The di(μ-mercaptide) species $Fe_2(CO)_6S_2C_2R_2$ (R = CF_3 or Ph) could be obtained directly from iron carbonyls using appropriate sulphur derivatives[71], or by ligand exchange between $[Fe(S_2C_2R_2)_2]_n$ or $[Ni(S_2C_2R_2)_2]$ and $Fe(CO)_5$ [52]. Under certain conditions, reaction of $[Ni(S_2C_2Ph_2)_2]$ with $Fe(CO)_5$ afforded a novel compound, $Fe_2(CO)_6SC_2Ph_2$, whose structure indicated that the sulphur ligand could be described as a coordinated thioketocarbene (Figure 6.2)[72].

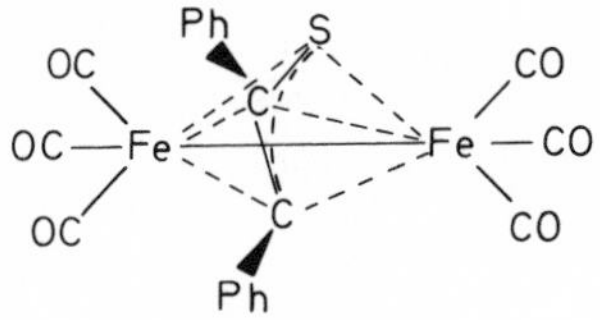

Figure 10.2

U.V. irradiation[56] of a mixture of $M(CO)_6$ (M = Mo or W) with $[Ni(S_2C_2R_2)_2]$ afforded $[M(CO)_4S_2C_2R_2]$ and cis-$[M(CO)_2(S_2C_2R_2)_2]$ (R = Me or Ph). Reaction of the latter compounds with PR_3' or diphos gave $[M(CO)(PR_3')(S_2C_2R_2)_2]$ and $[M(diphos)(S_2C_2R_2)_2]$, and with $Na_2S_2C_2-(CN)_2$ and $Na_2S_2C_2H_2$ the interesting mixed ligand species $[M(S_2C_2R_2'')-(S_2C_2R_2)_2]$ (R'' = H or CN). Polarographic examination of the monophosphine derivative showed that it could be reduced in three steps, the first certainly corresponding to the generation of a monoanion.

Reaction of $[Rh(CO)_2Cl]_2$ with $\{S_2C_2(CN)_2\}^{2-}$ and other 1,2-dithiolato ligands afforded $[Rh(CO)_2(S—S)_2]^-$ [73]. Treatment of $[Rh(CO)_2S_2C_2(CN)_2]^-$ with PPh_3 or $P(OPh)_3$ gave $[Rh(CO)(PR_3)S_2C_2(CN)_2]^-$, and with $S_2C_2(CF_3)_2$ $[Rh\{S_2C_2(CF_3)_2\}\{S_2C_2(CN)_2\}]_n^-$ ($n \geqslant 1$). With $S_2C_2(CF_3)_2$ $M(CO)(PPh_3)_2Cl$ afforded $[M_2(PPh_3)_2\{S_2C_2(CF_3)_2\}_3]$ (M = Rh or Ir) which underwent two one-electron reductions. With toluene-3,4-dithiol $[Ir(CO)(PPh_3)_2HCl\{SC_6H_3Me(SH)\}]$ was formed[74]. $[Rh(CO)_2Cl]_2$ reacted with $S_2C_2(CF_3)_2$ giving $[Rh(CO)\{S_2C_2(CF_3)_2\}]_3^-$ and the apparently related $[Rh_3(CO)(PPh_3)_2\{S_2C_2(CF_3)_2\}_3]^-$ and $[Ir(CO)(PPh_3)_2\{S_2C_2(CF_3)_2\}_3]^0$ were obtained in the reactions between $M(CO)(PPh_3)_2Cl$ and $S_2C_2(CF_3)_2$ [73].

10.2.6 Polymetallic dithiolene complexes

Early reports[52] of the formulation of a sulphur-rich dinuclear iron complex, obtained from the reaction between $[Fe_2(CO)_6S_2R_2Ph_2]$ and S_8, as $[FeS-(S_2C_2Ph_2)_2]_2$ have not been substantiated. However, by refluxing $Fe_2(CO)_6$ $S_2C_2R_2$ (R = CF_3 or Ph) in xylene with S_8, the tetrameric $[Fe_4S_4(S_2C_2R_2)_4]$ were obtained[75]. $[Fe_4S_4(S_2C_2(CF_3)_2)_4]$ underwent four voltammetric one-electron reduction processes, and on dissolution in basic solvents afforded the corresponding dianion. The latter, when reacted with an equimolar amount of the former, gave $[Fe_4S_4\{S_2C_2(CF_3)_2\}_4]^-$. The monoanion was paramagnetic ($S = 1/2$) whereas the other two were diamagnetic. Reaction of the neutral species with PPh_3 afforded $[Fe(PPh_3)\{S_2C_2(CF_3)_2\}_2]^-$. The corresponding diphenyldithiolene complex (R = Ph) could be reduced in three one-electron steps, and the monoanion was obtained from the neutral species using hydrazine. The trinuclear compound $[Fe_3S_2\{S_2C_2(CF_3)_2\}_4]$ was also obtained from the reaction which afforded the tetramers; it could be reduced in two one-electron steps.

Reaction of $[M(CO)_2(S_2C_2Ph_2)_2]$ (M = Mo or W) with S_8 afforded[56] the dinuclear $[M_2S_2(S_2C_2Ph_2)_4]$ which could be reduced polarographically giving mono-, di- and tetra-anionic species. The structure of these neutral dimers was believed to consist of two trigonal prismatic units, the sulphur atoms sharing a common edge. Reaction of paramolybdate ion with $(Ar_2C_2 S_2PS_2)_2$ (Ar = Ph, 4-MeC_6H_4 or 4-$MeOC_6H_4$) afforded[76] $[Mo_2S_4(S_2C_2 Ar_2)_2]$. These complexes could be reduced in two one-electron steps, and also oxidised to a monocation. Their structures were believed to consist of two fused trigonal prisms in which the four S atoms shared the common face.

Treatment of $[Fe_2(CO)_6(XR)_2]$ (X = S, R = Ph, Me or Et; X = Se, R = Ph) with $S_2C_2(CF_3)_2$ afforded[77] the tetramers $[Fe(CO)(XR)\{S_2C_2 (CF_3)_2\}]_4$. In dichloromethane, the compounds underwent an irreversible multi-electron voltammetric oxidation reaction, but could be reduced reversibly in a one-electron step. They appeared to dissociate in acetone forming dimeric species. Reaction of $Fe_2(CO)_9$ under carefully controlled conditions afforded the extremely reactive $[Fe(CO)_3\{S_2C_2(CF_3)_2\}]_2$ which is believed to contain 6-coordinate Fe, the sixth site being occupied by an S atom in the adjacent $Fe(CO)_3S_2C_2(CF_3)_2$ group[77]. Reaction of $Co_3(CO)_9CY$ (Y = Cl or Me) with $S_2C_2(CF_3)_2$ was reported[77] to give $[Co_3(CO)_3\{S_2C_2-(CF_3)_2\}_3CY]$.

The trinuclear complexes $[Co(CO)\{S_2C_2(CF_3)_2\}]_3$ and $[Fe(NO)\{S_2C_2 (CF_3)_2\}]_3$, obtained by reaction of $S_2C_2(CF_3)_2$ with $Co_2(CO)_8$ and $Hg[Fe-(NO)(CO)_3]_2$ respectively[80], underwent a one-electron reduction, and the former was oxidised to a monocation whereas the latter could be further reduced to a dianion[75]. Treatment of $[Fe(S_2C_2Ph_2)_2]_n$ with NO gave[79]

$[Fe(NO)(S_2C_2Ph_2)_2]$ and diamagnetic $[Fe_2(NO)_2(S_2C_2Ph_2)_3]$. The latter underwent two one-electron reductions, and the paramagnetic monoanion was isolated.

Addition of $NaMn(CO)_5$ to $[M\{S_2C_2(CN)_2\}_2]_2^{2-}$ (M = Fe or Co) gave[80] $[(CO)_5Mn\text{—}M\{S_2C_2(CN)_2\}_2]^{2-}$, and on treatment of the cobalt dithiolene with $Ir(CO)(PPh_2Me)_2Cl$, the similar $[Cl(Ph_2MeP)_2(CO)Ir\text{—}Co\{S_2C_2(CN)_2\}_2]^-$. An e.s.r. spectrum was obtained from the iron complex, and it was shown that the metal—metal bond was easily cleaved.

10.2.7 π-Cyclopentadienyl dithiolene complexes

There are three general classes of π-cyclopentadienyl dithiolene complexes — those containing only $\pi\text{-}C_5H_5$ and the sulphur ligand, those having also CO, and those containing NO.

The simple species are of four main types — dimeric, e.g. $[\pi\text{-}C_5H_5MS_2C_2R_2]_2$ (R = CF_3, M = V, Cr, Mo)[83], '7'-coordinate, e.g. $[\pi\text{-}C_5H_5M(S_2C_2R_2)_2]^z$ (z = 0 or −1; R = CN, M = Ti, Mo, W; R = H, M = W; R = CF_3, M = Mo or W)[84–86] or $[\pi\text{-}C_5H_5M(S_2C_6Cl_4)_2]^-$ (M = Ti or Mo)[41], '4'-coordinate, e.g. $[\pi\text{-}C_5H_5MS_2C_2R_2]$ (R = CN or H; M = Co, Rh; R = CF_3, M = Fe, Co, Ni, Rh) and bis-π-cyclopentadienyl species, e.g. $(\pi\text{-}C_5H_5)_2TiS_2C_2R_2$ (R = CN, H)[82, 83], $(\pi\text{-}C_5H_5)_2MS_2C_6H_3R$ (M = Ti, R = Me; M = Mo or W, R = H)[85, 86] and $(\pi\text{-}C_5H_5)_2TiS_2C_6Cl_4$[41].

All of these complexes, except $[\pi\text{-}C_5H_5NiS_2C_2(CF_3)_2]$ and $[\pi\text{-}C_5H_5VS_2C_2(CF_3)_2]_2$, were diamagnetic; the Ni species had one unpaired spin and for the vanadium complex, $\mu = 0.7$ BM (instead of 1.7 BM) per V atom suggesting that the V—V bond order was greater than unity. The structures of $[\pi\text{-}C_5H_5M\{S_2C_2(CF_3)_2\}]_2$ (M = V[87], Cr[88] and Mo[89]), $[\pi\text{-}C_5H_5Mo\{S_2C_2(CN)_2\}_2]$[90], $[\pi\text{-}C_5H_5CoS_2C_2R_2]$ (R = CF_3[91] or CN[92]) and $[(\pi\text{-}C_5H_5)_2MoS_2C_6H_3Me]$[93] have been reported. The binuclear species were of interest since the V and Mo complexes, which contain a metal–metal bond, are 7-coordinate, whereas the Cr complex contains essentially octahedral metal atoms (only one S atom in each S-ligand is bridging) with the Cr—Cr bond effectively rendering each Cr atom 7-coordinate. The anionic Mo species is a 'four-legged piano-stool' structure whereas $[(\pi\text{-}C_5H_5)_2MoS_2C_6H_3Me]$ has the expected 'oyster' structure. Of further interest are the structures of $[\pi\text{-}C_5H_5CoS_2C_2R_2]$ which suggest that coordinated $S_2C_2(CF_3)_2$ has more dithioketonic character than coordinated $S_2C_2(CN)_2$.

Many of the complexes have been studied electrochemically[94, 95] and most undergo at least one reduction process. The binuclear species undergo a two-electron reduction.

Reaction of $\pi\text{-}C_5H_5W(CO)_3H$ with $S_2C_2(CF_3)_2$ afforded $[\pi\text{-}C_5H_5W(CO)\{S_2C_2(CF_3)_2\}]_2$, and of $\pi\text{-}C_5H_5Fe(CO)_2Cl$ with $Na_2S_2C_2R_2$ (R = H or CN), $[\pi\text{-}C_5H_5Fe(CO)_2]_2S_2C_2R_2$. In the former[83, 84] there is a W—W bond, and each metal atom is 7-coordinate, one sulphur atom in each ligand being bridging. In the latter[82, 83], neither S atom bridges both metals.

Reaction of a series of 1,2-dithiolato ligands with $[\pi\text{-}C_5H_4RMn(NO)(CO)_2]$

$[PF_6]$ (R = H or Me) afforded[95] the diamagnetic $[\pi\text{-}C_5H_4RMn(NO)\text{-}$ $(S\text{—}S)_2]^-$ $(S\text{—}S = S_2C_2(CN)_2, S_2C_6Cl_4, S_2C_6H_3Me)$. All of these complexes underwent at least one voltammetric oxidation process, and oxidation of the monoanions chemically afforded the paramagnetic $[\pi\text{-}C_5H_4RMn\text{-}$ $(NO)(S\text{—}S)]$; voltammetric evidence was also obtained for monocationic species. The $E_{\frac{1}{2}}$ values for the couple $[\pi\text{-}C_5H_4RMn(NO)(S\text{—}S)]^z + e^- \rightleftharpoons$ $[\pi\text{-}C_5H_4RMn(NO)(S\text{—}S)]^{z-1}$ were dependent on the substituent on the sulphur ligand but not on the cyclopentadienyl ring. Interestingly, the 1,1-dicyanoethylene-dithiolato and dithiocarbamato analogues, $[\pi\text{-}C_5H_4R\text{-}$ $Mn(NO)S_2C{=}C(CN)_2]^-$ and $[\pi\text{-}C_5H_4RMn(NO)S_2CNR'_2]$ were also easily oxidised in a one-electron step.

The reactions between $[\pi\text{-}C_5H_5Mo(NO)I_2]_2$ and dithiolato and dithiete ligands proceeded[98] in stages giving $[\pi\text{-}C_5H_5Mo(NO)\text{-}(S\text{—}S)]_2$, $[\pi\text{-}C_5H_5Mo$ $(NO)(I)(S\text{—}S)]^-$, $[\pi\text{-}C_5H_5Mo(S\text{—}S)_2]^-$ and finally $[Mo(S\text{—}S)_3]^{2-}$. All of the species, except $[\pi\text{-}C_5H_5Mo(NO)(S\text{—}S)]_2$, could be oxidised in a one-electron step, whereas the dimers could apparently be reduced in an irreversible two-electron process.

10.2.8   Other organometallic dithiolene complexes

When treated with certain unsaturated hydrocarbons, Ni, Pd and Pt bis-diphenyldithiolenes[57, 97] and also $[Ni\{S_2C_2(CF_3)_2\}_2]$ [98] formed 1:1 adducts. These hydrocarbons included norbornadiene, butadiene and its 2,3-dimethyl homologue, 1,3-cyclohexadiene and norbornene.

The crystal structure of the norbornadiene adduct of $[Ni\{S_2C_2(CF_3)_2\}_2]$ showed[98] that two S—C bonds had been formed between the reactants and that one of the norbornadiene C—C double bonds had been transformed into a single bond. The S atoms were in the *exo* configuration, with one from each S-ligand being involved in the new bonds, which were about 0.1 Å longer than the S—C bonds within the S-ligands. The adduct is diamagnetic despite its extreme non-planarity, and underwent one one-electron oxidation, and a one-electron reduction process. The potentials of these processes corresponded very closely to those of the couples $[Ni\{S_2C_2$ $(CF_3)_2\}_2] + e^- \rightleftharpoons [Ni\{S_2C_2(CF_3)_2\}_2]^-$ and $[Ni\{S_2C_2(CF_3)_2\}_2]^- + e^- \rightleftharpoons$ $[Ni\{S_2C_2(CF_3)_2\}_2]^{2-}$, suggesting that the electrochemical behaviour of the neutral adduct was similar to that of the bisdithiolene monoanion, and, accordingly, that there was an increase of electron density within the ligand system of the former. However, electronic spectral studies indicated that the adduct should be regarded as a simple derivative of Ni^{2+} complexed by a tricyclic tetradentate sulphur ligand.

The reaction of norbornadiene with $[Ni(S_2C_2Ph_2)_2]$ has been reported[59] to give the Diels–Alder adduct (see Figure 6.3), 1, presumably via the intermediate 2. However, a norbornadiene adduct of the bisdiphenyldithiolene was obtained[99] which, on thermolysis, gave only $[Ni(S_2C_2Ph_2)_2]$, NiS, 2-phenylthionaphthene and norbornadiene, and it was suggested that an adduct similar to that obtained with $[Ni\{S_2C_2(CF_3)_2\}_2]$, 3, was obtained. This adduct did not react with tri-substituted phosphines, but, on alkylation with MeI, afforded a paramagnetic species, 4, which, on hydrolysis afforded the macrocyclic ligand, 5. The structure of 5 was confirmed by ^{1}H n.m.r. measurements, and a similar species derived from α,α'-dibromoxylene was also formed. It was shown that the norbornadiene adduct was formed from the dithiolene complex in an excess of the refluxing unsaturate, but that when the reaction was carried in cold dichloromethane, only the Diels–Alder adduct, 1, was produced. It was further shown that adducts of the type 3 were only formed with the nickel dithiolenes, whereas the Diels–Alder derivative, 1, was formed by reaction of norbornadiene with $[M(S_2C_2Ph_2)_3]$ when M = Cr, Mo or V but not when M = W, Re or Os. Small amounts of

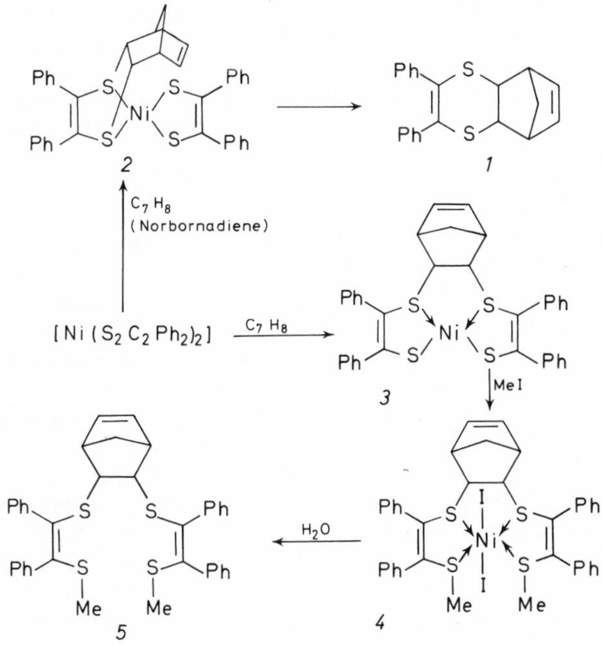

Figure 10.3

the Diels–Alder adduct containing CF_3 substituents were obtained as by-products of the reaction of the diolefin with $[Ni\{S_2C_2(CF_3)_2\}_2]$.

Alkylation of $[Ni(S_2C_2Ph_2)_2]^{2-}$ by MeI afforded[99] $[Ni(Me_2S_2C_2Ph_2)(S_2C_2Ph_2)]$, which, from electronic and n.m.r. spectroscopic measurements, was clearly planar. Similar alkylations with other alkyl halides were easily effected and the thioether could be readily displaced by α,α'-dipyridyl or diphos (L—L) giving $[Ni(L—L)(S_2C_2Ph_2)]$. The compounds exhibited no significant redox behaviour, and the dibenzyl derivative was easily photo-dealkylated. Attempts to akylate $[M(S_2C_2Ph_2)_3]^{2-}$ (M = Cr, Mo or V) with MeI were unsuccessful, although $Me_2S_2C_2Ph_2$ was obtained[102]. However, $[W\{(MeS)_2C_2Ph_2\}(S_2C_2Ph_2)_2]$ and $[Re\{(MeS)SC_2Ph_2\}(S_2C_2Ph_2)_2]$ were obtained by alkylation of the corresponding ionic species. These complexes were said to be derivatives of W(IV) and Re(V), and the tungsten compound reacted with diphos giving $[W(diphos)(S_2C_2Ph_2)_2]$.

10.3 ANALOGUES OF THE DITHIOLENE COMPLEXES

In 1965, interest in the electron transfer properties of complexes containing the $[MS_4]^z$ coordination unit was expanded to include compounds containing, as well as S donor atoms, NH or O donor atoms or groups. It was suggested[101] that if species derived from 1,2-disubstituted benzenes having the coordination units $[MN_2S_2]^z$ or $[MS_2O_2]^z$ could be prepared, then a five-membered electron transfer series containing $[MS_2X_2]^z$, $z = \pm 2, \pm 1$, 0, should exist. The terminal members of this series were expected to be defined as metal complexes of the potential stable anions $\{C_6H_4SX\}^{2-}$ and the thioquinonoidal species $\{C_6H_4SX\}$.

Nickel complexes of the ligands o-aminothiophenol, glyoxalbis(2-mercaptoanil), thiobenzoylhydrazine, biacetylthiobenzoylhydrazine and several diketone thiosemicarbazides have been investigated voltammetrically and polarographically[102, 103]. All readily underwent redox reactions, although the full extent of the proposed redox series were not realised. Several points of interest arose from these redox studies. Firstly, as expected, the $E_{\frac{1}{2}}$ values were dependent on the ligand substituents in the same way as in the 1,2-dithiolenes. Secondly, if there was full conjugation throughout the ligand system, $E_{\frac{1}{2}}$ values were generally more positive than those of their non-conjugated analogues. Thirdly, in fully conjugated ring systems, there was no apparent effect on the $E_{\frac{1}{2}}$ values when the central chelate ring was varied in size. The structures of two nickel complexes derived from thiosemicarbazide and benzene-1,2-dialdehyde and from hexane-2,5-dione have been elucidated[104]. The former was shown to be planar whereas the latter was slightly distorted, with one of the donor N atoms lying out of the plane of the other donor atoms and the metal.

An examination was made of the electrochemical properties of complexes derived from o-mercaptophenol, $[M(SOC_6H_4)_2]^z$. Evidence was obtained for mono- and dianionic species when M = Cu, Pd or Ni [109].

Reaction of various metal carbonyls and phosphine complexes with $Se_2C_2(CF_3)_2$ afforded a series of bis- and trisdiselenolene complexes[106]. The compound obtained included $[Ni\{Se_2C_2(CF_3)_2\}_2]$, $[M(Se_2C_2(CF_3)_2\}_2]^-$ (M = Ni, Co, Fe or Cu), $[M\{Se_2C_2(CF_3)_2\}_2]^{2-}$ (M = Ni or Cu), $[M\{Se_2C_2(CF_3)_2\}_3]^{0, -2}$ (M = Mo or W) and $[V\{Se_2C_2(CF_3)_2\}_3]^{-1, -2}$ The neutral Mo complex apparently has trigonal prismatic geometry. The complexes readily underwent redox reactions at potentials essentially identical to those of their dithiolene analogues. The complexes, particularly the more oxidised species, were generally less stable than their sulphur analogues, but from the e.s.r. spectral data obtained from $[Ni\{Se_2C_2(CF_3)_2\}_2]^-$ and $[Cu\{Se_2C_2(CF_3)_2\}_2]^{2-}$, it was apparent that the orbitals of the unpaired electron had considerably more ligand character than those of the corresponding dithiolenes.

4-Coordinate metal complexes of dithiotropolone have recently been described[107]. Spectral and electrochemical studies of the diamagnetic $[M(S_2C_7H_5)_2]$ (M = Ni, Pd or Pt) revealed that the compounds exhibited intense near-infrared absorption bands, and were reducible in two apparent one-electron processes, which, in contrast to those of the corresponding dithiolenes, were irreversible, and occurred at potentials considerably more negative. The Cu, Zn and Cd dithiotropolonates also exhibited irreversible redox behaviour. Comparison was made between the redox behaviour of $[Ni(S_2C_7H_5)_2]$, $[Ni(SOC_7H_5)_2]$, $[Ni\{S(NMe)C_7H_5\}_2]$ and $[Ni\{(NMe)_2-C_7H_5\}_2]$, and it was shown that all species were electrochemically active, each being reducible in at least one single electron step, and several species showing a one-electron oxidation. Generally, the $E_{\frac{1}{2}}$-values became more negative in the order S,S < S,O < S,N < N,N.

10.4 METAL COMPLEXES OF LIGANDS DERIVED FROM CARBON DISULPHIDE

10.4.1 Dithiocarbamates, xanthates and thioxanthates

Silver(I) dithiocarbamates are hexamers in the solid state[108] and in solution. The copper(II) complexes $[Cu(S_2CNR_2)_2]$, were monomeric in solution, but dimeric in the solid state[109, 110]. Each Cu atom in the dimers was 5-coordinate with respect to S. From e.s.r. spectral and electronic spectral work, it was concluded[110] that, in $[Cu(S_2CNR_2)_2]$ there was considerable σ-covalency in the Cu—S bond. Oxidation of $[Cu(S_2CNR_2)]_4$ with halogen (X_2), afforded[111] $[Cu(S_2CNR_2)X_2]$ ($S = 0$) and $[Cu(S_2CNR_2)X]$ ($S = 1/2$); the former (X = Br) was planar[111]. The analogous gold complexes were obtained similarly, but if a deficiency of X_2 was used $[Au(S_2CNR_2)_2]-[AuX_2]$ was obtained[112]; the latter complex (X = Br, R = Bu^n) had a planar cation and linear anion[112].

The zinc(II) bis-dithiocarbamates are dimeric but are not always iso-structural with $[Cu(S_2CNR_2)_2]$. Thus $[Zn(S_2CNMe_2)_2]$ contained[113] essentially tetrahedrally coordinated Zn atoms, with two of the sulphur ligands functioning as bridging groups between the two metal atoms. However, $[Zn(S_2CNEt_2)_2]$ was isomorphous and isostructural with $[Cd(S_2CNEt_2)_2]$ which has a structure similar to that of $[Cu(S_2CNEt_2)_2]$ [114].

Oxidation of $[Cu(S_2CNEt_2)_2]$ and its nickel analogue with Br$_2$ afforded $[Cu(S_2CNEt_2)Br_2]$ and $[Ni(S_2CNEt_2)_2Br_2]$ [115]. The structure of $[Ni\{S_2CN-(Bu^n)_2\}_3]^+Br^-$, obtained by oxidation of $[Ni\{S_2CN(Bu^n)_2\}_2]$ with Br$_2$ [116], and which is probably similar to $[Ni(S_2CNR_2)_2Br_2]$ [115], consists of an octahedral arrangement of S atoms, with Ni—S distances consistent with Ni^{IV}. While $[Ni(S_2CNEt_2)_2]$ was reluctant to form adducts with nitro-genous bases, $[Ni(S_2CNH_2)_2]$ and $[Ni\{S_2CNH(p–ClC_6H_4)\}_2]$ formed 6-coordinate adducts with pyridine and its derivatives[117]. However, nucleo-philic attack of either morpholine or piperidine occurred on the —CS$_2$ carbon atom of $[Ni(S_2CNH_2)_2]$ giving[118]

$$\left[Ni\left(S_2CN\diagup\!\!\diagdown O\right)_2\right] \quad \text{and} \quad \left[Ni\left(S_2CN\diagup\!\!\diagdown\right)_2\right]$$

respectively. Addition of $\{S_2COEt\}^-$ to planar $[Ni(S_2COEt)_2]$ gave $[Ni(S_2COEt)_3]^-$ identified magnetically, spectrally and crystallographically as an octahedral NiS_6 species[117]. Both $[Ni(S_2COEt)_2]$ and $[Ni(S_2CSEt)_2]$ formed 2:1 adducts with pyridine[117], but with piperidine the former gave

(Ref. 125)

The Pd(II) and Pt(II) dithiocarbamates and xanthates formed 5-coordinate adducts with $PMePh_2$ [120]. However, when the xanthates were treated with an excess of phosphine, dealkylation occurred and $[M(PR_3)_2S_2CO]$ was formed[119].

A number of dipositive metal bis-dithiocarbamates, e.g. $[M(S_2CNEt_2)_2]$, M = Fe, Mn or Cr, have been prepared[122], and are isomorphous with $[Cu(S_2CNEt_2)_2]$. Reaction of the ferrous compound with $S_2C_2(CF_3)_2$ afforded $[Fe(S_2CNEt_2)_2\{S_2C_2(CF_3)_2\}]$ [123]. An x-ray study of this compound confirmed its essentially octahedral geometry and magnetic studies over a temperature range were consistent with a temperature dependent equilibrium between singlet ground and triplet excited states. The compounds $[Fe(S_2CNR_2)_2\{S_2C_2(CF_3)_2\}]$ (R = Me, Et, $-(CH_2)_3-$, and $-(CH_2)_5-$) underwent two polarographic one-electron reductions (one near 0.0 V $v.$ S.C.E.) and a single oxidation step. Because of the paramagnetism of the complexes, their ^1H n.m.r. spectra were contact-shifted and, from a study over a temperature range, indicated that the molecules were stereochemically non-rigid. At temperatures above $+50\,°C$, the spectra revealed that rapid racemisation of the chiral complexes occurred, and suggested that the racemisation was intramolecular. This was confirmed by the observation that the exchange equilibrium $[Fe(S_2CNEt_2)_2\{S_2C_2(CF_3)_2\}]+[Fe(S_2CN Me_2)_2\{S_2C_2(CF_3)_2\}] \rightleftharpoons 2[Fe(S_2CNMe_2)(S_2CNEt_2)\{S_2C_2(CF_3)\}]$ was established immediately upon mixing.

A considerable amount of work has been carried out on the magnetic properties of the trisdithiocarbamates, $[Fe(S_2CNR_2)_3]$ [9]; these species exist at a magnetic 'crossover' point, between $^2T_2$ (low temp.) and $^6A_1$ states. The crystal structure determination of $[Fe\{S_2CN(Bu^n)_2\}_3]$ [124] ($\mu = 5.32$ BM at 264 K) revealed that the molecule could only be described as an intermediate between a trigonal prism and a trigonal antiprism: the angle between the two equilateral triangles formed by the two groups of three S atoms was only 32 degrees from the idealised trigonal prism. It was thought that this unusual coordination might be due to lattice forces since there were considerable differences between the compound's magnetic properties in the solid state and in solution. However, in $[Fe(S_2CNEt_2)_3]$, the six S atoms constituted[125] an essentially complete octahedral arrangement around the metal atom. From the magnetic data obtained from iron trisdithiocarbamates, it was estimated[126] that there was an increase of the Fe—S bond length on passing from the $^2T_2$ to the $^6A_1$ state. The compounds could be divided into four distinct classes according to the value of μ in solution. These classes were (i)

$\mu = 5.8$ BM; (ii) N,N-dialkyl compounds, $\mu = 4.3$ BM; (iii) N-alkyl, N-aryl, $\mu = 3.5$ BM; and (iv) N,N-s-alkyl, $\mu = 2.5$ BM. The R—N—R angle in $[Fe(S_2CNR_2)_3]$ was expected in increase from (i) to (iv), with a corresponding change in the C—N bond order, which was expected to affect the ligand field strength, and hence μ.

The magnetic properties of $[Fe(S_2COR)_3]$ have been described[126] as being characteristically low-spin ($\mu = 2.5$ BM). The ethyl xanthate $[Fe(S_2COEt)_3]$, had magnetic properties generally similar to those of $[Fe\{S_2CN(Pr^i)_2\}_3]$ [127], and its crystal structure determination revealed[128] that the angle between the two equilateral triangles formed by the two groups of three sulphur atoms was 41 degrees from the trigonal prismatic configuration. The tris-thioxanthates $[Fe(S_2CSR)_3]$, were described [3, 129] as exhibiting high-spin-low-spin magnetic behaviour, and $[Fe\{S_2CS(Bu^t)\}_3]$ had $\mu = 2.46$ BM[130] at room temperature in solution. Interestingly, this t-butyl xanthate exhibited reversible voltammetric one-electron oxidation and reduction behaviour.

The 5-coordinate $[Fe(S_2CNR_2)_2X]$ was prepared[131] by treating $[Fe(S_2CNR_2)_3]$ with HX (X = halide). The complexes were monomeric, had three unpaired electrons and adopted a square pyramidal configuration (X = Cl apical) [132].

The structure of $[Mo_2O_3(S_2COEt)_4]$ [133], consisted of a linear Mo—O—Mo group, and two distorted octahedrally coordinated Mo atoms, the M = O groups being mutually cis to the Mo—O—Mo bond. The structure of $MoO_2(S_2CNR_2)_2$, both in the solid state[134] and in solution[135], contained cis-MoO_2 groups. Reduction of $[Mo_2O_3(S_2CNR_2)_4]$ afforded[136] the monomeric, diamagnetic, square pyramidal $[MoO(S_2CNR_2)_2]$. Reaction of $ReOCl_3(PPh_3)_2$ with NaS_2CNEt_2 afforded $[Re_2O_3(S_2CNEt_2)_4]$ which contained[137] a linear O=Re—O—Re=O arrangement (cf. $[Mo_2O_3\{S_2P(OEt)_2\}_4]$ [138]; $[ReN(S_2CNEt_2)_2$, obtained from $ReNCl(PPh_3)_2$, had a square-pyramidal structure (apical Re≡N).

The square-pyramidal $[VO(S_2CNR_2)_2]$ were paramagnetic[139], and although the e.s.r. spectra obtained from $[V(S_2CNEt_2)_4]$ were shown to be time, temperature and solvent dependent, no clear structural information could be deduced[140]. Reaction of MCl_5 (M = Nb or Ta) with NaS_2CNEt_2 in acetonitrile afforded[141] $[Nb(S_2CNEt_2)_4]$ and $[Ta(S_2CNEt_2)_5]$. If these reactions were carried out in dichloromethane or benzene, however, the products were [142] $[M(S_2CNEt_2)_4]^+X^-$, $[M(S_2CNEt_2)_4]^+[MX_6]^-$ and $[M(S_2CNEt_2)_4S]$ (X = Cl, Br). In methanol, the monomeric $[M(S_2CNEt_2)_2(OMe)X]$ (X = Cl, Br, NCS) were formed[143].

Reaction of $M(NR_2)_4$ (M = Ti, Zr, Hf) and $M'(NR_2)_5$ (M' = Nb or Ta) with CS_2 gave $[M(S_2CNR_2)_4]$, $[Nb(S_2CNR_2)_4]$ and $[Ta(S_2CNR_2)_5]$ [144]. The crystal structure determination of $[Ti(S_2CNEt_2)_4]$ confirmed[145] the 8-coordination of the Ti atom and the general dodecahedral geometry; the V and Zr analogues were isomorphous.

The tripositive lanthanide complexes $[M(S_2CNEt_2)_3]$ and $[M(S_2CNEt_2)_4]^-$ (M = La−Lu), and $[U(S_2CNEt_2)_3]$, have been described[146]. Reaction of $[M'(S_2CNEt_2)_3]$ (M' = Np or Pu) with NaS_2CNEt_2 and $[Et_4N]Cl$ gave $[Et_4N][M'(S_2CNEt_2)_4]$, and the structure of the Np complex consisted of a metal atom 8-coordinate with respect to S having a grossly distorted dodecahedral configuration; the distortion was unusual in that it did not tend towards the alternative square antiprismatic arrangement[147]. The structure of $[Th(S_2CNEt_2)_4]$, too, was intermediate between an ideal dodecahedron and a square antiprism; the U^{IV}, Np^{IV} and Pu^{IV} complexes were structurally isomorphous[147].

The structure of $[Me_4N][UO_2(S_2CNEt_2)_3]$ [148] consisted of a linear OUO arrangement in a hexagonal prism.

A series of diselenocarbamato complexes have been reported[149]. These are of the type $[M(Se_2CNR_2)_2]$, $[M'(Se_2CNR_2)_3]$, $[U(Se_2CNR_2)_4]$ and $[Fe(Se_2CNR_2)_2Cl]$; M = Cu, Ni group, and M' = Au, Rh, Fe, Ru, Os, Co and Cr. $[Fe(Se_2CNEt_2)_3]$ was low spin ($\mu = 2.37$ BM) [150] at room temperature whereas $[Fe\{Se_2CN(Bu^n)_2\}_3]$ was high-spin ($\mu = 4.14$ BM)[151].

10.4.2   Metal complexes of dithioacid salts, $(S_2CR)^-$

The crystal structures of $[M(S_2CPh)_2]$ (M = Ni [152] or Pd [153]) revealed that the molecules were trimeric, containing one inner tetragonally distorted octahedral metal atom and two outer 5-coordinate square-pyramidal metal atoms. The structure of $[Ni(S_2CCH_2Ph)_2]$ was, however, completely different. Here the molecule was dimeric[154] and had a bridging acetate cage structure with a Ni—Ni bond; each nickel atom was therefore 5-coordinate. A similar structural arrangement was found in $[Ni(SOCPh)_2]_2 \cdot EtOH$ [155].

The complex $[NiS(S_2CPh)_2]$, first formulated as $[(PhCS_2)_2NiS_2Ni(S_2CPh)_2]$ [156], was reformulated as $[Ni(S_2CPh)(S_3CPh)]$ [157], and other similar compounds containing Ni, Pd or Pt were also prepared. While it was observed that nickel 1,1-dithiolato complexes readily absorbed two moles of sulphur, $[Ni(S_2CR)_2]$ took up only one[158]. The completely sulphurated species, $[Ni(S_3CR)_2]$ could be obtained only via metathetical reactions between the corresponding perthio–zinc complexes and $NiCl_2$ [158]. The mechanism of S-insertion into $[Ni(S_2CPh)_2]$ has been elegantly studied[159] using ^{35}S-enriched sulphur. By a mass spectrometric examination of $[Ni(S_2CPh)(S_2^{35}SCPh)]$ obtained by photolytic addition of ^{35}S to $[Ni(S_2CPh)_2]$, it was established that the added S appeared to be adjacent to the CS_2-carbon atom, and that this S atom is specifically removed by PPh_3 regenerating $[Ni(S_2CPh)_2]$. However, the mass spectral analysis of $[Zn(S_2CPh)(S_2^{35}SCPh)]$ revealed that ^{35}S scrambling occurred on photolysis of $[Zn(S_2CPh)_2]$ in the presence of ^{35}S. It was further established that S-scrambling occurred when $[Ni(S_2CPh)(S_2^{35}SCPh)]$ was prepared by heating together $[Ni(S_2CPh)_2]$ and enriched sulphur in refluxing dimethylformamide. The kinetic lability of the S atoms in the related Zn complex was apparent from ^1H n.m.r. spectral studies since the n.m.r. spectra of the Ni, Pd and Pt analogues clearly revealed the inequivalence of the two chelate ring systems, whereas no such inequivalence was observed, at −1 °C, in the zinc compound.

The interesting perthioiron complex, $[Fe\{S_3C(p\text{-}MeC_6H_4)\}_3]$ underwent[160] partial desulphurisation giving $[Fe\{S_2C(p\text{-}MeC_6H_4)\}\{S_3C(p\text{-}Me C_6H_4)\}_2]$, whose crystal structure determination revealed that the co-ordination polyhedron was a distorted octahedron. The desulphurated complexes were low-spin (μ = 2.2–2.4 BM at room temperature), as were $[Fe(S_2CPh)_3]$ and $[Fe(S_2CCH_2Ph)_3]$[151].

10.4.3 Metal complexes of 1,1-dithiolates, $(S_2C=X)^{2-}$

A large variety of metal complexes of the type $[M(S_2C=X)_2]^{2-}$ and $[M'(S_2C=X)_3]^{3-}$, X = S, N(CN), NPh, O, C(CN)$_2$, C(CN)(CO$_2$Et), C(CN)Ph, CH(NO)$_2$, etc., have been synthesised[3]. From a study of their electronic spectral and magnetic properties, it was apparent that most of the 4-coordinate species were planar, whereas the 6-coordinate compounds were octahedral (D_3 symmetry). The crystal structure determinations of salts of $[Au\{S_2C=C(CN)_2\}_2]^-$ [3] $[Ni\{S_2C=N(CN)\}_2]^{2-}$ [161] and $[Ni(S_2C=S)_2]^{2-}$ [162] confirmed the planarity of these species. The electronic spectra of these complexes did not exhibit the intense low-frequency absorptions so characteristic of the analogous 1,2-dithiolenes, nor did any of them undergo reversible oxidation or reduction reactions.

Repeated recrystallisation of $[R_4N]_2[Cu\{S_2C=C(CN)_2\}_2]$ led to the formation of a Cu (I) species, $[R_4N]_4[Cu_8\{S_2C=C(CN)_2\}_6]$ [163]. The crystallographic examination of this complex established that the S atoms were arranged at the apices of an icosahedron in which a cube of eight Cu atoms was inscribed (idealised T_h symmetry).

In attempts to oxidise $[Ni(S_2C=X)_2]^{2-}$ using iodine, the perthio-complexes $[Ni(S_3C=X)_2]^{2-}$ were formed[164], and these could also be formed by treatment of the 1,1-dithiolates with S_8. The 'extra' sulphur was extracted quantitatively with PPh$_3$, as Ph$_3$PS. In an effort to understand these sulphur addition reactions, radioactive ^{35}S was added to $[Ni(S_2C=S)_2]^{2-}$, and the product treated with PPh$_3$ [164]. It was found that S addition was specific, since the same sulphur that was added was extracted by the PPh$_3$. The data also implied that the inserted S atom lay adjacent to the C atom of the CS$_2$, and not next to the Ni atom. These results were similar to those obtained from studies of nickel and zinc dithiobenzoates[159]. The perthio-complexes, $[Ni(S_3CX)_2]^{2-}$ exhibited, voltammetrically, an apparent irreversible one-electron oxidation reaction[164].

10.4.4 Nitrosyl, carbonyl and organometallic complexes

It has been shown that the 5-coordinate mononitrosyls $[M(NO)(S_2CNR_2)_2]$, have square-pyramidal structures, with axial NO [165]. There has been considerable discussion about the M—N—O bond angle, but recent low temperature x-ray studies have established that when M = Fe and R = Me, this angle is 170 degrees [166]. Infrared and dipole moment data indicated that the 6-coordinate $[M(NO)_2(S_2CNR_2)_2]$ [167] and $[M(NO)_2\{S_2C=C(CN)_2\}_2]^{2-}$ [65] have cis-NO groups. ^{1}H n.m.r. data revealed that there was restricted rotation about the C—N bond of the S$_2$C—NR$_2$ group, caused by the partial double bond character of this C—N link, which is pronounced in dithiocarbamato complexes containing π-acceptor ligands[168]. The structure of $[Ru(NO)(S_2CNMe_2)_3]$ has been shown[169] to consist of two bidentate and one monodentate dithiocarbamate ligands, the latter being cis to the

NO. This structure persisted in solution[167], but both crystallographic and n.m.r. studies of $[Mo(NO)(S_2CNR_2)_3]$ [167, 170] revealed that all three S-ligands were bidentate, and that the molecule had a pentagonal bipyramidal structure (one S atom was in the apical position, *trans* to NO).

Reaction of the appropriate metal carbonyl halides with NaS_2CNR_2 has given $[Rh(CO)_2(S_2CNR_2)]$ [168], *cis*-$[Fe(CO)_2(S_2CNR_2)_2]$ [168] and $[M(CO)_n(S_2CNR_2)_2]$ (M = Mo, W; n = 2 or 3 [171]. It was shown that $[Mo(CO)_2(S_2CNR_2)_2]$ reversibly took up CO giving the tricarbonyl, and that these compounds were easily oxidised to $[Mo_2O_3(S_2CNR_2)_4]$. Reaction of the red solutions obtained by passing CO into $RuCl_3$ in refluxing ethanol with NaS_2CNR_2 or $[Me_2NCS_2]_2$ afforded[172] monomeric $[Ru(CO)(S_2CNR_2)_2]$, or, under modified conditions, *cis*-$[Ru(CO)_2(S_2CNR_2)_2]$ and *trans*-$[Ru(CO)_2\{S_2CN(CH_2Ph)_2\}_2]$ which, on standing in solution in $CHCl_3$ afforded the paramagnetic (μ = 1.8 BM) $[Ru(CO)_2\{S_2CN(CH_2Ph)_2\}_2]^+Cl^-$.

Treatment of $Rh(CO)(PPh_3)_2Cl$ and $Rh(PPh_3)_3Cl$ with NaS_2CNMe_2 gave[173] $[Rh(PPh_3)_2(S_2CNMe_2)]$ (bidentate S-ligand), $[Rh(CO)(PPh_3)(S_2CNMe_2)]$ (bidentate), $[Rh(CO)(PPh_3)\{S_2CNMe_2\}_3]$ (two monodentate, one bidentate) and $[Rh(PPh_3)(S_2CNMe_2)_3]$ (one monodentate and two bidentate). Both $[Ru(PPh_3)_2(S_2CNR_2)_2]$ and $[Ru(PPh_3)_2(S_2COR)_2]$, obtained from $Ru(PPh_3)_3Cl_2$, were believed to have octahedral structures with *trans*-PPh_3. Reaction of $Pt(PR_3)_4$ and $Ir(CO)(PR_3)_2Cl$ with $[(CF_3)_2C=CS_2]_2$ gave[174] $[Pt(PR_3)_2\{S_2C=C(CF_3)_2\}]$ and $[Ir(CO)(PR_3)_2(Cl)\{S_2C=C(CF_3)_2\}]$.

Reaction of $[\pi\text{-}C_5H_5M(CO)_n]_n$ (M = Fe, n = 2, M = Mo, n = 3) with $[Me_2NCS_2]_2$ gave[174] $[\pi\text{-}C_5H_5M(CO)_{n-1}S_2CNMe_2]$. However, reaction of $\pi\text{-}C_5H_5Fe(CO)_2Cl$ under mild conditions with NaS_2CNMe_2 afforded[173] $\pi\text{-}C_5H_5Fe(CO)_2SC(:S)NMe_2$. The 1,1-dithiolato complexes $[\pi\text{-}C_5H_4RMn(NO)S_2CX]^-$ and $[\pi\text{-}C_5H_4RMn(NO)S_2CNR_2]$, formed by addition of the appropriate sulphur ligand to $[\pi\text{-}C_5H_4RMn(NO)(CO)_2][PF_6]$ [95, 175], underwent facile, reversible one-electron transfer reactions generating neutral and, occasionally, monocationic species. The existence of these oxidation products was confirmed by e.s.r. spectral methods, and it was shown that the $E_{\frac{1}{2}}$ values and v_{NO} were dependent on the nature of X.

10.4.5 Carbon disulphide elimination and insertion reactions

Carbon disulphide insertion reactions occurred when $M(CO)_5R$ (M = Mn, Re; R = alkyl, aryl, benzyl, Ph_3C) was treated with CS_2 above 100 °C[176], and $M(CO)_4S_2CR$ was formed.

Dissolution of $[\pi\text{-}C_5H_5Ni(SR)]_2$ in CS_2 apparently afforded[177] the trithioxanthate $[\pi\text{-}C_5H_5NiS_2CSR]$, whose existence was established by ^1H n.m.r. spectroscopy. Reaction of $\pi\text{-}C_5H_5Fe(CO)_2Cl$ with S_2CSR^- afforded[178] $[\pi\text{-}C_5H_5Fe(CO)_2SC(:S)SR]$ which on u.v. irradiation, gave $[\pi\text{-}C_5H_5Fe(CO)S_2CSR)]$. Thermolysis of this compound afforded $[\pi\text{-}C_5H_5Fe(CO)SR]_2$.

The bistrithioxanthate $[Ni(S_2CSR)_2]$ decomposed thermally[119] with loss of CS_2 giving $[Ni(SR)(S_2CSR)]_2$ whose structure determination[179] revealed that the molecule was bent about the SR 'hinges', and that there was a Ni—Ni bond. In attempts to prepare $[Fe(S_2CSR)_3]$ (R = Et, Pr^n, Bu^n or benzyl), only $[Fe(SR)(S_2CSR)_2]_2$ was obtained[130]. These dimers were diamagnetic, and the ^1H n.m.r. spectra were consistent with a structure consisting of bridging ($\mu-$)SR groups and two bridging and two terminal trithioxanthate ligands, which was confirmed crystallographically[180]. Electrochemical studies revealed that the compound underwent, irreversibly, a two-electron reduction and a one-electron oxidation step.

10.5 COMPLEXES DERIVED FROM SULPHUR, SULPHIDE OR POLYSULPHIDE IONS

Reaction of cis-$Pt(PMe_2Ph)_2Cl_2$ with Na_2S gave[181] $[Pt_2(PMe_2Ph)_4S_2]$ and $[Pt_3(PMe_2Ph)_6S_2]Cl_2$. The former probably contained sulphide bridges and is a derivative of planar Pt^{II}. The latter may have consisted of a trigonal bipyramidal arrangement of Pt and S atoms, the latter in the apical positions. Reaction of $Pt(PPh_3)_2Cl_2$ with Na_2S gave $[Pt_2(PPh_3)_4S]$, which may be similar to $[Pt_2(PPh_3)_2(CO)S]$, obtained by thermal decomposition of $[Pt(PPh_3)_2S_2CO]$ [182], in that it contains a triangular $\underset{Pt\!-\!\!-\!Pt}{\overset{\diagup\;S\;\diagdown}{}}$ group[183].

Reaction of π-$C_5H_5V(CO)_4$ and $[\pi$-$C_5H_5Fe(CO)_2]_2$ with S_8, or of the latter with cyclohexane sulphide, afforded $[\pi$-$C_5H_5)_2V_2S_5]_n$ and $[\pi$-$C_5H_5FeS]_4$. The structure of the latter consisted[184] of an elongated tetrahedron of Fe atoms with an S atom above each face and a π-C_5H_5 ring projecting from each corner.

Treatment of $[\pi$-$C_5H_5Ni(CO)]_2$ with S_8 afforded[185] the paramagnetic $(S = 1/2) [(\pi$-$C_5H_5Ni)_3S_2]$ whose structure consisted of a regular trigonal bipyramid of Ni and S atoms, the latter in the apical positions; the π-C_5H_5 groups were bonded to the Ni atoms as expected. Decomposition of this complex in ether gave[186] $[(\pi$-$C_5H_5)_4Ni_5S_4]$. One Ni atom in this compound was coordinated by four S atoms in a square planar arrangement. The four remaining Ni atoms lay in a plane perpendicular to that of the four S atoms and were arranged in pairs such that each pair was bridged by two of the S atoms, and presumably by a metal–metal bond. Reaction of π-$C_5H_5Co(CO)_2$ with S_8 gave[187] $[(\pi$-$C_5H_5)_4Co_4S_6]$, whose structure is that of a tetrahedron of non-bonded Co atoms (each carrying a terminal π-C_5H_5 group), two of the tetrahedron faces being bridged by S_2 groups and the other two by S atoms; it was suggested that each S atom and S_2 group must be functioning as a four-electron donor. Reaction of $[(\pi$-$C_5H_5)_2NbCl_2(OH)]$ with H_2S in the presence of X$^-$ afforded[188] the monomeric, diamagnetic $[(\pi$-$C_5H_5)_2Nb(S_2)X]$ (X = Cl, Br, I, SCN). From preliminary x-ray data it is clear that this compound contains a three-membered NbS_2 ring.

The crystal structures of $[Co_3(CO)_9S]$ and of a 1:1 mixture of $[Fe(CO)_3S]_2$ and $Fe_3(CO)_9S_2$ have been determined. The first consisted of a Co_3S tetrahedral arrangement with terminal CO groups[189]. The second complex was described[190] as being derived from the intersection of two basal planes of two distorted tetragonal pyramids along the S—S bond, there being a 'bent' Fe—Fe bond. The third compound contained two iron atoms and two S atoms at alternate corners of a basal plane, with the remaining $Fe(CO)_3$

group bonded to all four atoms, thereby forming a square-pyramidal molecule[191]. The structure of $[Co_3(CO)_7S]_2S_2$ showed[192] that there were two
essentially identical $Co_3(CO)_7S$-groups (derived from $Co_3(CO)_9$), which
were linked by a disulphide bridge (Figure 6.4). The structure of $[\pi$-C_5H_5-
$MoO]_2S_2$ could be described ideally[193] as two octahedra (assuming π-C_5H_5

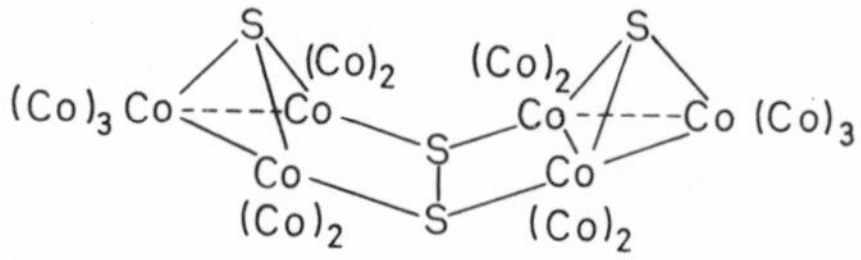

Figure 10.4

occupies three sites) sharing a common edge bisected by a direct Mo—Mo
bond which was proposed to account for the diamagnetism of the compound;
the S atoms constituted the bridges.

The crystal structure of $[(\pi$-$C_5H_5)_2TiS_5]$ confirmed[194] that the compound, obtained from the reaction of $[(\pi$-$C_5H_5)_2TiCl_2]$ with $(NH_4)_2S_5$ [195],
contained the S_5^{2-} chelate group; the chelate ring was puckered; $[(\pi$-$C_5H_5)_2$
$TiSe_5]$ has also been prepared[201]. Similar reactions of S_5^{2-} or Se_5^{2-} with
$[(\pi$-$C_5H_5)_2MCl_2]$ (M = Mo, W), afforded $[(\pi$-$C_5H_5)_2MX_4]$ (X = S, Se) [196].
Reaction of $M(PPh_3)_4$ or $M(diphos)_2$ (M = Pd, Pt) with S_8 gave
$[M(PPh_3)_2S_4]$ and its diphos analogue[181]. The Pt^{IV} complex, $[Pt(S_5)_3]^{2-}$,
was obtained by reaction of $[PtCl_6]^{2-}$ and $(NH_4)_2S_5$, and had an octahedral
geometry[197].

10.6   MISCELLANEOUS METAL SULPHUR COMPLEXES

10.6.1   Metal dithioacetylacetonates and their analogues

The simple bis- and tri-dithioacetylacetonato complexes, $[M(SacSac)_2]$
M = Ni, Pd, Pt, Co, and $[M(SacSac)_3]$, M = Co, Rh, Ir, Fe, Ru, Os (Figure
6.5; X = S, n = 2,3; $R^1 = R^2$ = Me) were obtained by reaction of the
appropriate metal salts with acetylacetone and H_2S in methanolic HCl [198, 199].
The structure of $[Co(SacSac)_2]$ (S = 1/2) is planar[200] and, from a study of the
^1H n.m.r. spectrum, it was deduced[201] that the unpaired electron was in an

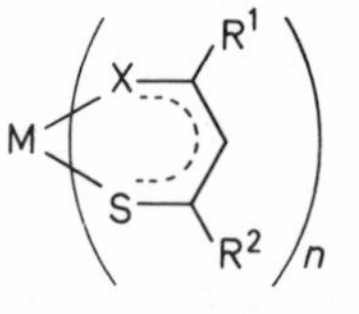

Figure 10.5

orbital of d_{z^2} character; $[Ni(SacSac)_2]$ is isomorphous with its Co analogue.
The iron complex, $[Fe(SacSac)_3]$ was low spin (μ = 2.00 BM at room temperature), and had a distorted octahedral structure[202]. The triad $[M(SacSac)_3]$,
M = Fe, Ru, Os, have been examined polarographically[199] and all underwent
a one-electron reduction giving $[M(SacSac)_3]^-$, the ease of reduction increasing in the order Fe < Os < Ru. The electronic spectra of these compounds were dominated by charge-transfer bands which obscured the d–d
transitions[199].

A variety of analogues of $[Ni(SacSac)_2]$ with $X = S$, $R^1 = Me$, $R^2 = Ph$, CF_3, etc. (Figure 6.5) have been prepared[203], and a polarographic examination of these, and of $[Ni(SacSac)_2]$ and $[Co(SacSac)_2]$, established that they underwent several irreversible reduction processes, the first of which was believed to correspond to the reduction of Ni^{II} to Ni^0 [204]. Halogenation of $[Ni(SacSac)_2]$ and other similar compounds afforded[205] the corresponding dithiolium salts stabilised by the appropriate tetra- or hexahalometallate anion.

References

1. McCleverty, J. A. (1968). *Progr. Inorg. Chem.*, **10**, 49
2. Schrauzer, G. N. (1968). *Trans. Met. Chem.*, **4**, 299; (1969) *Acc. Chem. Res.*, **2**, 72
3. Coucouvanis, D. (1970). *Progr. Inorg. Chem.*, **11**, 233
4. Eisenberg, R. (1970). *Progr. Inorg. Chem.*, **12**, 295
5. Livingstone, S. E. (1965). *Quart. Rev. (London)*, **19**, 386
6. Harris, C. M., Livingstone, S. E. (1963). 'Bidentate Chelates', *Chelating Agents and Metal Chelates*, Ed. by Dwyer, F. P. and Mellor, D. P. London: Academic Press
7. Jorgensen, C. K. (1963). *Inorganic Complexes*, London: Academic Press
8. Gray, H. B. (1965). *Trans. Met. Chem.*, **1**, 240; Gray, H. B., Eisenberg, R., Stiefel, E. I. (1967). *Advan. Chem. Ser.*, **62**, 641
9. Martin, R. L., White, A. H. (1968). *Trans. Met. Chem.*, **4**, 113
10. Abel, E. W., Crosse, B. C. (1967). *Organomet. Chem. Rev.*, **2**, 443
11. Schrauzer, G. N., Mayweg, V. P. (1962). *J. Am. Chem. Soc.*, **84**, 3221
12. Gray, H. B., Williams, R., Bernal, I., Billig, E. (1962). *J. Am. Chem. Soc.*, **84**, 3596
13. Davison, A., Edelstein, N., Holm, R. H., Maki, A. H. (1963). *J. Am. Chem. Soc.*, **85**, 2029
14. Gray, H. B., Billig, E. (1963). *J. Am. Chem. Soc.*, **85**, 2019
15. Eisenberg, R., Ibers, J. A., Clark, R. J. H., Gray, H. B. (1964). *J. Am. Chem. Soc.*, **86**, 113
16. Fritchie, C. J. (1966). *Acta Cryst.*, **20**, 107
17. Sartain, D., Truter, M. J. (1967). *J. Chem. Soc., A*, 1264
18. Forrester, J. D., Zalkin, A., Templeton, D. H. (1964). *Inorg. Chem.*, **3**, 1500
19. Eisenberg, R., Dori, Z., Gray, H. B., Ibers, J. A. (1968). *Inorg. Chem.*, **7**, 741
20. Forrester, J. D., Zalkin, A., Templeton, D. H. (1964). *Inorg. Chem.*, **3**, 1507
21. Weiher, J. F., Melby, L. R., Benson, R. E. (1964). *J. Am. Chem. Soc.*, **86**, 4329
22. Davison, A., Holm, R. H. (1967). *Inorg. Syn.*, **10**, 8
23. McCleverty, J. A. (1971). 'Redox Reactions and Electron Transfer Chains of Inert Transition Metal Complexes', *Reactions of Molecules on Electrodes*, Ed. by Hush, N. S., (London: John Wiley)
24. Balch, A. L., Dance, I. G., Holm, R. H. (1968). *J. Am. Chem. Soc.*, **90**, 1139
25. Olson, D. C., Mayweg, V. P., Schrauzer, G. N. (1966). *J. Am. Chem. Soc.*, **88**, 4876
26. Taft, R. W., Jr. (1965). *Steric Effects in Organic Chemistry*, Chap. 13, Ed. by Newman, M. S., (New York: John Wiley & Sons)
27. Schrauzer, G. N., Mayweg, V. P. (1965). *J. Am. Chem. Soc.*, **87**, 3585
28. Baker-Hawkes, M. J., Billig, E., Gray, H. B. (1966). *J. Am. Chem. Soc.*, **88**, 4870
29. Davison, A., Edelstein, N., Holm, R. H., Maki, A. H. (1963). *Inorg. Chem.*, **2**, 1227
30. Davison, A., McCleverty, J. A., Shawl, E. T., Wharton, E. J. (1967). *J. Am. Chem. Soc.*, **89**, 830
31. Maki, A. H., Edelstein, N., Davison, A., Holm, R. H. (1964). *J. Am. Chem. Soc.*, **86**, 4580
32. Shupack, S. I., Billig, E., Waters, J. H., Williams, R., Gray, H. B. (1964). *J. Am. Chem. Soc.*, **86**, 4594
33. Schmitt, R. D., Maki, A. H. (1968). *J. Am. Chem. Soc.*, **90**, 2288
34. Enemark, J. H., Lipscomb, W. N. (1965). *Inorg. Chem.*, **4**, 1729
35. Davison, A., Howe, D. V., Shawl, E. T. (1967). *Inorg. Chem.*, **6**, 458
36. Hamilton, W. C., Bernal, I. (1967). *Inorg. Chem.*, **6**, 2003
37. Baker-Hawkes, M. J., Dori, Z., Eisenberg, R., Gray, H. B. (1968). *J. Am. Chem. Soc.*, **90**, 4253

38. Williams, R., Billig, E., Waters, J. H., Gray, H. B. (1966). *J. Am. Chem. Soc.*, **88**, 43
39. McCleverty, J. A., Atherton, N. M., Connelly, N. G., Winscom, C. J. (1969). *J. Chem. Soc., A,* 2242
40. King, R. B. (1963). *Inorg. Chem.,* **2**, 641
41. Wharton, E. J., McCleverty, J. A. (1968). *J. Chem. Soc., A,* 2258
42. Eisenberg, R., Ibers, J. A. (1966). *Inorg. Chem.,* **5**, 411
43. Smith, A. E., Schrauzer, G. N., Mayweg, V. P., Heinrich, W. (1965). *J. Am. Chem. Soc.,* **87**, 5798
44. Eisenberg, R., Stiefel, E. I., Rosenberg, R. C., Gray, H. B. (1966). *J. Am. Chem. Soc.,* **88**, 2874; (1967). Eisenberg, R., Gray, H. B., *Inorg. Chem.,* **6**, 1844
45. Dickinson, R., Pauling, L. (1923). *J. Am. Chem. Soc.,* **45**, 1466
46. Stiefel, E. I., Dori, Z., Gray, H. B. (1967). *J. Am. Chem. Soc.,* **89**, 3353
47. Brown, G. F., Stiefel, E. I. (1970). *Chem. Commun.,* 728
48. Sequeira, A., Bernal, I. (1967). *Abstracts, American Crystallographic Association Meeting,* **75**, Minneapolis, Minn.
49. Schrauzer, G. N., Mayweg, V. P. (1966). *J. Am. Chem. Soc.,* **88**, 3235
50. Stiefel, E. I., Eisenberg, R., Rosenberg, R. C., Gray, H. B. (1966). *J. Am. Chem. Soc.,* **88**, 2956
51. Balch, A. L. (1967). *Inorg. Chem.,* **6**, 2158
52. Schrauzer, G. N., Mayweg, V. P., Finck, H. W., Heinrich, W. (1966). *J. Am. Chem. Soc.,* **88**, 4604
53. McCleverty, J. A., Ratcliff, B. (1970). *J. Chem. Soc., A,* 1631
54. Genser, E. E. (1968). *Inorg. Chem.,* **7**, 13
55. Mayweg, V. P., Schrauzer, G. N. (1966). *Chem. Commun.,* 640
56. Schrauzer, G. N., Mayweg, V. P., Heinrich, W. (1966). *J. Am. Chem. Soc.,* **88**, 5174
57. Schrauzer, G. N., Mayweg, V. P. (1965). *J. Am. Chem. Soc.,* **87**, 1483
58. Khare, G. P., Eisenberg, R. (1970). *Inorg. Chem.,* **9**, 2211
59. Pierpoint, C. G., Eisenberg, R. (1970). *Inorg. Chem.,* **9**, 2218
60. Epstein, E. F., Bernal, I., Balch, A. L. (1970). *Chem. Commun.,* 136
61. McCleverty, J. A., Atherton, N. M., Locke, J., Wharton, E. J., Winscom, C. J. (1967). *J. Am. Chem. Soc.,* **89**, 6082
62. Rae, A. I. M. (1967). *Chem. Commun.,* 1245
63. McCleverty, J. A., Ratcliff, B. (1970). *J. Chem. Soc., A,* 1627
64. James, T. A., McCleverty, J. A. (1970). *J. Chem. Soc.,* 3318
65. Connelly, N. G., Locke, J., McCleverty, J. A., Phipps, D. A., Ratcliff, B. (1970). *Inorg. Chem.,* **9**, 278
66. Connelly, N. G., Dahl, L. F. (1970). *Chem. Commun.,* 880
67. Wharton, E. J., Winscom, C. J. (1969). McCleverty, J. A., *Inorg. Chem.,* **8**, 393
68. Connelly, N. G., Locke, J., McCleverty, J. A. (1968). *Inorg. Chim. Acta.,* **2**, 411
69. King, R. B., Eggers, C. A. (1968). *Inorg. Chem.,* (1968). **7**, 1214
70. Hieber, W., Gscheidmeier, M. (1966). *Chem. Ber.,* **99**, 2312
71. King, R. B. (1963). *J. Am. Chem. Soc.,* **85**, 1584
72. Schrauzer, G. N., Rabinowitz, H. N., Frank, J. A. K., Paul, I. C. (1970). *J. Am. Chem. Soc.,* **92**, 212
73. Connelly, N. G., McCleverty, J. A. (1970). *J. Chem. Soc., A,* 1621
74. Singer, H., Wilkinson, G. (1965). *J. Chem. Soc., A,* 2516
75. Balch, A. L. (1969). *J. Am. Chem. Soc.,* **91**, 6962
76. McCleverty, J. A., Locke, J., Ratcliff, B., Wharton, E. J. (1969). *Inorg. Chim. Acta.,* **3**, 283
77. Jones, C. J., McCleverty, J. A., Orchard, D. G. (1971). *J. Organomet. Chem.,* **26**, C19
78. King, R. B. (1963). *J. Am. Chem. Soc.,* **85**, 1587; (1963). *Inorg. Chem.,* **2**, 1275
79. Locke, J., McCleverty, J. A., Wharton, E. J., Winscom, C. J. (1967). *Chem. Commun.,* 1289
80. Johnson, R. W., Muir, W. R., Sweigart, D. A. (1970). *Chem. Commun.,* 643
81. King, R. B. (1963). *J. Am. Chem. Soc.,* **85**, 1587
82. Locke, J., McCleverty, J. A. (1966). *Inorg. Chem.,* **5**, 1157
83. King, R. B., Eggers, C. A. (1968). *Inorg. Chem.,* **7**, 340
84. King, R. B., Bisnette, M. B. (1967). *Inorg. Chem.,* **6**, 469
85. Kopf, H., Schmidt, M. (1965). *J. Organomet. Chem.,* **4**, 426
86. Green, M. L. H., Lindsell, W. E. (1967). *J. Chem. Soc., A,* 1455

87. Baird, H. W., private communication quoted in ref. 90
88. Watkins, S. F., Dahl, L. F. (1965). *Abstract-52, 150th A.C.S. Meeting*, Atlantic City
89. Baird, H. W., private communication quoted in ref. 86.
90. Churchill, M. R., Cooke, J. (1970). *J. Chem. Soc., A,* 2046
91. Baird, H. W., White, B. M. (1966). *J. Am. Chem. Soc.,* **88,** 4744
92. Churchill, M. R., Fennessey, (1968). *Inorg. Chem.,* **7,** 1123
93. Knox, R., Prout, C. K. (1969). *Acta Cryst.,* **B25,** 2013
94. Dessy, R. E., Stary, F. E., King, R. B., Waldrop, M. (1966). *J. Am. Chem. Soc.,* **88,** 471;
    Dessy, R. E., King, R. B., Waldrop, M. (1966). *J. Am. Chem. Soc.,* **88,** 5112
95. McCleverty, J. A., James, T. A., Wharton, E. J. (1969). *Inorg. Chem.,* **8,** 1340
96. James, T. A., McCleverty, J. A. (1970). *J. Chem. Soc., A,* 3308
97. Schrauzer, G. N., Ho, R. K. Y., Murillo, R. O. (1970). *J. Am. Chem. Soc.,* **92,** 3508
98. Wing, R. M., Tustin, G. C., Okamura, W. H. (1970). *J. Am. Chem. Soc.,* **92,** 1935
99. Schrauzer, G. N., Rabinowitz, H. N. (1968). *J. Am. Chem. Soc.,* **90,** 4297
100. Schrauzer, G. N., Rabinowitz, H. N. (1969). *J. Am. Chem. Soc.,* **91,** 6522
101. Balch, A. L., Rohrscheid, F., Holm, R. H. (1965). *J. Am. Chem. Soc.,* **87,** 2301
102. Holm, R. H., Balch, A. L., Davison, A., Maki, A. H., Berry, T. E. (1967). *J. Am. Chem. Soc.,* **89,** 2866
103. Jones, C. J., McCleverty, J. A. (1970). *J. Chem. Soc., A,* 2829
104. Bailey, N. A., Hull, S. E., Jones, C. J., McCleverty, J. A. (1970). *Chem. Commun.,* 43
105. Balch, A. L. (1969). *J. Am. Chem. Soc.,* **91,** 1948
106. Davison, A., Shawl, E. T. (1970). *Inorg. Chem.,* **9,** 1820
107. Forbes, C. E., Holm, R. H. (1970). *J. Am. Chem. Soc.,* **92,** 2297
108. Hesse, R., Nilson, L. (1969). *Acta Chem. Scand.,* **23,** 825; Akerstrom, S. (1959). *Arkiv. Kemi,* **14,** 387
109. Bonamico, M., Dessy, G., Mugnoli, A., Vaciago, A., Zambonelli, L. (1968). *J. Chem. Soc., A,* 1351; O'Connor, B. H., Masten, E. N. (1966). *Acta Cryst.,* **21,** 828
110. Weeks, M. J., Fackler, J. P. (1968). *Inorg. Chem.,* **7,** 2548
111. Beurskens, P. T., Blaauw, H. J. A., Steggerda, J. J. (1968). *Inorg. Chem.,* **7,** 810
112. Beurskens, P. T., Blaauw, H. J. A., Cras, J. A., Steggerda, J. J. (1968). *Inorg. Chem.,* **7,** 805
113. Klug, H. P. (1966). *Acta Cryst.,* **21,** 536
114. Bonamico, M., Mazzone, G., Vaciago, A., Zambonelli, J. (1965). *Acta Cryst.,* **19,** 898;
    Domenicano, A., Torelli, L., Vaciago, A., Zambonelli, L. (1968). *J. Chem. Soc., A,* 1351;
    Shugam, E. A., Agre, V. M. (1966). *Acta Cryst.,* **21A,** 152
115. Nigo, Y., Masuda, I., Shinra, K. (1970). *Chem. Commun.,* 476
116. Brinkhoff, H. C., Cras, J. A., Steggerda, J. J., Willemse, J. (1969). *Rec. Trav. Chim.,* **88,** 633
117. Coucouvanis, D., Fackler, J. P. (1967). *Inorg. Chem.,* **6,** 2047
118. Fackler, J. P., Seidel, W. C. Unpublished work quoted in Ref. 3
119. Fackler, J. P., Seidel, W. C. (1969). *J. Am. Chem. Soc.,* **8,** 1631
120. Fackler, J. P., Seidel, W. C., Fetchin, J. A. (1967). *J. Am. Chem. Soc.,* **90,** 2707
121. Fackler, J. P., Fetchin, J. A., Seidel, W. C. (1969). *J. Am. Chem. Soc.,* **91,** 1217
122. Fackler, J. P., Holah, D. G. (1966). *Inorg. Nucl. Chem. Letters,* **2,** 251
123. Pignolet, L. H., Holm, R. H. (1970). *J. Am. Chem. Soc.,* **92,** 1791
124. Hoskins, B. F., Kelly, B. P. (1968). *Chem. Commun.,* 1517
125. Ewald, E. H., Martin, R. L., Ross, I. G., White, A. H. (1964). *Proc. Roy. Soc. (London),* **A280,** 235
126. Ewald, E. H., Martin, R. L., Sinn, E., White, A. H. (1969). *Inorg. Chem.,* **8,** 1837
127. White, A. H., Roper, R., Kokot, E., Waterman, H., Martin, R. L. (1964). *Australian J. Chem.,* **17,** 294
128. Hoskins, B. F., Kelly, B. P. (1970). *Chem. Commun.,* **45**
129. Ewald, A. H., Sinn, E. (1968). *Australian J. Chem.,* **21,** 927
130. Coucouvanis, D., Lippard, S. J., Zubieta, J. A. (1970). *J. Am. Chem. Soc.,* **92,** 3342
131. Martin, R. L., White, A. H. (1967). *Inorg. Chem.,* **6,** 712
132. Hoskins, B. F., Martin, R. L., White, A. H. (1966). *Nature,* **211,** 627
133. Blake, A. B., Cotton, F. A., Wood, J. S. (1964). *J. Am. Chem. Soc.,* **86,** 3024
134. Moore, F. W., Larson, M. L. (1967). *Inorg. Chem.,* **6,** 998
135. Moore, F. W., Rice, F. R. (1968). *Inorg. Chem.,* **7,** 2510
136. Jowitt, R. N., Mitchell, P. C. H. (1969). *J. Chem. Soc., A,* 2632

137. Fletcher, S. R., Rowbottom, J. F., Skapski, A. C., Wilkinson, G. (1970). *Chem. Commun.,* 1572
138. Knox, J. R., Prout, C. K. (1969). *Acta Cryst.,* **B25,** 2281
139. McCormick, B. J. (1968). *Inorg. Chem.,* **7,** 1965
140. Bradley, D. C., Moss, R. H., Sales, K. D. (1969). *Chem. Commun.,* 1255
141. Smith, J. N., Brown, T. M. (1970). *Inorg. Nucl. Chem. Letters,* **6,** 441
142. Heckley, P. R., Holah, D. G. (1970). *Inorg. Nucl. Chem. Letters,* **6,** 865
143. Pantaleo, D. C., Johnson, R. C. (1970). *Inorg. Chem.,* **9,** 1248
144. Bradley, D. C., Gitlitz, M. H. (1965). *Chem. Commun.,* 289; (1969). *J. Chem. Soc., A,* 1152
145. Colapietro, M., Vaciago, A., Gradley, D. C., Hursthouse, M. B., Rendall, I. F. (1970). *Chem. Commun.,* 743
146. Brown, D., Holah, D. G., Rickard, C. E. F. (1970). *J. Chem. Soc., A,* 786
147. Bibler, J. P., Karraker, D. G. (1968). *Inorg. Chem.,* **7,** 982; Brown, D., Holah, D. G., Rickard, C. E. F. (1970). *J. Chem. Soc., A.,* 423
148. Bowman, K., Dori, Z. (1968). *Chem. Commun.,* 636
149. Lorenz, B., Kirmse, R., Hoyer, E. (1970). *Z. Anorg. Allg. Chem.,* **378,** 144
150. Cervone, E., Camessei, F. D., Luciani, M. L., Furlani, C. (1969). *J. Inorg. Nucl. Chem.,* **31,** 1101
151. Furlani, C., Cervone, E., Camessei, F. D. (1968). *Inorg. Chem.,* **7,** 265
152. Bonamico, M., Dessy, G., Fares, V. (1969). *Chem. Commun.,* 324
153. Bonamico, M., Dessy, G. (1968). *Chem. Commun.,* 483
154. Bonamico, M., Dessy, G., Fares, V. (1969). *Chem. Commun.,* 1106
155. Bonamico, M., Dessy, G., Fares, V. (1969). *Chem. Commun.,* 697
156. Hieber, W., Bruck, R. (1952). *Z. Anorg. Allg. Chem.,* **269,** 13
157. Fackler, J. P., Coucouvanis, D. (1967). *J. Am. Chem. Soc.,* **89,** 1745
158. Fackler, J. P., Coucouvanis, D., Fetchin, J. A., Seidel, W. C. (1968). *J. Am. Chem. Soc.,* **90,** 2784
159. Fackler, J. P., Fetchin, J. A., Smith, J. A. (1970). *J. Am. Chem. Soc.,* **92,** 2910
160. Coucouvanis, D., Lippard, S. J. (1968). *J. Am. Chem. Soc.,* **90,** 3281; (1969). *J. Am. Chem. Soc.,* **91,** 307
161. Cotton, F. A., Harris, C. B. (1968). *Inorg. Chem.,* **7,** 2140
162. McKechnie, J. S., Miesel, S. L., Paul, I. C. (1967). *Chem. Commun.,* 152
163. McCandlish, L. E., Bissel, E. C., Coucouvanis, D., Fackler, J. P., Knox, K. (1968). *J. Am. Chem. Soc.,* **90,** 7357
164. Coucouvanis, D., Fackler, J. P. (1967). *J. Am. Chem. Soc.,* **89,** 1346
165. Davies, G. R., Mais, M. H. B., Owston, P. G. (1968). *Chem. Commun.,* 81; Colapietro, M., Domenicano, A., Scarmuzza, L., Vaciago, A., Zambonelli, L. (1967). *Chem. Commun.,* 583; Alderman, P. R. H., Owston, P. G., Rowe, J. M., *J. Chem. Soc.* (1962). 668
166. Davies, G. R., Jarvis, J. A. J., Kilbourn, B. T., Mais, M. H. B., Owston, P. G. (1970). *J. Chem. Soc., A,* 1275
167. Johnson, B. F. G., Al-Obadi, K. H., McCleverty, J. A. (1969). *J. Chem. Soc., A,* 1668
168. Cotton, F. A., McCleverty, J. A. (1964). *Inorg. Chem.,* **3,** 1398
169. Domenicano, A., Vaciago, A., Zambonelli, L., Loader, P. L., Venanzi, L. M. (1966). *Chem. Commun.,* 476
170. Brennan, T. F., Bernal, I. (1970). *Chem. Commun.,* 138
171. Colton, R., Scollary, G. R., Tomkins, I. B. (1968). *Australian J. Chem.,* **21,** 15; Colton, R., Rose, G. G. (1970). *Australian J. Chem.,* **23,** 1111
172. Kingston, J. V., Wilkinson, G. (1966). *J.Inorg. Nucl. Chem.,* **28,** 2709
173. O'Connor, C., Gilbert, J. D., Wilkinson, G. (1969). *J. Chem. Soc., A,* 84
174. Green, M., Osborn, R. B. L., Stone, F. G. A. (1970). *J. Chem. Soc., A,* 944
175. McCleverty, J. A., Orchard, D. G. (1970). *J. Chem. Soc., A,* 3315
176. Lindner. E., Grimmer, R., Weber, H. (1970). *Angew. Chem. Int. Edn. Engl.,* **9,** 639
177. Bladon, P., Bruce, R., Knox, G. R. (1965). *Chem. Commun.,* 557
178. Ahmad, M., Bruce, R., Knox, G. R. (1966). *J. Organometal. Chem.,* **6,** 1; Bruce, R., Knox, G. R. (1966). *J. Organometal. Chem.,* **6,** 67
179. Villa, A., Manfredotti, A. G., Nardelli, M., Pelizzi, C. (1970). *Chem. Commun.,* 1322
180. Coucouvanis, D., Lippard, S. J., Zubieta, J. A. (1970). *Inorg. Chem.,* **9,** 2775
181. Chatt, J., Mingos, D. P. (1970). *J. Chem. Soc., A,* 1243
182. Baird, M. C., Wilkinson, G. (1967). *J. Chem. Soc., A,* 865

183. Skapski, A. C., Troughton, P. G. H. (1969). *Chem. Commun.*, 170
184. Schunn, R. A., Fritchie, C. J., Prewitt, C. T. (1965). *Inorg. Chem.*, **5**, 892; Wei, C. H., Wilkes, G. R., Treichel, P. M., Dahl, L. F. (1966). *Inorg. Chem.*, **5**, 900
185. Vahrenkamp, H., Uchtmann, V. A., Dahl, L. F. (1968). *J. Am. Chem. Soc.*, **90**, 3272
186. Vahrenkamp, H., Dahl, L. F. (1969). *Angew. Chem. Int. Edn. Engl.*, **8**, 144
187. Uchtmann, V. A., Dahl, L. F. (1969). *J. Am. Chem. Soc.*, **91**, 3756
188. Treichel, P. M., Werber, G. P. (1968). *J. Am. Chem. Soc.*, **90**, 1753
189. Wei, C. H., Dahl, L. F. (1967). *Inorg. Chem.*, **6**, 1229
190. Wei, C. H., Dahl, L. F. (1965). *Inorg. Chem.*, **4**, 1
191. Wei, C. H., Dahl, L. F. (1965). *Inorg. Chem.*, **4**, 493; Dahl, L. F., Sutton, P. W. (1963). ibid., **2**, 1067
192. Stevenson, D. L., Magnuson, V. R., Dahl, L. F. (1967). *J. Am. Chem. Soc.*, **89**, 3727
193. Stevenson, D. L., Dahl, L. F. (1967). *J. Am. Chem. Soc.*, **89**, 3721
194. Epstein, E. F., Bernal, I. (1970). *Chem. Commun.*, 410
195. Kopf, H., Block, B., Schmidt, M. (1968). *Chem. Ber.*, **101**, 272; Kopf, H., Block, B. (1969). ibid., **102**, 1054
196. Kopf, H., Kahl, W., Wirl, A. (1970). *Angew. Chem. Int. Edn., Engl.*, **9**, 801
197. Jones, P. E., Katz, L. (1967). *Chem. Commun.*, 842
198. Barraclough, C. G., Martin, R. L., Stewart, I. M. (1969). *Australian J. Chem.*, **22**, 891
199. Heath, G. A., Martin, R. L. (1969). *Chem. Commun.*, 951; Heath, G. A., Martin, R. L. (1970). *Australian J. Chem.*, **23**, 1721
200. Beckett, R., Hoskins, B. F. (1967). *Chem. Commun.*, 909
201. Fitzgerald, R. J., Brubaker, G. R. (1969). *Inorg. Chem.*, **8**, 2265
202. Beckett, R., Heath, G. A., Hoskins, B. F., Kelly, B. P., Martin, R. L., Roos, I. A. G., Weickhardt, P. L. (1970). *Inorg. Nucl. Chem. Letters*, **6**, 257
203. Ouchi, A., Nakatani, M., Takahashi, Y. (1968). *Bull. Chem. Soc., Japan*, **41**, 2044
204. Furuhashi, A., Kawai, S., Hayakawa, Y., Ouchi, A. (1970). *Bull. Chem. Soc., Japan*, **43**, 553
205. Furuhashi, A., Watanuki, K., Ouchi, A. (1969). *Bull. Chem. Soc., Japan*, **42**, 260